Learning and Analytics in Intelligent Systems

Volume 14

The main aim of the series is to make available a publication of books in hard copy form and soft copy form on all aspects of learning, analytics and advanced intelligent systems and related technologies. The mentioned disciplines are strongly related and complement one another significantly. Thus, the series encourages cross-fertilization highlighting research and knowledge of common interest. The series allows a unified/integrated approach to themes and topics in these scientific disciplines which will result in significant cross-fertilization and research dissemination. To maximize dissemination of research results and knowledge in these disciplines, the series publishes edited books, monographs, handbooks, textbooks and conference proceedings.

More information about this series at http://www.springer.com/series/16172

George A. Tsihrintzis · Maria Virvou
Editors

Advances in Core Computer Science-Based Technologies

Papers in Honor of Professor Nikolaos Alexandris

Editors
George A. Tsihrintzis
Department of Informatics
University of Piraeus
Piraeus, Greece

Maria Virvou
Department of Informatics
University of Piraeus
Piraeus, Greece

ISSN 2662-3447 ISSN 2662-3455 (electronic)
Learning and Analytics in Intelligent Systems
ISBN 978-3-030-41198-5 ISBN 978-3-030-41196-1 (eBook)
https://doi.org/10.1007/978-3-030-41196-1

This Springer imprint is published by the registered company Springer Nature Switzerland AG
The registered company address is: Gewerbestrasse 11, 6330 Cham, Switzerland

The Assembly of the Department of Informatics of the University of Piraeus, with its unanimous decision of December 16, 2019, wholeheartedly dedicates this volume in honor of Prof. Nikolaos Alexandris.

Foreword—The Knot: A Tool for Craft and Science

Nikolaos Alexandris is a mathematician, who stands out for knowing how to solve knotty problems without boasting.

The knot, just like the wheel, is one of the fundamental tools humans created, which effectively promoted the development of civilization. The first knot in everyone's life is the umbilical cord, which is severed when the new-born leaves the mother's womb (Fig. 1).

Fig. 1 Knot

A knot (the Ancient Greek *komvos*, meaning not only the actual knot, but also a nod and a nodal point, and when turned into a verb meaning 'boast'—as in the first sentence of this text—which means 'creating my own public image', similar to 'blowing my own trumpet') can be used to restrict or connect. In other words, it is directly related to a bond, bonding or commitment, and the activities of knitting, weaving and needlework. To cover their nakedness human invented weaving. Knitting produces braids and rope, two prevalent tools for daily life. Many objects are joined/linked with the help of knots. The loom operates using knots and is directly reflected in the computer. Needlework highlights the aesthetic aspect of

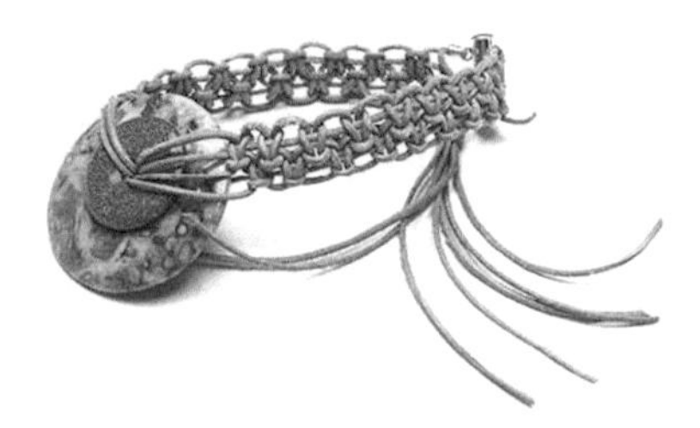

Fig. 2 Macramé

Fig. 3 Celtic knotwork

knots both in the craftsmanship and the mechanics of making them. This is also true in jewellery-making and the decorative arts (see Figs. 2 and 3).

A French knot, braids, bows, neckties and the way they are tied are all important for both men's and women's appearance. Swaddling and girdles also use knots. Buttons (*komvion* or *koumbi*, in Ancient and Modern Greek, respectively) and the button-hole (Ancient Greek *komviodochi* = button receptacle) replaced knots in garments.

Knots led to webs, and the intersection points are called nodal (*komvika* in Ancient Greek).

In ancient Egypt, 'rope stretchers' often used ropes with knots to form right angles to mark field borders following the River Nile floods.

The knot is also a kind of 'Swiss knife' for sailors (ropes/nets), while naval knots are particularly important in the training of Sea Cadets. A fisherman is considered ready to join the trade when they can tie the hook and disentangle the numerous vertical lines dropped with '*katheti*' angling or with net fishing. A ship is tied to the dock with an ordinary knot. Knots were also used to define the unit to measure the speed of marine vessels.

There is a list classifying all knots, from the simplest to most complex one [8].

When referring to weaving, Plato underlined: Of course, no man of sense would wish to pursue the discussion of weaving for its own sake, but most people, it seems to me, fail to notice that some things have sensible resemblances which are easily perceived' (Statesman/*Politicus* 285d), concluding that a politician's art resembles the craft of weaving wool.

A knot may be necessary, inevitable or undesirable. It may be born of need, carelessness, rush, intention or just by coincidence, while its purpose determines the material used to make it (yarn, rope, fishing line, etc.). The knot holds together, protects, stops, blocks or prevents a hazard.

Experience teaches that it is difficult to unravel a knot if one does not know the way it was created. The time and cost spent to create a knot and the techniques used are the basic parameters that will help unravel it.

Playing with knots can be an interesting leisure activity or used as a juggling trick.

Undoing a knot is necessary when its existence makes it difficult for the system containing it to operate. However, the security certain systems require makes it imperative to use knots that cannot be easily undone.

A knot is associated with the loss of life or with a person's rescue. Execution by hanging or rescuing a mountaineer or a speleologist, both rely on the help of knots and loops.

In Medicine, a knot is a basic tool for therapeutic interventions and operations (surgery, trauma care, physiotherapy, etc.). In the paper '*Iatrikai Synagogai*' (*Medical Collections*) by Oreibasius of Pergamum (Pergamenus), an eminent physician and philosopher of the fourth century AD, includes the orthopaedic knots and slings Heraclas, a physician used in the first century AD (see Figs. 4 and 5) Finally, the pharyngeal knot (*globus pharyngeus*/lump in the throat) is a known annoying symptom.

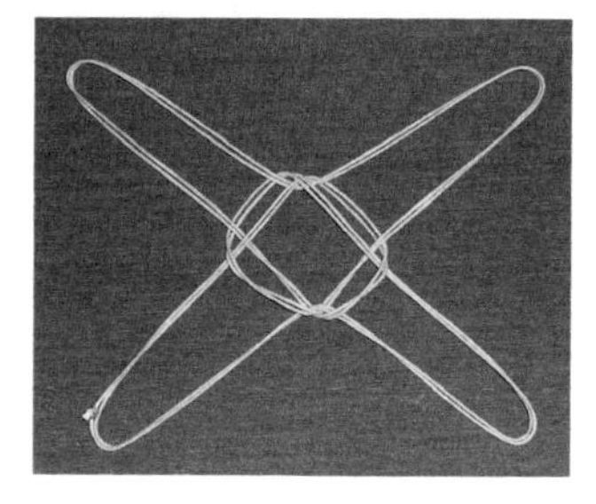

Fig. 4 *Plinthios Brokhos* [4-loop bandage noose]

Fig. 5 Crossed noose

Linking would be inconceivable without knots. Sorcerers promised that by tying and using knots on effigies they could harness harmful forces. A prayer rope is made up of successive cross-shaped knots, to accompany the devout during their prayers. The worry-beads (*komboloi* = literally a string of knots) have replaced prayer knots with beads, so that a sound is also produced during their use.

The Incas kept accounting records for their daily needs with the help of a quipu, i.e., a knotted yarn system (like prayer beads), which they also put to other uses.

The knot, as a symbol, describes typical aspects of human behaviour. Indicatively, the Ancient Greeks used the verb *kompazo* and the Mediaeval Greek the verb *komborrhemono* to express boasting. Aeschylus' works '*But his boast is too proud for a mere human*' [*Seven against Thebes, 425*] characterises a bragging persons way of thinking. Similarly, Sophocles [Antigone 127] notes that '... *Zeus detests above all the boasts of a proud tongue*'.

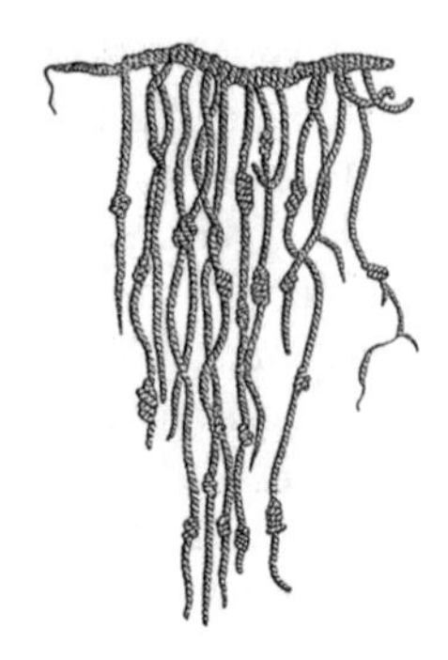

Fig. 6 Quipu [Talking Knot]

Daily expressions like 'the knot hit the comb' [that's the last straw] or 'this is the knot (crux) of the matter' indicate life problems and difficulties, while if you are 'buttoned up' (*Koumbomenos*), you are cagey. Finally, 'I turned my sorrows into knots' implies a stoic approach to misfortunes, while 'I got mixed up' implies being entangled in a ball of yarn (Fig. 6).

'*Kombodema*' (money tied in a kerchief knot) describes the arduous way of saving pennies to achieve a specific goal. The term '*kombogiannitis*' (charlatan—a person trying to restore health by tying knots, like a sorcerer) refers to anyone who lacks the qualifications to undertake the tasks of their trade, while 'tying a knot in one's kerchief' is a reminder for something one should not forget.

In Fine arts, knots appear in various forms, as in Escher's works. In cinematic art, there is a well-known film by the Russian director Aleksandr Sokurov called 'The Knot' [Uzel or Dialogues with Solzhenitsyn].

In music, knots are used for tuning string instruments.

Aeschylus, in '*Prometheus Bound*', presents the titan's divine punishment using the chains made by Hephaestus to tie the tragic hero to the rock.

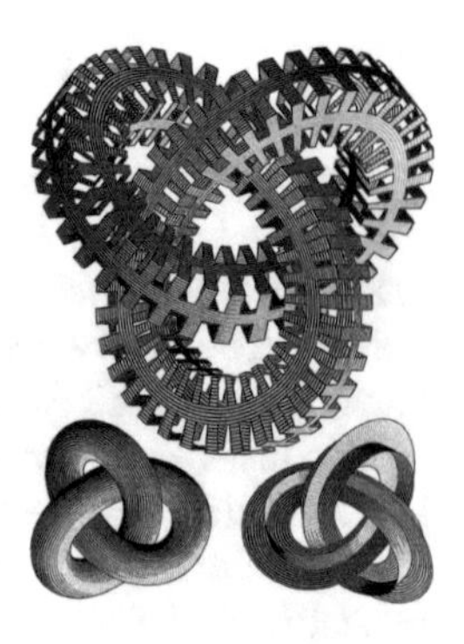

Fig. 7 Escher's knots

For psychiatrist R. D. Laing, 'Knots', his famous work [5], refers to the intertwining of patient's words to express their condition (Fig. 7).

The power and craftsmanship required to make or unravel a knot often need coordinated movements and a fair amount of patience. Of course, the way the Gordian knot was undone by Alexander the Great presents a violent solution.

Based on the definition of Zeno from Citium (the Stoic), 'Art is a skill that opens roads, that is, gradually and methodically creates something', easily explains why the craft of a knot is algorithmic, i.e., it requires a mathematical process following methodically predetermined steps. In fact, this is how the Knot Theory was established in Mathematics [1, 4]. The solution to any problem, mathematical or not, is comparable to that of undoing one or more knots.

A basic problem of systemic analysis is to determine the 'erroneous moves' that result in the formation of 'knots' in the corresponding systems.

A simple knot leads to the concept of a mathematical knot, which is defined as a closed, one-dimensional and continuous single curve in a three-D/R^3 space. This is how from the simple knot of Fig. 8 the closed knot of Fig. 9 emerges, which leads to the mathematical one, the diagram of which on a single level is presented in Fig. 10.

Fig. 8 Open noose knot

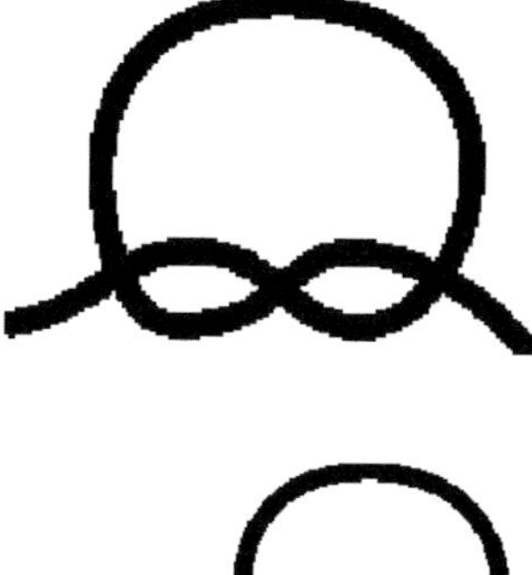

Fig. 9 Closed noose knot

In the diagram of the mathematical knot, solid curved lines correspond to preceding sections and dashed curved lines to ensuing ones. A similar picture is used to depict the multi-level junctions on major highways. A typical page is the one below from Gauss's notes (1794), where he uses knots in electrodynamics (Fig. 11).

Fig. 10 Mathematical knot

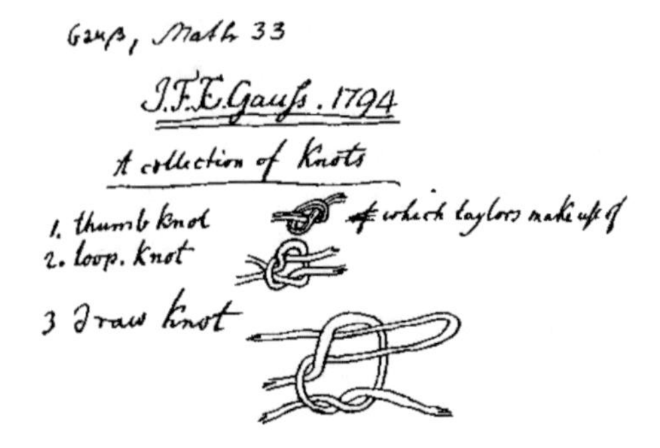

Fig. 11 The text by Gauss

Before Gauss, Leibniz (1679) had worked with knots in what is today known as Combinatorial Topology and Vandermonde (1771) with situational problems [10].

Lord Kelvin (1876) and Lord Maxwell (1873), with their papers on Chemistry and Electromagnetism, respectively, were influential in the development of the Knot Theory (Figs. 12 and 13).

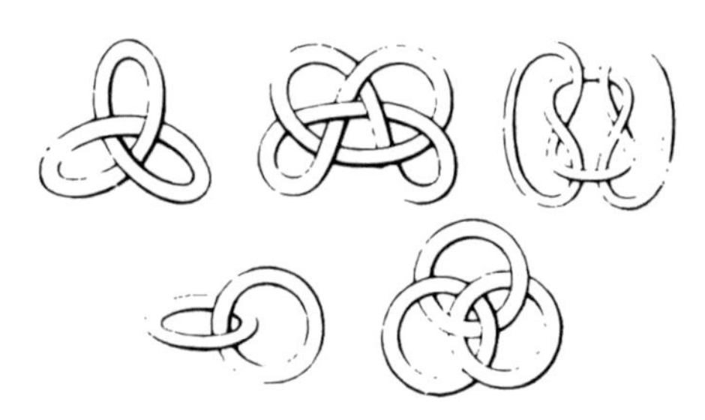

Fig. 12 Knots and links by Kelvin

Fig. 13 Link by Maxwell

Tait (1898), in his three-volume work 'On Knots' was the first to present the foundation of the Theory of Knots [9]. In the beginning of the twentieth century, Dehn (1910) with his work, followed by Alexander & Briggs (1927) [2] and Reidermeister's independent work [7], laid the foundations of the Combined Theory of Knots.

Mathematical knots are classified based on certain of their characteristic features. So, in the following table from Wikipedia, the classification is based on the number of crossings a knot has.

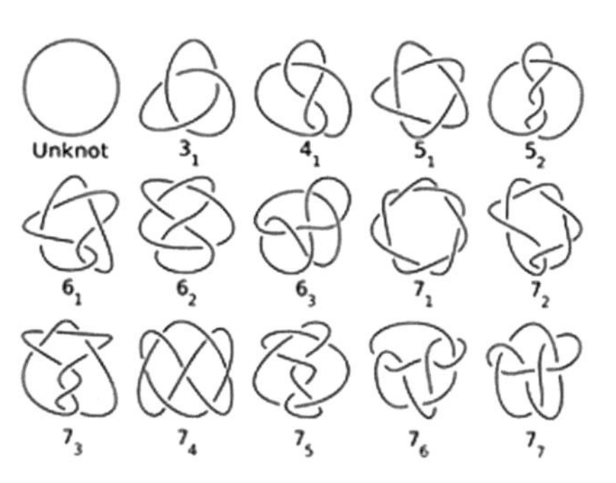

Fig. 14 Knot classification

The first knot, i.e., the simple circle, is called unknot, given that there is no physical knot (Fig. 14).

There are various codifications for the diagrams of mathematical knots. So, according to Gauss' code, the mathematical knot below is fully described by the following sequence:

1, -2, 3, -4, 5, 6, -7, -8, 4, -9, 2, -10, 8, 11, -6, -1, 10, -3, 9, -5, -11, 7

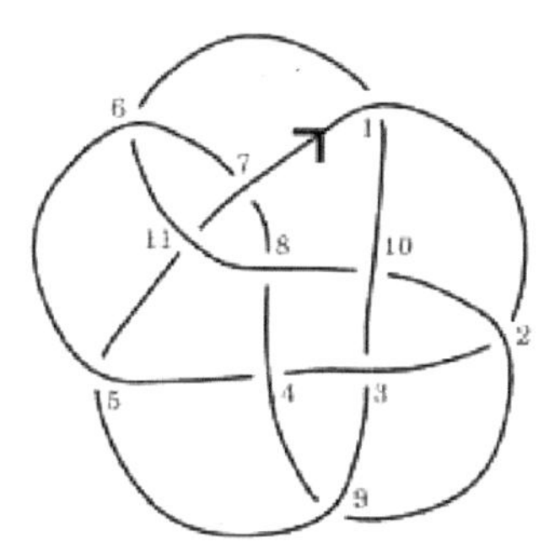

Fig. 15 Knot and Code by Gauss

Where positive signs indicate an over-crossing and the negative signs an under-crossing (Fig. 15).

The fundamental problem of the Knot Theory is unknotting, which raises the question whether the given knot may lead to an unknot. Reidermeister [7] solved this problem with the help of moves R1, R2, R3, named after their inventor (Figs. 16 and 17).

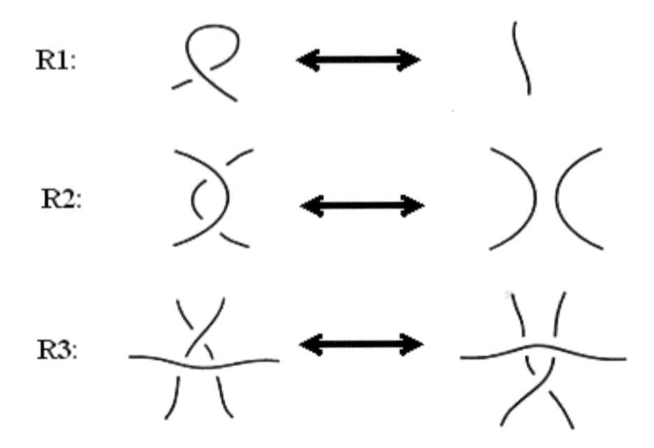

Fig. 16 Reidermeister's moves

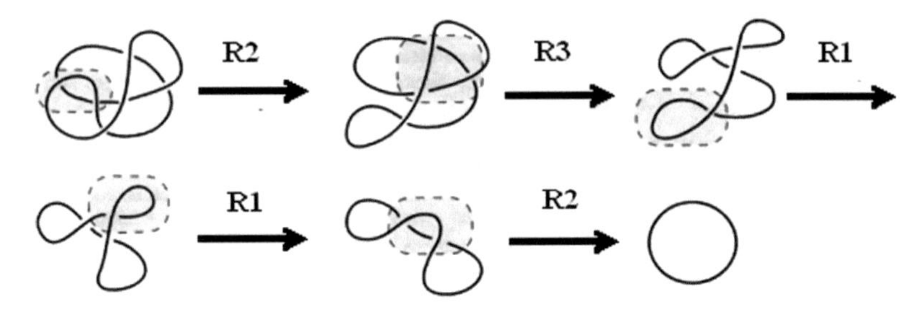

Fig. 17 Example of applying Reidermeister's moves

The following example presents an algorithmic application of these moves.

However, the general problem of unknotting comes under the class of NP-complete problems, which have not been solved within a satisfactory time limit.

In Physics [3], knots appear when dealing with issues of Statistical Mechanics, elastic materials, tubing, electrical grids, Quantum Mechanics, etc. A solanium is formed due to knots created by field dynamics (Fig. 18).

Fig. 18 Elastic Bands/Rings

Fig. 19 Coiled pipes/tubes

In Chemistry, the construction of chiral compounds owes a lot to the Theory of Knots (Fig. 19).

Brain neurons and synapses are directly related to braids, the theory of which was created by E. Artin (1936) (Fig. 20).

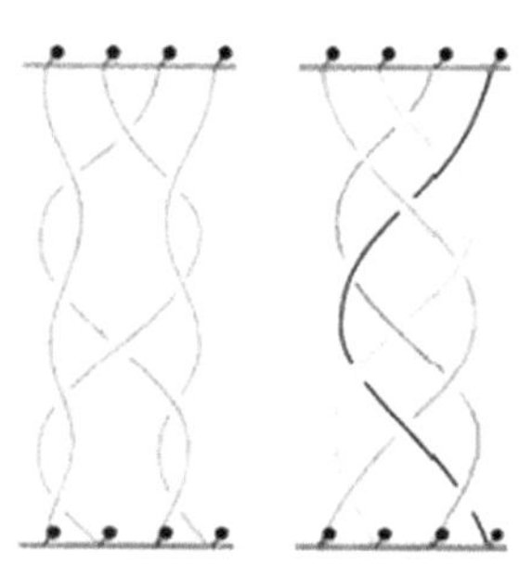

Fig. 20 Braids

It is easily comprehended that by joining the two ends of a braid a knot is created. The mathematical theory of knots helps study contemporary problems of proteins and RNA (Figs. 21 and 22).

Fig. 21 Protein

Fig. 22 RNA

Finally, knots are directly related to meandric colliers, wherever these are needed in applications [6].

Folk wisdom and the mathematical Theory of Knots coexist in the classic saying that ‘a knot cannot be solved or be unravelled by stretching it’.

Antonios Panayotopoulos
Emeritus Professor
Department of Informatics
University of Piraeus
Piraeus, Greece

References

1. C.C. Adams, *The Knot Book. An Elementary Introduction to the Mathematical Theory of Knots* (W.H. Freeman, New York, 1994)
2. J.W. Alexander, C.B. Briggs, On the types of knotted curves. Ann. Math. **28** (1927)
3. L.H. Kauffman, *Knots and Physics* (World Scientific Publishers, 1991), 3rd edn. (2002)
4. A. Kawauchi, *A Survey of Knot Theory* (Birkhäuser, Basel, 1996)
5. R.D. Laing, *Knots* (Penguin Books, 1969)
6. A. Panayotopoulos, On meandric colliers. Math. Comput. Sci. (2018)
7. K. Reidmeister, *Knotentheorie* (Springer, 1932)
8. P. Suber, Knots on the web, http://www.earlham.edu/~peters/knotlink.htm
9. R.G. Tait, *On Knots I, II, III*, Scientific papers (1898)
10. J.C. Turner, P. Van De Griend (eds.), *History and Science of Knots.*
11. A.T. Vandermonde, Remarques sur les problèmes de situation. *Mémoires de l’ Académie Royale des Sciences* (Paris) (1771)
12. https://www.researchgate.net/figure/a-Dessin-original-de-Kelvin-Zapato-2009-b-Exemples-danneaux-tourbillonnaires_fig2_278676613
13. http://katlas.org/wiki/Knot_Atlas:About

Preface

What is the best way to honour an academician who withdraws from his teaching duties, (without, however, retiring from his research activities) after having devoted almost five decades serving as advisor to generations of undergraduate, graduate and doctoral students? Such a task is certainly difficult to fulfil! When, additionally, this same academician has served his University and his profession as a top-level administrator and mentor to faculty and colleagues, the task of appropriately honouring him becomes even harder! **In our opinion, the best way to honour this person is to ask his former doctoral students, colleagues and also fellow researchers from around the world sharing similar research interests with him, to include some of their recent research results in a high-quality volume edited in his honour, indicating that they are pursuing and extending further what they have learned from him or from research areas where he made his contributions**.

Professor **Nikolaos Alexandris** has been serving computer science from various posts for almost 50 years now. He holds a B.Sc. in Mathematics from the National and Kapodistrian University of Athens, a M.Sc. from the Department of Computation of Manchester Institute of Science and Technology (UMIST) and a Ph.D. from the Department of Control Center of UMIST. He is a Professor (now Emeritus) in the Department of Informatics of the University of Piraeus. He has served as Head of the same Department and Vice Rector of the same University for several terms. He has also served as Head of the Hellenic Mathematical Society for six terms, President of the Monitoring Committee of the Integrated Mediterranean Programmes for Information Technology in Greece, Head of the Research, Documentation, and Education Technology Department of the Hellenic Pedagogical Institute and as a Member of the Board of Governors of the Ionian University. In his long academic career, he has supervised the research of over 15 doctoral students and tens of undergraduate and graduate students. Many of his ex-students currently hold academic positions.

We have been former colleagues and mentees of Prof. Nikolaos Alexandris. We have also served for several terms as Heads of the Department of Informatics of the University of Piraeus (home Department of Professor Nikolaos Alexandris).

Thus, we have proposed and undertaken with pleasure the task of editing a book to honour him. The response from former colleagues and mentees of his has been great! Unfortunately, page limitations have forced us to limit the works to be included in the book. We apologise to those authors whose works are not included in the current book.

While honouring Prof. Nikolaos Alexandris, this book also serves the purpose of exposing its reader to some of the most significant advances in core computer science-based technologies. As such, the book is directed towards professors, researchers, scientists, engineers and students in computer science-related disciplines. It is also directed towards readers who come from other disciplines and are interested in becoming versed in some of the most recent computer science-based technologies. We hope that all of them will find it useful and inspiring in their works and researches.

We are grateful to the authors and reviewers for their excellent contributions and visionary ideas. We are also thankful to Springer for agreeing to publish this book in its **Learning and Analytics in Intelligent Systems** series. Last, but not least, we are grateful to the Springer staff for their excellent work in producing this book.

Piraeus, Greece

George A. Tsihrintzis
Maria Virvou

Contents

Chapter 1
Advances in Core Computer Science-Based Technologies

George A. Tsihrintzis and Maria Virvou

Abstract At the dawn of the 4th Industrial Revolution, the field of computer science-based technologies is growing continuously and rapidly, developing in both itself and towards applications of many other disciplines. The book at hand aims at exposing its reader to some of the most significant advances in core computer science-based technologies. As such, the book is directed towards professors, researchers, scientists, engineers and students in computer science-related disciplines. It is also directed towards readers who come from other disciplines and are interested in becoming versed in some of the most recent computer science-based technologies. An extensive list of bibliographic references at the end of each chapter guides the reader to probe further into the application areas of interest to him/her.

1.1 Editorial Note

At the dawn of the 4th Industrial Revolution [1, 2], the field of core computer science-based technologies is growing continuously and rapidly, developing in both itself and towards applications of many other disciplines. The advances of computer science already affect many aspects of everyday life, the workplace and people's homes as well as human relationships both on professional and social interactions levels and are expected to affect yet more aspects in the foreseeable future. In this respect, researchers need to respond quickly, creatively and as accurately as possible to new challenges that arise perpetually in many areas of computer science which are often interrelated and, sometimes, involved in disciplines other than computer science, in which machine learning and artificial intelligence are expected to play an important role [3, 4].

Papers in Honor of Professor Nikolaos Alexandris.

G. A. Tsihrintzis (✉) · M. Virvou
Department of Informatics, University of Piraeus, Piraeus 185 34, Greece
e-mail: geoatsi@unipi.gr

G. A. Tsihrintzis and M. Virvou (eds.), *Advances in Core Computer Science-Based Technologies*, Learning and Analytics in Intelligent Systems 14,
https://doi.org/10.1007/978-3-030-41196-1_1

For example, the *Internet*, which is an achievement of the area of networks and telecommunications, has provided the means for the growth of e-services and m-services in disciplines such as *education* [5–7], *health* [8, 9], *entertainment* [10] or *commerce* [11, 12]. Similarly, the development of user interfaces within software engineering has expanded enormously in a way that the area of *human-computer interaction* requires a major part of the whole software product development effort in all kinds of computers, including handheld devices [7]. Additionally, the *Internet of Things* (IOT), an active stage of the 4th Industrial Revolution, is expected to "invade" everyday life with new technologies, such as smart homes [13], smart cities [14, 15], smart industries [16] or smart health [17]. At the same time, the more the computers expand in everyday life of people, the greater the need becomes for protecting all kinds of data from malicious abuse [18–21], a task that is now strongly required under the recently-established European *General Data Protection Regulation* (GDPR) [22–26].

Thus, one can identify many current core computer science-based technologies which form the foundation for many areas of modern and future everyday life, including:

1. *Computer Science-based Technologies in Education.*
2. *Computer Science-based Technologies in Risk Assessment and Readiness.*
3. *Computer Science-based Technologies in IoT, Blockchains and Electronic Money.*
4. *Computer Science-based Technologies in Mobile Computing.*
5. *Computer Science-based Technologies in Scheduling and Transportation.*
6. *Computer Science-based Technologies in Medicine and Biology.*
7. *Theoretical Advances in Computer Science with Significant Potential Applications in Technology.*

The book at hand aims at exposing its reader to some of the most significant advances in core computer science-based technologies. As such, the book is directed towards professors, researchers, scientists, engineers and students in computer science-related disciplines. It is also directed towards readers who come from other disciplines and are interested in becoming versed in some of the most recent computer science-based technologies.

More specifically, the book at hand consists of an editorial chapter (this chapter) and an additional seventeen (17) chapters. All chapters in the book were invited from authors who work in the corresponding chapter theme and are recognized for their significant research contributions. In more detail, the chapters in the book are organized into seven parts, as follows.

The *first part* of the book consists of four chapters devoted to *Advances in Computer Science-Based Technologies in Education*.

Specifically, Chap. 2, by Andreas Triantafyllou, George A. Tsihrintzis, Maria Virvou and Efthimios Alepis is on "*A Bimodal System for Emotion Recognition* via *Computer of Known or Unknown Persons in Normal or Fatigue Situations*." The authors summarize and present research on the detection via computer of emotions and/or the mood of a single person or groups of people by using a visual-facial and a keyboard-based modality, with special interest in applications in the educational environment.

Chapter 3, by Georgios Feretzakis, Dimitrios Kalles and Vassilios S. Verykios is on "*Knowledge Hiding in Decision Trees for Learning Analytics Applications*." The authors focus on preserving the privacy of sensitive patterns when inducing decision trees and demonstrates the application of a heuristic to an educational data set.

Chapter 4, by Stylianos Karagiannis and Emmanouil Magkos, is on "*Engaging Students in Basic Cybersecurity Concepts Using Digital Game-Based Learning: Computer Games as Virtual Learning Environments*." The authors present a commercial computer game is evaluated for the effectiveness of using gamification to the learning process. The result of this approach is a learning experience, featuring positive outcomes in terms of engagement and distinct impact in terms of perceived learning.

Chapter 5, by David Martín Santos Melgoza, José Armando Landa Hernández, Franco Ariel Ulloa González, and Abel Valdés Ramírez, is on "*Didactics for the development of Mathematical Thinking and the Sense of Academic Agency.*" The authors propose to validate the structural relationships between the measurements of the academic agency components of student self-report and those observed in a mathematical learning episode, in order to provide evidence of how the metacognitive processes involved in success interact before and during an academic learning process.

The *second part* of the book consists of three chapters devoted to *Advances in Computer Science-Based Technologies in Risk Assessment and Readiness*.

Specifically, Chap. 6, by Dimitris Gritzalis, George Stergiopoulos, Efstratios Vasilellis and Argiro Anagnostopoulou, is entitled "*Readiness Exercises: Are Risk Assessment Methodologies Ready for the Cloud?*" The authors compare and examine whether widely used and accepted risk management met-hods and tools (e.g. NIST SP800, EBIOS, MEHARI, OCTAVE, IT-Grundschutz, MAGERIT, CRAMM, HTRA, Risk-Safe Assessment, CORAS) are suitable for cloud computing environments. Based upon existing literature, they point out the essential characteristics that any risk assessment met-hod addressed to cloud computing should incorporate and suggest three new ones that are more appropriate based on their features.

Chapter 7, by Spiridon Papastergiou, Eleni-Maria Kalogeraki, Nineta Polemi and Christos Douligeris, is on "*Challenges and Issues in Risk Assessment in Modern Maritime Systems*." The authors present challenges and risks concerning cyber security in a supply chain environment.

Chapter 8, by Ioannis Stellios, Panagiotis Kotzanikolaou, Mihalis Psarakis and Cristina Alcaraz, is on "*Risk assessment for IoT-enabled Cyber-Physical Systems*." The authors review risk assessment methodologies for IoT-enabled cyber-physical systems and present a high-level risk assessment approach to enable an assessor to identify and assess non-obvious (indirect or subliminal) attack paths introduced by IoT technologies.

The *third part* of the book consists of three chapters devoted to *Advances in Computer Science-Based Technologies in IoT, Blockchains and Electronic Money*.

Specifically, Chap. 9, by Stefanos Gritzalis, Maria Sideri, Aggeliki Kitsiou, Eleni Tzortzaki and Christos Kalloniatis, is on "*Sustaining Social Cohesion in Information and Knowledge Society: The Priceless Value of Privacy*." The authors, based on a literature review, present the issue of privacy in social network sites focusing on factors that affect people's privacy concerns and behavior while relating these to social cohesion.

Chapter 10, by Georgios Spathoulas, Lydia Negka, Pankaj Pandey and Sokratis Katsikas, is entitled "*Can Blockchain Technology Enhance Security and Privacy in the Internet of Things*?" The authors discuss the convergence of the blockchain and internet of things technologies, analyse possible use cases, where blockchain technology can enhance internet of things security and privacy, and propose enhancements of blockchain technology to make it appropriate for application in the internet of things domain.

Chapter 11, by Pankaj Pandey and Sokratis Katsikas, is on "*The Future of Money: Central Bank Issued Electronic Money*." The authors present the key aspects of currency in modern society, the advancements in the field of financial technologies and how could these be harnessed to launch a central bank-backed electronic currency, while also discussing the position of various central banks and governments on cryptocurrencies, blockchain technology, and the initiatives related to issuing electronic currency.

The *fourth part* of the book consists of two chapters devoted to *Advances in Computer Science-Based Technologies in Mobile Computing*.

Specifically, Chap. 12, by Mike Burmester and Jorge Munilla, is on "*Lightweight stream authentication for mobile objects*." The authors propose a lightweight stream authentication scheme for mobile objects that approximates continuous authentication, only requires the user and object to share a loosely synchronized pseudo-random number generator and is provably secure.

Chapter 13, authored by Constantinos Kapetanios, Theodoros Polyzos, Efthimios Alepis and Constantinos Patsakis, is entitled "*From no Content to User Profiling, Surveillance and Exploitation*." The authors showcase that unprivileged apps, without actually using any permissions, can harvest a considerable amount of valuable user information via monitoring and exploiting the file and folder metadata of the most well-known messaging apps in Android.

The *fifth part* of the book consists of two chapters devoted to *Advances in Computer Science-Based Technologies in Scheduling and Transportation*.

Specifically, Chap. 14, by Konstantina Chrysafiadi, is on "*A Fuzzy Task Scheduling Method*." The author presents a rule-based fuzzy task scheduling method for use in operating systems, which considers both the execution time and the waiting time for each task.

Chapter 15, by Manolis N. Kritikos and Pantelis Z. Lappas, is on "*Computational Intelligence and Combinatorial Optimization Problems in Transportation Science*." The authors highlight the use of computational intelligence algorithms for solving a special class of combinatorial optimization problems in transportation science called routing problems.

The *sixth part* of the book consists of two chapters devoted to *Advances in Computer Science-Based Technologies in Medicine and Biology*.

Specifically, Chap. 16, by Athanasios Sofronis and Panagiotis Vlamos, is on "*Homeodynamic Modelling of Complex Abnormal Biological Processes.*" The authors outline the concept of Homeodynamic modelling of complex abnormal biological processes and then proceed to present a case study on the application of bifurcation theory and stability analysis, both topological approaches of dynamical systems, on three biological mechanisms of great importance in the context of Homeodynamics, namely protein folding, protein dynamics and epigenetics.

Chapter 17, by Marios Poulos, Sozon Papavlasopoulos, is on "*Metadata Web Searching EEG Signal.*" The authors investigate the problem of developing appropriate information search and retrieve mechanisms and tools in the web environment for the dataclass of uniform diagnostic EEG features and propose a suitable metadata schema.

Finally, the *seventh part* of the book consists of one chapter devoted to *Theoretical Advances in Computer Science with Significant Potential Applications in Technology.*

Specifically, Chap. 18, by Doru Stefanescu, is on "*Algorithmic Methods for Computing Bounds for Polynomial Roots.*" The author presents some basic results concerning the evaluation of the absolute values of roots of univariate polynomials with complex or real coefficients and discusses classical and modern algorithmic methods and their computational efficiency.

In this book, we have presented some of the most significant advances in core computer science-based technologies, while honouring Professor Nikolaos Alexandris who has inspired, advised and mentored computer science students and colleagues for almost fifty years. The book is directed towards professors, researchers, scientists, engineers and students in computer science-related disciplines. It is also directed towards readers who come from other disciplines and are interested in becoming versed in some of the most recent computer science-based technologies. We hope that all of them will find it useful and inspiring in their works and researches.

On the other hand, societal demand continues to pose challenging problems, which require ever more efficient tools, methodologies, systems and computer science-based technologies to be devised to address them. Thus, the reader may expect that additional related volumes will appear in the future.

References

1. J. Toonders, Data is the new oil of the digital economy. Wired. https://www.wired.com/insights/2014/07/data-new-oil-digital-economy/
2. K. Schwabd, The fourth industrial revolution—what it means and how to respond, *Foreign Affairs*, December 12, 2015. https://www.foreignaffairs.com/articles/2015-12-12/fourth-industrial-revolution
3. G.A. Tsihrintzis, D.N. Sotiropoulos, L.C. Jain (eds.), *Machine Learning Paradigms—Advances in Data Analytics*, volume 149 in Intelligent Systems Reference Library Book Series, Springer 2018

4. G.A. Tsihrintzis, M. Virvou, E. Sakkopoulos, L.C. Jain (eds.), *Machine Learning Paradigms—Applications of Learning and Analytics in Intelligent Systems*, volume 1 in Learning and Analytics in Intelligent Systems Book Series, Springer 2019
5. K. Chrysafiadi, M. Virvou, *Advances in Personalized Web-based Education*, volume 78 in Intelligent Systems Reference Library Book Series, Springer 2015
6. M. Virvou, E. Alepis, G.A. Tsihrintzis, L.C. Jain, *Machine Learning Paradigms—Advances in Learning Analytics*, volume 158 in Intelligent Systems Reference Library Book Series, Springer 2020
7. E. Alepis, M. Virvou, *Object-Oriented User Interfaces for Personalized Mobile Learning*, volume 64 in Intelligent Systems Reference Library Book Series, Springer 2014
8. S. Bahri, N. Zoghlami, M. Abed, J. Manuel, R.S. Tavares, Big data for healthcare: a survey. IEEE Access **7**, 7397–7408 (2019)
9. Qiong Cai, Hao Wang, Zhenmin Li, Xiao Liu, A survey on multimodal data-driven smart healthcare systems: approaches and applications. IEEE Access **7**, 133583–133599 (2019)
10. B.H. Thomas, A survey of visual, mixed, and augmented reality gaming. Comput. Entertain **10**(1), 3:1–3:33 (2012)
11. Peihai Zhao, Zhijun Ding, Mimi Wang, Ruihao Cao, Behavior analysis for electronic commerce trading systems: a survey. IEEE Access **7**, 108703–108728 (2019)
12. B. Yoo, M. Jang, A bibliographic survey of business models, service relationships, and technology in electronic commerce. Electr. Commer. Res. Appl. 33 (2019)
13. Hongbo Jiang, Chao Cai, Xiaoqiang Ma, Yang Yang, Jiangchuan Liu, Smart home based on WiFi sensing: a survey. IEEE Access **6**, 13317–13325 (2018)
14. Du Rong, Paolo Santi, Ming Xiao, Athanasios V. Vasilakos, Carlo Fischione, The sensable city: a survey on the deployment and management for smart city monitoring. IEEE Commun. Surv. Tutor. **21**(2), 1533–1560 (2019)
15. B.P.L. Lau, M.S. Hasala, Y. Zhou, N.U. Hassan, C. Yuen, M. Zhang, U.-X. Tan, A survey of data fusion in smart city applications. Inf. Fus. **52**, 357–374 (2019)
16. X. Liu, J. Cao, Y. Yang, S. Jiang, CPS-based smart warehouse for industry 4.0: a survey of the underlying technologies. Computers **7**(1), 13 (2018)
17. M.M. Dhanvijay, S.C. Patil, Internet of Things: A survey of enabling technologies in healthcare and its applications. Comput. Netw. **153**, 113–131 (2019)
18. N. Panwar, S. Sharma, S. Mehrotra, L. Krzywiecki, N. Venkatasubramanian, Smart home survey on security and privacy, (2019), https://arxiv.org/abs/1904.05476v2
19. M. Khawla, T. Mazri, A survey on the security of smart homes: issues and solutions, ICSDE 2018, pp. 81–87
20. David Eckhoff, Isabel Wagner, Privacy in the smart city—applications, technologies, challenges, and solutions. IEEE Commun. Surv. Tutor. **20**(1), 489–516 (2018)
21. Abdullah Algarni, A survey and classification of security and privacy research in smart healthcare systems. IEEE Access **7**, 101879–101894 (2019)
22. Omer Tene, Katrine Evans, Bruno Gencarelli, Gabe Maldoff, Gabriela Zanfir-Fortuna, GDPR at year one: enter the designers and engineers. IEEE Secur. Priv. **17**(6), 7–9 (2019)
23. Paul Breitbarth, The impact of GDPR one year on. Netw. Secur. **7**, 11–13 (2019)
24. A. Brombacher, Quality, reliability, data, and the impact of the GDPR .... Qual. Reliab. Eng. Int. **35**(4), 869 (2019)
25. D. Huth, F. Matthes, "Appropriate technical and organizational measures": identifying privacy engineering approaches to meet GDPR requirements, in *The 25th Americas Conference on Information Systems (AMCIS 2019)*, Cancún International Convention Center, Cancún, México, August 15–17, 2019
26. D. Rösch, T. Schuster, L. Waidelich, S. Alpers, Privacy control patterns for compliant application of GDPR, in *The 25th Americas Conference on Information Systems (AMCIS 2019)*, Cancún International Convention Center, Cancún, México, August 15–17, 2019

Part I
Computer Science-Based Technologies in Education

Chapter 2
A Bimodal System for Emotion Recognition via Computer of Known or Unknown Persons in Normal or Fatigue Situations

Andreas M. Triantafyllou, George A. Tsihrintzis, Maria Virvou, and Efthimios Alepis

Abstract The recognition of emotion/mood in groups of people is particularly interesting as it constitutes a whole new research area that can provide solutions to various high level problems and applications, such as, for example, the development of educational technologies for groups of learners in which teaching is adjusted to the mood of the group. The purpose of this chapter is to summarize and present research on the recognition via computer of emotions and/or the mood of a single person or groups of people by using a visual-facial modality combined with a keyboard-based modality. Additionally, this chapter presents research about a program that recognizes the emotions/mood of groups of people in live pictures collected directly from webcams. New algorithmic approaches are proposed and conclusions are drawn for use in similar systems. The implemented system successfully recognizes emotions in the following situations:

1. The system is only trained to recognize emotions/the mood of a specific person.
2. The system is trained with one set of sample emotions/mood from which it aims at generalizing and recognizing emotions/mood of unknown people.
3. The system is trained with several sets of sample emotions/mood from which it aims at generalizing and recognizing emotions/the mood of unknown people.

The chapter is complemented with illustrations of use of the implemented system, as well as suggestions for future related work in this area.

A. M. Triantafyllou · G. A. Tsihrintzis (✉) · M. Virvou · E. Alepis
Software Engineering Lab, Department of Informatics, University of Piraeus, Piraeus 185 34, Greece
e-mail: geoatsi@unipi.gr

M. Virvou
e-mail: mvirvou@gmail.com

E. Alepis
e-mail: talepis@unipi.gr

Pattern Recognition and Machine Learning-Multimedia Systems Lab, Department of Informatics, University of Piraeus, Piraeus 185 34, Greece

G. A. Tsihrintzis and M. Virvou (eds.), *Advances in Core Computer Science-Based Technologies*, Learning and Analytics in Intelligent Systems 14,
https://doi.org/10.1007/978-3-030-41196-1_2

2.1 Introduction

Computer science-based technologies are a fast-growing discipline in which new fields constantly emerge for research and development. One of these emerging fields is the subject of this chapter on recognizing through a computer the emotion and or disposition of a single person, as well as of groups of people.

Emotion recognition can find use in many problems and applications of modern information sciences [1–6]. Specific areas that can benefit from automated emotion recognition include (without being limited to) the following:

- *Robotics*, in which robots could deduce the emotional state of humans and interact with them accordingly.
- *Educational technologies*, in which recognizing the emotional state and mood of learners can make the teaching experience more effective and provide better results.
- *Human-machine interaction,* in which software systems self-adapt according to their users' emotional state.

In addition to the above indicative examples, there are many more application areas in which emotion recognition could prove very useful. Ultimately, the research fields of human-machine interaction and affective computing focus on studying and developing software systems and hardware devices that can recognize, interpret, process, and simulate human affects [7–16].

The aim of this chapter is to summarize and present new research on the recognition via computer of emotions and/or the mood of a single person or groups of people by using a visual-facial modality. Some of the theoretical aspects analyzed in this chapter include:

- Relationship of emotions/mood with samples for computational processing.
- Suggestions for researching new potential samples.
- Suggestions for sample collection experiments.
- Algorithm suggestions and data-sample processing methods.
- Proposals for continuation of related research and development.
- Conclusions and future work.

Regarding the practical aspects of the chapter, we have developed and present real-time software that recognizes the emotional state/mood of a single person and/or a group of people originally based on the image of each person's face. This program processes images and video selected via the computer camera and recognizes faces and mood automatically based on training samples. The program also supports direct sampling for its training, again by using the camera and determining the emotion/mood of the sample.

The practical part of our work also contributes to the presentation of results and their comparison and analysis, as the samples in the training set will be different from experiment to experiment. This helps us seek optimized processing and come up with novel algorithms, which are essential for better understanding the problem of emotion/mood recognition.

Additionally, the software provides a function that supports recognition of the users' "writing style" at any time by presenting frequency-related and time-related results between the keys users press on the keyboard, as well as information on time gaps between key pressing and the mistakes the users make. This feature allows us to extract conclusions which, in the future, can be integrated into algorithms that use a combination of data to achieve more accurate results about human emotion/mood recognition via computer.

2.2 What is Emotion?

At this point, we have to define the term "emotion" with regard to relevant literature. In computer science it firstly appeared in [1], while a more recent and thorough study on the topic of visual affect recognition can be found in [2]. The definition of the term is not sufficient to take us further, as some distinct grouping/sorting of emotions is necessary. Grouping/sorting of emotions is done in a variety of ways in the literature. Ekman and his co-authors, influenced by Tomkins's interpretation [17] of Darwin's ideas on human feelings and their expressions [18], suggested that there are six basic emotions: anger, disgust, fear, happiness, sadness and surprise [19, 20].

Today, a vast literature exists of psychological works investigating human emotion [e.g. 17–24]. However, the research in the field of human emotion is, to a large extent, incomplete and several issues remain under investigation. Thus, in this chapter, we will specifically concentrate on only the following six emotional states, which appear to be the most common in human-computer interaction:

- Neutral
- Happiness—Joy
- Sadness
- Surprise
- Anger
- Disgust.

2.3 Features to Draw Conclusions

After specifying the emotional states we will be dealing with, we identified features which allow the recognition of these emotional states from facial images of people.

One of the difficulties we faced is that of pretense. For example, if a person shows a smiling face and no other information is available, then we probably think the person is in a state of happiness. But is this person really happy or just pretending to be? A "fake" smile can easily lead both humans and an emotional state recognition program to erroneous conclusions. Research in emotion recognition based on speech and face expressions has yielded a significant number of fruitful results, while the majority of existing surveys concentrate on documenting these basic modalities [7–16].

Assuming that we talk to a smiling person and observe him/her further, additional information is also available, such as the timbre of his/her voice, the way he/she looks at us, his/her posture and even the content of our conversation. Certainly, this additional information allows us to draw a more reliable conclusion about the emotional state of this person. Similarly, it is conceivable that an electronic program can identify the emotional state of a person with higher reliability if such additional information is available.

Several factors affect the ability of humans to deduce the emotional state of another person. Such factors include previous experiences of humans or how well they know the other person. Currently, the ability of computers to deduce the emotional state of a person is still inferior to that of humans. In some cases, however, it is feasible for a computer to deduce the emotional state of a person. For example, in a human-computer interaction environment, a human is directly related and interacting with a computer and the computer can deduce relatively reliably its user's emotional state by identifying facial expressions of the user or his/her typing and mouse use patterns or lingual and para-lingual information or a combination of all these.

More generally, in the currently-feasible automated emotion recognition, we can identify two different data-collection frameworks to be used for building corresponding systems, namely *specialized sensor-based* (which require additional sensors and other hardware/software than what usually accompanies a common desktop computer) and *software-based* (which requires common hardware, such as keyboard, mouse, camera, microphone) frameworks. An implemented system that utilizes these frameworks successfully in mobile devices has been shown in [5].

Specifically, *specialized sensor-based* frameworks include:

- Cardiac pulse recognition and monitoring.
- Brain wave recognition and monitoring.
- Body posture recognition and monitoring.
- Walk mode recognition and monitoring.

On the other hand, *software-based* frameworks include:

- Face recognition and monitoring.
- Text typing recognition and tracking.
- Sound recognition and tracking for lingual and para-lingual information extraction.
- Mouse movement recognition and tracking.
- User errors recognition and tracking.
- Recognition and tracking of a specific user's habits (e.g. musical genre to which he/she usually listens).

In this chapter, we present software that has been developed to implement a *bi-modal* system for drawing conclusions about a computer user's emotional state.

Specifically, our system consists of (1) a visual-facial module and (2) a keyboard-based module. Our software provides emotion recognition via facial image and typing pattern recognition and tracking, in combination with assessment of the level of concentration of its user based on the mistakes he/she makes while typing.

The visual-facial module uses a commonly-available camera (possibly an embedded laptop camera) to collect and process facial images of the user. On the other hand, the keyboard-based module tracks and processes information of the keys pressed by the user and corresponding frequencies, as well as the frequencies of mistyped words and use of the backspace key.

2.4 Suggestions for Sample Collection Experiments

The ultimate goal of our research work was to implement a fully functional software system for human-computer interaction, which would respond taking into account its user's emotional state. Thus, a number of experiments were designed to help us to collect visual-facial and keyboard data samples to create databases for later use in emotion recognition algorithms.

The first set of experiments related to the collection of samples of facial images from people in various emotional states. Towards this goal, we invited people to participate in this process through the internet and via an online platform.

Initially, we created an online platform to inform interested users about the goal of the experiment, provide them with information about the field of emotion recognition, and enable them to send us their facial images in various emotional states. Users of the platform could send images anonymously, without a need to create an account first. The images sent samples were checked before they were properly tagged and entered into the corresponding database.

The second set of experiments related to the collection of additional data samples under the following two conditions:

- Real-time collection of simultaneous data samples from several users for higher reliability.
- Data samples after some emotional stimuli that would attempt to bring the people involved in the experiment into a specific emotional state.

In order to simultaneously collect different visual-facial and keyboard-based data samples from both a single human and a group of people, all the people who participated in the experiment were connected simultaneously and data collection and extraction programs ran in parallel. More specifically, we asked participants to watch two short movies that would invoke an intense emotional response. Data samples were collected before, during and after each film.

For the collection of keyboard-based data samples, we asked participants to type a given text before, after and in between the two films. In this way, we were be able to observe and compare differences in keyboard stroke patterns of the participants according to the emotional stimuli they had received and link them to the text typing style and to the mistakes that can be made.

2.5 Emotion Recognition Application

With the software requirements set as previously, we have developed a functional emotion recognition and sample collection program which, based on facial images, is able to recognize in real time the emotional state of an individual user or groups of users. Additionally, through the program one can collect samples of facial images, as well as samples of keyboard-based data. Specifically, the functional and hardware requirements of our software were set as analyzed in the following.

2.5.1 Functional Requirements

The functional requirements for our emotion recognition application are:

- To be able to collect samples of facial images.
- To be able to recognize in real time people's emotions from their facial images based on the samples it has gathered.
- To have a text typing option.
- To collect and display metrics from samples of the text typing.
- To collect and display metrics about the mistakes that someone might make when typing text.

2.5.2 Hardware Requirements

The hardware requirements to support the resulting application are:

- (Desktop) computer.
- Keyboard.
- Screen.
- (Web)camera.

2.5.3 Use Case Diagram—UML

The corresponding use case diagram for our emotion recognition software is illustrated in Fig. 2.1.

Key functions include:

- Recognition and storage of the facial image sample.
- Recognizing emotions of an individual or a group of people.
- Text typing and recognize samples of typing style and mistakes.
  - Sample recognition of electronic form of typing style.
  - View metrics of sample for typing style recognition.
  - View metrics of sample for recognition of mistakes in text typing.

2.5.4 Measurements, Samples and Conclusions

Having specified the functions of the emotion recognition program that we have developed, we now present its usability with regard to making measurements, collecting data samples and recognizing emotions.

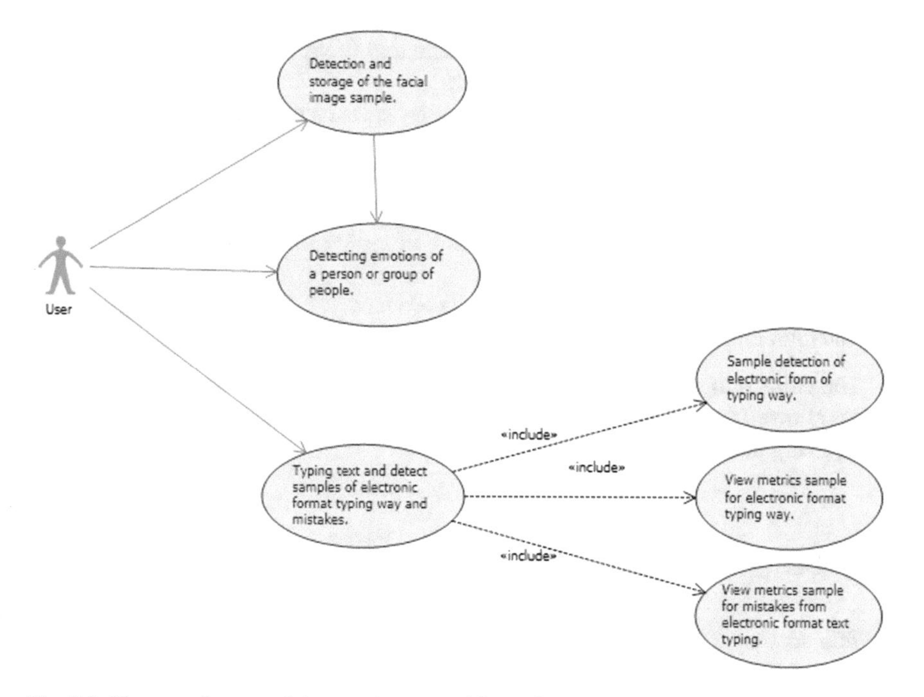

Fig. 2.1 Use case diagram of the emotion recognition software

As we have mentioned in Sect. 2.2, the visual-facial data samples for recognizing emotions are (black and white) facial of the six emotional states: *Neutral, Happiness—Joy, Sadness, Surprise, Anger* and *Disgust*. More specifically, we have collected samples of these six emotional states from which we can draw conclusions to be incorporated in the program to use in human emotion recognition. Of course, the program can be easily expanded to incorporate the recognition of additional emotional states.

On the other hand, and with regard to sample recognition for typing style module and the mistakes a user may make when typing as a result from his/her emotional state at the time of typing, the following parameters were recorded:

- Timestamps from each key pressed by the user.
- Information about each key.

Moe specifically, the following measurements are recorded:

- Average time delay between typing two keys (in milliseconds).
- Average time delay between typing two words, with one or more spaces between words (in milliseconds).
- The ratio of the average time delays typing keys and words.
- Average time for typing one word.
- The ratio between the average time for one word typing time and the total number of words.
- Total time for typing the entire text (not-counting the time spent on typing spaces).
- The ratio of the total time typing the entire text (not-counting the time spent on typing spaces) and the number of words.
- Total time of typing the entire text (counting the time spent on typing spaces).
- The ratio of the total time of typing the entire text (counting the time spent on typing spaces) and the number of words.
- The ratio between the total time of typing the entire text not-counting the time spent on typing spaces and the total time of typing the entire text counting the time spent on typing spaces.
- The total number of mistakes made and corrected.
- The total number of mistakes made and not corrected.

The collection of a sufficient number of data samples of the previous measurements allowed us to study them and draw conclusions so that proved them useful in recognizing emotions.

2.6 Using the Emotion Recognition Program

In this section, a user guide is presented for the emotion recognition program developed in the framework of this chapter.

At the program start, we face the starting menu illustrated in Fig. 2.2.

In this menu, the user can perform the following actions:

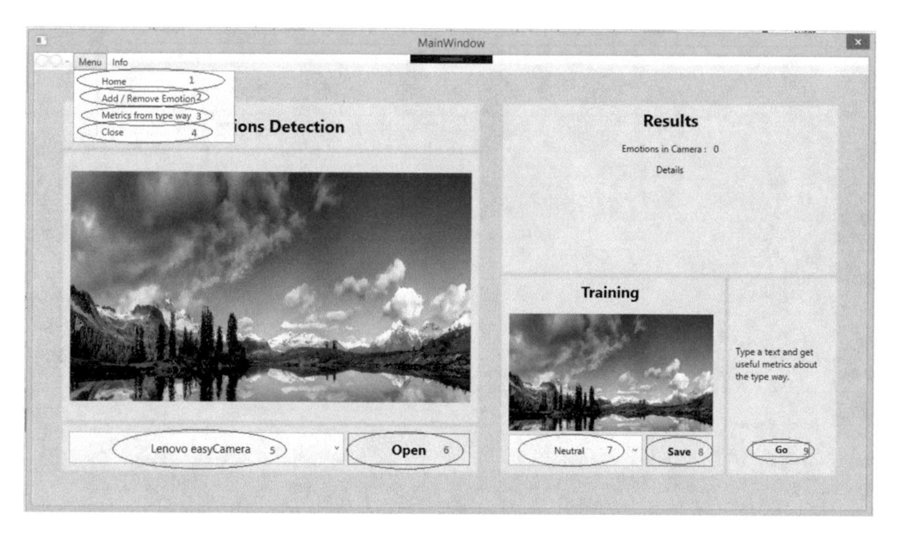

Fig. 2.2 Starting menu of the emotion recognition program

1. Add or delete types of emotions (emotional states).
2. Collect metrics with regard to the typing style of a piece of text.
3. Terminate the program.
4. See the list of cameras connected to the computer and select a camera.
5. Activate the selected camera.
6. We select a type of emotion (emotional state) so a sample of it can be stored.
7. Store the selected sample.
8. The user can be directed to the page where metrics can be collected regarding the style of typing a piece of text.

When the user selects to add or delete a type of emotion (emotional state), the screen in Fig. 2.3 appears:

In this screen, the user needs to follow the steps:

1. Type the name of the emotion he/she wishes to add (The program adds the emotion you have typed in Box 1 in Fig. 2.3) or
2. Double-click on an emotion if the emotion is already listed and the user wishes to delete it.

When the user selects to use the collection of metrics from typing a piece of text, the screen in Fig. 2.4 appears and the user follows the steps:

1. The user types in Box 1 in Fig. 2.4 the text from which he/she wishes to collect metrics.
2. Step 1 can be repeated as many times as the user wishes by clicking on Box 2 in Fig. 2.4.
3. The user sees the collected metrics by clicking on Box 3 in Fig. 2.4.

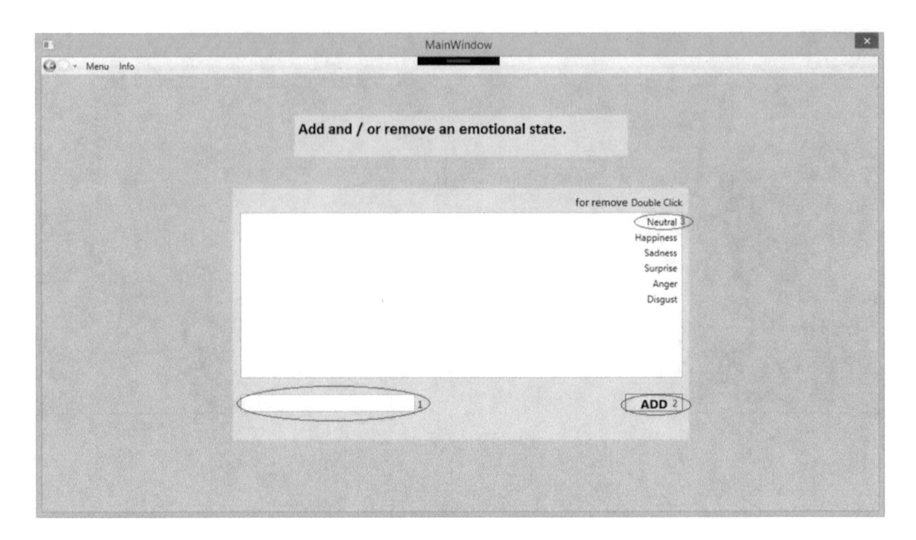

Fig. 2.3 Screen when adding or deleting an emotional state

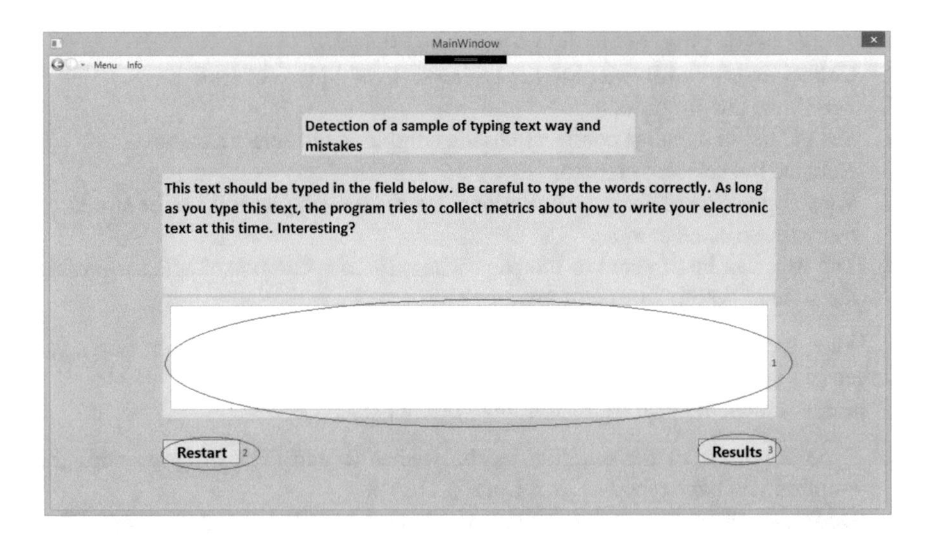

Fig. 2.4 Collection of metrics from typing a piece of text

More specifically, Step 3 in the previous screenshot leads the user to the screen shown in Fig. 2.5.

At this point, the user can follow the steps:

1. Click on the "Home" button to return to the home page or
2. Click on the "Type again" button, to return to the page where metrics are collected from text typing style.

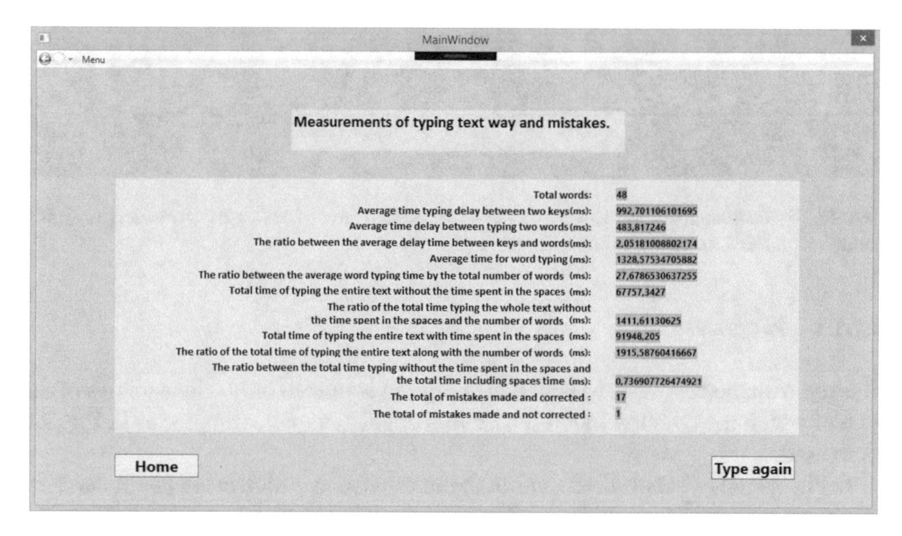

Fig. 2.5 Results of collection of metrics from typing a piece of text

2.7 Comparison of the Results—Experiments

In this section, we present some results from using the emotion recognition program. The following results are obtained in relation to recognition of emotion via facial images, which were extracted from several people who had formed facial expressions corresponding to various emotional states. Measurements on text typing style were also exported by different people at different times, whether in the same day or on different days.

2.7.1 Emotion Recognition

After developing the program for real-time emotion recognition, several experiments were made for sample collection and recognition. In these experiments, four different cases were investigated, as follows:

1. The program is trained only with samples of the face of which the emotion/mood it tries to recognize.
2. The program is trained with samples of one face other than the face of which the emotion/mood it tries to recognize.
3. The program is trained with samples of more than one faces other than the face of which the emotion/mood it tries to recognize.
4. The program is trained with samples of several faces including the face of which the emotion/mood it tries to recognize.

Fig. 2.6 Typical samples used in first case experiments. Emotional states (from left to right): neutral, happiness, sadness, surprise, anger, disgust

2.7.1.1 First Case

Starting with the first case, in which the program is trained only with samples of the face of which the emotion/mood it tries to recognize, typical samples as in Fig. 2.6 were used.

In Fig. 2.7a–f, we show the results of the emotion recognition in images of the same face expressing various emotions. Clearly, our software has correctly recognized the emotion via its visual-facial modality.

2.7.1.2 Second Case

We continue with the second case, in which the program is trained with samples of a face other than the face of which the emotion/mood it tries to recognize, typical samples as in Fig. 2.6 were used.

In Fig. 2.8a–f, we show the results of the emotion recognition. Clearly, our software has correctly recognized the Neutral (Fig. 2.8a), Happiness (Fig. 2.8b), Sadness (Fig. 2.8c) and Surprise (Fig. 2.8d) emotions via its visual-facial modality. However, the system has failed to recognize the Anger (Fig. 2.8e) and the Disgust (Fig. 2.8f) emotions via its visual-facial modality.

2.7.1.3 Third Case

Continuing with the third case, the program is trained with samples of several faces other than the face of which the emotion/mood it tries to recognize.

In Fig. 2.9a–f, we show the results of the emotion recognition. Clearly, our software has correctly recognized the Neutral (Fig. 2.9a), Happiness (Fig. 2.9b), Sadness (Fig. 2.9c) and Disgust (Fig. 2.9f) emotions via its visual-facial modality. However, the system has failed to recognize the Surprise and Anger emotions via its visual-facial modality.

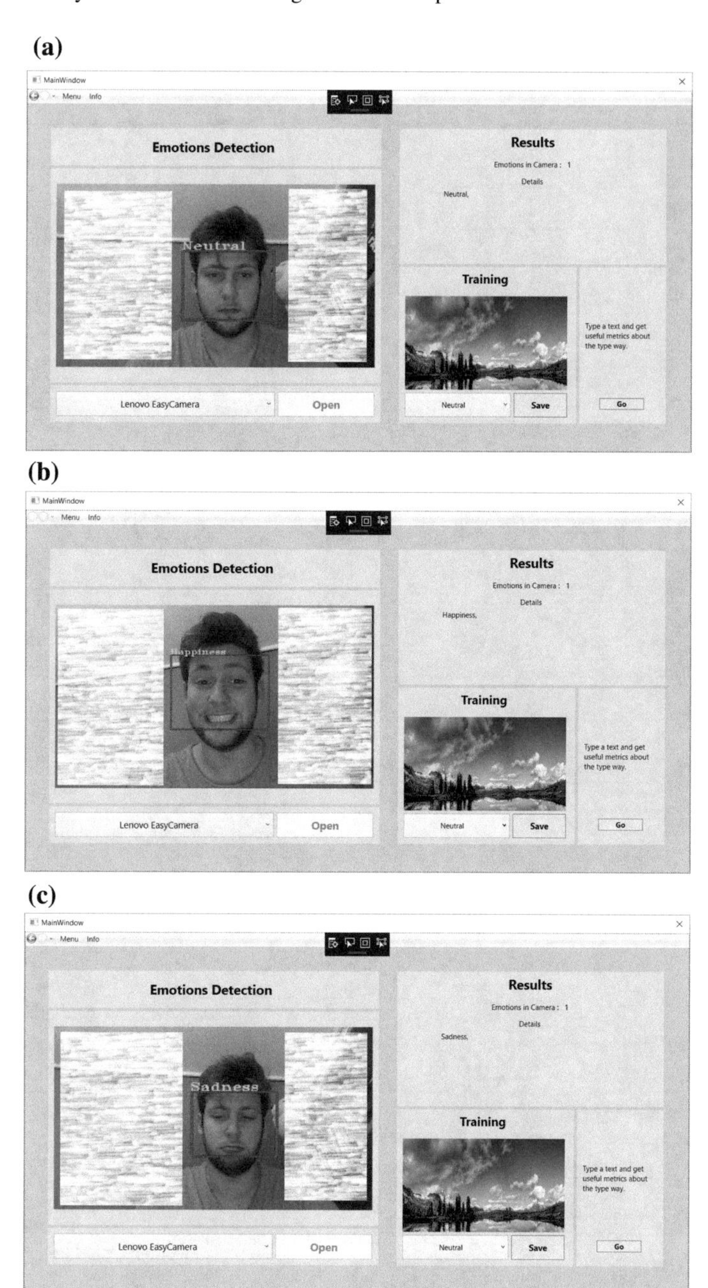

Fig. 2.7 **a** Sample of face expressing neutral emotion. **b** Sample of face expressing happiness. **c** Sample of face expressing sadness. **d** Sample of face expressing surprise. **e** Sample of face expressing anger. **f** Sample of face expressing disgust

(d)

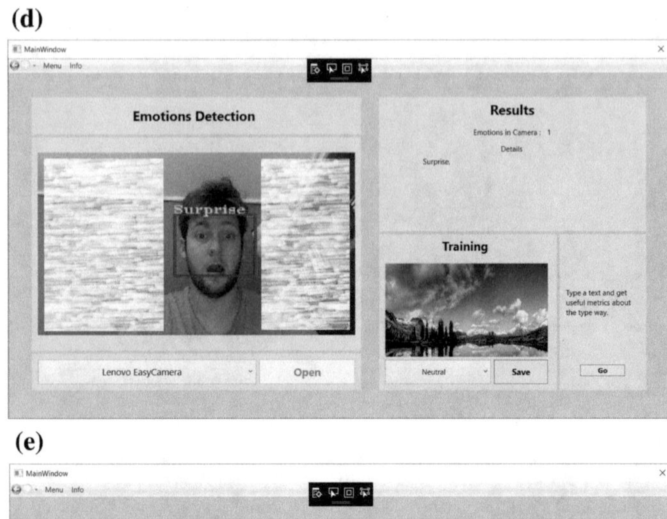

(e)

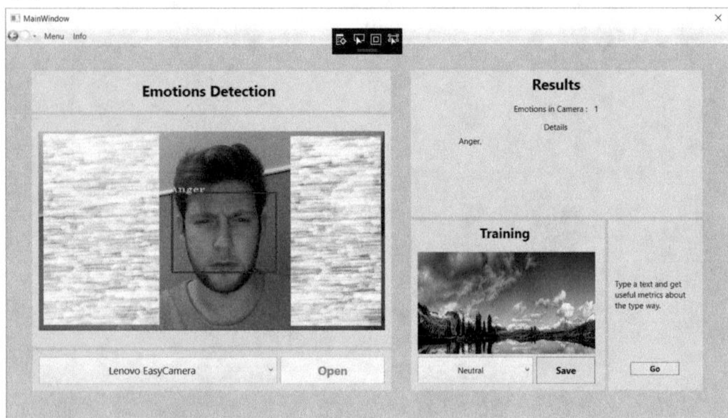

(f)

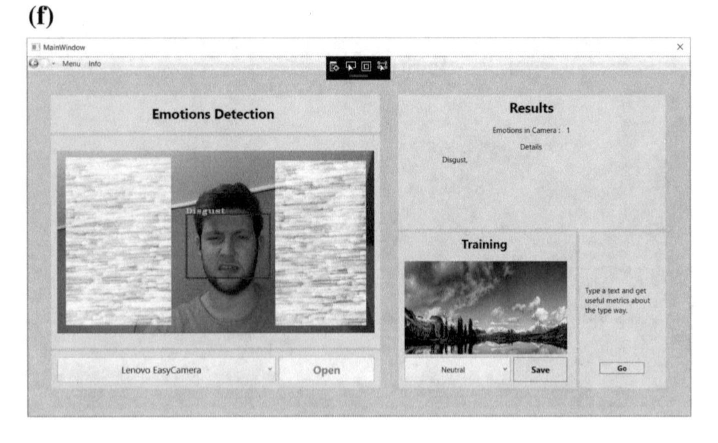

Fig. 2.7 (continued)

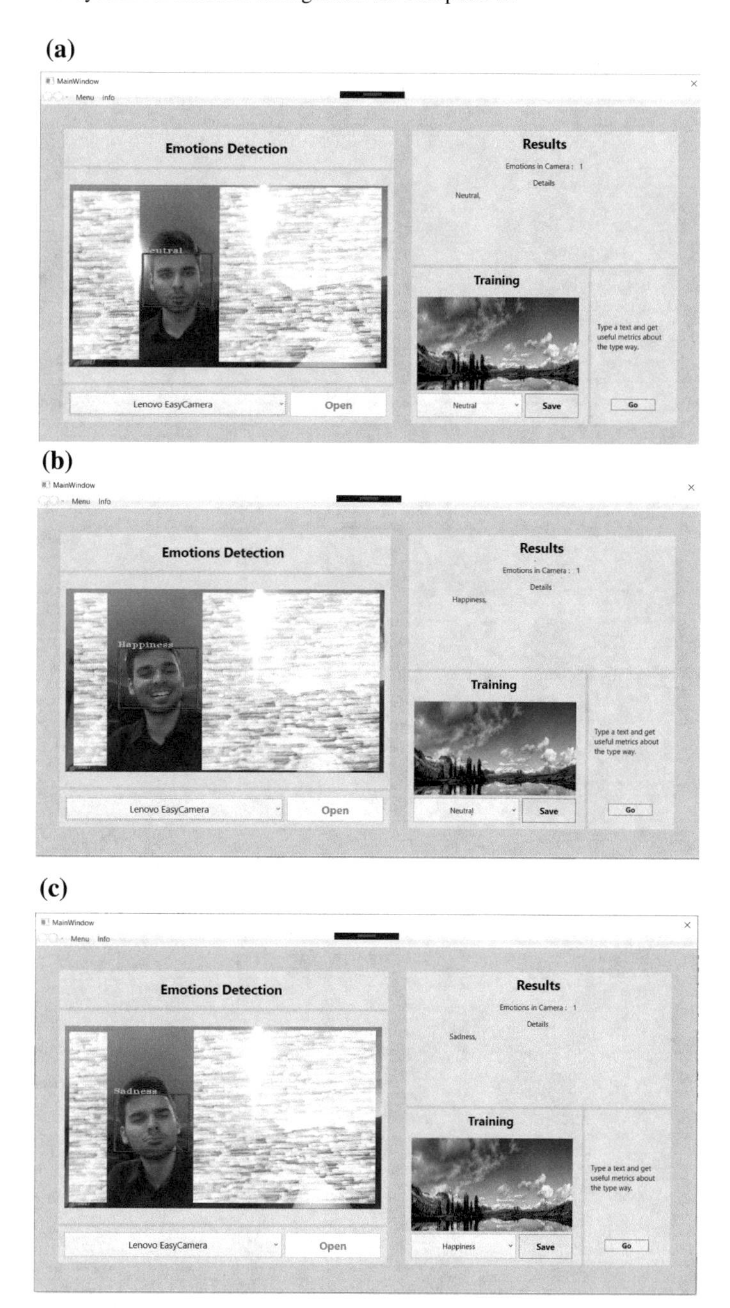

Fig. 2.8 **a** Successful recognition of neutral emotion of a face other than the specific user. **b** Successful recognition of happiness of a face other than the specific user. **c** Successful recognition of sadness of a face other than the specific user. **d** Successful recognition of surprise of a face other than the specific user. **e** Unsuccessful recognition of anger of a face other than the specific user. **f** Unsuccessful recognition of disgust of a face other than the specific user

(d)

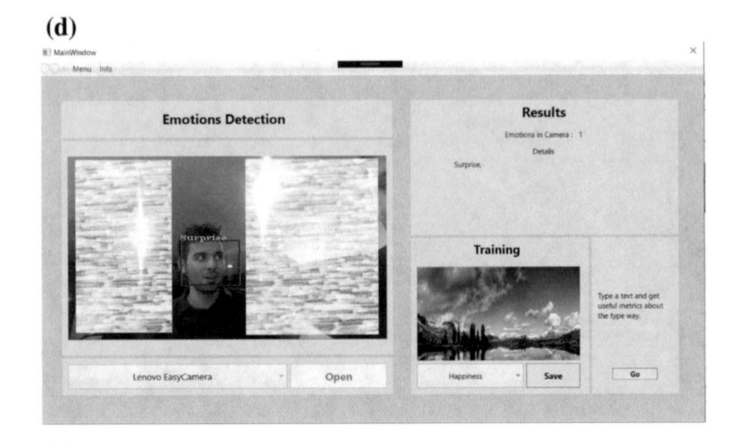

(e)

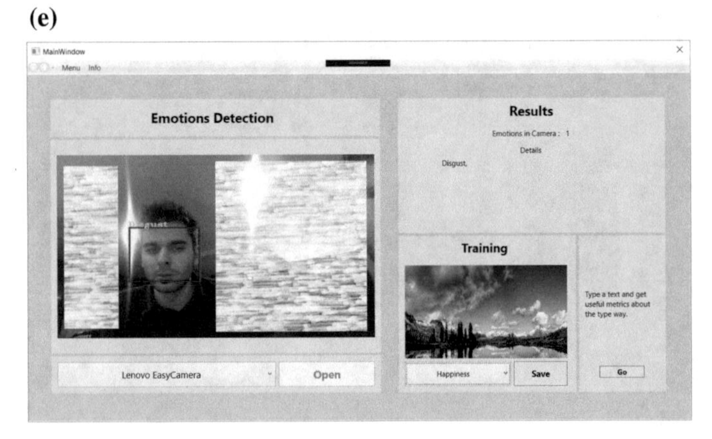

(f)

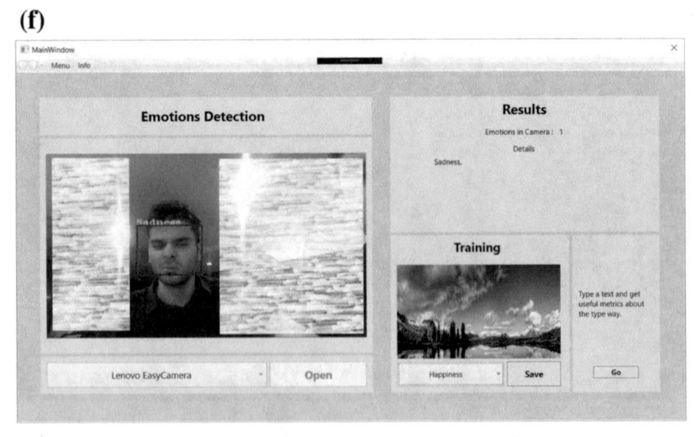

Fig. 2.8 (continued)

(a)

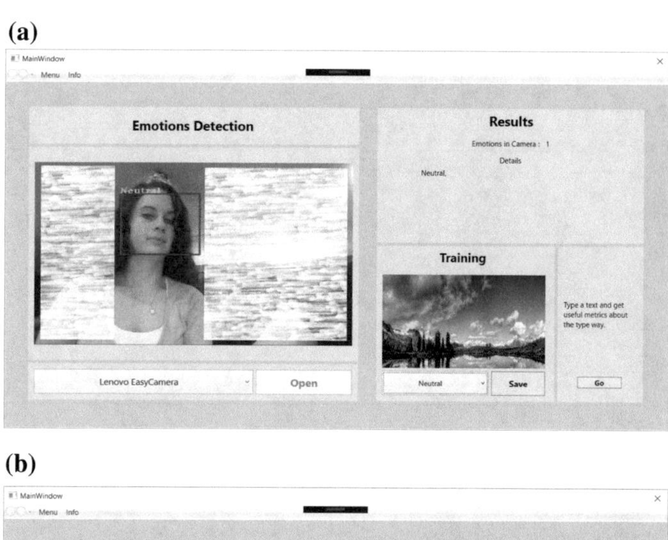

(b)

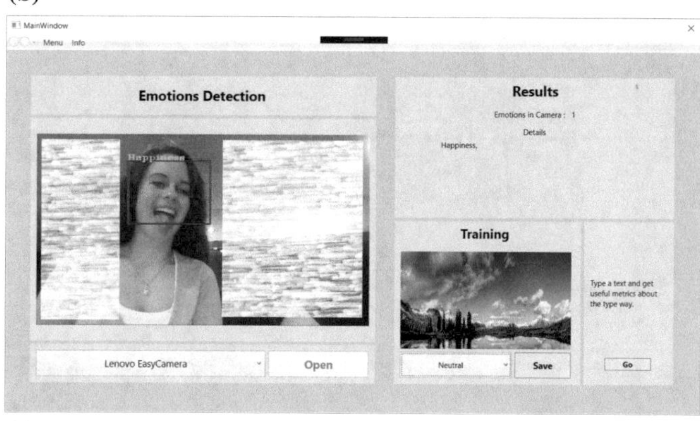

(c)

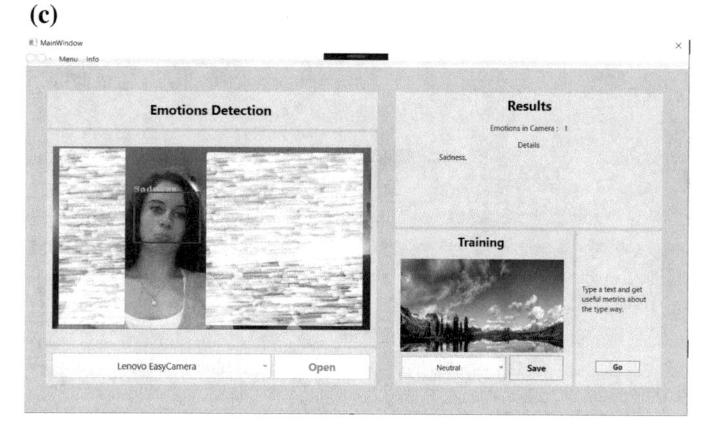

Fig. 2.9 **a** Successful recognition of neutral emotion of an unknown face. **b** Successful recognition of happiness emotion of an unknown face. **c** Successful recognition of sadness emotion of an unknown face. **d** Unsuccessful recognition of surprise emotion of an unknown face. **e** Unsuccessful recognition of anger Emotion of an unknown face. **f** Successful recognition of disgust emotion of an unknown face

(d)

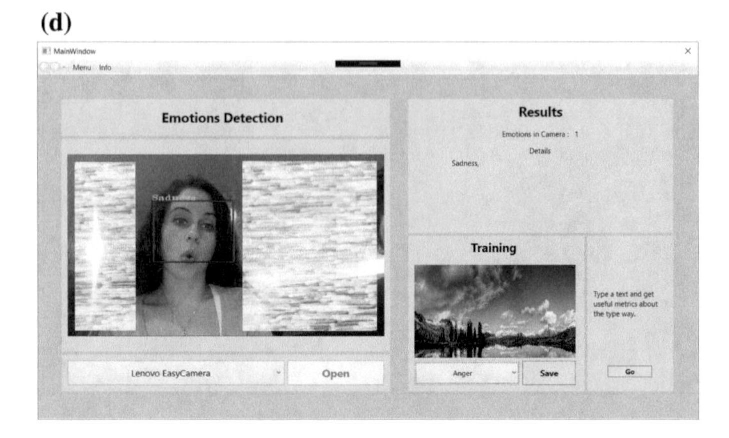

(e)

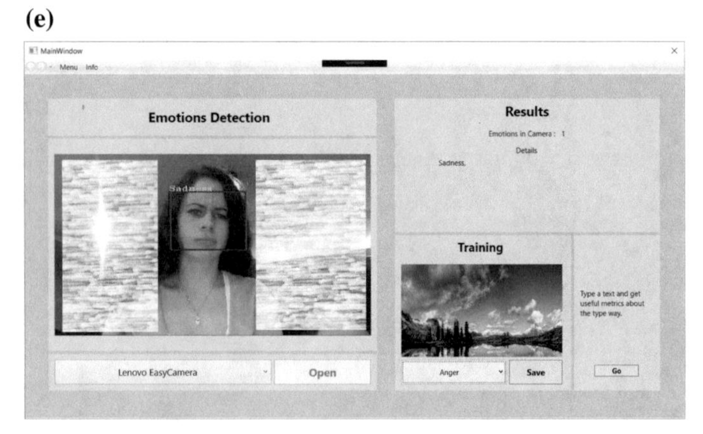

(f)

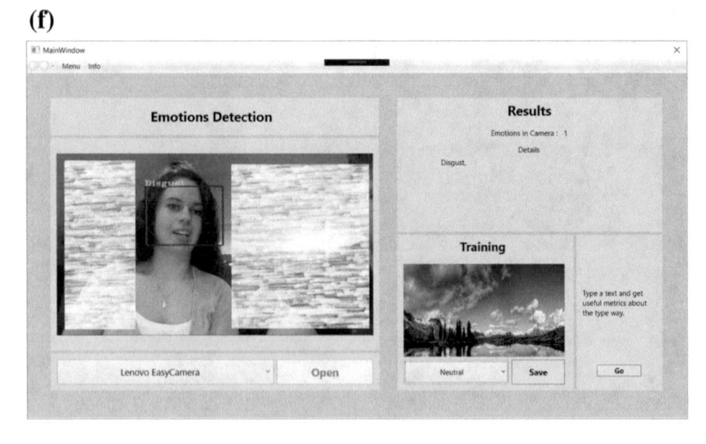

Fig. 2.9 (continued)

2.7.1.4 Fourth Case

Finally, we examine the fourth case, in which the program is been trained with samples from several faces including samples of the face of which the emotion/mood the program is to be recognized.

In Fig. 2.10a–f, we show the results of the emotion recognition. Clearly, our software has correctly recognized the Neutral (Fig. 2.10a), Happiness (Fig. 2.10b), Sadness (Fig. 2.10c), Surprise (Fig. 2.10d), Anger (Fig. 2.10e) and Disgust (Fig. 2.10f) emotions via its visual-facial modality.

2.7.2 Measurements for Text Typing Style

We continue the experimental demonstration of our system, presenting results from its keyboard-based modality. This experiment involved three (3) people who were asked to type into the program a specific text. Measurements we recorded and processed as analyzed in Sect. 2.5.4. In order to achieve diverse results, each measured sample was collected at a specific time on the same or different day and before and/or after "fatigue" due to work.

Specifically, the text used in the experiment was the following and the participants used the form shown in Fig. 2.11 to import it into our system:

> This text should be typed in the field below. Be careful to type the words correctly. As long as you type this text, the program tries to collect metrics about how to write your electronic text at this time. Interesting?

For the purposes of identifying and comparing the samples, we will use the three participants were identified as *A*, *B*, and *C*.

Typical measurements from typing this text are shown in Fig. 2.12. We see here that the measurements recorded are those analyzed in Sect. 2.5.4. Some of the measurements appear to have "Not number" as value. These measurements correspond to typing nothing. Other measurements have a zero value. A total of 48 words were typed, in which 279 errors were made and not corrected. The measurements for the average word typing time and the ratio of total time typing text without the time spent typing spaces with the number of words, will identify whether the number of words typed by the user matches the total number of words of the entire text. The difference in the two measurements lies in that the average word typing time refers to words written by the user, while the second measurement (net text typing time) refers to the number of words the user is asked to write.

Beginning with participant A, we present results from several measurements in Fig. 2.13. Specifically, in Fig. 2.13a–d, we present samples with measurements that were successively exported with only a short time interval between them. This means that the user had started to get "bored" with the test and perhaps was responding in an automated manner.

(a)

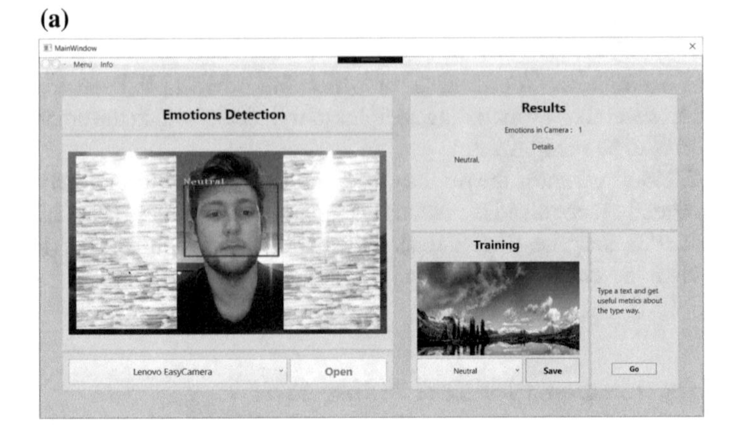

(b)

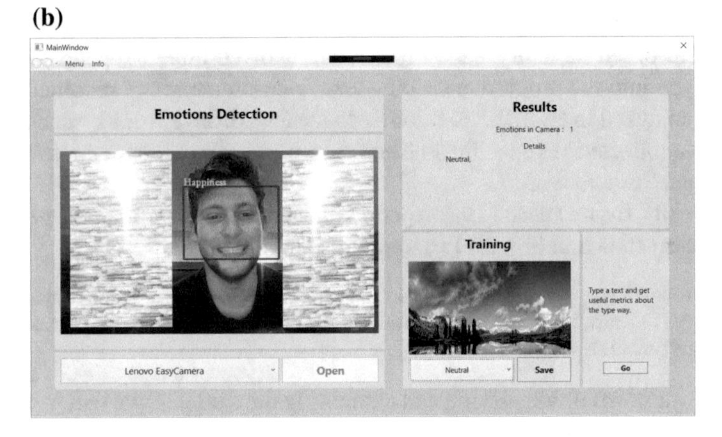

(c)

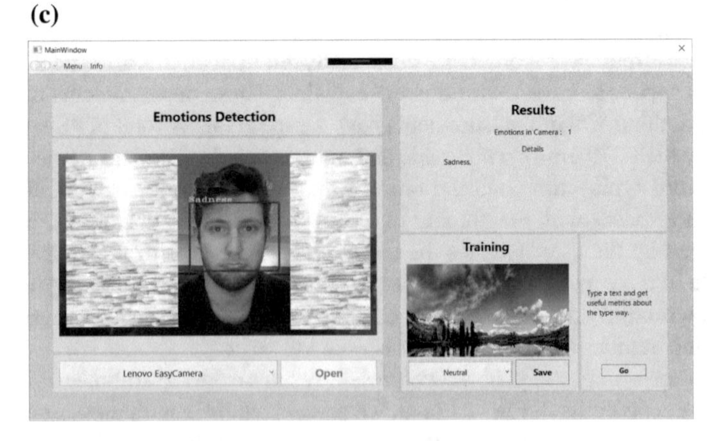

Fig. 2.10 **a** Successful recognition of neutral emotion of a known face. **b** Successful recognition of happiness emotion of a known face. **c** Successful recognition of sadness emotion of a known face. **d** Successful recognition of surprise emotion of a known face. **e** Successful recognition of anger emotion of a known face. **f** Successful Recognition of disgust emotion of a known face

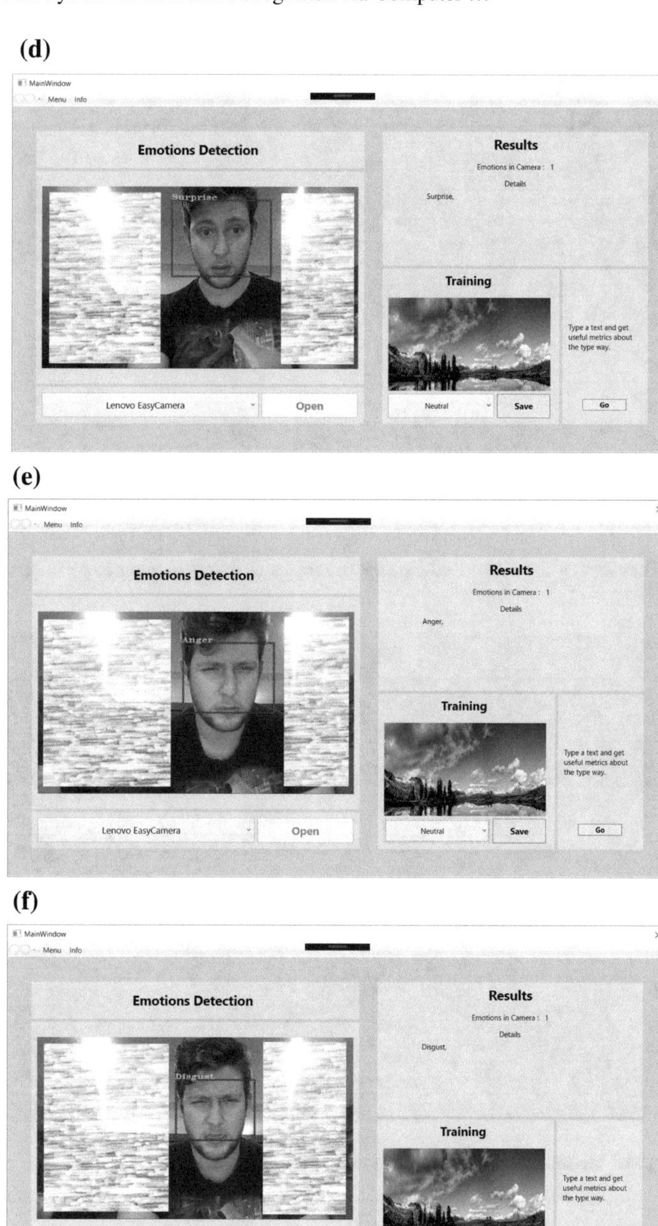

Fig. 2.10 (continued)

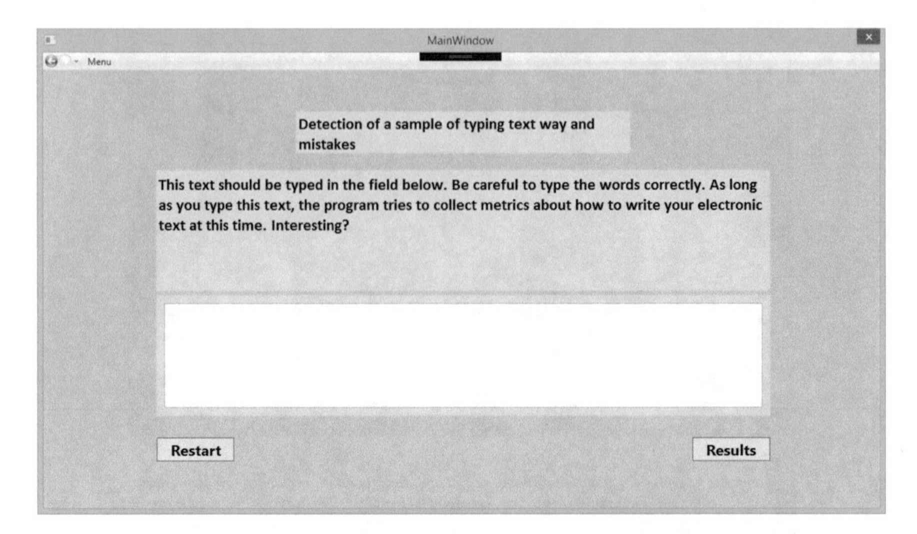

Fig. 2.11 Screenshot of the form used by experiment participants to enter experiment text

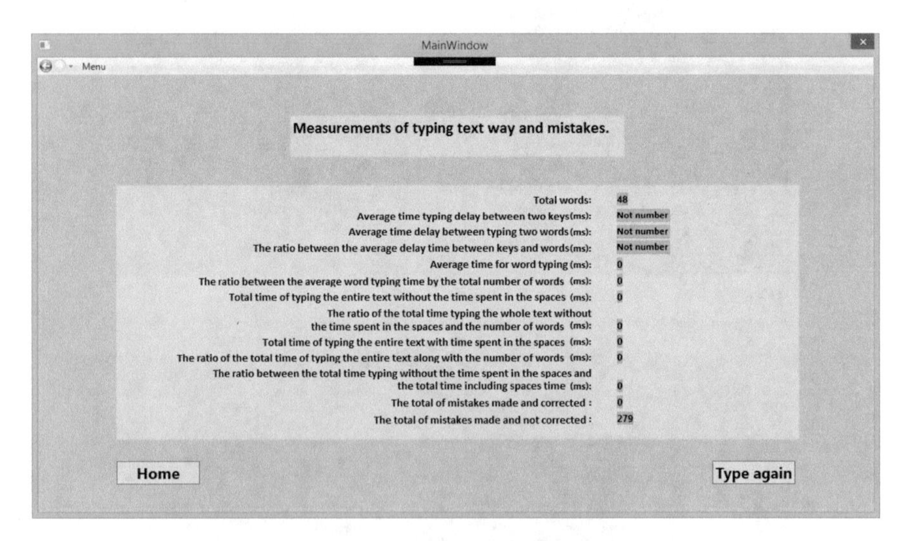

Fig. 2.12 Typical measurements from text typing

Then two additional samples of the same participant were measured in succession, in which the participant had spent several hours working. The corresponding results are shown in Fig. 2.13e–f. Finally, an additional two samples of the same participant were measured at the start of the following day and with the participant still fresh before work. The results are shown in Fig. 2.13g–h.

From these figures, we see that, even though there are differences in the measurements of the various experiments, there are still typing patterns of the user which are

(a)

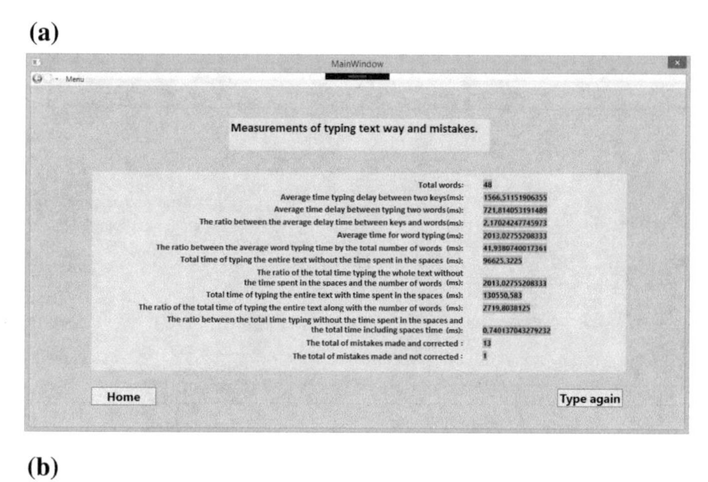

(b)

MainWindow
Menu
Measurements of typing text way and mistakes.
Total words: 48
Average time typing delay between two keys(ms): 1327,84555239852
Average time delay between typing two words (ms): 791,20134
The ratio between the average delay time between keys and words(ms): 1,67826504489808
Average time for word typing (ms): 1720,53240588235
The ratio between the average word typing time by the total number of words (ms): 35,844425122549
Total time of typing the entire text without the time spent in the spaces (ms): 87747,1527
The ratio of the total time typing the whole text without the time spent in the spaces and the number of words (ms): 1828,06568125
Total time of typing the entire text with time spent in the spaces (ms): 127307,2197
The ratio of the total time of typing the entire text along with the number of words (ms): 2652,23374375
The ratio between the total time typing without the time spent in the spaces and the total time including spaces time (ms): 0,689255117712699
The total of mistakes made and corrected : 3
The total of mistakes made and not corrected : 9
Home
Type again

(c)

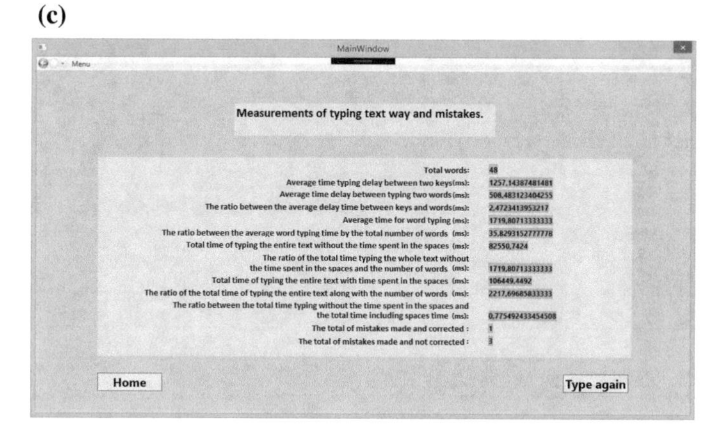

Fig. 2.13 **a** Example of measurements of users' typing texts. **b** Example of measurements of users' typing texts. **c** Example of measurements of users' typing texts. **d** Example of measurements of users' typing texts. **e** Example of measurements of users' typing texts under fatigue conditions. **f** Example of measurements of users' typing texts under fatigue conditions. **g** Example of measurements of users' typing texts under fatigue conditions. **h** Example of measurements of users' typing texts under fatigue conditions

(d)

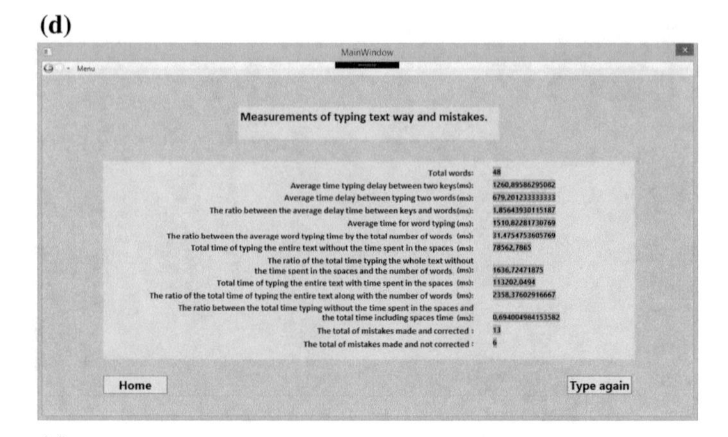

(e)

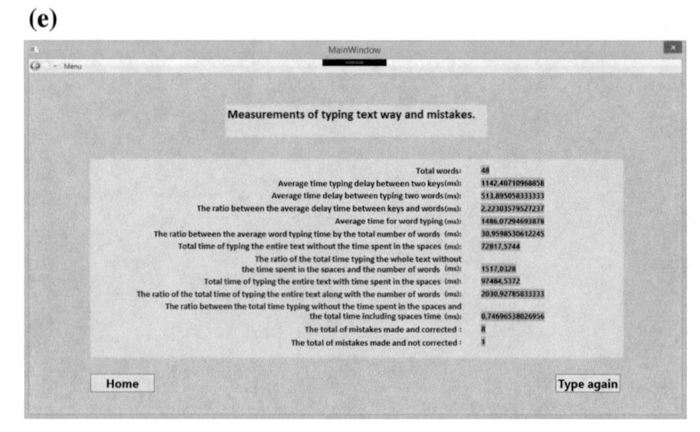

(f)

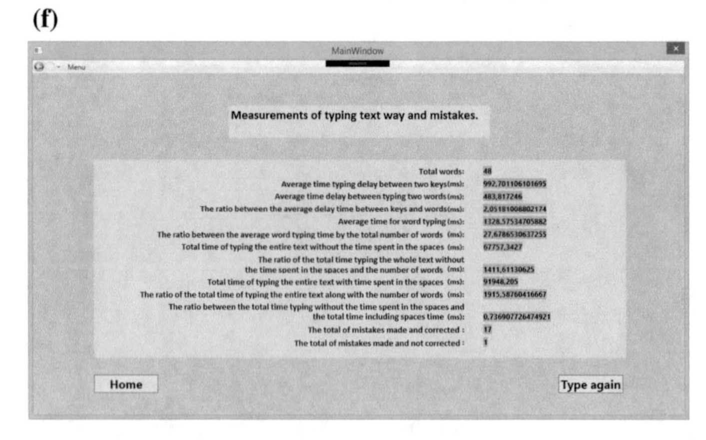

Fig. 2.13 (continued)

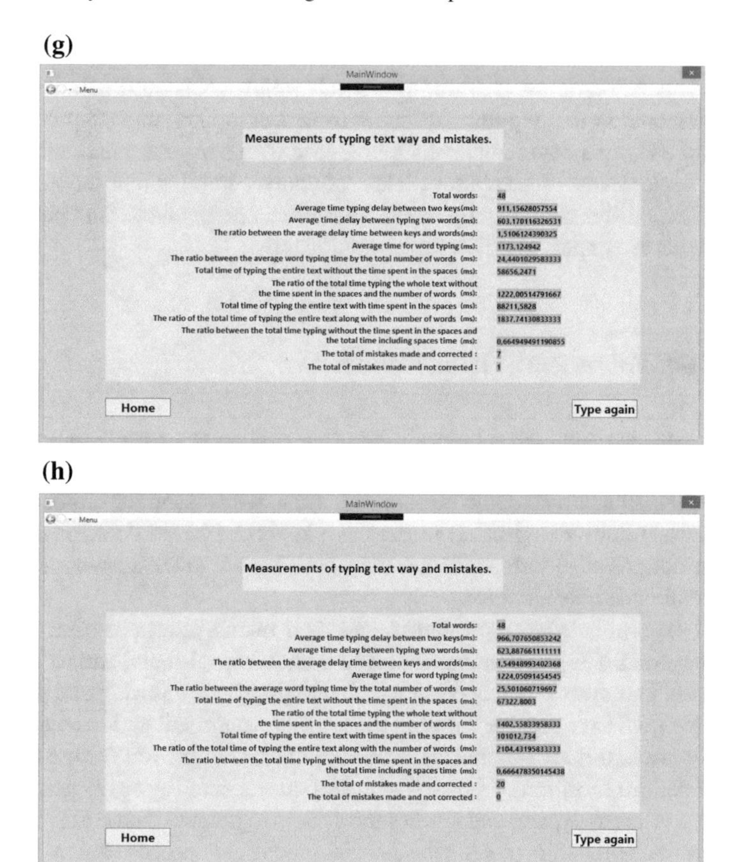

Fig. 2.13 (continued)

consistent. Specifically, the errors made by the user or the various time measurements are consistent from experiment to experiment. For example, we observe that in none of the experiments has the user been able to complete typing the text without making and correcting errors. This could be an indication that participant A was not fully concentrated at the time of testing.

Additionally, we observe differences between samples of the participant recorded at the *start* of a day, when the participant is fresh, and at a *later* time in the day, when the participant is tired. More specifically, the above measurements show that participant A typed words faster at the start of the day and the average values and time delays between successive key pressures were shorter. These conclusions are drawn via looking at the *speed of typing* of each word (average time of typing each word and time delay when hitting consecutive keys in words) at the start of the second day and comparing it with the speed of typing of each word during the experiments of the first day.

Similar observations were made regarding the experiments and corresponding measurements of participants B and C. At this point, we should mention that the three participants in the experiments came from a computer science/programming background and, thus, had a relatively better ease with computers than other people without a background in computers. This, of course, implies that the actual measurements may differ significantly when other users are involved, but "normalized" measurements are expected not to differ significantly.

2.8 Conclusions and Future Work

In the previous sections we referred to emotion as a means for interaction among humans and outlined several modalities to record measurements from which the emotional state of a human could be deduced. Then, we presented a bimodal system, which recorded both visual-facial images and keyboard-based data for use in recognizing emotions of its users whether they were known to the system or unknown under normal or fatigue situations.

Several experiments were presented, and their results analyzed from the actual use of our system. The experiments involved both its visual-facial and its keyboard-based modes. Our system can be used in collecting, properly storing and processing data with the goal to recognize emotion in various settings and various applications.

Future work in this area includes the development of improved feature extraction and classification algorithms to be used with the data collected by our system. Another step in our work is to expand and modify the system to process data corresponding to several simultaneous users (e.g. a visual-facial modality to process the facial images from many users participating in an event or a keyboard-based modality to process data collected from keyboard strokes of several simultaneous users). A third avenue for future work is to expand our system to include other modalities as well, such as an acoustical modality to process audio signals or a modality based on processing mouse use patterns. This and other similar works are currently under way and preliminary results have been obtained [25–27], which will be finalized and reported in full in the near future.

References

1. R.W. Picard, *Affective Computing* (MIT Press, 1997)
2. I.-O. Stathopoulou, G.A. Tsihrintzis, *Visual Affect Recognition*, Frontiers in Artificial Intelligence, vol. 214 (IOS Press, 2010)
3. P. Ekkekakis, *The Measurement of Affect, Mood, and Emotion—A Guide for Health-Behavioral Research* (Cambridge University Press, 2013)
4. E. Alepis, M. Virvou, *Object-Oriented User Interfaces for Personalized Mobile Learning*, Intelligent Systems Reference Library, vol. 64 (Springer, 2014)

5. E. Alepis, M. Virvou, S. Drakoulis, Human smartphone interaction: exploring smartphone senses, in *Proceedings of the 5th International Conference on Information, Intelligence, Systems and Applications (IISA 2014)*, Chania, Greece, pp. 44–48 (2014)
6. S. Bhattacharya, A predictive linear regression model for affective state detection of mobile touch screen users. Int. J. Mob. Hum. Comput. Interact. (IJMHCI) **9**(1), 30–44 (2017)
7. M. Pantic, L.J.M. Rothkrantz, Automatic analysis of facial expressions: the state of the art. IEEE Trans. Pattern Anal. Mach. Intell. **22**(12), 1424–1445 (2000)
8. M. Pantic, L.J. Rothkrantz, Toward an affect-sensitive multimodal human-computer interaction. Proc. IEEE **91**(9), 1370–1390 (2003)
9. A. Jaimes, N. Sebe, Multimodal human-computer interaction: a survey. Comput. Vis. Image Underst. **108**(1), 116–134 (2007)
10. I.-O. Stathopoulou, E. Alepis, G.A. Tsihrintzis, M. Virvou, On assisting a visual-facial affect recognition system with keyboard-stroke pattern information. Knowl.-Based Syst. **23**(4), 350–356 (2010)
11. M.A. Nicolaou, H. Gunes, M. Pantic, Continuous prediction of spontaneous affect from multiple cues and modalities in valence-arousal space. IEEE Trans. Affect. Comput. **2**(2), 92–105 (2011)
12. H. Gunes, B. Schuller, M. Pantic, R. Cowie, Emotion representation, analysis and synthesis in continuous space: a survey, in *Proceedings of 2011 IEEE International Conference on Automatic Face & Gesture Recognition and Workshops (FG 2011)*, pp. 827–834 (2011)
13. M. El Ayadi, M.S. Kamel, F. Karray, Survey on speech emotion recognition: features, classification schemes and databases. Pattern Recogn. **44**(3), 572–587 (2011)
14. M. Virvou, G.A. Tsihrintzis, E. Alepis, I.-O. Stathopoulou, K. Kabassi, Emotion recognition: empirical studies towards the combination of audio-lingual and visual-facial modalities through multi-attribute decision making. Int. J. Artif. Intell. Tools **21**(02) (2012)
15. C.-H. Wu, J.-C. Lin, W.-L. Wei, Survey on audiovisual emotion recognition: databases, features, and data fusion strategies. APSIPA Trans. Sig. Inf. Process. **3**, e12 (2014)
16. E. Sariyanidi, H. Gunes, A. Cavallaro, Automatic analysis of facial affect: a survey of registration, representation, and recognition. IEEE Trans. Pattern Anal. Mach. Intell. **37**(6), 1113–1133 (2015)
17. S.S. Tomkins, *Affect Imagery Consciousness: Volume I: The Positive Affects* (Springer, 1962)
18. C. Darwin, *The Expression of Emotions in Man and Animals*, 3rd edn. (Oxford University Press, 1872/1998)
19. P. Ekman, An argument for basic emotions. Cogn. Emot. **6**(3–4), 169–200 (1992)
20. P. Ekman, *Emotions Revealed: Recognizing Faces and Feelings to Improve Communication and Emotional Life* (Macmillan, 2007)
21. D. Watson, L.A. Clark, Measurement and mismeasurement of mood: recurrent and emergent issues. J. Pers. Assess. **68**(2), 267–296 (1997)
22. J.A. Russell, L.F. Barrett, Core affect, prototypical emotional episodes, and other things called emotion: dissecting the elephant. J. Pers. Soc. Psychol. **76**(5), 805 (1999)
23. J.A. Russell, Core affect and the psychological construction of emotion. Psychol. Rev. **110**(1), 145 (2003)
24. C. Beedie, P. Terry, A. Lane, Distinctions between emotion and mood. Cogn. Emot. **19**(6), 847–878 (2005)
25. A.M. Triantafyllou, G.A. Tsihrintzis, Group affect recognition: visual-facial data collection, in *Proceedings of 29th IEEE International Conference on Information, Intelligence, Systems and Applications (ICTAI2017),* Boston, MA, USA, pp. 677–681, Nov 6–8, 2017
26. A.M. Triantafyllou, G.A. Tsihrintzis, Group affect recognition: optimization of automatic classification, in M. Virvou, F. Kumeno, K. Oikonomou (eds.), *Knowledge-based Software Engineering: 2018, Proceedings of 12th Joint Conference on Knowledge-based Software Engineering (JCKBSE2018)*, vol. 108 in Smart Innovation, Systems and Technologies (Springer, 2018)
27. A.M. Triantafyllou, G.A. Tsihrintzis, Group affect recognition: evaluation of basic automated sorting, in *Proceedings of 9th IEEE International Conference on Information, Intelligence, Systems and Applications (IISA2018),* Zakynthos, Greece, July 18–20, 2018

Chapter 3
Knowledge Hiding in Decision Trees for Learning Analytics Applications

Georgios Feretzakis, Dimitris Kalles, and Vassilios S. Verykios

Abstract Nowadays there is a wide range of digital information available to educational institutions regarding learners, including performance records, educational resources, student attendance, feedback on the course material, evaluations of courses and social network data. Although collecting, using, and sharing educational data do offer substantial potential, the privacy-sensitivity of the data raises legitimate privacy concerns. The sharing of data among education organizations has become an increasingly common procedure. However, any organization will most likely try to keep some patterns hidden if it must share its datasets with others. This chapter focuses on preserving the privacy of sensitive patterns when inducing decision trees and demonstrates the application of a heuristic to an educational data set. The employed heuristic hiding method allows the sanitized raw data to be readily available for public use and, thus, is preferable over other heuristic solutions, like output perturbation or cryptographic techniques, which limit the usability of the data.

3.1 Introduction

In today's information age, all-around computing generates enormous amounts of data and information. Diverse fields, such as health care, finance, banking, education, e-commerce, and many others, have strongly invested on and benefitted from the analysis of this data [1]. However, a lot of the data collected may be related to sensitive or private information, which raises concerns about privacy. The Universal Declaration of Human Rights [2] of 1948 has recognized privacy as a right but to a limited extent: the right to privacy at home, with family and in correspondence. Data

G. Feretzakis (✉) · D. Kalles · V. S. Verykios
School of Science and Technology, Hellenic Open University, Patras, Greece
e-mail: georgios.feretzakis@ac.eap.gr

D. Kalles
e-mail: kalles@eap.gr

V. S. Verykios
e-mail: verykios@eap.gr

G. A. Tsihrintzis and M. Virvou (eds.), *Advances in Core Computer Science-Based Technologies*, Learning and Analytics in Intelligent Systems 14,
https://doi.org/10.1007/978-3-030-41196-1_3

protection is difficult to define because of the broad range of areas in which privacy applies [3].

In the domain of information, Westin [4] defined data protection as "*the claim of individuals, groups, or institutions to determine for themselves when, how, and to what extent information about them is communicated to others*". In terms of data control, Bertino et al. [5] have provided a similar definition but explicitly incorporate the risks of a violation of privacy. These authors define privacy as "*the right of an individual to be secure from unauthorized disclosure of information about oneself that is contained in an electronicrepository.*" Therefore, one can conclude that the main idea of privacy is to have control over personal data collection and processing.

Some information technology advantages can only be achieved by collecting and analyzing data. This can, however, lead to unwanted privacy breaches. In order to avoid information leakage, privacy conservation methods, to protect the exposure of the owner, have been developed by modifying the original data [6, 7]. However, data transformation can also decrease its usefulness, leading to inaccurate or even invaluable data extraction.

Privacy-preserving data mining (PPDM) methodologies are designed to guarantee a certain privacy level while maximizing the usefulness of the data so that data extraction can still be carried out effectively on the transformed data. Based on the variety of techniques developed by researches, several methods for assessing the level of privacy and data quality/utility of different techniques were proposed [8, 9]. In recent years, PPDM has attracted extensive attention among researchers, resulting in numerous privacy techniques under different assumptions and conditions. Various works focused on measurements to evaluate and compare these techniques in terms of the data protection, data utility, and complexity which were achieved. PPDM has therefore been applied effectively in many areas of scientific interest. The vast majority of PPDM surveys focus on the methods for assessing such techniques [5, 9–13]. Some discuss the parameters of evaluation and the trade-off between privacy and utility [14] briefly, while others describe briefly some of the current metrics [8].

Data mining protection is a relatively new area of research in the data mining community, which covers about a decade [15]. Since the pioneering work by Agrawal and Srikant [16] and Lindell and Pinkas [17] in 2000, several approaches to providing privacy in data mining have been proposed in the research literature. Most of the approaches proposed can be classified into two main directions: (i) data hiding approaches and (ii) knowledge hiding approaches.

Along the first direction lie methodologies which examine how raw data or information can be maintained before the data is mined. The general approach in such methodologies aims to delete or apply methods such as disruption, sampling, generalization or deletion, transformation, to the original data before disclosure, in order to generate a sanitized version of the original dataset. An objective is to enable data holders to receive accurate data mining results when the actual data is not given or when certain rules on microdata (sets of records containing information on individual persons, households or business entities) publication are not adhered to (e.g., as in the case of patient-specific data publication). Along the second direction, there lie methodologies aimed at protecting the sensitive data mining results (i.e., knowledge patterns extracted) produced on the original database through the application of data mining tools, and not the raw data itself. This approach focuses primarily on distortion and blocking techniques which prevent the leakage of sensitive knowledge patterns into disclosed data as well as techniques which reduce the effectiveness of classifiers in classification tasks so that no sensitive knowledge is shown by the produced classifiers.

The significant progress in data collection and data storage technologies has led to enormous amounts of transactional data being stored in companies and public sector organizations, at low cost, in warehouses or clouds. In addition to using these data per se (such as maintaining up-to-date clients' profiles and their purchases, keeping a list of available products, their quantities, and their prices, etc.), the mining of these data sets using existing data mining tools can uncover invaluable knowledge that the data holder did not previously know. Extracted patterns of knowledge can provide information for data holders and are of paramount importance for tasks such as decision-making and strategic planning. Additionally, companies are often willing or required to collaborate with competitors, based on perceived mutual benefit or regulatory supervision. Important patterns of knowledge can be derived and shared among partners through collaborative data mining. Public sector organizations and civil federal agencies ordinarily need to share or even make this information and knowledge public in order to comply with specific regulations, and such requirements are increasingly made of private entities too.

As can be seen in the previous discussion, a broad range of application scenarios exists where data collected, or knowledge patterns extracted from data, are to be shared with other (possibly not trusted) entities for specific purposes. The sharing of data and/or knowledge can harm privacy, mainly because of two main reasons: (1) when the data refer to persons, then the disclosure of such information may then be infringing on the privacy of the individuals registered in the data where their identity is disclosed to untrusted third parties or sensitive information about them can be extracted from the data, and (2) if this information concerns business (or corporate) information, the disclosure of this information or any knowledge gathered from the data may potentially reveal delicate trade secrets, the knowledge of which could give businesses a significant benefit and thus cause data owners to lose business to their competitors [15].

The rest of this chapter is structured in four subsequent sections. In the following section we present some basic concepts on privacy preserving in educational data and the sharing of learning analytics. Then, in Sect. 3, we present a new technique regarding the hiding of decision tree rules by using a minimum local distortion approach. In Sect. 4, we present an application of the Local Distortion Hiding algorithm to an educational dataset and, finally, we conclude the chapter by also offering some directions for further work.

3.2 Privacy Preservation Issues in Learning Analytics

The use of information technology has greatly changed the way learning experiences are perceived and experienced. The extensive use of a wide variety of digital devices combined with the proliferation of cloud computing services allows for the realization of an unprecedented wealth of learning scenarios. Students now have access to a vast multitude of learning resources, can interact at will with applications which focus on a particular topic, can enhance their experience with virtual environments and connect via social networks with others.

The technological advancement evolves with the ability to record events in a learning environment [18]. Each accessed interaction and resource can be captured and stored. As a result, learning scenarios can now be analyzed through large-scale data analytics.

Nowadays there is a wide range of digital information available to educational institutions regarding learners, including performance records, educational resources, student attendance, feedback on the course material, evaluations of courses and social network data. New educational environments, technology, and regulations are being developed to further enrich institutions' types of data. In modern educational systems, there is a lot of loosely structured and complex data from different information sources [19].

Learning analytics is defined as "the measurement, collection, analysis, and reporting of data about learners and their contexts, for purposes of understanding and optimizing learning and the environments in which it occurs" [20]. The ability to collect data from events in an environment can be used for a learning experience.

Predictive models which characterize a student's current performance can help predict future performance (possibly to prevent failure and to promote success). Increasing volumes of learning data typically improve the potential for analyzing various aspects of the learning process and help boost the effectiveness of cooperation and collaboration among learning groups [21]. Additionally, visualizing student data can lead to better and more prompt feedback.

Although collecting, using, and sharing educational data do offer substantial potential, the privacy-sensitivity of the data raises legitimate privacy concerns. Many initiatives and regulations strengthen the protection of personal data in fields such as health, trade, communication, and education [22, 23]. In most legislation, absolute confidentiality is not enforced if it is deemed that it might be more harmful than beneficial [24], but, rather, a trade-off is reached by protecting "individually identifiable data" that can be traced back to individuals with or without external knowledge. This has led to a large number of studies focusing primarily on the de-identification of private data with as little impact as possible to their information content, in order to protect privacy without compromising usefulness.

Without a specific context, it is difficult to provide a broad definition of data privacy [25]. Data privacy research officially defined and enforced privacy in two main scenarios: (1) data sharing with third parties without infringing the privacy of persons whose sensitive information is contained within the data [26], (2) mining data without abusing identifiable and sensitive information. This is often referred to as data mining or disclosure control that safeguards privacy.

Data analytics are about data examination so that conclusions can be drawn for better decision making, or models and theories can be checked [27]. Many academics see large-scale data collection and large-scale data as promising research fields. The future of learning analytics lies, as a matter of fact to Mayer-Schonberger and Cukier, in big data [28]. Studies (e.g., the PAR project [29]) have already been conducted which combine educational information on the web into a single, federated data set and analyze the factors affecting student retention, progression, and completion.

There is no doubt that educational data contain private and sensitive information. Recent learning analytics (LA) papers call for cooperation and open learning initiatives [30], for which researchers from around the world can benefit. However, sharing sensitive information requires additional privacy care.

In several LA works, the ethical and privacy implications of education data are discussed. In [31] Heath points out that institutions collect large educational data and provides a perspective on privacy for philosophers, lawyers, and educational specialists. In [32] the definition of consent and students' ability to opt-out of data collection is emphasized by Slade and Prinsloo. The policy frameworks of two distant education institutions are evaluated in [33] by Prinsloo and Slade, according to several considerations, including who benefits under a variety of conditions on consent, detection, and opt-out. Authors conclude that current policies are mainly related to academic analytics (data security, demographic data integrity) rather than to learning analytics (student and departmental data). As such, data protection during the process of obtaining and sharing LA results is urgently required.

For example, Verbert et al. [34] showed that learners' activities in a single course are probably so diverse that a system recommending learning resources would be practically useless when only based on data of that scale (i.e. of the single course). This challenge, therefore, applies to learn science research as well as potential products and services based on data generated during learning. It is also common for the data required for learning analytics to be found in various software systems. The increasing use of cloud computing services, where expertise or technology is provided to the educational institution by a separate organization, has increased the extent to which data is not only distributed between different IT systems but also distributed across legal entities. The combined effect of the need for data at scale and for combining data sources leads to the idea that data sharing among organizations, including potentially private and public sector entities, is an important tool for effective learning analytics.

The boundaries between the worlds of formal and informal learning have always been rather permeable and undefined, and it is hardly practical to separate education from non-traditional sources in a world in which many digital devices are capable of capturing and transforming any activity or utterance into data.

3.3 Using Minimum Local Distortion to Hide Decision Tree Rules

The sharing of data among education organizations has become an increasingly common procedure. However, any organization will most likely try to keep some patterns hidden if it must share its datasets with others. For example, let two parts, an online library of instructional videos covering software, business, and creative skills, which serves hundreds of thousands of educational requests every day and on the other side, a large technological university which wants to expand its activities in the online education industry by offering online courses in specific subjects. The two parts make a deal to share their data, but the university does not want to share the part of its data that reveals valuable educational usage insight, namely (for example) that female students in the age group 45–50 tend to be heavily interested in software applications for image editing and photo retouching. For that reason, the university's data have to be sanitized in order to keep all the rules which can be inferred except the rule above. This section focuses on preserving the privacy of sensitive patterns when inducing decision trees by presenting some indicative published techniques. For the scope of this work, we limit our treatment to rules which are defined by terminal nodes (leaves).

The primary representative of statistical methods [35] uses a parsimonious downgrading technique to determine if the loss of functionality associated with data not downgrading is worth further confidentiality. The reconstruction of public datasets [36, 37] includes the reconstruction of non-sensitive rules produced by the algorithms C4.5 [38] and RIPPER [39]. Perturbation-based techniques involve the modification of transactions to only support non-sensitive rules [40], the removal of sensitive rules-related tuples [41], the suppression of specific attribute values [42], and the redistribution of tuples supporting sensitive patterns to maintain the ordering of the rules [43]. Another useful approach is machine learning classification over encrypted data [44] in which a private decision tree classifier enables the server to traverse a binary decision tree using the client's input x such that the server does not learn the input x and the tree structure and thresholds at each node are not learned by the client. A recent work [45] proposes privacy-preserving decision tree evaluation protocols that hide the critical inputs from the counterparty using additively homomorphic encryption (AHE) like the ElGamal encryption method.

A series of techniques [46–49] have been presented to hide sensitive rules without compromising the information value of the entire data set and adequately protect against the disclosure of sensitive patterns of knowledge in classification rules mining. For these techniques, the class labels at the tree node corresponding to the tail of the sensitive pattern are modified after an expert selects the sensitive rules to eliminate the gain achieved by the information metric causing the split. By preserving the class balance of every node across the sensitive path, it can be assured that there will be no change in the hierarchy order of this path due to changes in entropy of the nodes along this path. The next step is to set the values of non-class attributes appropriately, adding new instances along the path to the root if necessary, so that non-sensitive patterns remain as unaffected as possible. This approach is critical because the sanitized data set may be subsequently published and even shared with the data set owner's competitors, as can be the case with retail banking [50].

A recent study [51] and an application of it in a medical dataset [52] proposed a technique which its objective is to allow the publishing or sharing of the original data set by hiding the critical rules which are produced by creating the corresponding decision tree and thus preserving to the maximum possible extent the privacy of data which have caused these critical rules to appear. This technique does not affect the class labels of the sensitive instances but instead modifies the attributes' values of these specific instances. While the new technique may need to modify more values of the initial data set, it does so by not requiring the addition of extra instances, and it thus saves on the size of the sanitized data set. This trade-off is an extra tool in the arsenal of the data engineer who might want to explore a range of possibilities when tasked with a data hiding mandate. In this method, we first identify the instances that contribute to the creation of a specific rule and then, by appropriately changing attribute values, we can successfully hide this rule with minimum impact to the rest of the decision tree.

Figure 3.1 shows a baseline problem assuming the representation of a binary decision tree with binary, symbolic attributes (X, Y, and Z) and binary classes (C1 and C2). Hiding R3 implies the suppression of splitting in node Z, hiding R2 as well.

Concerning the following figure (Fig. 3.1), hiding a rule is equivalent to hiding the leaf that corresponds to that rule (after all, a leaf corresponds to a conjunction of attribute tests which ends up in a class label). A straightforward approach to this hiding operation is the elimination of the last attribute test, which corresponds to

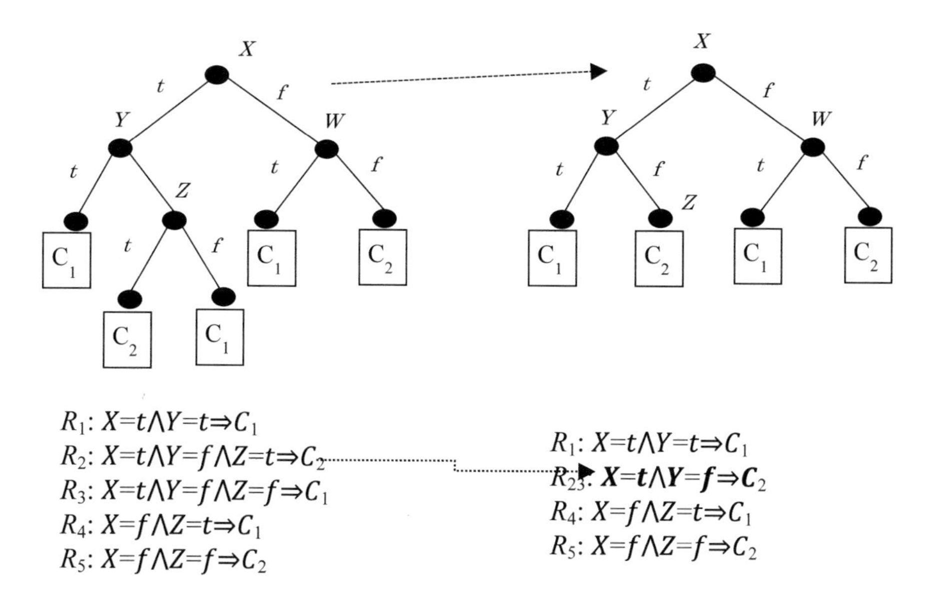

Fig. 3.1 A binary decision tree before (left) and after (right) hiding and the associated rule sets

the target leaf being merged with its sibling and their parent being turned into a leaf itself, thus decreasing the length of the original rule by one conjunction. Thus, the new terminal internal node is expected to be the parent of the target leaf; with the weaker version of our hiding technique, we merely strive to eliminate its original attribute test, and with the stronger version of our technique we aim to maximize the possibility that the new rule will be indeed shorter.

The first method of hiding R3 is to remove from the training data all instances of the leaf corresponding to R3 and to retrain the tree from the resulting dataset. This may, however, lead to a substantial restructuring of the decision tree, affecting other parts of the tree as well.

Another approach would be to turn the direct parent of the R3 leaf into a new leaf. However, the actual dataset would not be changed. This allows an opponent to recover the original tree.

To achieve hiding by minimally modifying the original data set, we can interpret "minimal" changes in data sets or whether the sanitized decision tree generated through hiding is syntactically close to the original with minimum modification of the initial data set. The minimum measurement of how decision-making trees are changed has been examined in the context of heuristics to ensure or approximate the effect of changes [53–55]. In our examples, we use the measure of the kappa statistic to compare the efficiency of the deduced decision tree after the proposed modification with the original one.

The information gain metric is used to select the test attribute at each node of the decision tree. The decision tree induction algorithm ID3 [56] used as a splitting criterion, the information gain, and its successor C4.5 use an improvement of information gain known as the gain ratio.

The split information value represents the potential information generated by splitting the training data set T into n partitions T_i, corresponding to n outcomes on attribute A.

$$SplitInfo_A(T) = -\sum_{i=1}^{n} \frac{|T_i|}{|T|} \times \log_2\left(\frac{|T_i|}{|T|}\right) \tag{3.1}$$

The gain ratio is defined as

$$GainRatio(A, T) = \frac{Gain(A, T)}{SplitInfo_A(T)} \tag{3.2}$$

where

$$Gain(A, T) = Info(T) - Info(A, T) \tag{3.3}$$

The attribute with the highest gain ratio is selected as the splitting attribute. Therefore, if we would like to suppress a particular attribute test at a node, a rule-of-thumb approach would be to try to modify the values (for that attribute) of the instances

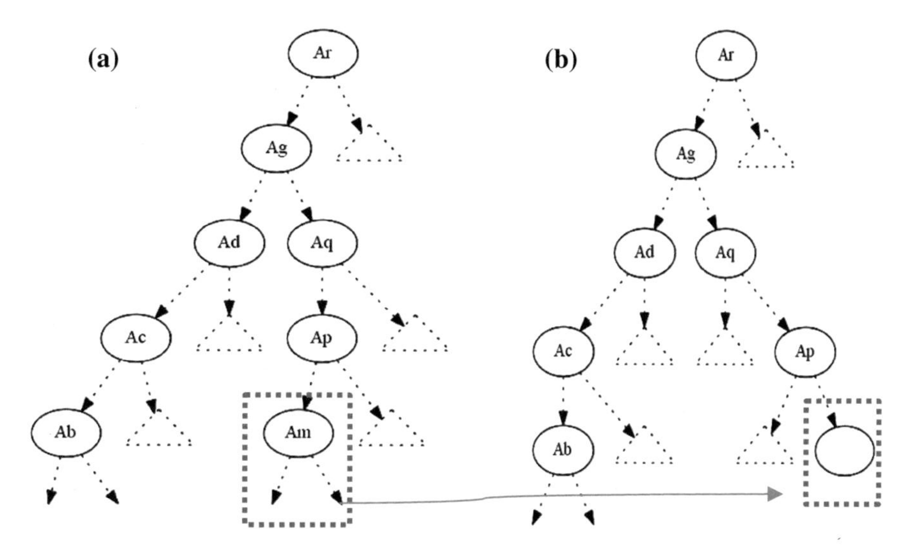

Fig. 3.2 **a** Original decision tree; **b** modified decision tree with the absence of A_m.

which would arrive at that node. By this modification, the resulting information gain due to that attribute would be reduced, becoming equal to zero if possible.

In the figure above (Fig. 3.2a), if we want to hide the terminal internal node A_m, we try to decrease its gain ratio at this particular splitting. The parent of the node A_m is A_p, and the critical moment to measure the gain ratio is when the splitting test is performed to decide which node will be under the true or false value of the node A_p. The decrease of the gain ratio of A_m could be accomplished by changing its attribute values only in the instances that correspond to this particular path (from the root A_r until the terminal internal node A_m). If we decrease the gain ratio of A_m and ideally became equal to zero, then it will not be displayed in its initial position. One can say that since the node A_m is not displayed any more at this position the hiding has been successfully done.

However, we have accomplished only half of our target since other nodes (attributes) could take the A_m position in this part of the decision tree. For that reason, we calculate for every node (attribute) the gain ratio under the conditions of the certain rule-path. Apart from the ancestors of A_m, all other attributes can be candidates to replace the initial position of A_m. Having calculated the gain ratios of all these competitors, we then proceed to decrease their gain ratios by shifting their values appropriately. We change only the attributes' values and never their class values. This is very important since every change in a class value can affect the total entropy of the entire data set, and then every change will not only have a local impact but cause a total distortion to the whole decision tree. On the other side, a local distortion in the critical instances will only affect the entropy at this specific subdomain. By completing the procedure above we end up (ideally) with a tree which looks like the tree in Fig. 3.2b; in other words, the final tree will be the same as the original one apart from the absence of the terminal internal node A_m.

The algorithm LDH locates the parent node of the leaf to be hidden and ensures that the attribute tested at that node will not generate a splitting which would allow that leaf to re-emerge. It does that by re-directing all instances following the branch from X to L towards L's sibling by means of simply manipulating their attribute values for the attribute tested at X. If we want to pursue the stronger approach to hiding, then we extend this instance manipulation for all attributes available at X and not just the one that is being tested at X.

This methodology allows one to specify which decision tree leaves should be hidden and then judiciously change some attribute values in specific instances in the original data set. Consequently, the next time one tries to build the tree, the to-be-hidden nodes will have disappeared as the instances corresponding to those nodes will have been absorbed by neighboring ones.

3.4 Local Distortion Hiding with Educational Data: A Case Study

The Student Performance Data Set from the UCI—Machine Learning Repository [57] deals with student achievement in secondary education in the context of two Portuguese schools. The data attributes include student grades, demographic, social and school-related features), and it was collected by using school reports and questionnaires. Two datasets are provided regarding the performance in two distinct subjects: Mathematics (mat) and Portuguese language (por). For our example we choose to use the Mathematics dataset and, for the sake of simplicity, we have kept for simplicity reasons twenty (20) attributes from the thirty-three (33) that was contained in the original dataset. The number of instances in the modified dataset is 363.

The 20 attributes are listed below:

1. Sex—student's sex (binary: 'F'—female or 'M'—male)
2. Address—student's home address type (binary: 'U'—urban or 'R'—rural)
3. Famsize—family size (binary: 'LE3'—less or equal to 3 or 'GT3'—greater than 3)
4. Pstatus—parent's cohabitation status (binary: 'T'—living together or 'A'—apart)
5. Guardian—student's guardian (binary: 'mother' or 'father')
6. Schoolsup—extra educational support (binary: yes or no)
7. Famsup—family educational support (binary: yes or no)
8. Paid—extra paid classes within the course subject (Math or Portuguese) (binary: yes or no)
9. Activities—extra-curricular activities (binary: yes or no)
10. Nursery—attended nursery school (binary: yes or no)
11. Higher—wants to take higher education (binary: yes or no)

12. Internet—Internet access at home (binary: yes or no)
13. Romantic—with a romantic relationship (binary: yes or no)
14. Famrel—quality of family relationships (numeric: from 1—very bad to 5——excellent)
15. Freetime—free time after school (numeric: from 1—very low to 5—very high)
16. Goout—going out with friends (numeric: from 1—very low to 5—very high)
17. Dalc—workday alcohol consumption (numeric: from 1—very low to 5—very high)
18. Walc—weekend alcohol consumption (numeric: from 1—very low to 5—very high)
19. Health—current health status (numeric: from 1—very bad to 5—very good)
20. G3—final grade in Mathematics (numeric: from 0 to 20, output target).

As we can observe, there are many attributes that may contain sensitive information about a student and, for that reason, this is an interesting dataset to demonstrate the LDH algorithm.

Since the LDH algorithm can be applied to binary data, multivalued nominal attributes must be converted to binary ones. The *NominalToBinary* filter of the WEKA—Data Mining Software in Java [58] transforms all specified multivalued nominal attributes in a data set into binary ones [59]. The binary values for the 43 attributes are "0" and "1", and the corresponding values for the class are "Pass" (greater or equal to 10) or "Fail" (less than 10).

The WEKA—Data Mining Software in Java workbench is a collection of machine learning algorithms and data preprocessing tools. We chose for our experiments to use the classification algorithm J48, which is the implementation of Quinlan's C4.5 algorithm. This algorithm uses the gain ratio for feature selection and to construct the decision tree. It handles both continuous and discrete features. The C4.5 algorithm is widely used because of its quick classification and high precision. It is implemented in WEKA as a classifier called J48.

In the first example in the Student Performance Data Set, we try to hide the terminal internal node $Walc = 3$, which is shown in Fig. 3.3.

Firstly, we have to determine the nine specific instances which correspond to that node. Next, we measure the information gain ratio (IGR) of all nodes apart those nodes which are located along the critical path from the root to the parent of $Walc = 3$. By measuring the IGRs of all nodes, we find (as it is expected) that node $Walc = 3$ has the highest IGR. By applying the LDH algorithm we first try to decrease the IGR of the node $Walc = 3$ and then if this is not adequate, we try to increase the IGR of any other competitor attribute to replace the node $Walc = 3$.

The above procedure can be achieved by changing the attributes' values in these specific nine instances. At this point, it is essential to note that if we change an attribute value which is located in the critical path, in our case, then this change also affects the gain ratios of all other values.

In this way, we have succeeded in eliminating the contribution of the node $Walc = 3$ when applying the splitting criterion in node $health = 3$. The node $Walc = 3$ is replaced by the nodes $farmel = 3$, $Dalc = 2$, and the result is shown in Fig. 3.4.

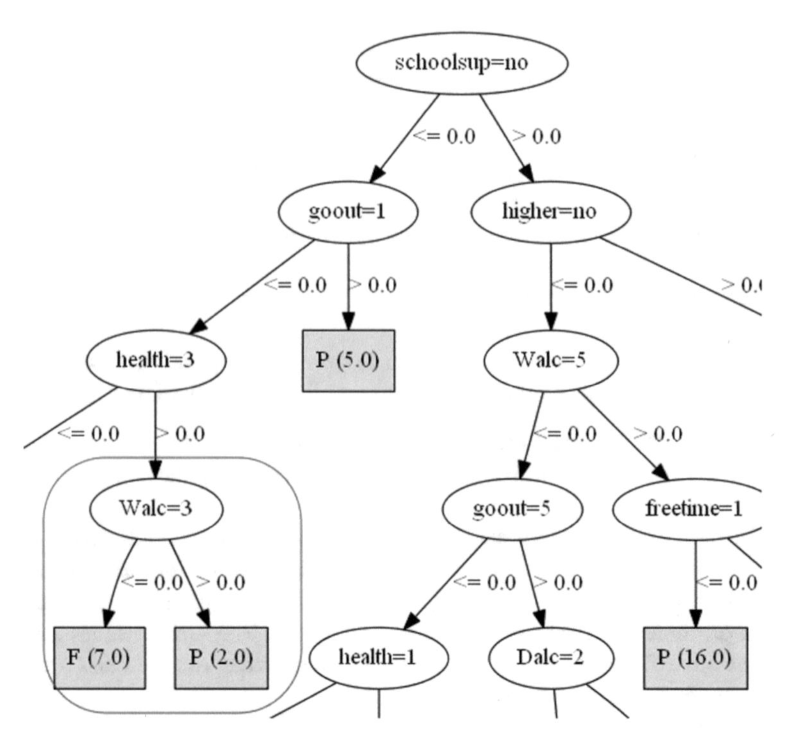

Fig. 3.3 The Original decision tree with the terminal node $Walc = 3$

Since the size of the above decision tree is too large to fit into one page, only the tree section around the critical node is presented. The reader can find all the WEKA data set files (.arff) and the corresponding model files (.model) on the website [60] before and after our distortion hiding method has been applied. In addition to these data sets, the two induced decision trees are also available in full-scale deployment (in a Portable Network Graphic format) as extracted from WEKA and then processed by Graphviz software [61].

Cohen's kappa coefficient is a statistical measure of inter-rater agreement for qualitative items. In WEKA, the kappa statistic is calculated as one of the measures of accuracy for classification [62]. The kappa statistic values corresponding to the original and modified data sets are the same ($K = 0.9936$). We can also conclude from all other WEKA statistics presented in Table 3.1 that the node $Walc = 3$ has been successfully hidden without affecting the effectiveness of the decision tree since there is not any change (in 4 d.p. accuracy) to the errors' numerical values after the distortion.

Furthermore, the decision tree induced by the modified data set produced by the LDH algorithm is identical to the original one apart from the replacement of the node $Walc = 3$ by the nodes $farmel = 3$, $Dalc = 2$ in the certain decision tree rule which is described above.

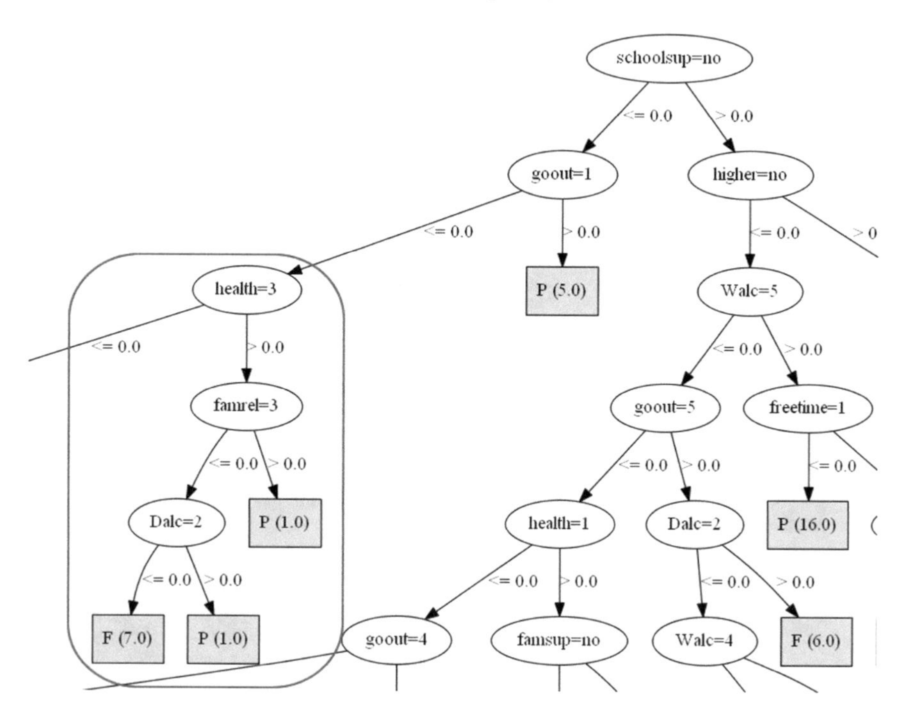

Fig. 3.4 Final decision tree without the terminal node *Walc = 3*

Table 3.1 Weka output for the first example

Weka statistics	Original	Modified
Correctly classified instances	362	362
Incorrectly classified instances	1	1
Kappa statistic (K)	0.9936	0.9936
Mean absolute error	0.0041	0.0041
Root-mean-squared error	0.0455	0.0455
Relative absolute error (%)	0.9583	0.9583
Root relative squared error (%)	9.7933	9.7933

In the second example in the Student Performance Data Set, we try to hide the terminal internal node *address* = *R*, which is shown in Fig. 3.5.

Firstly, we have to determine the two specific instances that correspond to that node. Next, we measure the information gain ratio (IGR) of all nodes apart those nodes which are located along the critical path from the root to the parent of *address* = *R*. By measuring the IGRs of all nodes, we find (as it is expected) that node *address* = *R* has the highest IGR. By applying the LDH algorithm we first try to decrease the IGR of the node *address* = *R* and then if this is not adequate, we try to increase the IGR of any other competitor attribute to replace the node *address* = *R*.

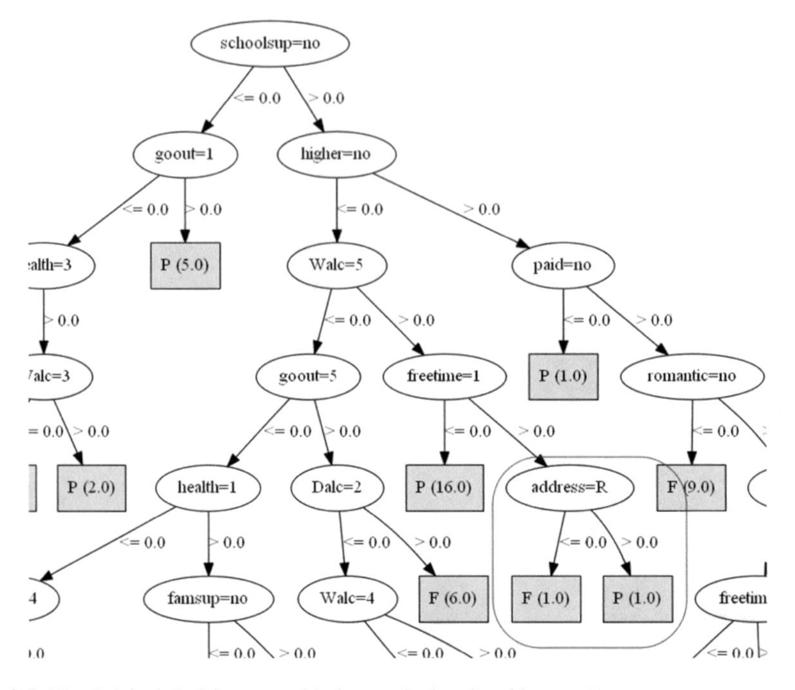

Fig. 3.5 The Original decision tree with the terminal node *address* = *R*

The above procedure can be achieved by changing the attributes' values in these specific nine instances. At this point, it is essential to note that if we change an attribute value which is located in the critical path, in our case, then this change also affects the gain ratios of all other values.

In this way, we have succeeded in eliminating the contribution of the node *address* = *R* when applying the splitting criterion in node *freetime* = *1*. The node *address* = *R* has disappeared, and the result is shown in Fig. 3.6.

We can also conclude from all other WEKA statistics presented in Table 3.2 that the node *address* = *R* has been successfully hidden without affecting the effectiveness of the decision tree since there is not any change to the errors' numerical values after the distortion. The decision tree induced by the modified data set produced by the LDH algorithm is identical as the original one apart from the absence of the node *address* = *R* in the certain decision tree rule, which is described above.

3.5 Conclusions

Our new methodology allows one to specify which decision tree leaves should be hidden and then change some attribute values in the original data set in specific instances. As a consequence, the next time one tries to build the tree; the hidden nodes

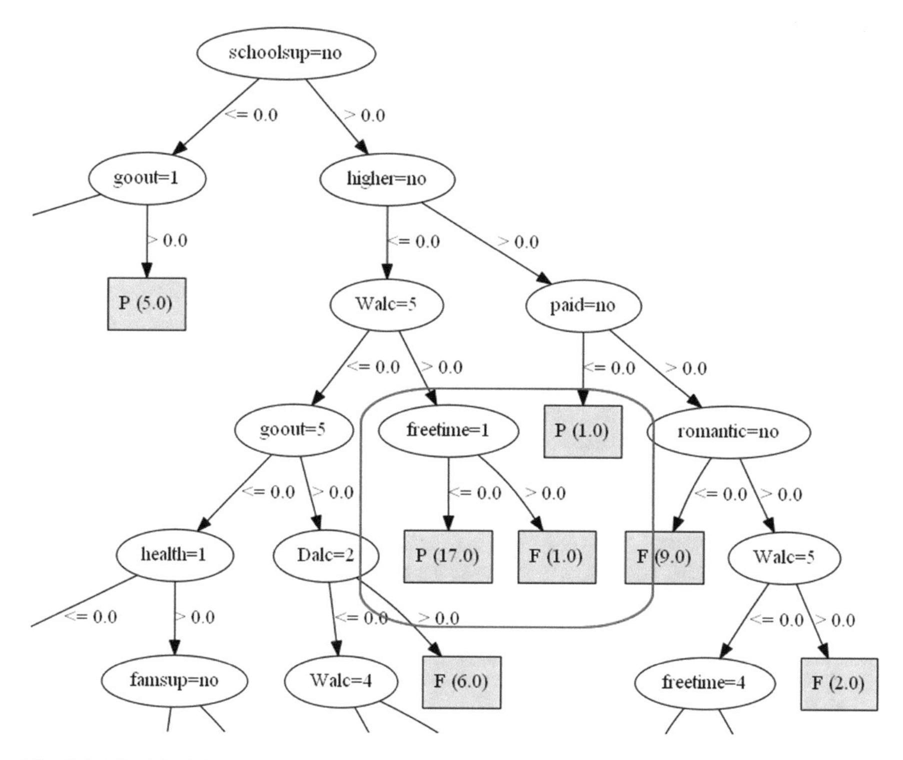

Fig. 3.6 Final decision tree without the terminal node *address* = *R*

Table 3.2 Weka output for the second example

Weka statistics	Original	Modified
Correctly classified instances	362	362
Incorrectly classified instances	1	1
Kappa statistic (K)	0.9936	0.9936
Mean absolute error	0.0041	0.0041
Root-mean-squared error	0.0455	0.0455
Relative absolute error (%)	0.9583	0.9583
Root relative squared error (%)	9.7933	9.7933

will disappear as neighboring nodes will have absorbed instances corresponding to these nodes. The above-detailed results have shown that the proposed method has successfully hidden the internal terminal nodes selected in the two selected different data sets. Also, we demonstrated by the proposed technique did not cause any extra change in the structure of the modified decision trees (compared with the original ones), besides the hidden node.

The medium-term development goal is to implement the technology proposed to accommodate hiding requirements as a standard data engineering service, combined with a suitable environment where the importance of each hiding request can be specified. Another important research and development aspect regarding the improvement of our technique is the ability to handle numerical attributes. Nonetheless, we believe that the apparent rise in the interest in solutions which protect privacy suggests that systems with theoretical backing are increasingly likely to appear.

References

1. L. Cranor, T. Rabin, V. Shmatikov, S. Vadhan, D. Weitzner, Towards a privacy research roadmap for the computing community, in *Computing Community Consortium committee of the Computing Research Association*, Washington, DC, USA, White Paper (2015)
2. Universal Declaration of Human Rights, *United Nation General Assembly* (New York, NY, USA, 1948), pp. 1–6. http://www.un.org/en/documents/udhr/
3. S. Yu, Big privacy: challenges and opportunities of privacy study in the age of big data. IEEE Access. **4**, 2751–2763 (2016)
4. S. Laughlin, A. Westin, Privacy and freedom. Mich. Law Rev. **66**, 1064 (1968)
5. E. Bertino, D. Lin, W. Jiang, A survey of quantification of privacy preserving data mining algorithms, in *Privacy-Preserving Data Mining* (Springer, New York, NY, USA, 2008), pp. 183–205
6. C.C. Aggarwal, P.S. Yu, A general survey of privacy-preserving data mining models and algorithms, in *Privacy-Preserving Data Mining* (Springer, New York, NY, USA, 2008), pp. 11–52
7. C.C. Aggarwal, *Data Mining: The Textbook* (Springer, New York, NY, USA, 2015)
8. S. Dua, X. Du, *Data Mining and Machine Learning in Cybersecurity* (CRC Press, Boca Raton, FL, USA, 2011)
9. S. Fletcher, M. Islam, Measuring information quality for privacy preserving data mining. Int. J. Comput. Theory Eng. **7**, 21–28 (2014)
10. R. Mendes, J. Vilela, Privacy-preserving data mining: methods, metrics, and applications. IEEE Access. **5**, 10562–10582 (2017). https://doi.org/10.1109/ACCESS.2017.2706947
11. A. Shah, R. Gulati, Privacy Preserving data mining: techniques, classification and implications—a survey. Int. J. Comput. Appl. **137**, 40–46 (2016)
12. Y. Aldeen, M. Salleh, M. Razzaque, A comprehensive review on privacy preserving data mining. SpringerPlus **4** (2015)
13. E. Bertino, I.N. Fovino, Information driven evaluation of data hiding algorithms, in *Proceedings of the International Conference on Data Warehousing and Knowledge Discovery* (2005), pp. 418–427
14. V.S. Verykios, E. Bertino, I.N. Fovino, L.P. Provenza, Y. Saygin, Y. Theodoridis, State-of-the-art in privacy preserving data mining. ACM SIGMOD Rec. **33**(1), 50–57 (2004)
15. A. Gkoulalas-Divanis, V.S. Verykios, Association rule hiding for data mining, in *Advances in Database Systems* (Springer US, 2010). https://doi.org/10.1007/978-1-4419-6569-1
16. R. Agrawal, R. Srikant, Privacy-preserving data mining. ACM SIGMOD Rec. **29**, 439–450 (2000)
17. P. Lindell, Privacy preserving data mining. J. Cryptol. **15**, 177–206 (2002)
18. A. Pardo, G. Siemens, Ethical and privacy principles for learning analytics. Br. J. Edu. Technol. **45**, 438–450 (2014)
19. L.P. Macfadyen, S. Dawson, A. Pardo, D. Gasevic, Embracing big data in complex educational systems: the learning analytics imperative and the policy challenge. Res. Pract. Assess. **9** (2014)
20. G. Siemens, P. Long, Penetrating the fog: analytics in learning and education. Educ. Rev. **48**(5), 31–40 (2011)

21. Y. Lou, P. Abrami, J. Spence, C. Poulsen, B. Chambers, S. d'Apollonia, Within-class grouping: a meta-analysis. Rev. Educ. Res. **66**, 423–458 (1996)
22. EUP, *Directive 2002/58/EC of the European Parliament and of the Council of 12 July 2002 concerning the processing of personal data and the protection of privacy in the electronic communications sector* (European Union, European Parliament, 2002)
23. T.W. House, Consumer data privacy in a networked world. Retrieved 13 April 2013 (2012)
24. M. Crook, The risks of absolute medical confidentiality. Sci. Eng. Ethics **19**, 107–122 (2011)
25. H. Nissenbaum, Privacy as contextual integrity. Wash. Law Rev. **79**(1), 101–139 (2004)
26. H. Drachsler, S. Dietze, E. Herder, M. d'Aquin, D. Taibi, The learning analytics & knowledge (LAK) data challenge 2014, in *Proceedings of the Fourth International Conference on Learning Analytics and Knowledge* (ACM, 2014), pp. 289–290
27. M. Gursoy, A. Inan, M. Nergiz, Y. Saygin, Privacy-preserving learning analytics: challenges and techniques. IEEE Trans. Learn. Technol. **10**, 68–81 (2017)
28. V. Mayer-Schonberger, K. Cukier, *Learning with Big Data: The Future of Education* (Houghton Mifflin Harcourt, 2014)
29. P. Ice, S. Díaz, K. Swan, M. Burgess, M. Sharkey, J. Sherrill, D. Huston, H. Okimoto, The PAR framework proof of concept: initial findings from a multi-institutional analysis of federated postsecondary data. Online Learn. **16** (2012)
30. G. Siemens, R.S. d Baker, Learning analytics and educational data mining: towards communication and collaboration, in *Proceedings of the 2nd International Conference on Learning Analytics and Knowledge* (ACM, 2012), pp. 252–254
31. J. Heath, Contemporary privacy theory contributions to learning analytics. J. Learn. Anal. **1**(1), 140–149 (2014)
32. S. Slade, P. Prinsloo, Learning analytics. Am. Behav. Sci. **57**(10), 1510–1529 (2013)
33. P. Prinsloo, S. Slade, An evaluation of policy frameworks for addressing ethical considerations in learning analytics, in *Proceedings of the Third International Conference on Learning Analytics and Knowledge* (ACM, 2013), pp. 240–244
34. K. Verbert, H. Drachsler, N. Manouselis, M. Wolpers, R. Vuorikari, E. Duval, Dataset-driven research for improving recommender systems for learning, in *Proceedings of the 1st International Conference on Learning Analytics and Knowledge* (ACM Press, New York, USA, 2011), pp. 44–53. https://doi.org/10.1145/2090116.2090122
35. L. Chang, I. Moskowitz, Parsimonious downgrading and decision trees applied to the inference problem, in *Proceedings of the 1998 Workshop on New Security Paradigms—NSPW '98*, Charlottesville, VA, USA, 22–26 September (1998)
36. J. Natwichai, X. Li, M. Orlowska, Hiding classification rules for data sharing with privacy preservation, in *Proceedings of the 7th International Conference*, DaWak 2005, Copenhagen, Denmark, 22–26 August (2005), pp. 468–467
37. J. Natwichai, X. Li, M. Orlowska, A reconstruction-based algorithm for classification rules hiding, in *Proceedings of 17th Australasian Database Conference*, (ADC2006), Hobart, Tasmania, Australia, 16–19 January (2006), pp. 49–58
38. J. Quinlan, *C4.5* (Morgan Kaufmann Publishers, San Mateo, California, 1993)
39. W.W. Cohen, Fast, effective rule induction, in *Proceedings of the Twelfth International Conference on Machine Learning*, Tahoe City, CA, USA, 9–12 July (1995)
40. A. Katsarou, A. Gkouvalas-Divanis, V.S. Verykios, Reconstruction-based classification rule hiding through controlled data modification, in *Artificial Intelligence Applications and Innovations III*, vol. 296, ed. by L. Iliadis, I. Vlahavas, M. Bramer (Springer, Boston, MA, USA, 2009), pp. 449–458
41. J. Natwichai, X. Sun, X. Li, Data reduction approach for sensitive associative classification rule hiding, in *Proceedings of the 19th Australian Database Conference*, Wollongong, NSW, Australia, 22–25 January (2008)
42. K. Wang, B.C. Fung, P.S. Yu, Template-based privacy preservation in classification problems, in *Proceedings of the Fifth IEEE International Conference on Data Mining (ICDM'05)*, Houston, Texas, 27–30 November (2005)

43. A. Delis, V. Verykios, A. Tsitsonis, A data perturbation approach to sensitive classification rule hiding, in *Proceedings of the 2010 ACM Symposium on Applied Computing—SAC '10*, Sierre, Switzerland, 22–26 March (2010)
44. R. Bost, R. Popa, S. Tu, S. Goldwasser, Machine learning classification over encrypted data, in *Proceedings of the 2015 Network And Distributed System Security Symposium*, San Diego, CA, USA, 8–11 February (2015)
45. R. Tai, J. Ma, Y. Zhao, S. Chow, Privacy-preserving decision trees evaluation via linear functions. Comput. Secur. ESORICS, 494–512 (2017). https://doi.org/10.1007/978-3-319-66399-9_27
46. D. Kalles, V.S. Verykios, G. Feretzakis, A. Papagelis, Data set operations to hide decision tree rules, in *Proceedings of the Twenty-second European Conference on Artificial Intelligence*, Hague, The Netherlands, 29 August–2 September (2016)
47. D. Kalles, V. Verykios, G. Feretzakis, A. Papagelis, Data set operations to hide decision tree rules, in *Proceedings of the 1St International Workshop on AI for Privacy and Security—Praise '16*, Hague, The Netherlands, 29–30 August (2016)
48. G. Feretzakis, D. Kalles, V. Verykios, On using linear diophantine equations for in-parallel hiding of decision tree rules. Entropy **21**, 66 (2019)
49. G. Feretzakis, D. Kalles, V. Verykios, On using linear diophantine equations for efficient hiding of decision tree rules, in *Proceedings of the 10th Hellenic Conference on Artificial Intelligence—SETN '18*, Patras, Greece, 9–12 July (2018)
50. R. Li, D. de Vries, J. Roddick, Bands of privacy preserving objectives: classification of PPDM strategies, in *Proceedings of the 9th Australasian Data Mining Conference*, Ballarat, Australia, 1–2 December 2011 (2011) pp. 137–151
51. G. Feretzakis, D. Kalles, V. Verykios, Using minimum local distortion to hide decision tree rules. Entropy **21**, 334 (2019)
52. G. Feretzakis, D. Kalles, V. Verykios, Hiding decision tree rules in medical data: a case study, in *Proceedings of the 17th International Conference on Informatics, Management and Technology in Healthcare—ICIMTH '19*, Athens, Greece, 5–7 July (2019)
53. D. Kalles, T. Morris, Efficient incremental induction of decision trees. Mach. Learn. **24**, 231–242 (1996). https://doi.org/10.1007/bf00058613
54. D. Kalles, A. Papagelis, Stable decision trees: using local anarchy for efficient incremental learning. Int. J. Artif. Intell. Tools **9**, 79–95 (2000). https://doi.org/10.1142/s0218213000000070
55. D. Kalles, A. Papagelis, Lossless fitness inheritance in genetic algorithms for decision trees. Soft. Comput. **14**, 973–993 (2009). https://doi.org/10.1007/s00500-009-0489-y
56. J.R. Quinlan, Induction of decision trees, in *Machine Learning 1* (Kluwer Academic Publishers, Boston, MA, USA, 1986), pp. 81–106
57. D. Dua, C. Karra Graff, UCI machine learning repository (The University of California, School of Information and Computer Science, Irvine, CA, 2019). http://archive.ics.uci.edu/ml. Accessed 16 April 2019
58. M. Hall, E. Frank, G. Holmes, B. Pfahringer, P. Reutemann, I. Witten, The WEKA data mining software. ACM SIGKDD Explor. Newsl. **11**, 10–18 (2009)
59. I.H. Witten, E. Frank, M.A. Hall, *Data Mining: Practical Machine Learning Tools and Techniques*, 3rd edn. (Morgan Kaufmann Publishers Inc., San Francisco, CA, USA, 2011)
60. G. Feretzakis, Local Distortion Hiding in Financial Technology Application: A Case Study with a Benchmark Data Set. http://www.learningalgorithm.eu/datafiles_GermanCredit.html. Accessed 16 April 2019
61. J. Ellson, E. Gansner, L. Koutsofios, S.C. North, G. Woodhull, Graphviz—Open source graph drawing tools (2001). Graph Drawing. https://doi.org/10.1007/3-540-45848-4_57
62. S.M. Vieira, U. Kaymak, J. M.C. Sousa, Cohen's kappa coefficient as a performance measure for feature selection, in *International Conference on Fuzzy Systems*, Barcelona (2010), pp. 1–8. https://doi.org/10.1109/fuzzy.2010.5584447

Chapter 4
Engaging Students in Basic Cybersecurity Concepts Using Digital Game-Based Learning: Computer Games as Virtual Learning Environments

Stylianos Karagiannis and Emmanouil Magkos

Abstract Teaching various topics using gamification elements or Game-Based Learning (GBL) methods is a top trend nowadays. Gamification has shown great results towards this direction, however, the usage of GBL methods has not been sufficiently studied for the effectiveness of the learning process. This study examines how instructional design could be applied and how computer games could be a learning environment for acquiring the basic skills and experience in fundamental cybersecurity topics. Towards this direction, this research aspires to discover how specific computer games, designed as simulations, could be converted into virtual learning environments and enhance the learning process, by increasing the levels of motivation and engagement of undergraduate students in the topics of cybersecurity. Computer games are appropriate for creating effective virtual learning environments specific to cybersecurity, providing positive learning outcomes. More specifically, in this study a commercial computer game is evaluated for the effectiveness of using GBL to the learning process. The result of this approach is a learning experience, featuring positive outcomes in terms of engagement and distinct impact in terms of perceived learning. For this study, the ARCS motivation model was used, for evaluating motivation levels and for investigating potential attributes which are related to perceived learning, knowledge and skill acquisition.

Keywords Game based learning · Gamification · Cybersecurity · Instructional design · Cybersecurity training

---

S. Karagiannis · E. Magkos (✉)
Department of Informatics, Ionian University, Plateia Tsirigoti 7, 49100 Corfu, Greece
e-mail: emagos@ionio.gr

S. Karagiannis
e-mail: skaragiannis@ionio.gr

G. A. Tsihrintzis and M. Virvou (eds.), *Advances in Core Computer Science-Based Technologies*, Learning and Analytics in Intelligent Systems 14,
https://doi.org/10.1007/978-3-030-41196-1_4

4.1 Introduction

Playing computer games has been correlated to different behavioural, motivational, cognitive and perceptual outcomes [10, 19]. Using computer games along with the official educational material could be a solution for maintaining self-paced learning tasks resulting in obtaining high levels of engagement and motivation during the learning process. Within a game, students are usually called to solve complex tasks and gradually acquire the desired skills. Using gamification elements in the learning process has shown great results in the past [13, 19, 21, 38]. Approaches like Problem and Challenge Based Learning (PBL and CBL) [33, 44, 61], introduce rewarding systems and other fun elements to the teaching material [14, 81]. CBL and PBL methods follow the learning by experience method, based on theories of *social constructivism* [12, 17, 45, 69], and often require high potential and effort from the participants in order to complete the tasks [70]. Maintaining balance between learning, competitiveness, social collaboration and creativity, while presenting sufficient academic and educational context still remains a research challenge [31]. Towards this direction, computer games could be a solution [41, 70, 75, 80].

Gamification turns the learning process into a game by embedding various gamification elements, while *Game-Based Learning (GBL)* is using some of the elements of a game, such as a computer game as part of the learning process [2, 65]. Through GBL, instructional material is able to be presented to the participants, while they experience a gamified learning experience. Active participation together with fun elements are important for enhancing the learning experience in order to convert it into a more interesting and engaging [71] process. Finally, GBL has been used successfully in various knowledge topics [70].

The effectiveness of gamification and GBL in terms of skills and knowledge acquaintance has been extensively studied [50, 70] with respect to how such methods could be effective in terms of the learning process [31, 55, 60, 72]. However, maintaining the participants motivated during the learning process, while achieving high levels of engagement is usually difficult. Especially in cybersecurity topics which usually require advanced technical skills, more engaging methods are required in order to achieve the required learning outcomes. Most of the issues derive from the complex concepts that participants are called to understand and usually require strong background knowledge [29, 81, 84]. Even if it is applicable to enhance the learning experience using *virtual learning environments* [62], the set of tasks are usually difficult to be followed by undergraduate students.

Computer games have already been used for teaching various concepts in computer science. For instance, particular approaches for teaching Boolean algebra are computer games focused on hardware design such as *"MHRD"*[1] and *"TIS-100"*[2] for teaching assembly language and computer architecture [43]. Another interesting approach is *"Shenzen I/O"*,[3] a circuit deisng game, focused on acquiring familiar-

[1] https://store.steampowered.com/app/576030/MHRD/.

[2] http://www.zachtronics.com/tis-100/.

[3] https://store.steampowered.com/app/504210/SHENZHEN_IO/.

ity in various concepts of engineering and computer architecture. These approaches tend to be able to improve coding skills, however, sufficient empirical data to quantify or to evaluate the effectiveness of the learning outcomes do not exist [1]. On the other hand, regarding tabletop games, empirical data exist which usually evaluate the positive impact of these approaches in terms of general introductory knowledge [83].

For integrating educational context and maintaining balance between acquiring knowledge while having fun, it is important to highlight specific indicators which affect the learning outcomes. It seems that using computer games as virtual learning environments could enhance the total learning experience. One of the most popular approaches for presenting cybersecurity challenges are *virtual machines* which are based on systems that maintain various vulnerable services and technologies. These approaches are usually used in *Capture The Flag (CTF)* challenges, where participants are called to solve puzzles and try to exploit the vulnerabilities of the systems. Through CTF challenges participants can actively participate in the learning experience and achieve high levels of engagement [15]. Towards this direction, practical customized courses and exercises could be developed, in order to train people in terms of cybersecurity threats and attacks [26, 63]. Cybersecurity topics often include complex processes and as an outcome, educational context must be presented as subtasks and reward the participants, in order to increase their motivation and persistence towards finding appropriate solutions. On the other hand, gamification has been previously applied as well as table-top games. Some approaches combine storytelling and puzzles [76, 77], while the most popular approaches of table-tops for cybersecurity are *Control-Alt-Hack* [24, 25] and *d0x3d!* [35]. However, further research is needed for providing sufficient empirical data, related to the learning outcomes and to the acquired skills. Furthermore, cybersecurity training using *serious games* is a young and developing field [39]. Serious games related to cybersecurity such as *CyberCIEGE* have been mentioned together with commercial computer games like *Hacknet-labyrinths* [36, 73]. Furthermore, specific studies relate education directly to games [82]. Moreover, CTF challenges and online cybersecurity challenges have been described along with serious games in a relative context. In order to discover how to adopt real-world challenges [8, 22, 51] and security scenarios in a computer game, a specific learning and educational methodology is required.

Embedding computer games, gamification and GBL in cybersecurity education, will continue to evolve and the enhancement of learning processes will most probably engage students in cybersecurity topics, together with the positive outcomes of increasing security and privacy awareness [4, 11]. A few studies show that empirical evidence towards this direction currently exists, however, no clear evidence is mentioned in terms of using these methods as an assessment method [6, 15, 26, 30, 46, 57].

4.1.1 Our Contribution

The main purpose of this research is to evaluate how and if *Digital Game-Based Learning (DGBL)* could enhance the learning experience regarding cybersecurity curriculum in academia, achieving high levels of engagement and active participation. This paper seeks answers to the following questions:

1. How can education be more interesting with the adoption of gamification elements and how can DGBL could be used in order to create virtual learning environments to this aspect?
2. What are the potential benefits and limitations regarding the learning effectiveness of DGBL in conducting cybersecurity training labs?
3. Which specific attributes derive from DGBL and possibly affect positively the learning outcomes?

More specifically, in this study, the commercial computer game, Nite Team 4[4], is evaluated in terms of the effectiveness of adopting DGBL into the learning process. This study aspires to make a contribution in providing empirical data and evidence on how and whether computer games can provide sufficient background knowledge and skills acquaintance in cybersecurity courses. With the implementation and evaluation of DGBL [60] as a main virtual learning environment in cybersecurity, this study aspires to fill the gap between theory and practice presenting a self-learning experience with the option to integrate with a collaborative learning process inside the classroom.

Our approach is highly correlated with approaches like CTF challenges and more specific Classroom Capture the Flag Challenges [28]. Using the recommended approach, students were able to acquire technical experience and skills in basic cybersecurity topics, in ethical hacking and penetration testing. Furthermore, the students were introduced to technical skills in topics of IT, while learning by experience the concepts that were already instructed according to the official academic curriculum.

4.1.2 Outline

This paper is organized as follows: Sect. 4.2 presents the methodology used for this research. Section 4.3 discusses background concepts and principles that were used in our approach, while Sect. 4.4 presents our approach. In Sect. 4.5, results and discussion of this research are given. Section 4.6 concludes the paper.

[4]https://www.niteteam4.com.

CTF	Tabletop & Computer Games	Instruction Design	Feedback from Students
Maintained a research, for approaches already used in cybersecurity, however focused on skills and knowledge acquaintance.	Tested various Computer and Table Top games in order to learn more about the current approaches. Related researches were also in the scope.	After selecting the computer game for evaluating the effectivenes in teaching, the educational material was structured and correlated with the game.	Observation was conducted on the behavior of students during the course, focusing on active learning capabilities, self-learning experience and collaborative learning. Finally a questionnaire about the total learning experience was given to the students.

Fig. 4.1 Methodology steps

4.2 Methodology

For this study a few computer and tabletop games were examined outside the classroom, including Hacknet and Hacknet Labyrinths, TIS-100, Shenzen I/O, [d0x3d!] and Control-Alt-Hack [2, 24, 35, 36, 43]. The chosen approaches include attributes which focus on skills and experience acquaintance and on discovering elements which enhance the self-learning experience.

For integrating educational material while maintaining balance between fun and knowledge, it is important to specify the key-elements which might affect the learning outcomes. The specific game has been chosen to be presented in the lab in order to quantify the effectiveness in terms of the learning outcomes. To ensure that DGBL methods enhance active participation during the learning process, participants were introduced to the in-game context in order to be familiar with the main concepts. Specific directions were given before starting the learning process.

In Fig. 4.1, the main steps which were followed in our approach are presented. During our study, active learning and instructional design guidelines were first analyzed for collecting a concrete and conceptual index, in order to maintain balance between fun and learning.

The specific computer game, *NITE Team 4*, was selected mostly because of the high correlation between the in-game context and of real-world tools and methods.

More importantly, the computer game is presenting extra information by providing various website links.

The following principles were taken into consideration for creating our methodology [23, 27] deriving from Vygotsky's theories on Zone of Proximal Development:

1. Provide related context to what have been already instructed before, in order for the participants to recall any required information [27].
2. Organize the learning process in four different stages and embed various subgoals in order to build up gradually from simple to complex concepts [27].
3. Emphasize on team collaboration [58] and on the importance of setting questions during the learning process [23].
4. Enhance the process with self-learning capabilities, in order to bypass the knowledge and experience gap between the participants [23, 27].
5. Maintain the concepts and scenarios in difficulty levels which will be comfortable and at the same time challenging [23, 27].

Expected learning goals include improvement in terms of background knowledge, skills acquaintance that are mostly related to basic concepts and methods, used in ethical hacking and penetration testing. Most of the presented processes are similar to real-world processes. The presented context inside the computer game helps for creating a real-world learning environment, providing rich context and a variety set of challenges.

The experiment was conducted on undergraduate students for enhancing the learning process of a specific course in the Department of Informatics, Ionian University, Corfu, Greece. Most of the participants did not have any significant experience and knowledge in cybersecurity. Specifically the observation was focused on discovering actions that highlight attributes of collaborative learning, indicators that present self-learning experience elements and connection to theoretical concepts. During the learning process, a variety of educational and informative material according to cybersecurity concepts has also been instructed. For analyzing the motivational aspects of learning environment, *the ARCS model of Motivational Design* [49] was taken into consideration. The questionnaire was based on the four key elements of the above model, namely Attention, Confidence, Relevance and Satisfaction [49]. During the learning process other various observations were performed related to the students behavior and are described in Sect. 4.4.

4.2.1 Common Approaches in Cybersecurity Topics

To begin with, it is important to present approaches of computer games and other gamification approaches, which were previously tested before this study, focusing on the expected learning outcomes. These approaches are worth-mentioning and were previously tested for discovering the key elements which might enhance our approach and might be used in future research.

4.2.1.1 Computer Games

Various approaches are presented on commercial computer games, mostly in the form of simulations.

1. **Hacknet Labyrinths**[5]: *Hacknet Labyrinths* is a simulation-based hacking computer game, featuring real-world networks and real-like system infrastructure. A set of tool-kits for penetration testing and ethical hacking similar to the real ones is presented [36].
2. **Uplink**[6]: The main positive learning outcomes of this game is to introduce network commands and familiarity in UNIX systems along with other basic topics of cybersecurity. However this approach is mostly a computer game featuring only themes deriving from cybersecurity [56].
3. **NITE Team 4**[7]: NITE team 4 is also a simulation computer game featuring topics of cybersecurity and adopting real terminology from NSA leaks. Tools that already exist in the real world are also presented in the computer game. The high engagement and the similarities with real ethical hacking tools and methods are the main attributes of this game, as well as the ability to create custom puzzles and challenges. Related information about network protocols and topology is presented, as well as real tools and services. The main difference with Hacknet is that methods presented in NITE Team 4 are similar to the real ones and the steps are clearly stated in terms of learning basic concepts of cybersecurity. Moreover, the in-game context is enhanced with educational material regarding real-world information such as real ethical hacking tools and penetration testing methods.
4. **CyberCIEGE**[8]: CyberCIEGE focuses on cyber defense where participants access a real VPN network. The target group for this game is to prepare workforce such as system engineers, software developers, system designers and network administrators in order to increase security awareness.

4.2.1.2 Vulnerable Virtual Images

Well known approaches for enhancing the learning process include the use of virtual systems that are vulnerable. These approaches are also used on CTFs, however the challenges usually are able to be hosted individually, focusing on self-learning experiences. Some approaches are published in ***Vulnhub***[9] and others are published in official websites like the well-known virtual machine of ***Metasploitable***[10] from Rapid7.

[5]http://www.hacknet-os.com/.

[6]https://www.introversion.co.uk/uplink/.

[7]https://www.niteteam4.com/.

[8]https://my.nps.edu/web/c3o/cyberciege.

[9]https://www.vulnhub.com/.

[10]https://sourceforge.net/projects/metasploitable/.

Furthermore, some other challenges are published in ***Over the Wire***.[11] Finally, various conferences publish the challenges and the solutions of conducted past CTF challenges.

1. **Vulnhub**: This website maintains various vulnerable virtual systems, focused on learning the basics in cybersecurity. Using walkthroughs the participants are able to learn and get help from others. Afterwards, it is possible to try and solve the challenges without using the walkthrough, since the website provides also vulnerable images for which the solutions are not provided.
2. **Metasploitable**: A well-known virtual image featuring various vulnerabilities. Focused on web exploitation methods this image features a large variety of vulnerabilities.
3. **Over the Wire**: This website maintains a variety of challenges which are presented and accessed mostly through SSH. Focused on step-by-step learning experience and featuring walkthroughs it is considered as a nice approach for beginners.
4. **Hack the Box**[12]: This approach is one of the most known in conducting individual CTFs. Hosting a lot of challenges, this platform maintains many vulnerable images and CTF challenges that participants can access using a VPN. This platform mostly focuses on various levels of challenges, however most of the challenges are for advanced users. This platform features options for workforce acquaintance and a large community. Moreover, this platform has been presented in various conferences for providing CTF challenges.

4.2.1.3 Tabletop Games

Furthermore, in this study the following tabletops were also studied:

1. **Control-Alt-Hack**[13]: Control-Alt-Hack is a card game for increasing security awareness. The game mechanics and designed content is engaging, encouraging interest in computer security. The findings were positive in terms of increasing security awareness and it was conducted on 22 educators representing 450 students [24].
2. **[d0x3d!]**[14]: [d0x3d!] is an open source card game designed for informal computer security education [35]. The game is cooperative and players assume the role of white-hat hackers, with the main task of retrieving digital assets from an adversarial network. The card game is released in three different forms.[15]

Most of the table-top games focus on acquiring familiarity related to cybersecurity terminology.

---

[11] http://overthewire.org/wargames/.

[12] https://www.hackthebox.eu/.

[13] http://www.controlalthack.com/.

[14] http://d0x3d.com/.

[15] https://github.com/TableTopSecurity/d0x3d-the-game.

4.3 Background: Gamification and GBL

4.3.1 Active Participation and High Levels of Engagement

A balance between challenges, instructive material and assessments together with gamification elements could result in positive learning outcomes [20, 70]. GBL provides fast response between action and feedback resulting in self-learning and self-assessment elements, similar to approaches like learning from mistakes [70]. These attributes are important for achieving high motivation and for developing self-paced elements during the learning process. Games have been highlighted with the method of "try and error", which is important for motivating players to keep trying until they succeed.

As a result, computer games could provide a sufficient learning environment as already applied in serious games [19, 48, 55, 59]. Context which is related to real-world cases and challenges could be presented inside a computer game. When carefully designed, the pedagogic content could be embedded and with the integration of quizzes and interactive exercises, computer games could provide the opportunities for achieving high levels of engagement through interaction and exploration [6, 9, 19, 78, 79].

4.3.2 Gamification, GBL and Expected Learning Outcomes

In approaches like gamification and GBL, the expected learning outcomes have to be described and specifically depicted. To enforce appropriate instructional context, the expected learning outcomes have to be specified. Entertainment elements alone could not guarantee sufficient learning outcomes. The differences between Gamification and GBL are presented in Table 4.1 in order to distinct such indicators.

In order to be effective in terms of the learning outcomes, pedagogical principles and directive instruction material has to be carefully embedded and organized [37]. Towards this approach, implementations that provide simulations focused on network infrastructure do exist [34].

Table 4.1 Differences between gamification and GBL [42, 46]

Gamification	Game based learning
Gamification is turning the learning process as a whole into a game, using gamification elements like reward system, badges, storytelling, theme-related context, different levels, Leader-boards	GBL is using a game as part of the learning process, using learning games to achieve an instructional goal. Serious games and classical computer games or table-top games might be included
Game design and game elements in non-game context. Convert instruction to a game-like approach	Educational content is included in the game. Play a digital or non-digital game in order to achieve the required learning outcomes

4.3.2.1 Basic Principles of Gamification

The use of digital and physical games is often presented [3] together with specific exercises from suggested gamification approaches [47]. However, in order to accept the broad use of gamification in academia, more careful approaches have to be considered as well. Gee [32], for example, indicated that 16 learning principles could be offered from games such as: interaction, well-structured problems, challenge and consolidation, pleasantly frustrating, system thinking, exploration and cross-functional teams, among other principles [17, 32, 53, 54]. These principles are directly related to cybersecurity concepts in order for the participants to discover design and configuration flaws.

4.3.2.2 Role of Engagement

The role of engagement in the learning process is also pinpointed during the design phase of a game [66]. When playing games, participants are called to solve complex challenges and participate without experiencing fatigue, while it is a comfortable learning task [30]. For achieving the expected learning outcomes it is important to highlight the attributes that have potential positive impact on the learning process [5, 9, 64]. Towards this direction, careful and comprehensive methods are required in order to introduce the instruction material and to define the learning outcomes [40].

4.3.2.3 Learning Outcomes and Effectiveness

Specific studies argue that not all games could be effective as a learning environment and not all the participants will accept these methods [67]. Educational games are still in early stages of evolution and even nowadays more careful approaches have to be developed in order to achieve and to accept computer games as a successful learning approach. Main issues include difficulties in terms of integrating educational context inside games [35].

4.3.2.4 The Positive Impact of Games

Despite the limitations of using GBL, games in general have been associated with the benefits below which result in positive learning outcomes [2]:

1. Games can be used for examining individual characteristics such as self-esteem
2. Games are fun, which consequently leads to undivided attention and focus for long periods of time, helps developing various IT skills and engages participants to make mistakes and learn from them.

4.3.3 GBL and Education in Cybersecurity Topics

Educational domains that use GBL are interdisciplinary topics where skills such as critical thinking, group communication, debate and decision making are of high importance. Such topics require high collaboration and active participation in order to achieve the sufficient learning outcomes [70]. Towards this direction computer games could enhance this perspective and improve the attributes of collaborative learning and active participation.

4.4 Using a Computer Game as a Virtual Cybersecurity Learning Environment

Integrating a computer game as a virtual learning environment is difficult, especially for presenting complex topics such as cybersecurity. Computer games maintain the ability to create self-learning experiences, by presenting interactive walkthroughs and step-by-step guidelines. Towards this direction, students were firstly introduced to basic methods of ethical hacking and penetration testing. It was mentioned in the classroom that a computer game will be used as a virtual learning environment, highlighting the correlation between real commands and software tools. Finally, it was mentioned that approaches such as CTF challenges already exist in official training and that this proposed approach is a gamified version similar to CTF challenges.

4.4.1 Introduction to the Game

Participants were guided through a structured guideline and were called to fulfill the first steps and to learn the basic tactics, tools and commands. The first directions and information were presented with the following message: *"some of the most advanced technology used to hack through corporate systems and networks based on Real-World Events. The world today has evolved from regular warfare, to cyberwarfare. As such, we need experienced personnel to repel attacks from various sources at any given time. Join us for a better world"*. The message depicts the aspect of being involved in real-world cases and acquiring skills for repelling attacks. In each step, participants were rewarded with badges as virtual certification credits. These features are important for the participants to be motivated and at the same time important in order to control and to evaluate their own progress.

Computer games maintain the ability to present step-by-step guides. The in-game guidelines were presented in a well-structured way which was effective in terms of enhancing the self-learning experience. It is important to maintain the attribute of self-learning, in order to bypass issues deriving from knowledge and skills gap between the participants. After each set of challenges, solutions were presented in the class for those who were incapable to follow.

4.4.1.1 First Section, the Academy

This section includes practices important for learning the basic commands and getting familiar with the required tools. The basic topics are the following:

1. Basic Terminal Operations
2. Digital Forensics
3. Network Intrusion
4. Command and Control
5. Elite Training
6. Signal Intelligence
7. StingerOS Advanced.

Most of the missions were highly correlated to cyberthreats and cyberattacks, encapsulating information and software tools from the real-world.

4.4.1.2 Academy Level 1—Basic Terminal Operations

This topic includes a set of practices and missions, which helps the participants to learn the basic tools and commands and get familiar with the environment. For example, 6 training missions were included in the first academy level (Fig. 4.2), containing a subset of missions. More specifically the sub-modules were the following: Stinger OS Basics, Basic OSINT, Fingerprint, Advanced OSINT, Exploit Database and Foxacid. All the above tasks included a code number (ex. SOPS.01, OSINT.01). The used terminology was similar to the official cybersecurity curriculum. For example, even from the beginning the tasks and missions related to official terminology such as *Open-source Intelligence (OSINT)* methods.

4.4.1.3 Academy Level 2—Digital Forensics

Topics like intelligence gathering, directory enumeration and password attacks were presented. Moreover, concepts such as Xkeyscore (used from NSA) were mentioned deriving from elements of *Alternate Reality Games (ARG)* and in order to present real-world context.

4.4.1.4 Academy Level 3—Network Intrusion

Tools like Intelligence gathering focused on network infrastructure and exploitation tools were introduced. Concepts like *Man In The Middle (MITM)* were introduced, together with the Social Engineering Toolkit (SET Toolkit[16]).

---

[16]https://github.com/trustedsec/social-engineer-toolkit.

4.4.2 Game Components

NITE Team 4 maintains a variety of tools, software components, modules and sub-modules in order to create more engaging and customized experience.

4.4.2.1 Gameplay Option—Section 1: Basic Software and Tools

The main tools and software components of the game are presented in Table 4.2. The following tools were available for further processes:

- Stinger OS Cluster
- Satellite Feed (Drone and Geolocation Intelligence)
- Hivemind Network.

4.4.2.2 Gameplay Option—Section 2: Missions

Using elements of storytelling which derive from real-world events and cases, the learning process could be transformed in an immersive and educational process. Missions are divided into four different sections. Every section and topic included a different set of challenges, featuring storytelling and multimedia elements. Every scenario or story, is called *Operation*, maintaining a sub-set of challenges and sub-tasks. The attributes of every mission are the following:

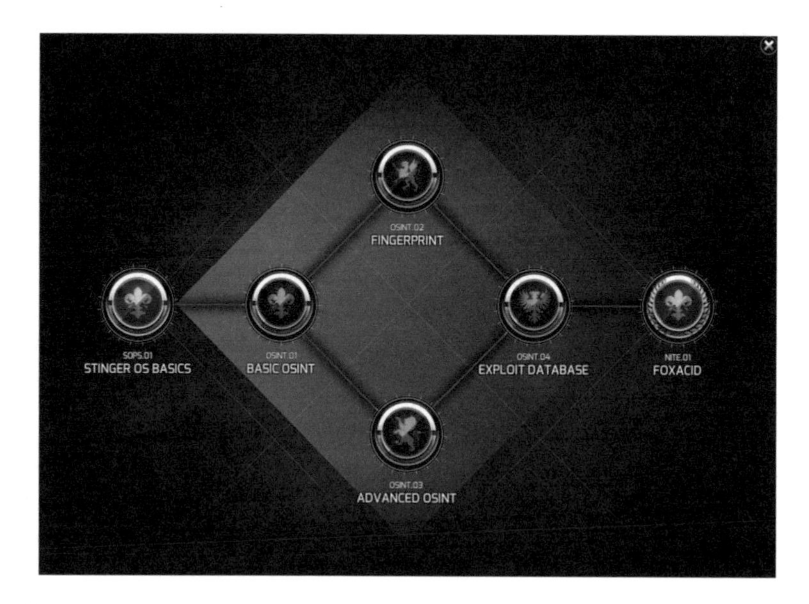

Fig. 4.2 Academy level 1 steps and progress tracking

Table 4.2 Main tools and software components

Information gathering	Network intrusion
Host fingerprint	Phone CID backdoor
Exploit database	Password attack
WMI scanner	MITM
Air crack	Social engineering toolkit
Active directory	Hydra terminal
Data forensics	Advanced tools
Xkeyscore forensics	Turbine C2 registry
File browser	Satellite feed
TBW archive	Hivemind network
Notepad	Command center

1. **Type**: The type of mission such as *Basic OSINT, enumeration or exploitation.*
2. **Real Life**: Correlation with real-world cases.
3. **Level**: Level of difficulty.
4. **Ambiance**: The role of the player in each mission.

Each mission has a unique name and a description. For example, a specific operation has the name *Operation Castle Ivy* and the description is the following: *"Military grade malware was stolen from NITE Team 4 as of yet unknown means. Assess the scope of the leak"*.[17]

4.4.2.3 Gameplay Option—Section 3: Multiplayer/Hivermind Network

Hivemind Network is a module which includes challenges and puzzles created from other users, using an extra feature/software of Nite Team 4, called *Network Administrator* (Fig. 4.3). Currently, 51 different challenges are presented. Users are able to execute network mapping processes similar to *arp-scan* or *netdiscover*. Through this process, the user is possible to extract information related to network topology and discover the running services. Focusing on methods of reconnaissance, enumeration and OSINT the participants are able to get familiar with network topology and services. Every time a user discovers some information, using tools like *sfuzzer* or *fingerprint*, the discovered information is included in the network topology.

Custom challenges could integrate storytelling elements including interesting names and other related information and context. Every challenge is hosted in a virtual domain *.hvm.* which includes multiple services and assets. Sub-domains and other system resources could be enumerated using reconnaissance methods in order to exploit the vulnerabilities and attack the systems. Each challenge holds informa-

[17] NITE Team 4—In-game statement.

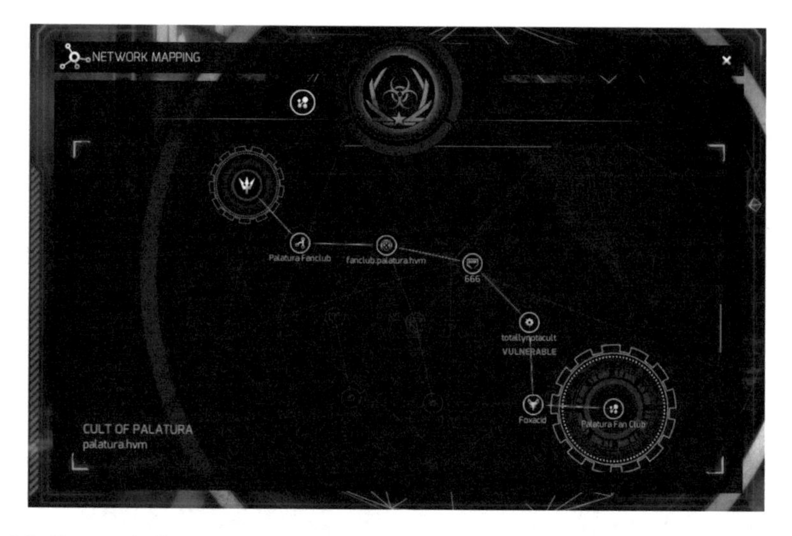

Fig. 4.3 Custom challenge created by another player

tion related to the duration of time needed for solving the challenge and indicators related to content quality, fun and difficulty level among others.

4.4.2.4 Gameplay Option—Section 4: Bounties

Participants were called to test their skills (Fig. 4.4) through challenges that have a time limit, maintain a rewarding system and give reputation points. This set of challenges are very similar to official CTF challenges.

4.4.2.5 Gameplay Option—Section 5: Open WorldCampaign

This option is a combination of ARG elements and of NITE Team 4 challenges. The potential of creating a real-world puzzle combined with storytelling elements could create real-world challenges and enhance the learning curve, since the engagement levels might increase.

4.4.2.6 Extra Module: Network Administrator

This is an extra software component, where users are able to create their own challenges and puzzles in order to make them public to the NITE Team 4 network universe. Network Administrator is still in beta version, however it is already up and running. All services, port numbers and other network and application components are highly correlated to real software components. Through this option, users are free to create a

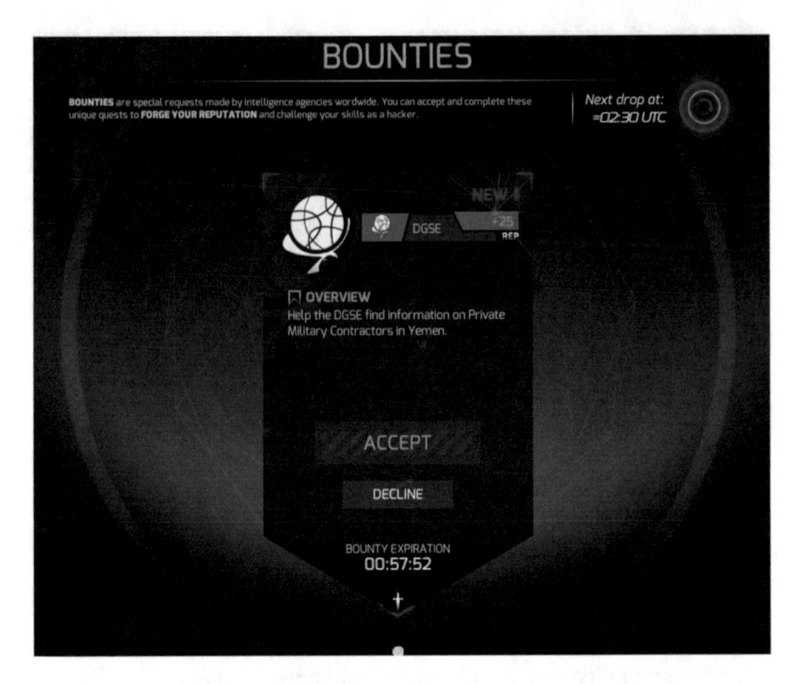

Fig. 4.4 Special challenges with time expiration

network topology and actually get familiar with threat modelling tools. Towards this direction, this tool could help the participants in terms of acquiring basic knowledge in topics of security modelling and systems' design. Moreover, this could be very important in terms of getting familiar with threat modelling tools and methods.

4.4.3 Comparison Between Game and Real-World

In conclusion, it seems that correlation exists between the in-game concepts and real-world information. Towards this direction, it is important to mention that some users mentioned that some challenges required skills in cryptography in order to complete the missions. This issue could be solved and enhance the learning experience in terms of real skills, if the users were introduced to basic principles and set of challenges between the main challenges. Basic methods and tools for executing the attacks are presented in Table 4.3.

NITE Team 4 has been already announced as a partnership with the *International Air Transportation Association (IATA)* for setting practices in order to improve skills related to the topics of cybersecurity. Featuring real cases including modern cyberthreats, the main focus is to maintain threat cases for educational purposes.

Table 4.3 Comparison between in-game commands and real commands

Action	NITE Team 4	Real-world
Port Scanning	Fingerprint	Nmap
DNS enumeration/sudomains (OSINT)	Osintscan	The Harvester, sublist3r
DNS enumeration/sudomains (Wordlist)	Sfuzzer	Dirb, Dirbuster, DNSRecon
Search vulnerabilities	Searchsploit	Searchsploit
Discover running services	Netscan	Nikto, Wpscan
Packet Capturing (802.11)	Airodump	Airodump-ng

Moreover, showcases related to the game are also presented in events such the RSA Conference, The Black Hat and SecTor conventions.[18]

4.4.3.1 Real-World Cases

The following methods are presented and are highly correlated to real cases used in ethical hacking and penetration testing:

1. Reconnaissance and Intelligence Gathering: Processes related to network scanning, sub-domain enumeration and network mapping using in-game tools such as *sfuzzer* and *OSINTscan*. Afterwards participants are invited to execute port scanning and vulnerability analysis using the related software components and tools.
2. Network Infiltration: Participants are called to attempt connecting to the network using various rootkits and exploits.
3. Network Scanning (Insider): Participants have to execute commands such as *netscan* and *airodump* for discovering any devices connected to the network and also for discovering and enumerating folders across the network, such as shared folders.
4. Password Attacks: Participants have to launch password attacks in order to access services and enumerate devices in the network. Basic tools are introduced which are similar to the real ones. Generic tools like *John the Ripper* and text files such as *RockYou.txt* are presented in this section. For example RockYou.txt is a popular wordlist used for bruteforce attacks and John the Ripper is a well-known software tool for executing password attacks.

Moreover, basic malware, hacking tools and exploits such as rootkits *Assassin* and *AfterMidnight* are presented. An in-game software tool called *Foxacid Server* is mostly used for exploits like *Metasploit*. Furthermore, commands such as *searchsploit* is presented in order to discover information related to exploits and vulnerabilities known as *Common Vulnerabilities and Exposures Exploits(CVEs)*.

[18]https://aliceandsmith.com.

Finally, the game provides more information regarding each concept and tool, redirecting the user to various websites, blogs and Wikipedia pages in order to learn more information regarding to real commands and software tools. For instance, information is provided related *Kali Linux* and *Metasploit* during the in-game exploitation phases. It is considered important for the students to follow such links, in order to learn more about the type of attacks and to understand the tools.

4.5 Results and Discussion

The learning outcomes from our approach were evaluated using the following indicators [49]:

1. **Attention**: Refers to elements of perceptual stimulation, active participation and the ability to present similar context to the participants' interests.
2. **Relevance**: Includes elements which help the participants understand the relevance between past knowledge and includes specific goal orientation, familiarity and context related to the learner's needs and motives.
3. **Confidence**: Attributes which enhance the learners' positive expectation for success, personal responsibility and self-control elements, that participants have during the learning process.
4. **Satisfaction**: Enjoyable and fun elements, and features such as extrinsic rewards while enhancing the extrinsic and intrinsic reinforcement related to the effort.
5. **Perceived Learning**: Refers to attributes like self-report capabilities, knowledge acquaintance or more accurately as *"Self-report of knowledge gain, generally based on some reflection and introspection"* [7].

For each of indicator, participants were called to answer in a 7-point Likert scale if they disagree (grade-1) or totally agree (grade-7).

Deriving from Keller's model of Motivation (ARCS), the concepts of Attention, Confidence, Relevance and Satisfaction [49] are presented, along with the statements which were included in this research. The indicator of actual learning is not evaluated, since specific assessment methods have to be integrated in order to achieve such results.

Attention. The main objective is to achieve high attention levels, in order to increase levels of engagement during the learning process. In Table 4.4 the statements related to the element of Attention are presented.

During the process, participants indicated high levels of attention, active participation and collaborative learning, achieving high levels of engagement. Regarding the statements in Table 4.4, average score for ***Attention*** **was 73.46%**. Some participants scored very low in evaluating the total process as a good approach for cybersecurity training and it was mentioned that approaches like CTF would be more appropriate. However, it was highlighted that gamification elements would enhance the approach, for instance if gamification elements were embedded in a specific CTF challenge.

Table 4.4 Questionnaire items to evaluate attention

Item code	Statement
ATT1	The presented process included various self learning capabilities
ATT2	During the lab I was focused and absorbed in the process
ATT3	It is an "eye-catchy process"
ATT4	The way in which learning objectives were presented helped me focus
ATT5	Storytelling is exciting and helps me to go on

Table 4.5 Questionnaire items to evaluate relevance

Item code	Statement
REL1	It reflects, to a good extent, possible real case scenarios
REL2	I can correlate the content with concepts that I am already familiar with such as Databases, Networks, Programming, and Operating Systems
REL3	The content is related to my general interests in the scientific field
REL4	This methodology corresponds to my IT needs
REL5	I am aware of most topics and I can discover related information in topics of Programming, Databases, Network and Web Infrastructure

Relevance. In Table 4.5, it is perceived that participants which have strong background knowledge would understand most of the presented cybersecurity topics. As a result participating in such challenges could improve skills related to other IT topics such as Databases and Operating Systems. Therefore, it is still difficult to understand how this process could result in achieving better grades in terms of evaluation processes such as exams. However, acquiring skills and background knowledge through this method may eventually have a positive impact in other topics of IT. The average score for ***Relevance*** **was 76.40%**.

Confidence. Participants mentioned that self-learning features are very helpful and improve their ability to maintain control during the entire process (Table 4.6). Towards this direction, computer games could be used as self-assessment methods in evaluating every step of each challenge. Elements of high interactivity enhance participants' confidence and together with elements of collaborative learning, the learning process achieve high engagement levels. Most of the participants wanted to continue the process after acquiring sufficient familiarity related to basic in-game concepts. The average score for ***Confidence*** **was 74.76%**.

Satisfaction. Some participants mentioned that this process was the best and most enjoying process they ever had during their undergraduate studies (Table 4.7).

Collaborative learning could enhance the engagement levels and make the learning environment more entertaining. Average score on ***Satisfaction*** **was 78.47%**.

Perceived Learning. For ***Perceived Learning*** **the average score was 78.70%**. The statements for assessing perceived learning are presented in Table 4.8.

Table 4.6 Questionnaire items to evaluate confidence

Item code	Statement
CON1	How much despite the difficulties this methodology increases the feeling of perseverance in order to discover the solution
CON2	I want to finish the lab (to see everything—it is interesting)
CON3	I would like to explore hidden sub-challenges or optional context
CON4	Gradually and during the lab, I think I can cope better and I feel self-confident
CON5	This methodology does correspond to my IT needs
CON6	During the process I feel I have the control
CON7	Using this method as an assessment method is a good idea. I am not scared of that as much as of the exams

Table 4.7 Questionnaire items to evaluate satisfaction

Item code	Statement
SAT1	I would like to repeat this process without caring for academic rewards and marks. I would like to learn more
SAT2	Really, I had a lot of fun
SAT3	I did not feel tired when I played
SAT4	I feel satisfied after the lab
SAT5	I really enjoyed this process as a virtual learning environment
SAT6	I felt that time passed very quickly
SAT7	I felt happy during the process

Table 4.8 Questionnaire items to evaluate perceived learning

Item code	Statement
PER1	Could create the right environment for learning more information and acquire skills
PER2	It is like a virtual learning environment, however it requires customization
PER3	It might help me to improve my grades (as a result of familiarity with other fields)
PER4	It is perceived by me that I have developed some skills
PER5	I can perceive that through this methodology I acquired knowledge and skills

Real-world tools and case scenarios. It is important to mention the importance of including real-world scenarios, tools and commands during the learning process. Towards this direction the virtual learning environment is important to be a representation of real cases and scenarios. Since in this study a computer game was used, it is important to compare it with the commands and tools used in the real world. In Table 4.9 the statements for evaluating how much the computer game matches reality are presented.

From the participants' responses on the statement of how much the process reflects a real case scenario, the answers were positive. However it was mentioned that some

Table 4.9 Questionnaire items to evaluate real-world tools and case scenarios

Item code	Statement
REA1	It feels that the scenarios are similar to real-world cases
REA2	This method is good for acquiring basic familiarity with some of the real penetration tools
REA3	Multi-faceted learning, helps me to understand most of the methods
REA4	It is possible that the presented scenarios and cases could be real cases
REA5	Cases are very real. If I did not know that the system is a game I would possibly think that this is real

Table 4.10 Questionnaire items to evaluate impact on other related fields

Item code	Statement
REL1	It would be nice to have similar learning processes with the right customization context in other courses as well
REL2	After my participation, my learning ability in other fields seems to improve

commands and software tools do not exist in reality. High correlation was indicated with Kali and other penetration tools and methods. The **correlation between real-world software tools and cases in contrary to the in-game context scored 72.20%**.

Related topics and Perceived learning. Since topics of cybersecurity require strong background knowledge in other IT topics, it is inevitable that we have to identify if perceived learning is achieved in other topics as well. Towards this direction it is important to identify any direct or indirect impact to the learning outcomes in general.

In Table 4.10 the possibility for our approach to present other IT topics, relevant to cybersecurity scored 80.51%.

4.5.1 Summary

The ARCS model itself provides information about the impact of the process on the engagement levels during the learning process. In our research high levels of students' engagement were achieved during the learning process. The average scores are summarized in Table 4.11.

Our approach takes into account the ARCS model [49] for evaluating motivation levels and for discovering the elements which are related to perceived learning. Most of the participants scored this method sufficient in terms of using this method as a learning process. Attributes such as actual learning and the usage of this method as an assessment method are not evaluated. Towards this direction more work is required regarding the impact of the specific approach in other I.T topics. **Total score of this**

Table 4.11 Summary scoring table

Attribute	Average score (%)
Attention	73.46
Relevance	76.40
Confidence	74.76
Satisfaction	78.47
Perceived learning	78.70
Relevance with other topics	80.51
Real-world	72.07

Attention

	N	%
Cases Valid	11	91.67
Excluded	1	8.33
Total	12	100.00

Reliability Statistics

Cronbach Alpha	N of Items
.86	7

Relevance

	N	%
Cases Valid	11	91.67
Excluded	1	8.33
Total	12	100.00

Reliability Statistics

Cronbach's Alpha	N of Items
.80	6

Confidence

	N	%
Cases Valid	11	91.67
Excluded	1	8.33
Total	12	100.00

Reliability Statistics

Cronbach's Alpha	N of Items
.84	6

Satisfaction

	N	%
Cases Valid	11	91.67
Excluded	1	8.33
Total	12	100.00

Reliability Statistics

Cronbach's Alpha	N of Items
.91	7

Perceived Learning

	N	%
Cases Valid	11	91.67
Excluded	1	8.33
Total	12	100.00

Reliability Statistics

Cronbach's Alpha	N of Items
.86	5

Real-world elements

	N	%
Cases Valid	11	91.67
Excluded	1	8.33
Total	12	100.00

Reliability Statistics

Cronbach's Alpha	N of Items
.39	5

Fig. 4.5 Reliability check—Cronbach's Alpha

approach acceptance from students indicates 76.49% acceptance as a sufficient learning method.

Reliability check for each construct is presented in Fig. 4.5. Most of the constructs achieve sufficient reliability scores, except for the construct related to the ability of this approach to present Real-world events and incidents or to be used to teach other IT topics. Therefore, it is perceived that some participants were confused about the relevance with other IT topics.

Considering the answers related to the acceptance of this approach in other topics, it is clear that it might be difficult to accept it as an official assessment method in academia. Students for example are confused in how this method could enhance their skills and knowledge in matters of the official academic curriculum.

4.6 Conclusions

In this study, a method for introducing students to the basic concepts of cybersecurity was proposed. The impact of GBL on the learning outcomes was also discussed, focused on the ARCS motivation model [49] and how it applies to our approach. The main purpose was to increase the engagement levels of the academic course and to maintain balance between fun, engagement and perceived learning in the complex topics of cybersecurity.

The in-game challenges and software tools are compared with tools and ethical hacking methods that exist in reality, in order to uncover the potential of computer games as virtual learning environments. In our proposal we mentioned the importance of presenting real software tools used for ethical hacking and penetration testing in contrary to the in-game context.

High correlation levels are indicated regarding the relevance between in-game context and real-case scenarios. It is perceived that the skills were gradually developed, enhanced with the self-learning elements which computer games could provide. Basic concepts and methods were presented in the game and even participants with insufficient knowledge in ethical hacking could follow the procedure.

Finally, through this research, we were able to extract empirical data on how Gamification or GBL could enhance the learning process. By focusing on GBL, we combined the fun elements deriving from computer games with an actual learning process enhanced by gamification elements.

4.6.1 Limitations

Students were informed that the main virtual learning environment would be a computer game and not a real platform. As a result, some participants did not directly correlate the process with a real virtual learning environment.

The second limitation was the small set of participants. However, using the correct methodology, this method could be used in order to enhance the learning process in the official academic curriculum and to collect more empirical data.

Finally, this approach could be an appropriate method for conducting assessments and exams. However not sufficient empirical data are presented in this research in order to support this approach and to create an appropriate assessment method.

4.6.2 Future Work

More work is required in terms of analyzing the features which affect attributes such as perceived learning, actual learning and skills acquaintance. In order to convert a computer game into a virtual learning environment more customization and research is required. We plan to study the interconnections and possible extensions of the proposed method, using real-case scenarios and defining specific learning goals.

We recognize the importance of increasing the data-set of this research in order to better investigate the impact of this approach on the learning outcomes. Towards this direction we plan to create attack scenarios, both inside the computer game and also in the lab, in order to collect more details. Scenarios might be created using the software component of Nite Team 4—*"Network Administrator"* in order to create custom challenges deriving from real-case scenarios. Furthermore, the potential of Gameplay Option—Section 3: Multiplayer/Hivemind Network (presented in Sect. 4.4.2.3) has to be further analyzed in order to highlight the importance of creating custom challenges which are highly related to real-world infrastructure and acquiring familiarity related to threat modelling tools.

References

1. F. Agalbato, D. Loiacono, Robo 3: a puzzle game to learn coding, in *2018 IEEE Games, Entertainment, Media Conference (GEM)* (IEEE, 2018)
2. R. Al-Azawi, F. Al-Faliti, M. Al-Blushi, Educational gamification vs. game based learning: comparative study. Int. J. Innov. Manage. Technol. **7**(4), 132–136 (2016)
3. P.D. Allen, K.A. Straub, Using games to enrich continuous cyber training. Johns Hopkins APL Tech. Dig. **33**(2) (2015)
4. F. Alotaibi et al., A review of using gaming technology for cyber-security awareness. Int. J. Inf. Secur. Res. (IJISR) **6**(2), 660–666 (2016)
5. A. Amory, R. Seagram, Educational game models: conceptualization and evaluation: the practice of higher education. South Afr. J. High. Educ. **17**(2), 206–217 (2003)
6. L.A. Annetta et al., Investigating the impact of video games on high school students engagement and learning about genetics. Comput. Educ. 53(1), 74–85 (2009)
7. D.R. Bacon, Reporting actual and perceived student learning in education research, pp. 3–6 (2016)
8. G. Barata et al., Engaging engineering students with gamification, in *2013 5th International Conference on Games and Virtual Worlds for Serious Applications (VS-GAMES)* (IEEE, 2013)
9. K. Becker, Video game pedagogy, in *Games: Purpose and Potential in Education* (Springer, Boston, MA, 2009), pp. 73–125
10. B. Bediou et al., Meta-analysis of action video game impact on perceptual, attentional, and cognitive skills. Psychol. Bull. **144**(1), 77 (2018)
11. K. Boopathi, S. Sreejith, A. Bithin, Learning cyber security through gamification. Indian J. Sci. Technol. **8**(7), 642–649 (2015)
12. A. Bruckman, Community support for constructionist learning. Comput. Support. Coop. Work (CSCW) **7**(1–2), 47–86 (1998)
13. P. Buckley, E. Doyle, Gamification and student motivation. Interact. Learn. Environ. **24**(6), 1162–1175 (2016)
14. S. Cass, Some assembly (language) required-Three games that make low-level coding fun [Resources-Geek Life]. IEEE Spectr. **54**(5), 19–20 (2017)

15. C. Cheong, F. Cheong, J. Filippou, Quick quiz: a gamified approach for enhancing learning. PACIS (2013)
16. R.S. Cheung et al., Effectiveness of cybersecurity competitions, in Proceedings of the International Conference on Security and Management (SAM). The Steering Committee of The World Congress in Computer Science, Computer Engineering and Applied Computing (WorldComp) (2012)
17. L. Cifuentes et al., An architecture for case-based learning. TechTrends **54**(6), 44–50 (2010)
18. B.D. Coller, D.J. Shernoff, Video game-based education in mechanical engineering: a look at student engagement. Int. J. Eng. Educ. **25**(2), 308 (2009)
19. T.M. Connolly et al., A systematic literature review of empirical evidence on computer games and serious games. Comput. Educ. **59**(2), 661–686 (2012)
20. D.I. Cordova, M.R. Lepper, Intrinsic motivation and the process of learning: beneficial effects of contextualization, personalization, and choice. J. Educ. Psychol. **88**(4), 715 (1996)
21. K. de Beer, M. Holmner, The design of an alternate reality game as capstone course in a multimedia post-graduate degree (2013)
22. M. Denk, M. Weber, R. Belfin, Mobile learning-challenges and potentials. IJMLO **1**(2), 122–139 (2007)
23. V.P. Dennen, Cognitive apprenticeship in educational practice: research on scaffolding, modeling, mentoring, and coaching as instructional strategies. Handb. Res. Educ. Commun. Technol. **2**(2004), 813–828 (2004)
24. T. Denning et al., Control-Alt-Hack: the design and evaluation of a card game for computer security awareness and education, in Proceedings of the 2013 ACM SIGSAC conference on Computer & Communications Security (ACM, 2013)
25. T. Denning, A. Shostack, T. Kohno, Practical lessons from creating the control-alt-hack card game and research challenges for games in education and research, in *2014 USENIX Summit on Gaming, Games, and Gamification in Security Education (3GSE 14)* (2014)
26. A. DomNguez et al., Gamifying learning experiences: practical implications and outcomes. Comput. Educ. **63**, 380–392 (2013)
27. B.C. Dunphy, S.L. Dunphy, Assisted performance and the zone of proximal development (ZPD); a potential framework for providing surgical education. Aust. J. Educ. Dev. Psychol. **3**(2003), 48–58 (2003)
28. C. Eagle, J.L. Clark, *Capture-the-Flag: Learning Computer Security Under Fire* (Naval Postgraduate School Monterey CA, 2004)
29. R.M. Felder, R. Brent, Navigating the bumpy road to student-centered instruction. Coll. Teach. **44**(2), 43–47 (1996)
30. P. Fotaris et al., From hiscore to high marks: empirical study of teaching programming through gamification, in *European Conference on Games Based Learning. Academic Conferences International Limited* (2015)
31. P. Fotaris et al., Climbing up the leaderboard: an empirical study of applying gamification techniques to a computer programming class. Electr. J. E-learn. **14**(2), 94–110 (2016)
32. J.P. Gee et al., Playing to learn game design skills in a game context, in Proceedings of the 8th International Conference on International Conference for the Learning Sciences-Volume 3 (International Society of the Learning Sciences, 2008)
33. R.E. Gewurtz et al., Problem-based learning and theories of teaching and learning in health professional education. J. Perspect. Appl. Acad. Pract. **4**(1) (2016)
34. F. Gilberg, Using games to improve network security decisions (2006)
35. M. Gondree, Z.N.J. Peterson, Valuing security by getting [d0x3d!]: experiences with a network security board game, in *Presented as part of the 6th Workshop on Cyber Security Experimentation and Test* (2013)
36. H. Gonzalez, R. Llamas, F. Ordaz, Cybersecurity teaching through gamification: aligning training resources to our syllabus. Res. Comput. Sci. **146**, 35–43 (2017)
37. F.L. Greitzer, O.A. Kuchar, K. Huston, Cognitive science implications for enhancing training effectiveness in a serious gaming context. J. Educ. Resour. Comput. (JERIC) **7**(3), 2 (2007)

38. J. Hamari et al., Challenging games help students learn: an empirical study on engagement, flow and immersion in game-based learning. Comput. Hum. Behav. **54**, 170–179 (2016)
39. M. Hendrix, A. Al-Sherbaz, B. Victoria, Game based cyber security training: are serious games suitable for cyber security training? Int. J. Ser. Games **3**(1), 53–61 (2016)
40. A. Hirumi, C. Stapleton, Applying pedagogy during game development to enhance game-based learning, in *Games: Purpose and Potential in Education* (Springer, Boston, MA, 2009), pp. 127–162
41. R. Ibrahim, A. Jaafar, Educational games (EG) design framework: combination of game design, pedagogy and content modeling, in *2009 International Conference on Electrical Engineering and Informatics*, Vol. 1 (IEEE, 2009)
42. S. Isaacs, The difference between gamification and game-based learning. *ASCD In Service* (2015), http://inservice.ascd.org/the-difference-between-gamification-and-game-based-learning
43. C. Johnson et al., Game development for computer science education, in *Proceedings of the 2016 ITiCSE Working Group Reports* (ACM, 2016)
44. L.F. Johnson et al., *Challenge-Based Learning: An Approach for Our Time* (The New Media Consortium, 2009)
45. D.H. Jonassen, Instructional design models for well-structured and III-structured problem-solving learning outcomes. Educ. Technol. Res. Dev. **45**(1), 65–94 (1997)
46. K.M. Kapp, What is gamification, in *Game-Based Methods and Strategies for Training and Education, The Gamification of Learning and Instruction* (2012), pp. 1–23
47. M.N. Katsantonis, P. Fouliras, I. Mavridis, Conceptualization of game based approaches for learning and training on cyber security, in *Proceedings of the 21st Pan-Hellenic Conference on Informatics* (ACM, 2017)
48. M. Kebritchi, A. Hirumi, H. Bai, The effects of modern math computer games on learners math achievement and math course motivation in a public high school setting. Br. J. Educ. Technol. **38**(2), 49–259 (2008)
49. J.M. Keller, Development and use of the ARCS model of instructional design. J. Instr. Dev. 10(3), 2 (1987)
50. K. Kiili, Digital game-based learning: towards an experiential gaming model. Internet High. Educ. **8**(1), 13–24 (2005)
51. L.M. Kinczkowski, Hacker's challenge: test your incident response skills using 20 scenarios. Secur. Manage. **47**(1), 108–108 (2003)
52. A.V. Kirillov et al., Improvement in the learning environment through gamification of the educational process. Int. Electr. J. Math. Educ. **11**(7), 2071–2085 (2016)
53. D.R. Krathwohl, A revision of Bloom's taxonomy: an overview. Theory Pract. **41**(4), 212–218 (2002)
54. D.R. Krathwohl, L.W. Anderson, *A Taxonomy for Learning, Teaching, and Assessing: A Revision of Bloom's Taxonomy of Educational Objectives* (Longman, 2009)
55. R.N. Landers, Developing a theory of gamified learning: linking serious games and gamification of learning. Simul. Gaming **45**(6), 752–768 (2014)
56. A.D.E. Le Compte, T. Watson, A renewed approach to serious games for cyber security, in *2015 7th International Conference on Cyber Conflict: Architectures in Cyberspace* (IEEE, 2015)
57. R.S.N. Lindberg, T.H. Laine, L. Haaranen, Gamifying programming education in K12: a review of programming curricula in seven countries and programming games. Br. J. Educ. Technol. **50**(4), 1979–1995 (2019)
58. L. Lin, Exploring collaborative learning: theoretical and conceptual perspectives, in *Investigating Chinese HE EFL Classrooms* (Springer, Berlin, Heidelberg, 2015), pp. 11–28
59. J. Lee et al. More than just fun and games: assessing the value of educational video games in the classroom, in *CHI'04 Extended Abstracts on Human Factors in Computing Systems* (ACM, 2004)
60. S. Maraffi, F.M. Sacerdoti, E. Paris, Learning on gaming: a new digital game based learning approach to improve education outcomes. US-China Educ. Rev. **7**(9), 421–432 (2017)

61. D.K. Meyer, J.C. Turner, C.A. Spencer, Challenge in a mathematics classroom: students' motivation and strategies in project-based learning. Elem. School J. **97**(5), 501–521 (1997)
62. P. McClean et al., Virtual worlds in large enrollment science classes significantly improve authentic learning. in *Proceedings of the 12th International Conference on College Teaching and Learning, Center for the Advancement of Teaching and Learning* (2001)
63. J. Mirkovic et al., Evaluating cybersecurity education interventions: three case studies. IEEE Secur. Priv. **13**(3), 63–69 (2015)
64. P. Moreno-Ger et al., Educational game design for online education. Comput. Hum. Behav. **24**(6), 2530–2540 (2008)
65. C.I. Muntean, Raising engagement in e-learning through gamification, in *Proceedings of 6th International Conference on Virtual Learning ICVL*, Vol. 1 (2011)
66. A. Nagarajan et al., Exploring game design for cybersecurity training, in *2012 IEEE International Conference on Cyber Technology in Automation, Control, and Intelligent Systems (CYBER)* (IEEE, 2012)
67. D. Oblinger, Games and learning. Educ. Q. **3**, 5–7 (2006)
68. S. Papert, I. Harel, Situating constructionism. Constructionism **36**(2), 1–11 (1991)
69. J.M. Pittman, R. Pike, An observational study of peer learning for high school students at a cybersecurity camp. Inf. Syst. Educ. J. **14**(3), 4 (2016)
70. M. Pivec, O. Dziabenko, I. Schinnerl, Game-based learning in universities and lifelong learning: UniGame: social skills and knowledge training game concept. J. Univ. Comput. Sci. **10**(1), 14–26 (2004)
71. M. Prensky, The games generations: how learners have changed. Dig. Game-based Learn. **1** (2001)
72. M. Prensky, Digital game-based learning. Comput. Entertain. (CIE) **1**(1), 21–21 (2003)
73. R. Raman, A. Lal, K. Achuthan, Serious games based approach to cyber security concept learning: Indian context, in *2014 International Conference on Green Computing Communication and Electrical Engineering (ICGCCEE)* (IEEE, 2014)
74. J.W. Rice, The gamification of learning and instruction: game-based methods and strategies for training and education. Int. J. Gaming Comput. Med. Simul. **4**(4) (2012)
75. M. Šakić, V. Varga, Video games as an education tool, in *The Sixth International Conference on e-Learning. eLearning-2015* (2015)
76. M. Schiffman, *Hackers Challenge: Test Your Incident Response Skills Using 20 Scenarios* (McGraw-Hill Inc, New York, NY, USA, 2001)
77. Z.C. Schreuders, E. Butterfield, Gamification for teaching and learning computer security in higher education, in *2016 USENIX Workshop on Advances in Security Education (ASE 16)* (2016)
78. A.C. Siang, R.K. Rao, Theories of learning: a computer game perspective, in *Proceedings of Fifth International Symposium on Multimedia Software Engineering, 2003* (IEEE, 2003)
79. K. Siau, H. Sheng, F.F.-H. Nah, Use of a classroom response system to enhance classroom interactivity. IEEE Trans. Educ. **49**(3), 398–403 (2006)
80. K. Squire et al., Electromagnetism supercharged!: learning physics with digital simulation games, in *Proceedings of the 6th International Conference on Learning Sciences* (International Society of the Learning Sciences, 2004)
81. G. Surendeleg et al., The role of gamification in education-a literature review. Contem. Eng. Sci. **7**(29), 1609–1616 (2014)
82. J.-N. Tioh, M. Mina, D.W. Jacobson, Cyber security training a survey of serious games in cyber security, in *2017 IEEE Frontiers in Education Conference (FIE)* (IEEE, 2017)
83. R. Van Solingen, K. Dullemond, B. Van Gameren, Evaluating the effectiveness of board game usage to teach GSE dynamics, in *2011 IEEE Sixth International Conference on Global Software Engineering* (IEEE, 2011)
84. V. Vbensk, J. Vykopal, Challenges arising from prerequisite testing in cybersecurity games, in *Proceedings of the 49th ACM Technical Symposium on Computer Science Education* (ACM, 2018)

Chapter 5
Didactics for the Development of Mathematical Thinking and the Sense of Academic Agency

David Martín Santos Melgoza, José Armando Landa Hernández, Franco Ariel Ulloa González, and Abel Valdés Ramírez

Abstract This paper proposes to validate the structural relationships between the measurements of the academic agency components of student self-report and those observed in a mathematical learning episode, in order to provide evidence of how the metacognitive processes involved in success interact before and during an academic learning process. It is proposed to analyze the way in which learning styles, epistemic beliefs, and reading comprehension correlate with the degree of self-regulation and the ability to self-direct one's cognitive resources in a conscious and voluntary way in the achievement of learning goals in tasks that involve solving mathematical problems around the notion of infinite process, mediated by the use of a computational tool. The findings underline the importance and role that the use of technology can play in developing teaching strategies and assessing learning outcomes that result in the promotion of metacognitive and self-regulated learning skills.

Keywords Mathematical learning · Infinite processes · Asymptotes · Line: learning science · Mathematics and technology

---

D. M. Santos Melgoza (✉)
Humanistic Disciplines Area, Universidad Autónoma Chapingo, Carretera México-Texcoco Km. 38.5, Chapingo, Edo. Méx, Mexico
e-mail: dmsm21@correo.chapingo.mx

J. A. Landa Hernández · F. A. Ulloa González · A. Valdés Ramírez
Math Area, Universidad Autónoma Chapingo, Carretera México-Texcoco Km. 38.5, Chapingo, Edo. Méx, Mexico
e-mail: armanlanda@hotmail.com

F. A. Ulloa González
e-mail: arielulloa03@gmail.com

A. Valdés Ramírez
e-mail: abel_on1@yahoo.com.mx

G. A. Tsihrintzis and M. Virvou (eds.), *Advances in Core Computer Science-Based Technologies*, Learning and Analytics in Intelligent Systems 14,
https://doi.org/10.1007/978-3-030-41196-1_5

5.1 Introduction

Learning mathematics is one of the most studied topics in the field of education. This is undoubtedly due to its importance in the progress of scientific thought and in its use as a tool in all areas of technological development. Among all subjects in the upper intermediate level, mathematics has one of the highest failure rates and ends up being a factor with great impact on school dropout. The reasons why student academic success in this subject is not achieved in a satisfactory proportion is multifactorial. Among the factors that have been studied are: factors related to the context in which instruction takes place [16] and factors inherent to students, such as anxiety [14], self-efficacy [31], attitudes towards mathematics [12], self-regulation of emotions [11], to cite some examples that show the complexity and breadth of the problem implicit in the teaching and learning of mathematics. Definitely, for educational institutions, addressing the problem is by no means a simple matter. Regardless, it is essential to focus attention on the problem based on an explanatory model that takes into account the multiplicity of factors involved and that gives guidelines for achieving better results in an efficient manner, and to work on building a model that allows us to concentrate our efforts on increasing the quality of learning, meaning that it is easier to learn and that what is learned is what is actually required.

A theoretical perspective that has gained ground in explaining school problems is found within the framework of the Theory of Social Cognition, which focuses attention on the study of the sense of academic agency [1], due largely to the fact that the motivational factor of the student represents the possibility of concentrating intervention efforts on fostering better academic results by promoting the student's own will to learn. Human agency, according to Bandura, "human agency is embedded in a self theory encompassing self-organizing, proactive, self-reflective and self-regulative mechanisms."[1] This perspective, in which the student is involved as an active agent who voluntarily develops activities with an eagerness to learn, includes the development of metacognition with a self-regulatory perspective, discerns that it is the goals set by the students themselves that represent the guiding axis of the self-directed learning process [30], and represents an effort to develop an integrative explanatory model, since the motivational problem is the result of the way in which impact the rest of the factors involved. Now, given that academic success represents a development factor, both individual and social, it becomes essential to broaden our knowledge of the motivational dynamics of students and the mechanisms involved in self-regulated academic performance. It becomes increasingly important to understand the interaction of cognitive, communicative, and social factors implicit in the educational process, considering that the fact that the student achieves learning in a metacognitive sense is an interactive function of multiple elements of the educational context.

We consider it pertinent to focus attention on learning outcomes and use them as an axis of analysis to understand the mechanics of the process. It seems only natural that any analytical approach that seeks to better achieve a high development

[1] 1, p. 21.

of the process should start from the evaluation of the expected learning results, since the value of teaching concentrates on what is expected as the best result of the educational process. Thus, if we want learning to result in an active student with a high degree of academic agency, the attitudinal, affective-motivational, and self-referential objectives of students must be more clearly defined in terms of curricular learning goals.

However, in relation to the results that could be considered optimal as a result of the educational process, it is by no means easy or trivial to determine what has been learned and in what sense it represents "quality learning". Learning is an adaptive process, so student learning derives directly from the social demands originating in the context in which the learning episodes and the assessment processes involved take place. Therefore, it is the interaction of all these elements of the context with the agentive profile of the students that motivates us to analyze and better and more broadly understand how the didactic conditions stimulate the development of the components of academic agency as a result of the teaching process.

In order to narrow down the analysis and illustrate the theoretical approach of this paper, we note that, on the one hand, we have the fact that mathematics is a tool used in many professions mainly for the execution of algorithms; but on the other hand, we observe that the performance of an individual in the use of such algorithms does not necessarily correspond to a total understanding of them. Thus, it is pertinent to point out that comprehension in general, not only of mathematics, should not be interpreted in a discrete manner, but rather as a continuum; that is, it is not a matter of "one understands" or "one does not understand", but rather there are degrees of comprehension, so to speak. In the sense that an algorithm represents the synthesis of a generalization in a numerical procedure, and although it is necessary to memorize the algorithm, we often ignore what is generalized. If we add to this the fact that frequently the motivation to learn the proposed curricular contents is reduced to the accreditation of a course, the problem of student attitude towards the task of learning mathematics is exacerbated.

Although the traditional didactics of mathematics has been mainly directed to procedures, that is to say, to the learning of operativity from mathematical language, giving emphasis to the rules that are applied in algebraic operations, the focus in teaching has gradually pivoted towards the development of comprehension and not only the memorization of algorithms. It is indisputable that, beyond the need to learn mathematical language, it is crucially important to promote the development of mathematical thinking parallel to the capacity of students to apply procedures, in addition to beginning to consider the metacognitive abilities underlying the learning process [25].

In order to achieve educational development of greater efficacy in learning outcomes, it is necessary to be careful about the alignment between what you want to teach, the way you teach it and the mechanisms with which you evaluate it [18]. Therefore, it is necessary to objectively establish what learnings derived from the process of teaching mathematics are necessary to favor adaptive processes, as well as to broaden the analysis of empirical evidence regarding the type of learning outcomes being obtained and the factors involved in the learning environments in order to optimize strategies and implement tools that effectively favor the learning outcomes needed in the adaptive process of individuals to an environment that is quickly transformed and continually poses new and more complex challenges. In this sense, we agree with the idea that this discipline allows us to achieve a broader and deeper purpose than only becoming an instrumental support for approaching and solving problems of the discipline: the development of logical thinking and the clear and forceful contribution to the improvement of metacognitive processes, and therefore to the ability to learn to learn [21].

In short, the didactic structuring of learning episodes should be aimed at having the students themselves monitor and evaluate the results obtained, in addition to being articulated with the type and degree of commitment that students assume in their learning. It is possible to think that without considering a theoretical model that takes into account the complexity of student participation in their own learning process, it is not possible to get an appropriate view of the problem. It is necessary to understand the interaction between the cognitive and metacognitive performance factors of students, with the formal aspects of communication and social development of the knowledge structures involved in any educational process. It becomes fundamental, as Castañeda and Pérez [6] put it, to have a greater understanding of how students are "active agents of their own development, making things happen, intentionally, through their actions". It is proposed that the evaluation of the sense of academic agency and the orientation of cognitive engagement in relation to a specific mathematical learning task and its learning outcomes at different levels of performance, whether in terms of consciousness, abstraction, execution, attitude or resource management, are the nodal aspects in the objectivation, knowledge development and interpretation of the results of the educational process.

As we have already said, the act of learning is a complex process. In the subject of calculus, there are reported data on the most common learning difficulties. Research in the area of educational mathematics has addressed the problems arising from the erroneous concepts that students develop in learning the notion of limit [10, 29]. Bezuidenhout [3] and Cornu [9] have studied the fact that that students, in developing an incorrect idea of limit, not only have their understanding of limit in itself affected, but also affect their subsequent development of mathematical thinking on topics such as the notion of continuity and the differentiability of functions, as well as the understanding of infinite series [4, 27]. The development of incorrect notions is not easily detectable in didactics focused on acquiring the skills to calculate numerical values using the calculation operations. In this sense, we are interested in directing research efforts towards determining how to improve the self-regulatory thinking process of students and exploring other didactic mechanisms that favor it. There are many studies that show that the development of mathematical thinking is favored with the use of images in the didactic process aimed at enhancing the understanding of the notion of limit [19, 22]. The use of computer simulators is one of the elements with great potential for promoting these thinking skills [17]. However, as already outlined, a fundamental need is not to limit the evaluation of learning to the learning outcome that is circumscribed to the academic mathematical content, but rather to involve the attitudinal aspects of the willingness to solve problems and in particular the student's ability to problematize his or her own learning process. Thus, on the one hand, we have the fact that the will to solve a problem does not necessarily correspond to generating the learning of a mathematical abstraction [23]; and on the other hand, that behavioral control tends to be mostly developed by the immediate or closest consequences and in relation to the satisfaction of administrative goals such as accrediting the subject or avoiding being reprimanded by the teacher. As a result, the academic learning process is seen as a mechanism that moves between the students' ability to structure goals around the identification of ideas established as independent objects of their own belief system and the ability they have to identify beforehand the type of argumentative structure that characterizes these objects as knowledge, a process that we will identify as objectifying information [24] and that is immersed in the students' belief system regarding what is considered knowledge and that has been studied as what is known as personal epistemology [15].

5.2 Objective and Object of Study

Given the bibliographic review that supports the importance of knowing more about the meaning of academic agency, that the school learning process is made up of multiple factors that interact in a complex manner and that each field of knowledge requires specific considerations, we became interested in determining the extent to which students can manage this complexity in the conduction of their own learning in the development of mathematical thinking with the aid of a computer graphing tool. Therefore, in order to deepen knowledge of the role of the metacognitive processes involved in academic learning, which represent the basis of the agentic profile in a mathematical learning task, the objective was to fit a structural model that validates the relationships between components of the agentic behavior during a learning episode.

The research was developed in a context of semi-independent and interactive learning for the development of notions of infinite process, asymptote and continuity and discontinuity of a function. The self-report was compared with respect to the study profile, personal epistemology and the reading capacity of each student with the performance in the task, considering the degree of strategic behavior and the motivational orientation observed in the learning episode.

In summary, the interaction of the agentive profile (both in terms of self-perception and performance) of the students with a learning task mediated by an interactive graphing tool was proposed as an object of study. The results obtained in the task were evaluated in three levels: (1) mathematical knowledge, (2) metacognitive performance, and (3) interaction of the agentic profile with the graphing tool as a facilitator of the development of the mathematical notions involved.

5.3 Main Assumptions of the Three Levels of Analysis

With respect to mathematical knowledge, it is assumed, as Schoenfeld [26] puts it, that mathematics is inherently a social activity and what is important is the development of mathematical thinking, which means: "(a) developing a mathematical point of view—valuing the processes of mathematization and abstraction and having the predilection to apply them, and (b) developing competence with the tools of the trade, and using those tools in the service of the goal of understanding structure—mathematical sense-making." This perspective was used as the basis for the development of the tool and in the choice of practical problems. A type of function was used that has an oblique asymptote that results from having an algebraic expression in which a quadratic expression is divided over a linear one. Its operative and graphic algebraic characteristics as a whole made it possible to confront the student with intuitions and conceptual errors, as well as to test hypotheses regarding the reasons for when a hyperbola or a line is caused in the graph. The development of algebraic notions of one of the hyperbola's basic structures was explored in relation to its oblique asymptote in order to exhibit in a practical manner the notion of geometric and algebraic discontinuity, in two forms: (a) when the algebraic expression of the function can be simplified in a linear expression and (b) when the division of the quadratic expression by the linear one generates a residual that produces an infinite approximation to an asymptote. We start from the understanding that a mathematical notion is not reduced to a simple definition and that the conceptualization made of it comprises multiple aspects. However, the evaluation was developed with the purpose of distinguishing and grouping the students' answers with respect to four performance categories: 1. the total absence of notions, 2. mistaken ideas, 3. intuitive comprehension and 4. handling of mathematical abstraction and symbolism.

In relation to the students, we decided to assess some basic aspects that the student had before starting the learning episode, such as having taken and accredited calculus courses in which they have been prepared to apply knowledge and procedures that revolve around the notions of infinite process, asymptote and continuity and discontinuity of a function under a program that mainly focuses on the teaching of the ability to develop algebraic operations (operative ability). Thus, with the intention of having references to weight the learning and performance during the task, in addition to evaluating the students' previous self-report of their study profile for comparison with their effective self-regulated performance during a learning episode, it was necessary to assess their previous knowledge of mathematics, their belief system in relation to knowledge, i.e. their personal epistemology, and their ability to comprehend texts.

Regarding the tool provided to students, which we have called Micro-world (MW), we take up some of the considerations of Papert [20, p. 79] about being "a small world, a small slice of reality". This tool was designed for students to explore and manipulate the geometric elements of graphical representation and to discover mathematical relationships implicit in the notion of limit. With this tool, it is possible to modify the parameters represented by sliders that, when being manipulated, simultaneously modify both the algebraic expression and its geometric representation; it was proposed as an exploratory tool that allows experimentation, both in geometric and algebraic terms, but also the permanent process of approximation for each value of X to the graphical representation of the asymptote. It is assumed that it is an aid to a strategic behavior so that students can develop their concepts around the problem studied.

Regarding the learning episode, while the study focuses on voluntary and conscious activity, it was explained to the students that the intention was to explore their learning style during an activity that involves solving problems in order for them to learn. It was also made explicit to them that the exercise was intended to improve their understanding of the concepts involved. Thus, they were invited to use MW as an aid in their experimentation in order to broaden their understanding of the aforementioned notions from the solution of the problems and the answers they had to develop around these notions. In addition, since the activity from which the learning outcome was objectified was the argumentation, the episode was carried out in pairs to encourage discussion and reflection through dialogue and comparison of points of view; however, each student was asked to write his or her own conclusions.

5.4 Methods

5.4.1 Participants

The sampling was intentional and consisted of 40 undergraduate students from the third semester in the Economics program at the *Universidad Autónoma Chapingo* (UACh), involving 27 women and 13 men, with an average age of 20.75 ± 1.1 years.

5.4.2 Scenario

The learning episode took place in a computer room with 45 computers designed for computer-assisted classes given in UACh's Division of Economic and Administrative Sciences. The room was set up so that students could work in pairs.

5.4.3 Materials

(a) Prior to the learning episode, the following were applied:

- Inventory of Study Strategies and Self-Regulation (ISSSR) [7]. This instrument is composed of 52 items that measure cognitive skills (to learn and teach), as well as self-regulatory ones (to be aware of the nature of the task and for the administration of necessary resources available.

- Inventory of Personal Epistemology (IPR) [5]. This instrument is composed of 26 items that measure variables about a student's epistemological beliefs (individual conceptions about knowledge and knowing).
- Instrument for the evaluation of reading comprehension [8]. This instrument is composed by the narrative reading "The two kings and the two labyrinths" by Jorge Luis Borges that consists of 303 words and a low lexical-technical difficulty level and a high syntactic-semantic one. It also has a questionnaire with 20 items that are organized into two tasks with ten items each that measure the understanding of information through information retrieval.
- Two practical problems to evaluate the development of mathematical notions related to the notions of limit, infinite process, asymptote, and continuity and discontinuity of a function:

  Problem 1. A medical research team established that the mass M(t) of a tumor, as a function of the time t to which the patient is exposed to radiation during treatment, is given by $M(t) = \frac{t^2-5t+6}{t-3}$ where M(t) is in milligrams and t in seconds. Due to equipment malfunction, it is impossible to expose the patient for exactly 3 s of radiation therapy.
  Problem 2. A swampy region used to drain agricultural areas has been contaminated with selenium. It was determined that irrigating the area with clean water will reduce the selenium for a while, but at some point it will increase its concentration again. A biologist has found that the percentage of selenium in the soil x months after watering is given by $f(x) = \frac{x^2+36}{2x}$, 1 less than or equal to x less than or equal to 12.

- Pencil-paper instrument to assess the degree of knowledge and operative ability with respect to the notions of limit, infinite process, asymptote, and continuity and discontinuity of a function from a questionnaire of 6 free response questions regarding the notions, graphing and intuition related to the geometric and operative behavior of the functions of the type $f(x) = \frac{dx^2+fx+c}{ax+b}$, as well as for finding the algebraic solution to the two practical problems posed.
- A rubric that provided a rating system based on a 4-point, Licket-type scale with respect to: (a) the presence of mathematical notions, (b) conceptual interaction, (c) operative ability and (d) graphical representation.

(b) For the learning episode, the following was used:

- A computational tool, called Micro-world, which allows a graphical manipulation of functions of the form $f(x) = \frac{dx^2+fx+c}{ax+b}$ by modifying the parameters. A computer program designed through the GeoGebra mathematics software structures the tool.[2] Micro-world allows you to experiment with the

[2]The program can be downloaded for free from the following link: http://www.geogebra.org.

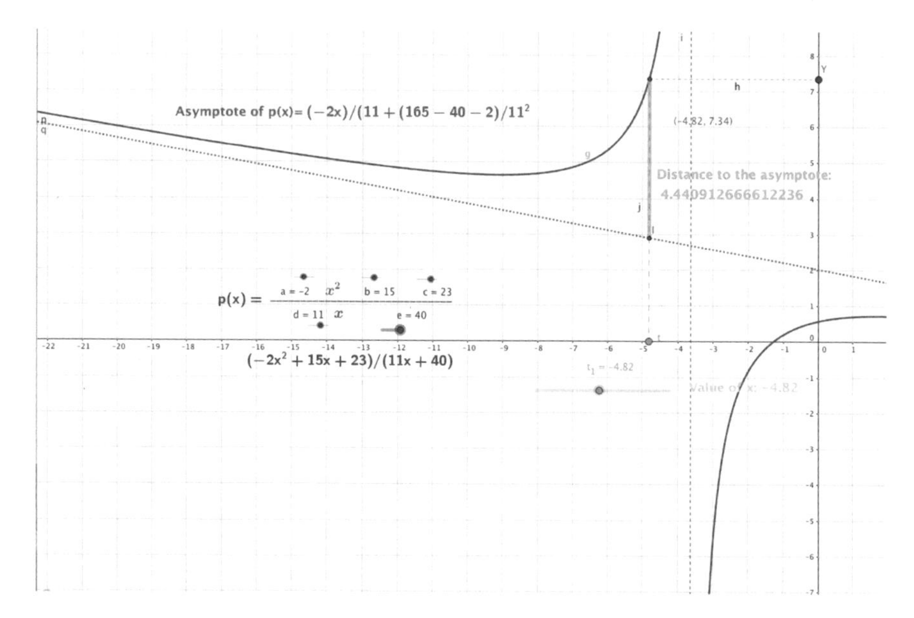

Fig. 5.1 Micro-world screen

variation of the parameters by means of objects called sliders and to visualize the properties of the geometric representation of functions presented by an oblique asymptote, through multiple objects that offer information to the student on the relations between them. That is, the variation of the parameters of a function, through the sliders, interactively shows its corresponding geometric representation, the distance to the asymptote, the general form of the function, as well as the function of the graph with the value of its parameters; one can also observe the value of f (x) given the value of x, manipulate the value of x in terms of t through a slider and observe the position of each pair of numbers in the graphical representation (see Fig. 5.1).

- Algebraic expressions of the type $f(x) = \frac{dx^2 + fx + c}{ax + b}$ have two forms of graphical representation—line or hyperbola—and it is their numerical, algebraic and graphical properties that we wish to highlight by requesting the argumentation of the notions of limit, infinite process, asymptote, and continuity and discontinuity of a function in relation to the solution of the two problems posed in the evaluation of the previous operative ability, supported by exploring the graphical representation of the problems. Students are expected to evaluate the answers offered in the previous evaluation based on the possibility of experimenting with the relationships between graphic, algebraic and conceptual elements during the learning episode aided by Micro-world (the representations of the problems are shown in Figs. 5.2 and 5.3 respectively).
- Rubric to evaluate change in student notions, their post-learning episode conceptual interaction and the attribution of didactic support to the tool based on

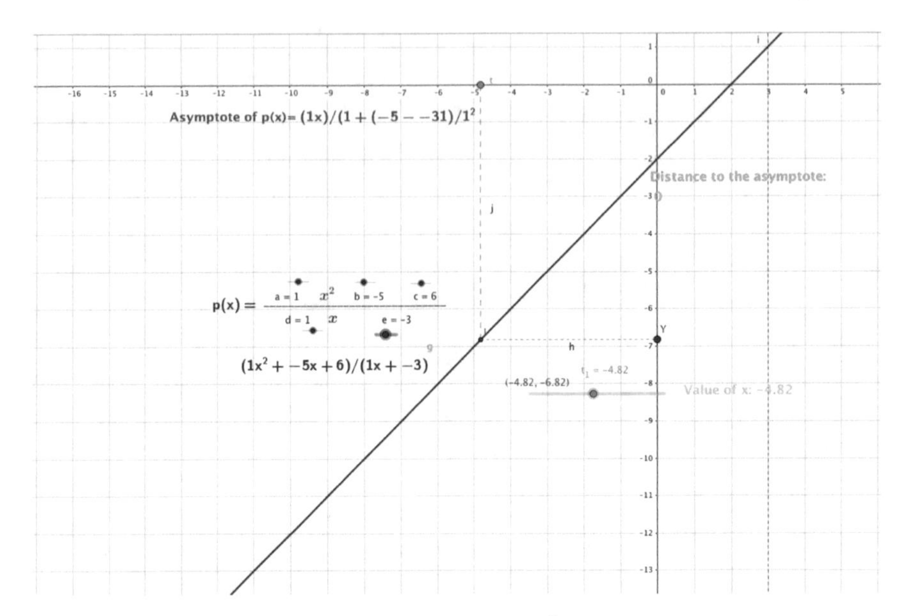

Fig. 5.2 Geometric representation of problem 1. $M(t) = \frac{t^2 - 5t + 6}{t - 3}$

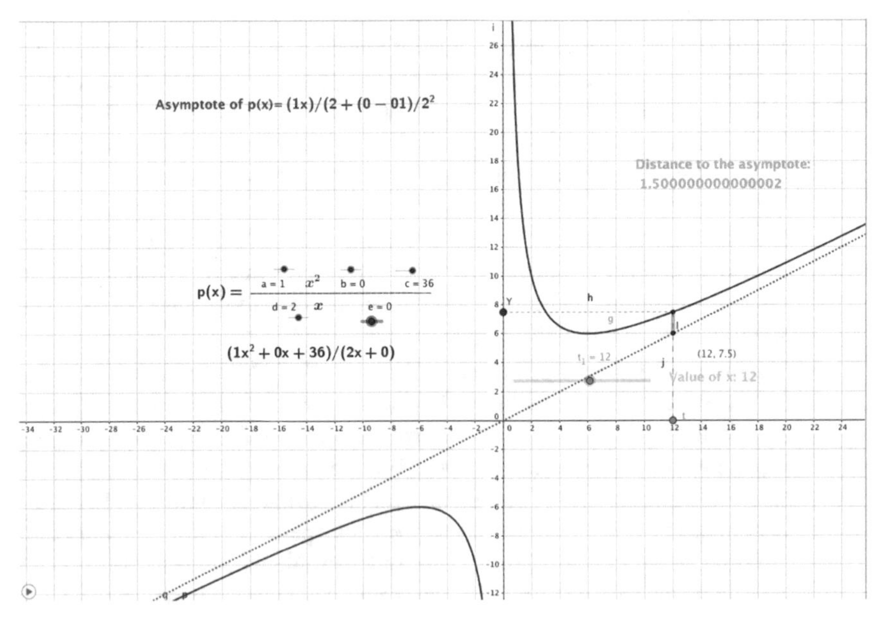

Fig. 5.3 Geometric representation of problem 2. $f(x) = \frac{x^2 + 36}{2x}$

performance in solving the same two problems of the previous evaluation. A scale was used regarding the modification of notions that allowed characterizing the students' behavior according to four conditions: (1) does not modify, reaffirms his or her conceptions although incorrectly (2) does not modify and does not relate the exercise to the need to generate greater understanding, (3) notices new elements of the mathematical thinking involved, (4) does not modify because his or her arguments are correct and can corroborate them. Conceptual interaction was characterized as: (1) does not integrate concepts in mathematical terms, (2) presents some isolated mathematical notions, (3) only integrates the notion of infinity and limit with clarity, (4) integrates all concepts. The support attributed to Micro-world by the students was classified as: (1) it is not useful, (2) it helps a little because it is better to do the algebraic procedures, (3) it helps to get the easiest solution, (4) it helps to better understand the mathematical ideas.

- Rubric to assess the self-regulated performance mediated by the use of the computer tool evaluated some metacognitive aspects of the self-regulated performance of students that are considered fundamental components of an agentive profile and that must become manifest in the students during the development of a learning episode and that basically refer to behavioral self-control when trying to achieve an academic learning goal. These elements were considered indicators of the degree to which students control their own cognitive resources in order to acquire new skills and better interpretations of problems they previously putted into perspective (see Table 5.1).

The 3 rubrics (1. to evaluate previous mathematical knowledge, 2. to evaluate change of notions during the learning episode and 3. to evaluate self-regulated performance) were constructed from a 4. point, Likert-type evaluation scale.

5.4.4 Validation by Judges of the Rubrics for Evaluating Performance During the Learning Episode

Once the evaluation rubrics had been constructed, judges using an assessment instrument in which 6 professors participated, 3 from UACh's mathematics area and 3 from its psychology area carried out validation. Each teacher was presented with the specifications matrix, complete with detailed definitions of what each of the rubric's elements assessed, and a table with instructions to establish the degree of agreement or disagreement in each element of the table, the reasons for this trial and a space for suggestions. Kendall's W coefficient was 0.85, indicating statistically significant agreement among the judges regarding the validity of the rubrics.

Table 5.1 Constructs and criteria for evaluating self-regulated performance in the learning episode

			1	2	3	4
Self-regulated performance	Cognitive commitment to task	Concentration	Very scattered	Scattered	Concentrated	Very Concentrated
		Reflection	Absent, shows no understanding of the task's intention	Seeks understanding from classmates	Answers questions without asking questions to oneself	Reflective, poses questions to generate greater understanding
		Motivational orientation	Disinterested problem solving	Problem solving with a desire to comply	Academic learning without strategy	Strategic academic learning
	Collaboration	Attitude	Indifferent	Difficult to collaborate and distracted at times	Listens carefully to the participations	Proactive searching for answers
		Involvement	Does not wish to participate	Makes an effort but not very convinced	Participates moderately	Participates assertively
		Clarity of goals	No clarity of goals	Identifies the need to solve problems numerically	Identifies the need to solve problems numerically and gives responses to theoretical approaches	Identifies the need to solve problems numerically, gives responses to theoretical approaches and identifies learning goals
	Argumentative capacity of mathematical knowledge		Null	Confused	Moderately coherent	Well argued
	Efficacy		Does not get the solutions	Gets the numerical solutions but shows little understanding	Gets the numerical solutions and shows some understanding	Gets the numerical solutions, shows little understanding, and identifies changes in his or her conceptual understanding
	Order and strategy		Disorderly	Orders unthinkingly	Strategic to solve the problem	Strategic, focuses on the tool to gain greater understanding

5.4.5 Model Specification

The specification of the model for trajectory analysis was generated from the 6 variables evaluated. It is proposed that students' evaluations in relation to Study Strategies and Self-Regulation, Personal Epistemology, text comprehension, as well as prior knowledge in mathematics have a predictive capacity on the self-regulated behavior and execution of students in the learning episode (Fig. 5.4).

5.5 Procedures

In the first 60-min session, the ISSSR, IPR, and text comprehension instruments were applied according to the standard procedure for their collective application.

In another 60-min session, the instrument for evaluating prior mathematical knowledge was applied, also collectively and through a standard procedure.

In a third session the learning episode with 20 pairs of students was carried out. Instructions for the episode were given in writing. In these, they were informed about the purpose of the study, that they had to work with the aid of the computer tool, and that they could manipulate the tool on their own but that they had to discuss their answers in pairs; however, they had to write their answers individually. They were instructed to re-solve the two practical problems they had already solved and to write the respective algebraic representations, but this time supported by Micro-world,

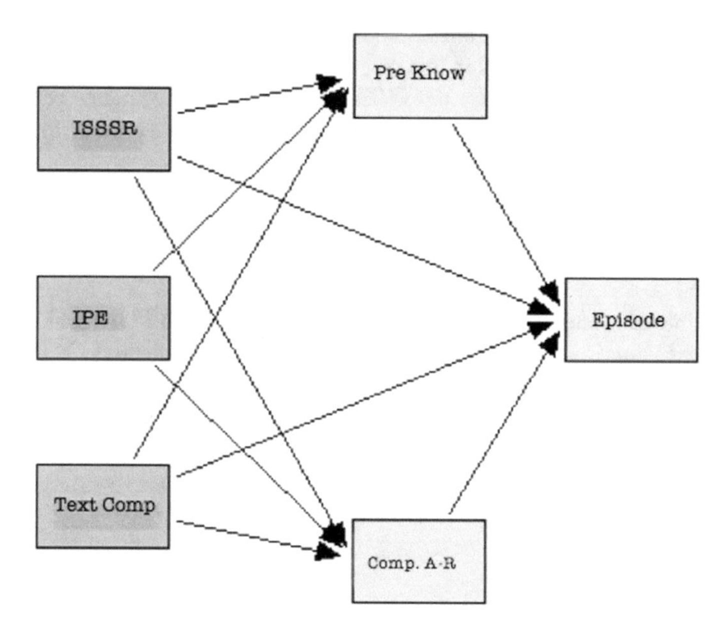

Fig. 5.4 Theoretical model of the relationship between academic agency variables

not necessarily through algebraic operability. They were asked to answer questions regarding the way in which the definition of limit, asymptote, infinite process, and continuity and discontinuity of a function was applied, the geometric characteristics of both graphical representations and their relationship with the type of function in each of the two problems; it was emphasized that they had to discuss their answers and that, in order to substantiate them, they could manipulate the graphs in Microworld to demonstrate what they were arguing. The activity of each pair's episode was recorded in audio and video through the Camtasia program [28] to later carry out the evaluation with greater calm and detail. The EQS program was used to fit the model [2].

5.6 Results

5.6.1 Agentive Profile

In the Inventory of Study Strategies and Self-Regulation, the criteria for assessing results indicate that scores above 76 do not represent a risk for the student; it is presumed that there is good development of learning strategies and motivational orientation. Scores between 56 and 75 suggest the need to reinforce the students' learning strategies or motivational orientations corresponding to the subscale(s) in question; it does not represent a critical deficiency. However, scores below 56 indicate a critical need to train students in learning strategies or motivational orientation in the corresponding subscale(s); it is considered a significant deficiency. According to the scores obtained by the students in this research, in the evaluation made with respect to the 13 subscales evaluated by the ISSSR, an average performance per subscale of more than 70 points is observed (Table 5.2). Nonetheless, in each of the scales at least one student had scores lower than 56.

No student had an average percentage below 56 on the instrument; however, 17 students scored below 56 on some of the scales, of which 11 students had a low score on only one subscale, 3 on 2 subscales, two on three subscales, and one student on four subscales.

Taking into consideration that 40 students were evaluated in 13 subscales grouped into six dimensions, we observed that 520 subscales were measured. Of these, only 27 had scores below 56; that is, only 5% are scores considered low and requiring attention, mainly concentrated in the area of divergent processing. On the other hand, there were 185 scores over 76, representing 35.6% of the scores. That is, there were 16 students (40%) with average scores above 76 (Table 5.3).

Table 5.2 Averages and standard errors

	Acquisition		Recovery		Processing		Self-regulated person				Self-regulated task		Self-reg. materials
	Selective	Generative	In the face of tasks	In the face of exams	Convergent	Divergent	Perceived efficacy	Perceived contingency	Perceived autonomy	External approval	Task	Task Achievement task	Materials
Min	50	50	44	50	**56**	44	44	50	50	50	50	50	50
Average	72	74	74	69	74	64	69	79	69	79	73	70	75
Max	88	100	100	100	94	81	88	100	94	100	94	94	94

Table 5.3 Number of students per score

No. of students related to scores	Acquisition		Recovery		Processing		Self-regulated person				Self-regulated task		Self-reg. materials
	Selective	Generative	In the face of e tasks	In the face of exams	Convergent	Divergent	Perceived efficacy	Perceived contingency	Perceived autonomy	External approval	Task	Task achievement	Materials
Below 56	1	2	3	2	**0**	6	2	1	4	1	1	1	1
56–76	31	25	22	30	29	36	30	18	30	18	29	28	24
Above 76	8	13	15	8	11	3	8	21	6	21	10	11	15

Table 5.4 Number of students per IPR score

No of students related to scores	Stability	Source	Usefulness	Nature
Below 56	1	11	0	0
56–76	31	26	14	25
Above 76	8	3	26	15

5.6.2 Personal Epistemology

The evaluation criteria regarding personal epistemology as measured by the IPR consider the same score limits as those used in the ISSSR.

With respect to the overall test score, as with the IPR, no student averaged less than 56. There were 12 (30%) who averaged scores greater than 76.

In relation to the IPR subscales, it can be seen that in the nature and usefulness subscales, no student scored below 56%; in the stability scale, only one student scored 41, while in the source of knowledge subscale, 11 students scored below 56% (Table 5.4). Of 120 dimensions assessed (40 students, 4 dimensions each), 12 scores (10%) fell below 56; 52 scores (43%) were placed above 76, yet only 3 students scored above 76 on their overall scores.

5.6.3 Assessment of Reading Comprehension

The average score obtained by the students in the study was 11.9, with a maximum score of 20 points being possible. There were 11 students (27%) who scored less than 8 (minus one standard deviation), 5 (12.5%) who scored above 16 (plus one standard deviation) and 24 (60%) who had an average performance.

Previous knowledge of the mathematical notion of infinite process and previous operative ability (PKMNIP)

Regarding the execution of the students in the instrument for evaluating previous knowledge and operative skills, the instrument contemplates a maximum score of 88 points. Students averaged 49 points, with the highest score being 64 and the lowest 36 (Fig. 5.5).

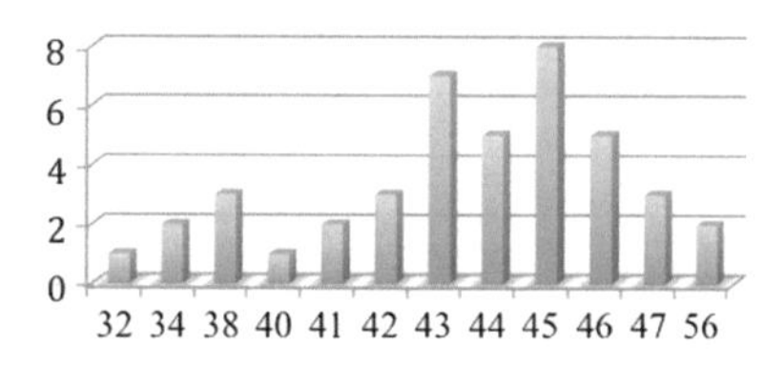

Fig. 5.5 CPNMPI score frequency

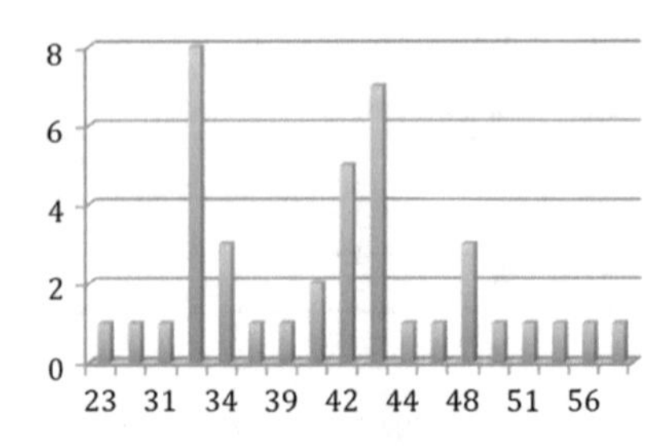

Fig. 5.6 Operative ability performances mediated by the tool (frequency of scores)

5.7 Learning Episode

5.7.1 *Evaluation of Self-regulatory Behavior During a Learning Episode*

The reliability assessment yielded a Cronbach's alpha of 0.778 for the previous knowledge, 0.92 for the post-performance scale, and 0.96 for the self-regulated behavior scale. The alpha of the three together is 0.92.

5.7.2 *Operative Performance Mediated by the Graphing Tool*

In relation to the operative performance mediated by the graphical tool, the rubric's maximum score is 84. The average performance score was 35, with the maximum being 59 and the minimum 23 (Fig. 5.6).

5.7.3 *Performance Self-regulation*

Self-regulation of performance, evaluated on the basis of cognitive commitment, collaborative capacity, and ability, which has a maximum score of 108, averaged 43 points, with a maximum score of 79 and a minimum of 35 (Fig. 5.7).

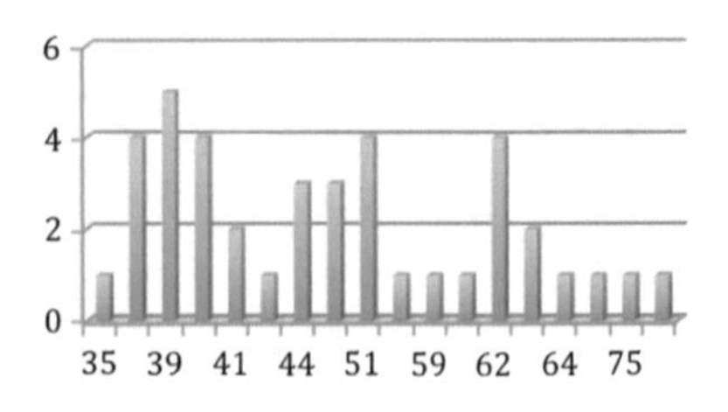

Fig. 5.7 Performance self-regulation

Table 5.5 Analysis of variance

Analysis of variance (one way)						
Summary						
Groups		Sample size		Sum	Mean	Variance
Previous knowledge		40		1,096	27.4	2,163
Post-argumentation		40		1,609	40.225	6,787
ANOVA						
Origen of variation	SS	df	MS	F	p level	F crit.
Between groups	3,289	1	3,289	7,350	0.00	396.347
Within groups	3,490	78	4,475			
Total	6,780	79				

The analysis of variance for measures of mathematical performance prior to the episode and in the mathematical learning episode was significant. The significance was less than 0.05, so the hypothesis of null differences is rejected (Table 5.5).

5.7.4 Structural Model Fit

The theoretical model showed statistical fit. While the chi-square value was not significant, it is feasible to assume that the distribution of the data corresponds to a normal distribution and that the assumptions tested by the model can be interpreted according to the theory. The practical fits showed adequate fit values (CFI = 1.00, RMSEA = 0.00), which allows assuming that the validation through the obtained factorial weights is appropriate and allow trusting that the evidence supports the assumptions made (Fig. 5.8).

5.8 Discussion and Conclusions

The results should be interpreted without losing sight of the context of the *Universidad Autónoma Chapingo*. It is necessary to take into account some relevant institutional characteristics that allow us to understand the data obtained. The university, according to its statutes, has two profession-oriented curricula: one of 7 years and another of 5 years. For the 7-year plan, students enter after finishing secondary school, with the first three years corresponding to higher secondary education. Those who enroll in the 5-year plan enter after having completed high school; however, the first year constitutes a preparatory year. In both plans, the highest academic dropout rate occurs during their passage through the Agricultural High School Department.

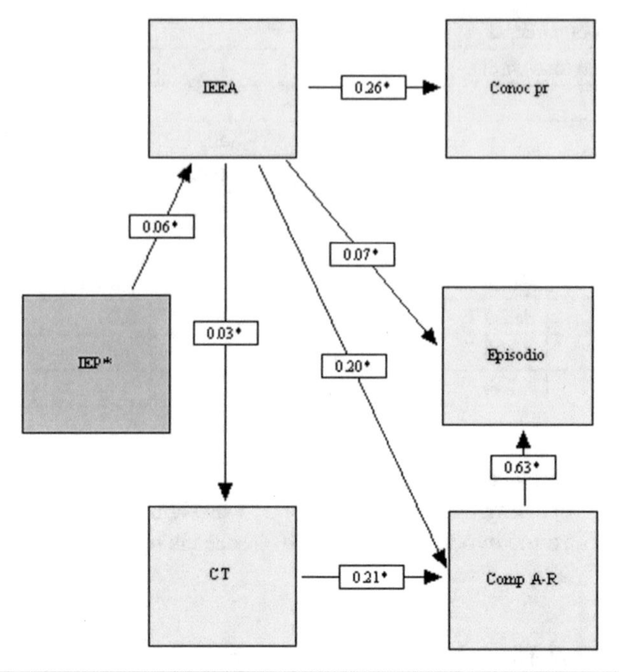

Fig. 5.8 Fitted model

However, according to ENLACE[3] [13] data, almost 90% of the students evaluated in their last year of high school in Chapingo, between 2012 and 2014, obtained results rated as good and excellent; in the case of the evaluation in mathematics, students assessed as excellent reached 75% in 2014. Thus, in light of the data, we observe that the average performance of students in terms of their agentive profile is good, even before the learning episode; better in relation to their motivational orientation and study strategies, good in their personal epistemology and a little less so in their comprehension of texts.

Regarding performance during the episode, it needs to be emphasized that the rubric that assesses mathematical performance and the tools available in Microworld considers a very broad conceptual domain of the evaluated field, in order to observe the tendency of students to investigate and develop knowledge by exploration. In this sense, the ratings of self-regulated performance and the efficacy with which they achieve solutions reflect that students in general in Chapingo have developed adaptive skills to achieve learning around completing assessments rather than investigating and generating more knowledge. Nevertheless, the fitted model, the

[3]ENLACE, an acronym for The National Evaluation of Academic Achievements in School Centers, was an exam that was applied each year in Mexico by the Secretary of Public Education (SEP) to all public and private elementary schools.

analysis of significant variances, and the reliability of the instruments allow us to conclude that students with motivational orientation and greater epistemic development tend to display greater investigative behaviors, which is a more critical agentive profile. The fact that performance in the text comprehension instrument has not been bad, but that it is not remarkable and that its influence on performance has not been greater, together with the fact that the majority of students work more to comply than to learn, leads us to think that it is necessary to broaden the perspective of evaluation in order to understand what is being taught and how it is possible to be more efficient by generating more adequate learning results.

Therefore, it seems important to emphasize that learning achieved from a technology-assisted episode, which is mediated by the tools and instructions of the task, interacts with the student in such a way that it could become a mechanism to foster self-directed learning and thus a better agentive profile. For this reason, it is essential to assess to what extent the achieved learning attains the objectives, both conceptual and practical required from the request to carry out the activity in order to learn, not just passing. The data shown here leads us to think that for the teaching of the use of mathematics as a tool it is necessary to complement it with teaching strategies that promote conceptual understanding and therefore a greater development of mathematical thinking aimed at solving problems. In this sense, the importance of using technology that facilitates the approach of students to complex concepts through activities that help in the development of more motivated schoolwork is evident, but this must be attached to evaluation mechanisms. However, all these conclusions require more empirical evidence and further inquiry into the interaction between the constructs underlying the sense of academic agency. This would require carrying out observations with a greater number of subjects that would allow the development of a model that considers a greater number of variables.

As a final conclusion, it is noteworthy that self-report variables regarding self-regulatory constructs of academic agency have proven to be a powerful predictor of academic performance; however, this work evidences the need to use these constructs in the daily evaluation of the learning process, as well as the usefulness of developing tools that promote learning outcomes oriented to a more critical, participatory and proactive profile. For this reason, we emphasize that the fundamental assumptions from which this proposal is constructed are relative to the importance of considering an explanatory model that takes into account the student as an active actor in the generation of his or her own learning. Therefore, the constructs involved are articulated from the notion of consciousness in a process identified as metacognitive, in which consciousness is considered as immediate knowledge of individuals regarding themselves, their environment and time, based on a learning goal. Consequently, the results are analyzed in a diachronic section of the learning process in terms of the evaluation of subjective-order psychological magnitudes and their interaction with a process we call the objectification of information that occurs during the solution of mathematical problems.

References

1. A. Bandura, Social cognitive theory: an agentic Albert Bandura. Asian J. Soc. Psychol. **2**(1), 21–41 (1999). https://doi.org/10.1111/1467-839X.00024
2. Bentler, P.M. (2006). EQS Estructural Equations Program. Versión 6.2 [CD-ROM] Encino, CA: Multivariate Software, Inc
3. J. Bezuidenhout, Limits and continuity: some conceptions of first-year students. Int. J. Math. Educ. Sci. Technol. **32**, 487–500 (2001)
4. Brousseau, G. (1997). Theory of didactical situations in mathematics (N. Balacheff, M. Cooper, R. Sutherland, & V. Warfield, Trans.) Dordrecht, The Netherlands: Kluwer Academic
5. S.Y. Castañeda, E. Peñalosa, Validando constructos en epistemología personal. Revista Mexicana de Psicología **27**(1), 65–75 (2010)
6. S. Castañeda, I. Pérez, Evaluando componentes de agencia académica en la WEB. Revista Mexicana de Psicología **22**, 82–83 (2014)
7. S. Castañeda, M. Pineda, N. Somoza, E. Castro, Construcción de instrumentos de estrategias de estudio, autorregulación y epistemología personal. validación de constructo. Revista Mexicana de Psicología **27**(1), 77–85 (2010)
8. S. Castañeda, E. Peñalosa, Y. Soto, Optimizando la Evaluación en Comprensión de Textos. Revista Mexicana de Psicología, 33, 1(January), 7–16 (2016)
9. B. Cornu, Limits, in *Advanced mathematical thinking*, ed. by D. Tall (Kluwer Academic, Dordrecht, The Netherlands, 1991), pp. 153–166
10. R.B. Davis, S. Vinner, The notion of limit: Some seemingly unavoidable misconception stages. J. Math. Behav. **5**, 281–303 (1986)
11. E. de Corte, F. Depaepe, P.O.T. Eynde, L. Verschaffel, Students' self-regulation of emotions in mathematics: An analysis of meta-emotional knowledge and skills. ZDM Int J Math Educ **43**(4), 483–495 (2011). https://doi.org/10.1007/s11858-011-0333-6
12. P. Di Martino, R. Zan, Attitude towards mathematics: a bridge between beliefs and emotions. ZDM Int J Math Educ **43**(4), 471–482 (2011). https://doi.org/10.1007/s11858-011-0309-6
13. Enlace (2014) http://www.enlace.sep.gob.mx/ms/informes_de_resultados/Geogebra
14. H.-Z. Ho, D. Senturk, A.G. Lam, J.M. Zimmer, S. Hong, Y. Okamoto, S.Y. Chiu, Y. Nakazawa, C.-P. Wang, The affective and cognitive dimensions of math anxiety: a cross-national study. J. Res. Math. Educ. **31**(3), 362–379 (2000). http://doi.org/10.2307/749811
15. B.K. Hofer, P.R. Pintrich, The development of epistemological theories: beliefs about knowledge and knowing and their relation to learning. Rev. Educ. Res. **67**(1), 88–140 (1997). https://doi.org/10.3102/00346543067001088
16. D.W.L. Hung, Meanings, contexts, and mathematical thinking: The meaning-context model. J. Math. Behav. **16**(4), 311–324 (1997). https://doi.org/10.1016/S0732-3123(97)90010-9
17. M. Kawski, How CAS and visualization lead to a complete rethinking of an introduction to vector calculus, in *Proceeding of the Third International Conference on Technology in Mathematics Teaching*, ed. by W. Fraunholz (Koblenz, Germany, 1997)
18. D. Leclercq, A. Cabrera, Conceptos y modelos para concebir, analizar y evaluar innovaciones curriculares basadas en competencias. en: Redes de colaboración para la innovación en la docencia universitaria: II Encuentro de Centros de Apoyo a la Docencia: ECAD: 29 y 30 de septiembre, 2011, UCM, Talca (2011)
19. M. Pinto, D. Tall, Building formal mathematics on visual imagery: a case study and a theory. Learn. Math. **22**, 2–10 (2002)
20. S. Papert, Microworlds: transforming education, in *Artificial Intelligence and Education volumen one: Learning Environments and Turoring Systems*, ed. by R. Lawler, M. Yazdani (Ablex Publishing, 355 Chestnut St. Norwood, NJ 07648, 1987)
21. R. Peñalva, Las matemáticas en el desarrollo de la metacognición. Política Y Cultura, (33), 135–151. Retrieved from http://www.scielo.org.mx/scielo.php?script=sci_arttext&pid=S0188-77422010000100008&lng=es&nrm=iso&tlng=es (2010)
22. K.H. Roh, Students' images and their understanding of definitions of the limit of a sequence. Educ. Stud. Math. **69**(3), 217–233 (2008). https://doi.org/10.1007/s10649-008-9128-2

23. D.M. Santos, S. Castañeda, Objetivación de información en aprendizaje matemático autorregulado. Revista Mexicana de Investigación …, 713–736. Retrieved from http://dialnet.unirioja.es/servlet/articulo?codigo=2748540 (2008)
24. D.M. Santos, Objetivar el conocimiento. Revista Mexicana de Psicología **27**(1), 103–110. Retrieved from http://www.redalyc.org/articulo.oa?id=243016325011 (2010)
25. A.H. Schoenfeld, *Mathematical Problem Solving* (Academic Press, New York, 1985)
26. A.H. Schoenfeld, Learning to think mathematically: problem solving, metacongnition, and sense-making in mathematics, in *Teaching and Learning*, ed. by D. Grouws (MacMillan, New York, 1992), pp. 334–370
27. A. Sierpińska, Humanities students and epistemological obstacles related to limits. Educ. Stud. Math. **18**, 371–397 (1987)
28. TECHSMITH, Camtasia Studio [CD-ROM]versión 6.0.0: TechSmith USA (2008)
29. S. Williams, Models of limit held by college calculus students. J. Res. Math. Educ. **22**, 219–236 (1991)
30. B.J. Zimmerman, Becoming a self-regulated learner : an overview **41**(2), 64–70 (2002). http://doi.org/10.1207/s15430421tip4102
31. A. Zuffianò, G. Alessandri, M. Gerbino, B.P. Luengo Kanacri, L. Di Giunta, M. Milioni, G.V. Caprara, Academic achievement: the unique contribution of self-efficacy beliefs in self-regulated learning beyond intelligence, personality traits, and self-esteem. Learn. Individ. Differ. **23**(1), 158–162 (2013). https://doi.org/10.1016/j.lindif.2012.07.010

Part II
Computer Science-Based Technologies in Risk Assessment and Readiness

Chapter 6
Readiness Exercises: Are Risk Assessment Methodologies Ready for the Cloud?

Dimitris Gritzalis, George Stergiopoulos, Efstratios Vasilellis, and Argiro Anagnostopoulou

Abstract Cloud computing is a type of service that allows the use of computing resources from a distance, rather than a new technology. Various services exist on-demand, ranging from data storage and processing to software as a service, like email and developing platforms. Cloud computing enables ubiquitous, on-demand access over the net to a shared pool of configurable resources, like servers, applications, etc. that can be accessed, altered or even restored rapidly with minimal service provider interaction or management effort. Still, due to the vast growth of cloud computing, new security issues have been introduced. Key factors are the loss of control over any outsourced resources and cloud's computing inherent security vulnerabilities. Managing these risks requires the adoption of an effective risk management method, capable of involving both the Cloud customer and the Cloud Service Provider. Risk assessment methods are common tools amongst IT security consultants for managing the risk of entire companies. Still, traditional risk management methodologies are having trouble managing cloud services. Extending our previous work, the purpose of this paper is to compare and examine whether popular risk management methods and tools (e.g. NIST SP800, EBIOS, MEHARI, OCTAVE, IT-Grundschutz, MAGERIT, CRAMM, HTRA, Risk-Safe Assessment, CORAS) are suitable for cloud computing environments. Specifically, based upon existing literature, this paper points out the essential characteristics that any risk assessment method addressed to cloud computing should incorporate, and suggests three new ones that are more appropriate based on their features.

D. Gritzalis (✉) · G. Stergiopoulos · E. Vasilellis · A. Anagnostopoulou
Information Security & Critical Infrastructure Protection (INFOSEC) Laboratory, Department of Informatics, Athens University of Economics & Business, Athens, Greece
e-mail: dgrit@aueb.gr

G. Stergiopoulos
e-mail: geostergiop@aueb.gr

E. Vasilellis
e-mail: vasilelliss@aueb.gr

A. Anagnostopoulou
e-mail: anagnostopouloua@aueb.gr

G. A. Tsihrintzis and M. Virvou (eds.), *Advances in Core Computer Science-Based Technologies*, Learning and Analytics in Intelligent Systems 14,
https://doi.org/10.1007/978-3-030-41196-1_6

Keywords Cloud computing · Cloud security · Risk assessment methods · Security risk assessment · Security risk · Risk analysis · Cloud computing risk · Cloud security risk assessment

6.1 Introduction

There is an emerging need for users to have ubiquitous and on-demand network access to computer resources, such as services, servers, storage or applications [1]. This trend induces significant changes to the way security experts operate and assess computing environments. Differences exist when cloud and traditional computing environments are compared in terms of features and security experts must face risks from both the Cloud Client (CC) and the Cloud Service Provider's (CSP) side [2]. It is important for CCs to evaluate security risks of the CSP's system in which their data will be stored. However, this is quite difficult, given the fact that CSPs do not make such information available to the public due to security restrictions. Moreover, another problem is estimating potential loss, since he does not know about the asset's real value. Only the client is aware of this value, which differentiates according to each CC [2].

In this paper, we analyze whether popular, traditional risk assessment methodologies are able to properly and fully evaluate cloud computing environments and, if not, which changes they should implement. As technology is advancing, RA methodologies should optimize their models and be adapted to the evolving threat landscape, in order to produce more accurate results. There is need for risk assessment methodologies to improve some features not only for their use in cloud computing environments, but also for being more accurate and efficient to traditional IT environments.

Section 6.2 describes the existing literature related to risk assessments addressed to cloud computing environments. In Sect. 6.3, cloud computing is defined, while Sect. 6.4 focuses on the presentation of features that risk assessment methodologies should incorporate in their implementation in order to be used effectively to cloud computing environments. Section 6.5 examines whether the traditional risk assessment methodologies have already included these features. In Sect. 6.6, we propose a quantitative business-process risk assessment methodology, which supports risks that are associated with cloud computing platforms. Finally, the importance of our findings is discussed in Sect. 6.7, while Sect. 6.8 concludes our work.

This paper extends our previous work [3] and tries to identify whether traditional risk assessment methods are suitable for assessing risks in cloud computing environments. Cloud computing environments have different structures when compared to traditional IT environments. Five essential characteristics of the cloud are examined to determine the appropriate criteria for analyzing methods and tools in the context of cloud computing. A minimum set of criteria is defined, for assessing if risk methodologies are suitable for cloud services.

6.2 Related Work

Current researcher interest focuses mostly on whether the existing risk assessment methodologies are sufficient for identification of risks in cloud computing ecosystems. Many state that traditional methodologies are unfit to this new type of environments and tend to propose [2] their own version of risk assessment. In this section these methodologies are referred.

In 2009, the European Network and Information Security Agency (ENISA) [4] published a report about the security benefits, as well as risks, of cloud computing. Their report included some practical recommendations. Afterwards, ENISA again [5] developed a set of questions that could be asked to a CSP, to ensure adequate protection of third-party information in the cloud. The purpose was the risk of adopting cloud services to be evaluated and help CC to compare potential offers made by different Cloud Provider. Finally, they developed a cloud security tool [6] for Small and Medium-sized Enterprises (SME), in order to help them not only to evaluate the opportunities and the risks, but also create some questions that support them to identify the main characteristics that the chosen CSP should have.

Cayirci et al. [7] recommended a qualitative and relative risk assessment model, called CARAM. The purpose of this model is to help CCs to choose the most appropriate CSP based on their risk profile. The interesting feature of this model is that it implements a risk assessment for every CC–CSP pair, and not a generic one.

Goettelmann et al. [8] suggested a technique that counts on two parts. The first is the security needs of the business process, while the second one is the conformance to guidelines of the CSP's side. The purpose of this technique is to give assistance to companies for deploying securely their business processes and applications in a multi-Cloud environment.

QUIRC [9] is a quantitative methodology for impact and risk assessment to cloud computing environments. It presents six key security criteria of cloud platforms, along with the typical attack vectors and events that can categorised based on these six categories.

Finally, COBIT was introduced in 1996 by the Information Technology (IT) Governance Institute and the Information Systems Audit and Control Association (ISACA). Its latest version is "COBIT 2019 Framework" and provides a common language to businesses in order they be able to express their goals, objectives and their results [10]. Even though it is a generic framework that used in IT environments, for some selected cases it can be successfully adapted for cloud computing use [11].

As we notice, there are plenty of new risk assessment methodologies that are focused on cloud computing environments. Our question is what happens with the methodologies that already exist. Are they sufficient to assess the risk of these environments? Our paper examines the characteristics of cloud computing and define the essential features that a risk assessment addressed to cloud computing environments should include.

6.3 Defining Cloud Computing

Cloud computing enables the provision of services that are more flexible in supporting daily operations, such as the on-demand network access to resources. Cloud computing successfully accomplishes this task because of its essential characteristics which have been defined by NIST [1] and are described at Table 6.1.

Moreover, cloud computing environments have the following three types of service models [1].

(1) **Software as a Service (SaaS)**:

- CC uses provider's applications which are running on a cloud infrastructure.
- Applications can be accessed by client devices through thin client interfaces, such as a web browser, or even program interfaces.
- CC does not manage or control the cloud infrastructure, and this may drive to limited user-specific application configuration settings.

(2) **Platform as a Service (PaaS)**:

- CC deploys the cloud infrastructure consumer-created or acquired applications created using material which is supported by the provider.
- CC does not manage the cloud infrastructure. He only controls the deployed applications and configuration settings for the application-hosting environment.

Table 6.1 Cloud computing characteristics

Characteristic	Description
On-demand self-service	A CC can unilaterally provide computing capabilities, e.g. network storage, as needed without the existence of human interaction with the CSP
Broad network access	Capabilities are available over the network and CCs have access to them via standard mechanisms for heterogeneous client platforms, such as mobile phones or tablets
Resource pooling	CSPs computing resources are pooled in order multiple CCs to be served. These resources are dynamically assigned based on CCs demand. They have no control or knowledge for the exact location of these provided resources but may be able to specify location at a higher level of abstraction (e.g., country, state, or datacenter)
Rapid elasticity	Capabilities can be elastically provisioned, as well as to automatically scale rapidly according to the demand. CC can use these capabilities in any quantity at any time
Measured service	Cloud systems automatically optimize resource use based on the type of service. The monitoring and reporting of the resource usage provides transparency for both the CSP and CC who make use of the service

(3) **Infrastructure as a Service (IaaS)**:

- CC is provided with processing, storage, networks, and other fundamental computing resources. He can deploy and run arbitrary software, including operating systems and applications.
- CC does not manage the cloud infrastructure. He only controls operating systems, storage, as well as deployed applications.
- CC may have limited control to the selection of networking components.

6.4 Criteria

We present two groups of evaluation criteria. Section 6.4.1 focuses on four criteria necessary for all risk assessment methodologies. It is worth-mentioning that these criteria should be implemented, whether a risk assessment methodology is addressed solely on traditional IT systems or not. Section 6.4.2 proposes five cloud-specific criteria. These are necessary for cloud-enabled risk assessment based on international literature and should be implemented by methodologies that aim to assess risk in environments with dynamic nature, such as cloud-enabled networks and systems.

6.4.1 Basic RA Criteria

6.4.1.1 Regular Updates

Risk assessment methodologies ought to analyze only necessary information, because they should optimize their services and balance data and timely analysis of existing threats with accurate results. Cloud computing's dynamic nature enables the provision of new types of services, thus risk assessment (RA) methods addressed to cloud should be able to handle such complexity [2]. For that reason, it is quite important for RA methodologies to be updated on a regular basis, and support potential new threats that come up or to improve some of its functions [3]. The life cycle of a RA method is quite important because an outdated one does not take into account recent treats, vulnerabilities, as well as safeguards and this slows the entire progress of risk assessment [3].

6.4.1.2 Types of Analysis

Risk assessments usually follow either qualitative or quantitative analysis of risks. The difference is mainly behind the method of risk calculation they use, as well as the factors that each methodology considers [12]. Specifically, the quantitative approach is based on mathematical or statistical calculation of the probability of a specific event to occur in the future [13]. The measurement is expressed with numbers defined discretely and always belong to an exact range. Finally, in order to work properly, they need great amounts of data [12]. As for the qualitative approach, the analysis of the risk is described values that are not numeric. The scale is relative, and thus users know only the measurements' arrangement. The distance between the measurement is not clearly specified [12, 13].

The ideal approach for risk assessment analysis is to combine the aforementioned types of analysis. It will be interesting if risk assessment's methodology could use inclusive metrics that will be both of qualitative and quantitative nature [14]. This will be beneficial for cloud computing environments that have complex services and are not easily classified in discrete ranges (for quantitative analysis) nor can be expressively described (as needed in qualitative scales). Also, literature suggests that if we separate these two types of analysis, risk assessment results are not as accurate as using their combination [14]. This solution will contribute to the increase of trust that a CC has to CSP and will also help cloud computing environments to secure their service provision for companies and organisations [15].

6.4.1.3 Ability to Analyze Cascading Risks

Unfortunately, not all existing risk assessment methodologies examine separated risks. It will be more beneficial if they study the overall risk, including cascading risks that come up after an incident occurs. The study of such risks can be defined with two different ways: (1) Methodologies should take into account that more than one risk could manifest simultaneously, and also (2) that the same risk is able to affect more than one system albeit with no apparent connection betweeen the two. Finally, due to the fact that most of the current methods undervalue hard-to-predict or rare event risks in information systems, they amy provide misleading information on what may occur to an organisation that uses cloud-enabled services [16].

6.4.1.4 Assets as Attack Vectors

Risk assessment methodologies should change the way they handle assets. The current consensus behind them is that assets are valuable entities for the organisation and must be protected. On the other end of the spectrum, they can provide the actual attack surfaces needed by malicious attackers. As technology is evolving, new types of services are developed, such as the Internet of Things, Cloud Computing etc, collectively named as "Industry 4.0". Thus, the way that systems should be protected changes. Risk assessments auditors should realize this fact and start thinking that maybe the assets previously thought solely as value entities, may now act as entry points and become attack vectors that actually cause security incidents. When analysts start thinking in that way, they will be able to identify more vulnerabilities that infrastructure should deal with [17].

6.4.2 Cloud-Specific Criteria

6.4.2.1 Provision of Security Assurance Recommendations

It is quite important for risk management frameworks to support the "*objective examination of evidence for the purpose of providing an assessment on risk management, control or governance processes*" [18] of CSPs. ENISA [4] published a set of recommendations that CCs can ask their CSPs in order to understand their security level. Such recommendations should assist CCs not only to evaluate the risk of the adoption of a cloud service, but also to decide which offer is more suitable for him among the offered ones.

CCs should be equipped with a set of questions that are targeted to security issues. This can be successfully implemented only if questions are constructed by security experts who understand the architecture of a cloud computing environment. Examining evidence to justify risk assessment of CSPs would benefit both the CCs and the CSPs. For once, if all clients desired a cloud infrastructure audit, then the CSP would have increase auditing load that could hinder operations. Also, multiple assessment by different parties would threaten to expose other customer data to unauthorised experts, since access to the infrastructure by third parties would inevitably increase [4].

6.4.2.2 Involvement of CC in CSP's RA

Cloud computing environments, according to NIST [19], include at least two teams of actors: the CSP and the CC. It is crucial that both these two groups remain active during the RA process. This will supply different types of information and provide a more targeted and effective assessment. For example, the CC is useful when the value of his assets is examined, while CSPs should only participate during the process of service risk management, since the number of CCs is often pretty large and joint procedures would be complicated. Generally, CCs should only participate when it is absolutely necessary. According to Albakri et al. [20], an imaginary concept would be that the CC does not participate in all steps of the risk assessment, but only during the following three: (1) when the regulatory and legal requirements are defined; (2) when the security risk factors are identified, and (3) when they get CSP's feedback and should apply the required security tasks. This will not allow the level of the complexity to be increased [2, 20].

6.4.2.3 Assessment of Cloud Threats

Cloud-enabled systems include threats and vulnerabilities that exist both in traditional information systems, as well as in new, cloud-specific systems [21]. OWASP [22] has published a top ten list of risks that are addressed specifically to cloud computing environments. These include (but are not limited to) Multi-tenancy and physical security, service and data integration and user identity federation. Moreover, Latif et al. [23] made a literature review and concluded to similar risk categories related to CCs and CSPs. From the perspective of CSPs, crucial risks that should be mitigated are classified in five major categories: (a) physical security; (b) organizational; (c) data security and privacy; (d) technological and (e) compliance audit. On the other hand, risks referred to the CC's perspective are classified to the following four groups: (a) physical security; (b) data security and privacy; (c) technological and (d) compliance audit. Finally, ENISA [4] published a report in which thirty-five risks were identified and grouped into three main categories: (i) technical; (ii) policy and organizational, as well as (iii) legal ones.

Sometimes, traditional vulnerabilities or threats can affect cloud computing environments with even greater impact than in traditional IT systems. An indicative example is the concurrency problem, where a resource is locked for use by a user and, at the same time, someone else is trying to gain access to this resource [24].

6.4.2.4 Risk Calculation Formula

Cloud computing ecosystems not only invalidate the assumption that assets exist within the organization, but also the fact that organizations themselves are responsible for managing their own security processes. This means that the traditional risk assessment formula will be inaccurate for such complex environments, due to the difficulty of calculating not only risk's likelihood, but also its impact [2]. On the one hand, CSPs will not be able to assess the value of the CC's assets, since the latter is the only one who can value the cost of a breach. On the other hand, CCs will not be able to assess the CSP's system. This happens so that CSP's may protect their systems from hackers impersonating to be CCs. However, it is mandatory for CSPs to provide guarantees to CCs concerning the protection of their processed data. Moreover, the assessment of probability is generally considered to be a strenuous issue, because a CC will have to assess his system through an open environment, i.e. the Internet. Also, the risk introduced by the client himself is quite high, since user security awareness varies to a great extend [2]. According to Theoharidou et al. [21], the assessment of risks in cloud computing environments still remains an open research issue. It is necessary for CC to trust CSP on providing all the appropriate assessment data and already implemented security controls.

Last but not least, it would be interesting if RA methods could incorporate the examination of the historical security status of an organisation. Such kind of record may contain incidents separated to adversarial and non-adversarial. The first category concerns incidents conducted by hackers, antagonists or cyber-criminal organisations. The latter category is for incidents that mostly include unintentional factors, such as system faults or earthquakes. In case of cloud computing environments, CCs should be able to consider the CSP's status, exposure and resilience, in order to understand the impact of threats on their services and data supported by the CSP's resources [25].

6.4.2.5 Service-Based Risk Assessment

One cloud computing characteristic that complicates assessments is the number of CCs that participate on a single RA. Each tenant has his own security requirements and thus, potential conflicts are often not avoided. A solution to this problem is for risk assessment to be implemented separately for every service that the CSP provides to his CCs [26]. However, all services should be examined because any exclusion may lead to serious organization risks going unnoticed [27].

No matter which side conducts the risk assessment, experts should use a service-based approach in order to avoid the aforementioned issues. Especially, when risk assessment is conducted by CSPs, the use of this approach is beneficial due to the large numbers of CCs they have to manage. A common issue is that a particular CC may participate to more than one service, and this makes the process quite complicated.

6.5 Comparison of Traditional Risk Assessment Methodologies

In this section, a further comparison among the traditional risk assessment methodologies that were analysed in our previous work [3] is presented, focused on cloud essential characteristics. Specifically, these methodologies were EBIOS, MEHARI, OCTAVE, IT-Grundschutz, MAGERIT, CRAMM, HTRA, NIST SP800, RiskSafe and CORAS.

The comparison is composed of two parts. In Sect. 6.5.1 we examine which of these risk assessment methodologies have incorporated four basic features in order to be more effective and accurate to their result. Section 6.5.2 checks which of them can be used to assess cloud computing environments.

6.5.1 Basic Criteria Comparison

Table 6.2 depicts the comparison of the aforementioned risk assessment methodologies.

The information about the criteria of *Regular Update* and *Type of Analysis* have been extracted from our previous work, with some minor changes based on recent information we came up with [3]. NIST SP800-30 is the only method which supports quantitative, qualitative as well as semi-qualitative type of analysis [28, 29].

As for the criterion of *Capture cascading risks*, EBIOS [30, 31] takes into account the cascading risks, since they consider as a significant issue to identify and comprehend the relationship among the media assets. This will contribute to the examination of the possible phenomena of propagation of incident or disaster. MEHARI and IT-Grundschutz [32] do not consider connections between assets, and thus they do not

Table 6.2 Comparison of basic RA criteria

	Regular update	Type of analysis	Capture cascading risks	Assets as an attack platform
EBIOS	2018	Qualitative	Yes	No
MEHARI	2017	Qualitative and quantitative	No	No
OCTAVE	2007	Qualitative	Yes	No
IT-Grundschutz	2015	Quantitative	No	No
MAGERIT	2014	Qualitative and quantitative	Yes	No
CRAMM	2005	Qualitative	Yes	No
HTRA	2007	Qualitative	Yes	No
NIST SP800	2012	Qualitative, quantitative, semi-qualitative	Yes	No
RiskSafe	2012	Qualitative	Yes	No
CORAS	2010	Qualitative and quantitative	Yes	No
COBIT	2019	Qualitative and quantitative	No	No

capture the cascading risks. OCTAVE [30, 33] considers threats which are outside the control of the organization, like natural disasters, as well as interdependency risks, such as unavailability of a Critical Infrastructure due to lack of power supply. MAGERIT is the only one that follows an asset dependency concept [32, 34]. Specifically, this method designates the assets relative to the examined organisation, as well as their values and the existing interdependencies. As for CRAMM and CORAS, not only analyse relationships among assets, but also among in relation to the existing vulnerabilities of the system [30, 35, 36]. HTRA methodology also focuses on the examination of the cascading effect of related vulnerabilities [37]. NIST SP 800-30, can assess single or multiple interrelated systems. In the second case, the domain of interest and all interfaces and dependencies should be well defined prior in order the methodology to be applied [28]. RiskSafe methodology is capable to identify the assets and potential relations that exist with other assets [30, 38]. Finally, as we notice in the description of its enabling process "*APO12. Manage Risk*" and more specifically in "*APO12.2 Risk Analysis*", COBIT 2019 Framework supports both quantitative and qualitative analysis for the assessment of risk [39, 40]. Moreover, identification of interdependencies among all potential risk events is crucial, in order to ensure completeness in the determination of likelihood and impact [40].

Last but not least, the criterion *Assets as an attack platform* refers to a new feature that traditional risk assessments fail to capture but should incorporate in their threat vectors. As such, none of these ten risk assessment methodologies yet takes into account this parameter.

6.5.2 Cloud-Specific Criteria Comparison

Table 6.3 presents the comparison of the aforementioned methodologies. The five criteria are focusing to characteristics that methodologies should incorporate in order to implement a sufficient risk assessment in cloud computing environments.

EBIOS is a generic approach and this makes it a real toolbox [41]. As such, actions as well as appropriate methods, according to the objective of the study, can be chosen for implementation. Consequently, after its final update by ClubEbios in 2018, the methodology can be used for assessment of cloud computing environments. One can assess whether a cloud provider meets the ANSSI essential security requirements (SecNumCloud) before migrating [42, 43]. Moreover, ANSSI has issued a case study on how a client can conduct an assessment after migration [44]. Finally, there is no relative information which supports the involvement of CC when the CSP while performs a risk assessment.

As for MEHARI, there is not much documentation available. What is currently known is that the methodology is still being further extended to cope with the evolutions of Information and Communication architectures, technologies as well as

Table 6.3 Comparison of cloud-specific RA criteria

	Provision of security assurance recommendations	Involvement of CC in CSP's RA	Assessment of cloud threats	Risk calculation formula	Service-based risk assessment
EBIOS	Yes	No	Yes	No	No, asset-based
MEHARI	N/A	N/A	Yes	No	Yes
OCTAVE	No	No	Yes	No	No, asset-based
IT-Grundschutz	Yes	No	Yes	No	Yes
MAGERIT	N/A	N/A	Yes	No	No, asset-based
CRAMM	N/A	N/A	N/A	No	No, asset-based
HTRA	No	No	NO	No	No, asset-based
NIST SP800	Yes	No	Yes	No	YES, if chosen
RiskSafe	N/A	N/A	N/A	N/A	N/A
CORAS	Yes	No	Yes, when combined with T&VA tool	No	No, asset-based
COBIT	Yes	No	Yes	No	No, threat-based

working changes, like BYOA, BYOD, Cloud, etc. [45]. According to its manual, MEHARI uses a service-based approach to identify assets lined to business processes [46, 47].

OCTAVE's Allegro approach can capture operational risks of cloud environments. The methodology evaluates the effectiveness of implemented security controls at the container level. In OCTAVE, the CSP is considered an external technical container, who stores, transfers or processes the client's information assets. Furthermore, OCTAVE does not incorporate a knowledge base for the identification and assessment of cloud risks. However, it focuses on information assets and their containers, as well as it uses worksheets. As a result, all details for cloud-specific threats and their mitigations can be captured, through appropriate definition of risk measurement criteria [48]. Finally, the methodology is considered as asset-centric in its identification of assets [48, 49].

The Federal Office for Information Security has published a report that incorporates directives in how cloud services can be used securely [50]. It is based on the IT-Grundschutz module M 1.17 "Cloud usage", but it can be used also by clients who do not use the aforementioned methodology. The publication describes all the appropriate steps for properly assessing the potential CSPs to assist decision making before migration. The module provides information about threat scenarios when using the cloud [50, 51]. Clients ought to have performed a risk assessment on the past in order to have a clearer picture of the desired security requirements that the CSP should cover. After migration, a risk assessment must be run again from time-to-time by the CC, in which the provider participates and gives information about implemented controls [50]. Cloud providers should undertake both tiers of the standard's risk assessment, because they are considered companies with special security requirements due to the nature of their business objectives [38]. Finally, IT-Grundschutz is a service-based methodology as its first step is the identification of business process, from which the assets of the organization will be derived [52].

Similar to MEHARI, there are not available resources for information about MAGERIT's employment on cloud environments. However, some information could be extracted from EAR/PILAR, which is a framework of tools used for analysis and management built around MEHARI. It supports the MAGERIT methodology and is specialized on Information and Communication Systems. Since its revision in 2013, the tool has updated its library with asset classes and protection measures for third party services, such as cloud [53, 54]. According to its manual, MAGERIT is considered an asset-oriented risk analysis approach [55].

CRAMM is a traditional risk assessment methodology which can assess information systems after identifying security risks. However, there is not relevant information which indicates its ability to be used in cloud computing ecosystems. According to the methodology's user guide version 5.1 and [38], it is an asset-based approach, which links the assets with the relevant end user services. Also, CRAMM is not fully supported since 2015, which prohibits any future incorporation of cloud computing in its assessment methodology.

On the one hand, the Harmonized Threat and Risk Assessment (TRA) Methodology has not been updated its documentation since 2007. As a result, no information relevant to the assessment of cloud services was found. On the other hand, it is quite interesting that its Threat Listing could be extended. Consequently, it could be enriched with cloud-oriented threats and be utilized in such environments. Finally, HTRA is an asset-centric approach as it relies in the fact that employees and other personnel use tangible assets to produce services [37].

NIST in [29] explains that the Risk Management Framework (RMF) introduced in SP800-37 Revision 1, could be used by CSPs to assess the layers of the functional stack that he manages. CC, on the other hand, will assess the upper functional layers that support service provision. Prior to acquisition and adoption of such a service, the provider has to undergo a risk assessment by the client. By doing so, associated risks as well as risk treatment of the cloud service will be identified and planned respectively. To do so, the CSP needs to support the client with all appropriate information about the ecosystem, such as implemented security controls, etc. After migration to the cloud, a periodic risk assessment of the systems' security is conducted by the CC. NIST does not define the analysis approach of the assessment (i.e. vulnerability-oriented, threat-oriented, asset-oriented) as it tailored to the need of the examined organization [28]. This is because different analysis approaches lead to different levels of detail and thus, the adverse events for which likelihoods are determined. Moreover, NIST SP800-53 contains the controls to address cloud security, through provision of a broad risk and security framework for evaluation of cloud ecosystems. The final version of NIST SP800-146 contains further guidance for using cloud service models. Finally, according to [29], each CC's specific security requirements are handled as black-box by CSPs and thus are projected as a generic core set.

Concerning RiskSafe Assessment, no information could be retrieved, since the official web site has suspended its operation. Only ENISA [56] still has some details about the methodology but is not relevant to our survey.

CORAS methodology has to be used by the CC both at service deployment phase to initially assess the offered services, as well as at service operation, where resources and data are managed by the provider [57]. In both circumstances, assessment data should be given by the CSP as requested. The methodology when combined with the Threat and Vulnerability Assessment tool (T&VA) [58] can be used for assessing cloud-specific threats, because the tool contains a threat model for distributed systems and software. More specifically, the T&VA provides a standard list of IT related threats, adopting the cloud relevant threats of the deployed cloud ecosystem under examination. Finally, CORAS is an asset-based approach, since it integrates CRAMM's asset value technique [38].

ISACA in 2011 published a document which contains useful guidance for enterprises which consider migrating their business processes into a cloud computing ecosystem [59]. This publication is aligned with the COBIT framework. Also, it is able to assess cloud-specific enterprise risks, because of IT involvement in almost all facets of a business's operation. Finally, COBIT's approach on risk event identification is threat-based, according to "*APO12.1 Collect data*" [39, 40].

Last but not least, as far as *Risk Calculation Formula* criterion is concerned, all surveyed methodologies rely on the traditional risk calculation. Moreover, none of them involve CCs in the RA process of CSPs, like when risk factors are defined. Thus, the assessment results are considered less realistic.

6.6 Cloud-Specific Risk Assessment Methodologies

In this section, three methodologies that can be used to cloud computing environments are presented. Each methodology has focused on different aspect for improvement, in order the results to be more subjective and complete.

6.6.1 A Quantitative Business-Process Risk Assessment Methodology

In order to improve the evaluation and decrease the subjectivity of the assessment that is implemented in cloud computing environments, Stergiopoulos et al. [12] developed a methodology that uses mathematical distributions over historical data. It estimates risks on asset-based processes that related to cloud computing platforms. They used Poisson distributions for the quantification of the likelihood of a threat occurrence in order to evaluate the security risks. The advantage of using this methodology is that the estimation of the threat likelihood is more objective, and the risk assessment is grounded on asset-based processes which is an appropriate approach for cloud services. Finally, it supports stakeholders to evaluate the risk of using cloud services to process data [12]. This methodology is incorporated as a top layer to a previous tool that was developed by Stergiopoulos et al. [60], named CIDA.

Finally, they presented an example of train routing times, modelling it as asset dependencies. Their purpose was to implement risk assessment to a real critical infrastructure with multiple services that are accessed remotely over a cloud server system. Some of the achievements of this methodology, contrary to the standard risk assessment methodologies, are the following [12]:

- Due to the fact that the methodology assesses the asset dependencies in terms of the business process, it is able to compute both the dependency risk measures and the representative dependency graph.
- Through a dynamic way, the methodology estimates the evolution of cascading failures of the assets that are participated in the same business processes.
- The tool can generate a catalogue with all asset-threat pairs for different time periods, including short, medium or even long-term. This catalogue is useful, in order risk assessment auditors to recognize every asset with eventual risk that will be above a designated threshold value for different time periods.
- The methodology considers realistic scenarios, including what-if scenarios. This will help to the evaluation of the scenarios involving common-caused failures

to targeted nodes. Moreover, it assesses the cascading effects, according to the geographical dependencies.
- Considering the risks of different time periods, auditors can make decisions, taking into account the dynamics of threats over time.

6.6.2 Risk Assessment Methodology for Cloud Clients

Drissi et al. [61] developed a methodology that is suitable for risk assessment to CCs. Specifically, this methodology is composed of the following six steps:

i. It is important for CC to conduct an archived risk assessment before he migrates to cloud. This helps him to evaluate the risk after he stores his data to the cloud.
ii. The expertise who will conduct the risk assessment should prioritize cloud providers, where the data are stored, the importance of asset, as well as the security objectives.
iii. With the use of AHP model, the weight of security objectives for each of the above mentioned locations will be defined, as well as its asset values. It is worth-mentioning that both qualitative and quantitative analysis is needed in order the decision-making methodology to be accurate. Drissi et al. [61] used AHP model, since it is appropriate when solution of complex group decision situations is needed.
iv. Then, the weight for all asset locations should be defined in order the asset value for each location to be estimated.
v. Next, risk assessment of each location is conducted, in order the threats and vulnerabilities to be identified. As such, the methodology should estimate which location and, thus, which provider is critical for the CC.
vi. Outsourcing services to cloud environment, entails new threat and vulnerabilities in CC's landscape. In order CC to be aware of the most critical location of his data, it is advisable to conduct again risk assessment for the identification of the current vulnerabilities which are present to the current cloud computing environment.

Drissi et al. [61] supported that this methodology will offer flexibility to Cloud Clients when they are to conduct risk assessment. This will be successfully accomplished with the described steps for the evaluation of risk.

6.6.3 Risk Assessment Conduct with the Intervention of the Cloud Client

Albakri et al. [20] proposed a risk assessment framework that consists of two basic parts, the actions that should be implemented by the CSP and the ones from the CC's perspective. Its main purpose was to support CCs to be more active during the process of risk assessment in order the results to be more accurate and realistic. On the other hand, this intervention should be controlled, otherwise this will bring the opposite results. CSPs will not be able to manage the process of the assessment in case that all the CCs are participating to all the steps of this procedure. Albakri proposed the participation of CCs only when this is necessary. Specifically, they defined that CCs' intervention is needed three times: (i) specify the regulatory and legal requirements; (ii) identify security risk factors, as well as (iii) when they get feedback from the CSP and they should implement the appropriate security actions. Albakri stated that when CSPs and CCs are combined, the results of a risk assessment will be more accurate since the process incorporates information from all actors of the cloud ecosystem.

6.7 Discussion

Since traditional IT environments are different comparing with these of cloud computing, traditional risk assessment methodologies are often not fully compatible with cloud-enabled networks and systems. Most traditional methodologies fail due to the rapid technological evolution and thus, are insufficient for cloud computing ecosystems. To begin with, the participation of the CCs is necessary, since only them are able to evaluate their data as data or service owners.

However, their participation should be controlled, in order not to interrupt the process of risk assessment. Moreover, examining each service as a flow seems to be more efficient whether the assessment is implemented at the cloud or not. Finally, it is advisable that risk assessment methodologies should combine both qualitative and quantitative ranking scales. This optimizes understanding of risks involved from the point of the organization, CC or CSP and helps them leverage the advantages of each type of analysis while eliminate the disadvantages of using only one of the two [14].

Some extra features often missing that would optimize risk assessment for both traditional IT and cloud environments are following: (1) Use of what-if scenarios in process of risk evaluation, even if it raises the cost of the implementation. This is useful for identifying the risk when an incident occurs. (2) Storing track records that include the prioritisation of implemented security controls, as well as those that are rejected along with relevant justification. This record can be used when an incident occurs to evaluate previous decisions on information security. Some of the controls may be rejected as inappropriate the moment the RA was performed but, under new light, may be deemed mandatory [27].

Finally, it is difficult to evaluate intangible assets that are mainly dynamic entities that change frequently. If a risk assessment methodology is not able to identify this kind of assets, the evaluation may be incomplete since these data compose the social and non-technical dimension of the examined infrastructure [16].

6.8 Conclusions

In this paper, we have outlined some criteria capable to optimize traditional risk assessment methodologies, in order to be used not only for traditional IT, but also for cloud environments. Incorporating the first group of suggested features will allow traditional risk assessment methodologies to evolve their assessment to support cloud computing. As new types of services come up, risk assessment methodologies should evolve to be capable to evaluate risk in cloud-enabled services.

From our analysis we concluded that the existing methodologies are not efficient enough for cloud ecosystems, unless they redesign their steps in a more generic way. Moreover, these frameworks have underestimated the importance of CCs involvement in the RA process. Also, they should optimize risk calculation formulas to consider the distributed nature of the cloud computing environments. It is a kind of challenge for them to change their concept and implementation from a static nature to a dynamic one. Cloud computing offers new opportunities, but also needs new ways of assessment since it brings new threats and new services.

References

1. P.M. Mell, T. Grance, Sp 800-145. The NIST Definition of Cloud Computing (2011)
2. S.H. Albakri, B. Shanmugam, G.N. Samy, N.B. Idris, A. Ahmed, Traditional security risk assessment methods in cloud computing environment: usability analysis, in *Proceedings of the 1st International Conference of Recent Trends in Information and Communication Technologies*, Universiti Teknologi Malaysia, Johor, Malaysia (2014), pp. 483–495
3. D. Gritzalis, G. Iseppi, A. Mylonas, V. Stavrou, Exiting the risk assessment maze: a meta-survey. ACM Comput. Surv. (CSUR) **51**(1), 11 (2018)
4. T. Haeberlen, L. Dupré, Cloud computing—benefits, risks and recommendations for information security, in *European Network and Information Security Agency (ENISA)* (2012)
5. D. Catteddu, G. Hogben, Cloud computing information assurance framework. Eur. Netw. Inf. Secur. Agency (ENISA) **13**, 14 (2009)
6. SME Cloud Security Tool—ENISA (2019), https://www.enisa.europa.eu/topics/cloud-and-big-data/cloud-security/security-for-smes/sme-guide-tool. Accessed 7 Jan 2019
7. E. Cayirci, A. Garaga, A. Santana, Y. Roudier, A cloud adoption risk assessment model, in *2014 IEEE/ACM 7th International Conference on Utility and Cloud Computing (UCC)* (IEEE, 2014), pp. 908–913
8. E. Goettelmann, K. Dahman, B. Gateau, E. Dubois, C. Godart, A security risk assessment model for business process deployment in the cloud, in *2014 IEEE International Conference on Services Computing (SCC)* (IEEE, 2014), pp. 307–314
9. P. Saripalli, B. Walters, QUIRC: a quantitative impact and risk assessment framework for cloud security, in *2010 IEEE 3rd International Conference on Cloud Computing (CLOUD)* (IEEE, 2010), pp. 280–288
10. COBIT 2019 Publications & Resources (2019), http://www.isaca.org/COBIT/Pages/COBIT-2019-Publications-Resources.aspx
11. S. Gadia, Cloud computing: cloud computing risk assessment: a case study. ISACA J. **4**, 11 (2011)
12. G. Stergiopoulos, D. Gritzalis, V. Kouktzoglou, Using formal distributions for threat likelihood estimation in cloud-enabled IT risk assessment. Comput. Netw. **134**, 23–45 (2018)

13. S. Taubenberger, J. Jürjens, Y. Yu, B. Nuseibeh, Problem analysis of traditional IT-security risk assessment methods—an experience report from the insurance and auditing domain, in *IFIP International Information Security Conference* (Springer, Berlin, Heidelberg, 2011), pp. 259–270
14. Y. Sivasubramanian, A.S. Zubair, P. Ved, Risk assessment for cloud computing. Int. Res. J. Electron. Comput. Eng. **3**, 7 (2017). ISSN Online: 2412-4370. https://doi.org/10.24178/irjece.2017.3.2.07
15. S. Drissi, S. Benhadou, H. Medromi, Evaluation of risk assessment methods regarding cloud computing, in *The 5th Conference on Multidisciplinary Design Optimization and Application* (2016)
16. G. Wangen, E. Snekkenes, A taxonomy of challenges in information security risk management, in *Proceeding of Norwegian Information Security Conference/Norsk informasjonssikkerhetskonferanse-NISK 2013-Stavanger, 18th–20th November 2013* (Akademika Forlag, 2013)
17. J.R. Nurse, S. Creese, D. De Roure, Security risk assessment in internet of things systems. IT Prof. **19**(5), 20–26 (2017)
18. Glossary (2019), https://www.isaca.org/Pages/Glossary.aspx?tid=1087&char=A. Accessed 7 Jan 2019
19. NIST Cloud Computing Standards Roadmap Working Group, *NIST Cloud Computing Standards Roadmap* (2013)
20. S.H. Albakri, B. Shanmugam, G.N. Samy, N.B. Idris, A. Ahmed, Security risk assessment framework for cloud computing environments. Secur. Commun. Netw. **7**(11), 2114–2124 (2014)
21. M. Theoharidou, N. Tsalis, D. Gritzalis, In cloud we trust: Risk-Assessment-as-a-Service, in *IFIP International Conference on Trust Management* (Springer, Berlin, Heidelberg, 2013), pp. 100–110
22. OWASP Cloud—10 Project—OWASP (2019), https://www.owasp.org/index.php/Category:OWASP_Cloud_%E2%80%90_10_Project. Accessed 7 Jan 2019
23. R. Latif, H. Abbas, S. Assar, Q. Ali, Cloud computing risk assessment: a systematic literature review, in *Future Information Technology* (Springer, Berlin, Heidelberg, 2014), pp. 285–295
24. S.V. Garde, A. Mudaliar, B. NCHSE, Concurrency Lock Issues in Relational Cloud Computing (2013)
25. F. Xie, Y. Peng, W. Zhao, D. Chen, X. Wang, X. Huo, A risk management framework for cloud computing, in *2012 IEEE 2nd International Conference on Cloud Computing and Intelligent Systems (CCIS)*, vol. 1 (IEEE, 2012), pp. 476–480
26. R. Alosaimi, M. Alnuem, Risk management frameworks for cloud computing: a critical review. Int. J. Comput. Scie. Inf. Technol. **8**(4) (2016)
27. A.B. Ruighaver, M. Warren, A. Ahmad, Does traditional security risk assessment have a future in Information Security? J. Inf. Warf. **10**(3), 16-IV (2011)
28. NIST, S. 800-30, *Guide for Conducting Risk Assessments* (2012)
29. M. Iorga, A. Karmel, Managing risk in a cloud ecosystem. IEEE Cloud Comput. **2**(6), 51–57 (2015)
30. G. Stergiopoulos, V. Kouktzoglou, M. Theocharidou, D. Gritzalis, A process-based dependency risk analysis methodology for critical infrastructures. Int. J. Crit. Infrastruct. **13**(2–3), 184–205 (2017)
31. EBIOS—Risk Management Methodology (2010), http://people.redhat.com/swells/anssi/EBIOS-1-GuideMethodologique-2010-01-25-english.pdf. Accessed 7 Jan 2019
32. B. Rahmad, S.H. Supangkat, J. Sembiring, K. Surendro, Threat scenario dependency-based model of information security risk analysis. IJCSNS **10**(8), 93 (2010)
33. R.A. Caralli, J.F. Stevens, L.R. Young, W.R. Wilson, *Introducing OCTAVE Allegro: Improving the Information Security Risk Assessment Process* (No. CMU/SEI-2007-TR-012) (Carnegie-Mellon University, Software Engineering Institute, Pittsburgh, PA, 2007)
34. F. Crespo, M. Gómez, J. Candau, J. Mañas, *MAGERIT—Version 2 Methodology for Information Systems Risk Analysis and Management. Book* (Ministerio de Administraciones Públicas, Madrid, 2006)

35. J. Viehmann, Reusing risk analysis results—an extension for the CORAS risk analysis method, in *2012 International Conference on Privacy, Security, Risk and Trust (PASSAT) and 2012 International Conference on Social Computing (SocialCom)* (IEEE, 2012), pp. 742–751
36. G. Brændeland, H.E. Dahl, I. Engan, K. Stølen, Using dependent CORAS diagrams to analyse mutual dependency, in *International Workshop on Critical Information Infrastructures Security* (Springer, Berlin, Heidelberg, 2007), pp. 135–148
37. R. CSE, *Harmonized Threat and Risk Assessment (TRA) Methodology*. TRA-1 Date: October 23 (2007)
38. L. Coles-Kemp, J.W. Bullee, L. Montoya, M. Junger, C. Heath, W. Pieters, L. Wolos, *Technology-supported Risk Estimation by Predictive Assessment of Socio-technical Security* (2015)
39. P. Bernard, *COBIT® 5-A Management Guide* (Van Haren, 2012)
40. COBIT Control Practices: Guidance to Achieve Control Objective for Successful IT Governance, 2nd Edition (2019), http://www.isaca.org/Knowledge-Center/Research/ResearchDeliverables/Pages/COBIT-Control-Practices-Guidance-to-Achieve-Control-Objective-for-Successful-IT-Governance-2nd-Edition.aspx. Accessed 7 Jan 2019
41. M. Grall, *EBIOS: The Risk Management Toolbox* (Club EBIOS, Viroflay, France, 2018), pp. 1–27, https://club-ebios.org/site/wp-content/uploads/productions/EBIOS-GenericApproach-2018-09-05-Approved.pdf
42. Agence nationale de la sécurité des systèmes d'information (ANSSI), *Fiches méthodes* (2018), p. 43, https://www.ssi.gouv.fr/uploads/2018/10/fiches-methodes-ebios_projet.pdf
43. Agence nationale de la sécurité des systèmes d'information (ANSSI), *Prestataires de services d'informatique en nuage (SecNumCloud)—référentiel d'exigences* (2018), https://www.ssi.gouv.fr/uploads/2014/12/secnumcloud_referentiel_v3.1_anssi.pdf
44. Agence nationale de la sécurité des systèmes d'information (ANSSI), *Etude De Cas: Securite D'un Service Du Cloud* (2011), https://julienlhonore.files.wordpress.com/2013/02/logiciel-ebios-etudedecassc3a9curitc3a9servicecloud-2011-07-e280a6.pdf
45. Mehari — ENISA, https://www.enisa.europa.eu/topics/threat-risk-management/risk-management/current-risk/risk-management-inventory/rm-ra-methods/m_mehari.html
46. D.F.C. Velasco, J.E.F. Quinayás, S.A. Donado, Adaptación De La Metodología Mehari A La Fase De Planeación De Un Sgsi Para Un Procedimiento De Estudio Propuesto/Adaptation of the Mehari methodology to the planning phase of an ISMS for a proposed study procedure. Rev. Teckne **14**(1) (2017)
47. Mehari 2007—Security Stakes Analysis and Classification Guide, Club de la Sécurité de l'Information Français (CLUSIF) (2007)
48. M. Masky, S.S. Young, T.Y. Choe, A novel risk identification framework for cloud computing security, in *2015 2nd International Conference on Information Science and Security (ICISS)* (IEEE, 2015), pp. 1–4
49. G. Wangen, C. Hallstensen, E. Snekkenes, A framework for estimating information security risk assessment method completeness. Int. J. Inf. Secur. 1–19 (2016)
50. Federal Office for Information Security, *Secure Use of Cloud Services*. Bonn, Germany, pp. 1–23, https://www.bsi.bund.de/SharedDocs/Downloads/EN/BSI/CloudComputing/SecureUseOfCloudServices/SecureUseOfCloudServices.pdf?__blob=publicationFile&v=6
51. Federal Office for Information Security, *IT-Grundschutz Catalogues*. Bonn, Germany (2016), pp. 132–136, https://www.bsi.bund.de/SharedDocs/Downloads/DE/BSI/Grundschutz/International/GSK_15_EL_EN_Draft.pdf?__blob=publicationFile&v=2
52. K.V.D. Kiran, L.S.S. Reddy, N.L. Haritha, A comparative analysis on risk assessment information security models. Int. J. Comput. Appl. **82**(9) (2013)
53. EAR—Tools—versions, https://www.pilar-tools.com/download/stable_en.html
54. PILAR—Manual de Usuario (6.2) (2016), https://www.pilar-tools.com/doc/v62/manual_std_risk_es_2016-08-21.pdf
55. MAGERIT v. 3: Metodología de Análisis y Gestión de Riesgos de los Sistemas de Información (2012)

56. RiskSafe Assessment—ENISA, https://www.enisa.europa.eu/topics/threat-risk-management/risk-management/current-risk/risk-management-inventory/rm-ra-methods/m_risksafe-assessment
57. A.U. Khan, M. Oriol, M. Kiran, M. Jiang, K. Djemame, Security risks and their management in cloud computing, in *2012 IEEE 4th International Conference on Cloud Computing Technology and Science (CloudCom)* (IEEE, 2012), pp. 121–128
58. Information Risk Analysis Methodology, IRAM, https://www.securityforum.org/iram#iramtva
59. ISACA, Information Systems Audit, & Control Association, *IT Control Objectives for Cloud Computing: Controls and Assurance in the Cloud*. ISACA (2011)
60. G. Stergiopoulos, P. Kotzanikolaou, M. Theocharidou, D. Gritzalis, CIDA: Critical Infrastructure Dependency Analysis Tool, Information Security and Critical Infrastructure Protection Laboratory, Department of Informatics, Athens University of Economics and Business, Athens, Greece (2014), http://github.com/geostergiop/CIDA
61. S. Drissi, H. Medromi, A new risk assessment approach for cloud consumer. J. Commun. Comput. **11**, 52–58 (2014)

Chapter 7
Challenges and Issues in Risk Assessment in Modern Maritime Systems

Spyridon Papastergiou, Eleni-Maria Kalogeraki, Nineta Polemi, and Christos Douligeris

Abstract In this paper, we present challenges and risks concerning cyber security in a supply chain environment. In particular, we focus on the MITIGATE (Multidimensional, IntegraTed, rIsk assessment framework and dynamic, collaborative risk manaGement tools for critical information infrAstrucTrurEs) Supply Chain Risk Assessment methodology, which is in compliance with ISO28001 and can be applied in order to assess the security risks of all the organizations involved in a supply chain. To validate the MITIGATE approach, we provide use cases based on real-life maritime scenarios and real-world data collection. To this end, a number of best practices in the form of guidelines for a successful application of the MITIGATE risk management system in supply chain environments are presented. The main advantages of the Mitigate Risk Assessment approach over existing maritime initiatives and efforts are also highlighted.

7.1 Introduction

7.1.1 Scope and Objectives

This paper describes the adoption of the MITIGATE (Multidimensional, IntegraTed, rIsk assessment framework and dynamic, collaborative risk manaGement tools for critical information infrAstrucTrurEs) approach as a best practice for identifying, classifying, assessing and mitigating risks affecting the Information Technology (IT) infrastructures of all business partners and organizations involved in dynamic Supply Chains. This is achieved by providing a framework capable of supporting information infrastructures while addressing their respective risk assessment and risk management challenges. In particular, the MITIGATE methodology, which is implemented through a set of distinct steps, integrates a number of activities that range from the asset identification and impact and threat analysis to the identification of the

S. Papastergiou · E.-M. Kalogeraki · N. Polemi · C. Douligeris (✉)
Department of Informatics, University of Piraeus, Piraeus, Greece
e-mail: cdoulig@unipi.gr

G. A. Tsihrintzis and M. Virvou (eds.), *Advances in Core Computer Science-Based Technologies*, Learning and Analytics in Intelligent Systems 14,
https://doi.org/10.1007/978-3-030-41196-1_7

existing controls and the disclosure, evaluation and treatment of the inherent and interdependent risks.

The MITIGATE methodology is implemented as a unified integrated system (MITIGATE system), which aims to provide an innovative, scalable Risk Assessment Toolkit that facilitates the Supply Chains' (SCs) operators to efficiently identify, assess and treat their security issues in a collaborative manner. The aforementioned tool has a plethora of potential possibilities which could have a positive impact in: the economy of the maritime and transport sector (as well as other sectors), their business value, the competitiveness and the reputation of the SC operators' facilities, the image of SC operators and their market differentiation, the offering of additional services/products and the reduction of security breaches (and their costs).

Despite the fact that the maritime sector has been selected as a suitable test bed for the Mitigate platform, there are no major differences between the various domains (e.g. transportation, airport, energy, telco and maritime domains) since all of them are facing similar challenges concerning cyber security.

Thus, the generic nature of the MITIGATE approach makes it applicable to various transport infrastructures (such as airports and railway infrastructures) and different types of port infrastructures (including infrastructures of different sizes and different business activities). In particular, the MITIGATE system can be used by various organizations coming from different sectors to evaluate the risks of their IT infrastructures.

In this context, this paper has a two-fold objective. Firstly, to illustrate the security needs and challenges of various transport and different types of port infrastructures and secondly to provide guidelines for the successful deployment of the MITIGATE system.

7.1.2 Structure of the Paper

The remainder of this paper is organized as follows:

Section 7.2 illustrates the common challenges and issues concerning cyber security facing all dynamic supply chains involving various transport infrastructures (such as airports and railway infrastructures) and different types of port infrastructures (including infrastructures of different sizes and different business activities). Section 7.3 documents a number of best practices derived from the MITIGATE system applicability that can be used by various SC operators to construct their incorporated risk management models, including various types of transport Critical Iinfrastructures (CIs) (such as airports and railway infrastructures), port infrastructures of different sizes serving miscellaneous business operations. Section 7.4 documents how the MITIGATE best practices can address SCS security challenges by illustrating and analyzing the mapping of the MITIGATE best practices with specific cyber aspects of SC security challenges. Section 7.5 draws conclusions and recommendations.

7.2 Supply Chains (SCs)

7.2.1 Introduction

According to [24], a supply chain (SC) can be defined as a globally distributed, interconnected set of entities (i.e., organizations, individuals or/and CIs), processes and services that relies upon an interconnected web of Information and Communication Technologies (ICT) infrastructures and cyber networks to leverage the flows of products, services and information from a source to a customer. Supply chain management is defined as the systemic, strategic coordination of the traditional business functions and the tactics across these business functions within a particular company and across businesses within the supply chain, for the purposes of improving the long-term performance of the individual companies and the supply chain as a whole [3, 31].

In the modern era, Global Supply Chains (GSCs) are a way of life for modern business; in particular they are the blood veins of global trade and economy where individual operators and organizations coming from different critical sectors (e.g. health, finance, energy and transportation) collaborate in exploring new labour markets and offering complex SC services towards building an innovative, vibrant and secure Digital Single Market. Over the last years, these SCs have evolved becoming more complex and integrated. The organizations/businesses that operate within the SCs have become smarter and are not only heavily dependent on Information and Communication Technologies ICT (e.g. IoT, cloud technologies, telecommunications, ICS), but are also interconnected in order to share a large amount of data.

Information sharing is associated with all the supply chain functions like transportation, distribution, logistics, warehousing, inventory management, sourcing, procurement, and order and production planning, required to deliver goods and services from the point of origin to the point of consumption. This is very prominent in the case of transportation, energy and health. Thus, today's SCs can be viewed as complex cyber systems composed by heterogeneous, interconnected ICT assets, owned by different national and E.U. organizations (e.g. industries, airports, telecommunications service providers, railways, energy providers, banks, transport companies), ensuring seamless and swift product/data exchange from the producer down to the end consumer during the provision of the SC services. Physical logistics become more dependent on information technologies, and these technologies can also become enablers of further cooperative arrangements. Firms are then faced with the management of an extended enterprise as a network of processes, relationships and technologies creating an inter-dependence and shared destiny. The truly strategic nature of supply chain management thus becomes apparent for participating companies, with successful implementation becoming a source of competitive advantage [31].

The established interconnections reflect the relationships that exist between the involved entities representing how one process, activity or resource relies upon another. For example, an entity could be dependent on receiving a process or information from another entity or organization as an input to one of its critical business processes. The increased usage of information technology in modern SCs means that they are becoming more vulnerable to the activities of hackers and other perpetrators of cyber-related crime. Several recent studies have shown that the cyber threats landscape is changing continuously and the nature of attacks of this sort are evolving and are becoming even more targeted, sophisticated, ingenious and coordinated, resulting in so-called Advanced Persistent Threats [4, 35, 39]; thus, the cyber criminals will continue to unexpectedly discover new ways to break into ICT processes and operations of the SCs.

7.2.2 The Concept of Security in Supply Chains

SC systems are increasingly prone to complexity and uncertainty. While globalization, extended supply chains, and supplier consolidation offer many benefits in efficiency and effectiveness, they can also make supply chains more vulnerable and can increase the risks of supply-chain disruption. In particular, the challenging new threat landscape has turned supply chains to an attractive target for both physical attacks (e.g., sabotage, cargo theft/altering, terrorism) and cyber attacks (e.g., disruption of SC management, loss of SC integrity, loss of privacy) [23]. Even worse, the management of a SC is becoming a very complex process due to the high volume of data that need to be managed and due to the high number of participants in a supply chain, that many times may belong to different sectors (such as industry, transportation, IT, government etc.). The mitigation of these risks in a SC network requires novel SC risk management methodologies that take into consideration innovative and disruptive technologies (e.g. [19, 30]).

The topic of security for supply chains and logistics functions has risen in importance both in practice as well as in research and has emerged as its own area of research within SC management and logistics [9]. SC security has been recognized as an important part of managing business risks. In fact, the increased global free flow of people, goods, and terrorist threats created significant challenges for businesses, countries, and the global economy. Industries must implement continuous improvement processes that enhance both supply chain execution and security. SC security management may be defined [6, 36] as "the application of policies, procedures, and technologies to protect supply chain assets (products, facilities, equipment, information, and personnel) from theft, damage, or terrorism, and to prevent the introduction of unauthorized contraband, people, or weapons of mass destruction into the supply chain". Enhancing supply chain security can reduce theft and losses, reduce the number of delayed shipments, improve planning, increase customer loyalty and employee commitment, reduce the number of safety incidents, lower inspection costs of suppliers and increase cooperation with them, reduce crime and vandalism, and improve security and communication between supply chain partners. Furthermore, security management facilitates international trade by reducing transit time [42].

According to [5], the ability of organizational and cultural tools to increase supply chain security has not been fully exploited yet. Tools to mitigate the negative effects on security due to the inadequacy of partners are not popular or they are not considered as powerful enough, despite the fact that this has been highlighted as the most relevant causal factor of lack of security in SC. In addition, organizational responses to several SC incidents vary; some companies are proactive while other companies are reactive. Many companies find it difficult to invest in transport logistics security because justifying investments on a pure financial basis may be untenable [22]. Therefore, a security culture is essential for the implementation of security initiatives and for better supply chain security operational performance [42]. Three major objectives exist for investing in supply chain security: heightening security of supply chains, improving efficiency of business processes, and improving response and resilience to security incidents [10]. The actions implemented by companies to protect their business continuity from threats like theft, smuggling and terrorism may include: (1) identification and assessment of threats along the SC; (2) identification of weak points; (3) identification, development and provision of suitable target processes that increase security without negatively affecting efficiency; (4) evaluation of expected impacts of the identified target processes on SC security and efficiency; (5) implementation and monitoring of the performance of the identified target processes [28].

7.2.3 Cyber-Security Threats in Modern Supply Chains

As discussed above, SCs are vulnerable to various cyber and physical attacks. The complexity and the cross-sectoral nature of supply chains make them vulnerable to various threats that cannot be addressed by traditional risk assessment methodologies [1, 2]. As an example, consider a typical supply chain service: transfer of goods from production to the consumers. This typical SC may involve entities from the industry, road transportations, port infrastructures, vessels, customs, logistics etc. The security needs and constraints of all those players and all those sectors must be addressed. In addition, various cyber security threats must be considered, since huge amounts of data is generated and managed. Although typical security controls such as data encryption and integrity, authentication and access control mechanisms can be applied [38], SC are also subject to cascading attacks, i.e. attacks that may originate from one node of the SC, that may also affect many other nodes (entities or systems) within the SC [7, 20, 21]. The "portfolio" of the attacks for potential adversaries is rich and various attacks at various different layers may trigger cascading effects: Replay Attacks, Brute Force Attacks, Keys Exposure, Single Point of Failure, Time Synchronization etc. are just a few among the existing security threats. The identification of the proper security controls that may successfully mitigate such cascading threats [37] may also require time-based risk assessment, since in many scenarios the time that an attack is realized, as well as the time frame that the proper mitigation controls must be activated, is of high importance for the successful mitigation of the threats [38].

In addition, the evolution of the ICT-empowered SCs has brought with it a certain degree of risk. The use of ICT systems offers a high degree of flexibility, scalability, and efficiency in the communication and coordination across the SCs but also increases the danger to businesses as well as to the SCs as a whole from cyberattacks [8]. The integrated nature of the SCs introduces new potential entry points for cybersecurity risks in terms of, for example, obsolete security infrastructure or outdated hardware/software, infected devices on a corporate network, and lack of appropriate security protocols across partners, that can be exploited by sophisticated attackers using any mean such as advanced persistent threats (APTs) to gain access to sensitive data or to interfere with the information flows [4, 17].

The research indicates not only that the threat of cyberattacks is rising but also how insidious they can be. In particular, the SCs' highly valuable interconnected cyber assets have become lately targets for cyberattacks attracting the attention of security researchers, cyber-criminals, hacktivists (e.g. Anonymous, LulzSec) and other such role-players (e.g. cyber-spies). These cyber actors have been creating attack toolkits that give them an asymmetric quantum leap in capability. In the past years, there have been a number of cybersecurity meltdowns and high-profile breaches with supply chain involvement. The ransomware attack, WannaCry or WanaCrypt0r 2.0, that took place a couple of years ago (mid-May 2017), affected more than 230,000 computers in over 150 countries, with the NHS, Spanish phone company Telefónica and German state railways among those hardest hit. Another wave of ransomware attacks (called Petya; NotPetya; Nyetya; Goldeneye) were also reported the first half of 2017, infecting networks in multiple countries, like the US pharmaceutical company Merck, Danish shipping company Maersk, and Russian oil giant Rosnoft. In most cases, in order for the adversaries to achieve their goal, they targeted the organizations' supply chains as a means of targeting the broadest possible victims for their malware [18, 33].

Obviously, the impact of a compromised IT infrastructure can extend far beyond the corporate boundaries, putting not just individual SCs' operators but also entire supply chains at risk. Cyber-attacks on the SCs cause not only a disruption of the SC services but tremendous damage to the SCs operations, national and EU safety, economies, societies and environment. For example, attacks in the Industrial Control Systems (e.g. supervisory control, SCADA, distributed control systems and programmable logic controllers) hosted in an energy provider or a telecommunications service provider may cause damage of critical mechanical devices to other organizations (e.g. hospitals, banks) and even worse they may cause loss of life. For this reason, SCs' operators need to understand how cybersecurity incidents at the other partners could affect them, and how these can be handled in a collaborative and privacy aware manner [25, 26].

In 2013 the House of Representatives in the USA released the National Cybersecurity and Critical Infrastructure Protection Act establishing the framework for the provision of shared situational awareness among federal entities to enable real-time, integrated, and operational actions to protect from, prevent, mitigate, respond to, and recover from cyber incidents. In 2016, the Commission introduced the E.U. Directive NIS 2016 that enforces all CIs to report to an appropriate Computer Security Incident Response Team (CSIRT) any incident having a substantial impact on the provision of their services. Unfortunately, this Directive provides only the legal basis and creates the assurance framework for boosting the cyber security culture across sectors which are vital for the EU economy and society and moreover rely heavily on ICTs (e.g. energy, transport, water, banking, healthcare, digital, financial market infrastructures).

Nowadays, the attackers have a range of capabilities and resources and use various advanced techniques and tools (e.g. social engineering techniques and zero-day exploits programs) to initiate advanced targeted attacks. These threats employ multiple technologies and malware, deployed in multiple stages, to bypass traditional security mechanisms in order to penetrate an organization's defense. The attack vectors vary significantly including Application-Layer, Social Engineering Unauthorized Access, Malicious Code, and Reconnaissance and Networking-based service attacks that target applications, host and client operating systems, and even networking equipment. In this vein, the attackers use these techniques to get valuable data assets, such as financial transaction information, user credentials, insider information etc. In addition, in many cases, the attackers take advantage of the interdependencies in the SCs in order to intrude to other organizations. Novel multi-stage attacks can be used to exploit vulnerabilities of the interconnected ICT systems to cross the organization's boundaries enabling the attackers to move within the supply chain across multiple businesses and functions. In this context, an attacker can use a SC entity as the stepping stone to attack and compromise the security of a dependent organization [40, 41].

Thus, there is an urgent, pressing challenge for the SCs security officers and operators to protect their interconnected cyber systems and infrastructures in the new digital era. In particular, they need to cooperate closely exchanging security-relevant information in order to timely reveal such SCs-related attacks. Information sharing is an important step to acquiring a thorough and common understanding of cyberattack situations, and is necessary to warn others about attacks occurring into their infrastructures. Consequently, an analysis of shared incident information is crucial in attempting to detect the presence of a threat, within an organization's infrastructure that has already been detected in other interdependent entities of the SCs [34].

7.2.4 Challenges and Issues in Modern Supply Chains

This Section illustrates and summarizes the common challenges and issues concerning cyber security, being faced by all dynamic supply chains, involving various transport infrastructures (such as airports and railway infrastructures) and different types of port infrastructures (including infrastructures of different sizes and different business activities).

7.2.4.1 High Interdependencies Within the SCS

Modern **SC operators** are highly dependent on the operation of complex, dynamic ICT systems of the ICT-based supply chains. They interact with various organizations coming from different sectors (including ministries, transportation companies, industries, agencies, insurance companies and other Critical Information Infrastructures (CIIs), e.g. railways, airports, energy networks and telecommunication. They are the nodes in the supply chain networks, exchanging a huge amount of data to deal with various operations (e.g. logistics procedures). The growing concerns about the security aspects of such data shifts in the way the SC operators assess cyber risks and vulnerabilities on their assets, according to the cyber risk management methodologies they use. The plethora of complex interdependencies within the SCS reveals threats with catastrophic impacts; beyond the provision of the SCS and the business partners involved, the whole SC ecosystem, the economic environment and, most importantly, the safety of people and involved nations may be affected. Good scenarios to take into account for deepening into the study of cascading effects in the supply chains' security incidents would be the study of whole supply chain services (SCS), e.g. the transport of containers, cruise passengers and dangerous goods. A starting point for modelling the cascading effects of supply chain threats in every supply chain service lies on the understanding of the various business partners interdependencies involved, including physical, cyber, geographic and others (logical dependencies). Moreover, techniques for **modelling the SCS assets and their interdependencies and understanding risks in the scope of Systems-of-Systems (SoS) are needed**.

The degree of interactions and interdependencies among transnational business partners increases as Supply Chain Services become global. **Modelling as well as visualised procedures are required in order to capture the interdependencies of all transnational business partners involved in the SCS**, which is the first step in any risk assessment process [15]. The problem becomes more complex when it comes to capture the interdependencies of all physical and cyber assets participating for the provision of the SCS. A cartography of the assets' dependencies is required, in order to recognize the specific threats of the SCS assets, identify the propagation of their vulnerabilities to their interconnected assets and estimate the risk levels for each SCS asset to various physical and/or cyber threats. **Advanced modelling standardised procedures are needed in order to dynamically generate the SCS assets interdependencies**.

7.2.4.2 Low Awareness of SCS Security

A literature review revealed a general lack of knowledge of the full meaning and implications of the concepts "supply chain risks assessment", "cascading effects" and the general risks (both physical and cyber) within their SCS.

SC operators do not adopt "Good ICT supply chain security", i.e. they do not cope effectively with interdependent external threats (e.g. masquerading identities, network traffic monitoring, theft/modification of personal data) causing tremendous damage and putting at risk (e.g. loss of reputation, loss of legal compliance, disruption of business operation) not only themselves but the whole value chain. Thus, they need to improve their security culture by realizing the value of information security as a means to support their supply chain services, their business models, and their current and future business plans.

SCs rely upon various Information and Communication Technology (ICT) systems to offer a plethora of electronic services, exchanging and storing electronic data, following complex processes and interacting with many entities. The criticality of these services and the technologies used (e.g. RFID) need to be evaluated and appropriate security measures need to be implemented, in order to avoid catastrophic consequences (e.g. commercial espionage, terrorist cyber attacks) in the whole European business eco-system. Cross border security aware services are not offered by the SC operators (helping them increase their trust in the antagonistic, new era, digital maritime markets and enhance their business opportunities) since certification and interoperability issues have not yet been resolved.

In general, SC operators consider the impact of physical and cyber risks along with their cascading effects to their organizations. Each SC operator has its own security/safety plan which usually is coordinated with other plans at upper or lower levels, but their analyses usually stop in the boundaries of the organization since their competences are not beyond these limits. Usually, they do not concentrate on their cyber risks and the risks risen from their interdependencies with other entities in the provision of their SCS.

The 2011 White Paper on Transport and the Single Market Act II emphasize the need for a well-connected SC infrastructure and reliable services. Security is the basic requirement for the adequate SC infrastructure, good performance of the providing services, improving SC operators' growth potential and create a sustainable and inclusive EU transport system which will be able to address the challenges of the global competitive, digital market.

7.2.4.3 Lack of Targeted Standards and Methodologies

Considering the aforementioned, it is obvious that there is a lack of specific tools or methodologies implemented for the specific analysis or assessment of supply chain risks and their cascading effects. European initiatives with this respect are mainly aimed at studying concrete aspects of goods/passenger transport within the supply chain, but an initiative which addresses the issue from a holistic point of view is not yet identified. It seems that only Large Government Administrations and important big companies are the interested stakeholders in the whole Supply Chain risk assessment and the cascading effects, especially in their economical and social dimensions.

For over a decade, significant efforts have been allocated in the introduction of risk management and assurance methodologies for Critical Information Infrastructures (CIIs) concentrating either on the cyber or on the physical threats ignoring the dual nature of the ports CII. At the same time, they tend to ignore the complex nature of the ICT systems, assets (e.g. SCADA) and e-services (e.g. cargo management) used in the various sectors along with their interrelationships. This is for example the case with several international standards and legislation (e.g., International Ships and Port Facilities Security Code (ISPS), the International Safety Management Code and EC Regulation No 725/2004 on enhancing ship and port facility security, the EC Directive 2005/65 on enhancing port security), as well as with related risk assessment methodologies e.g. MSRAM (Maritime Security Risk Analysis Model) and MARISA (MAritime RISk Assessment). The existing risk management methodologies do not adequately take into account the cyber nature of the ports and the security requirements of the business processes associated with supply chains, which are nowadays ICT enabled and therefore severely dependent on intentional and unintentional compromise of CIIs.

ISPS, the European Regulation (Marucheck, Greis et al.) 753/2004 on enhancing ship and port facility security and the Directive 2005/65/EC on enhancing port security do not adequately address the security of the ports' supply chain since they do not consider cyber security but they concentrate on safety (physical security).

In particular, ISPS (as well as the respective EU regulation EC725/2004) defines a set of measures to enhance the physical security of port facilities and ships. Methodologies for security assessment are described and a guideline for the implementation of the respective security measures is given. Additionally, roles and responsibilities concerning maritime security at various levels are defined. Nevertheless, due to the increased interaction and exchange of information of ports with other CIs in the maritime eco-system (e.g. port authorities, ministries, maritime companies, ship industry, etc.) the sole focus on physical security is not sufficient and the security of the port's ICT systems becomes equally important.

ISO 28001 [12] is the security management system, which has been specifically developed for supply chains, due to an increased demand from the transportation and logistics industry for a common SC security management standard. The standard specifies the requirements for a security management system, including those aspects critical to the security assurance of the SC. These aspects include, but are not limited to, financing, manufacturing, information management and the facilities for packing, storing and transferring goods between modes of transport and locations. **ISO 28001 is applicable to all sizes of organizations in manufacturing, service, storage or transportation at any stage of the SC; however, it does not consider the specificities of maritime SCS**. The standard defines the need for a SC security management system; assures conformance with the stated security management policy and demonstrates conformance; seeks certification by an Accredited third party Certification Body, or self-declaration of conformance with the standard.

ISO 28001 describes a generic SC risk assessment methodology. It is intended to assist organizations in establishing adequate level of security for the whole supply chain and to make better risk-based decisions. The methodology includes a process of identifying threats and consequences to the organization in the event that a threat is realized, the likelihood of such an occurrence and identification of countermeasures sufficient to reduce risk to levels acceptable to executive management. It should be noted that **ISO 28001 provides a generic guideline as an informative reference for organizations seeking to implement or refine a specific methodology. It does not directly addresses the ports' SCS needs and specificities**.

7.2.4.4 Economic Crisis: A Stumbling Block to SC Security

SC's activities were severely affected by the most recent economic crisis. Personnel is reducing and no budget is foreseen, in order to increase the security team of the SC operators and authorities. In particular, they are not willing to finance the security enhancements of their companies.

At the same time, SC operators remain at the core of the regional and E.U. economic recovery. Cross-border, efficient SCS services will accelerate their presence in the promising digital market, assuming that they have harmonised their digital security practices ensuring information sharing and knowledge creation. Thus, the trustworthy, cross border provisioning of SCS services needs to be guaranteed in order to promote the local, national, regional and E.U. economies and expand their commercial activities in the digital era. E.U. economies even more rely upon their maritime business activities. A prerequisite to build trustworthy SCS services is the operators' compliance with cyber security standards.

Furthermore, technologies used in the utilisation of SCS services need to be protected, e.g. Radio Frequency Identification (RFID) technologies, commonly used in logistics, container identification and cargo management face various threats.

7.2.4.5 The Policy Dimension in Enhancing Maritime Supply Chain Security

It was realised by the MITIGATE survey that the ports and port authorities do not question the security of their Ports' Single Window although they are not aware of their security mechanisms. They consider it as a secure black box. The further coordinated implementation of the Directive 2010/65/EU23 by establishing guidelines on the "national single windows" need to include security risk assessment in order to avoid security incidents with great impact at national and E.U. levels.

The IMO and the Commission need to become the drivers for improving maritime security awareness. In particular, they need to provide security guidelines for the implementation of various existing initiatives (e.g. the National Single Window) and for the upcoming ones: e.g. the "Blue Belt" initiative which aims at reducing the administrative burden for EU goods carried between EU ports, by digitalisation which will simplify customs procedures. This initiative is also a key action under the Single Market Act II22 and needs to be embraced by security guidelines as well as the "e-maritime" initiative, which promotes the use of electronic information for the reduction of administrative burden and the "e-Freight" initiative, which aims to facilitate the exchange of information along multimodal logistics chains. These initiatives which are highly related to the SCS provisioning will contribute to improve port efficiency and ports multimodality only if security is guaranteed.

The Commission (in line with Articles 151 and 154 of the TFEU) is willing to facilitate the Social Dialogue at Union level by providing technical and administrative support. The Commission expects that the EU social partners will have the capacity to tackle issues related to work organisation and working conditions. These efforts need to be upgraded and include physical and cyber security issues and provide support for mitigating risks and handling security incidents, securing cargo handling and passenger services.

The global port industry increasingly dependent on technological innovations across the entire logistics chain and their security will enhance the competitiveness of European ports. The ports critical roles as multi-modal hubs require secure cross-modal connections and use of security management tools in order to further increase their attractiveness and competitiveness in the global maritime market.

The European Cybersecurity Strategy needs to embrace the ports Critical Information Infrastructures (CIIs) in order to establish an open, safe and secure cyberspace, highly contributing to coordinated prevention, detection and mitigation of risks enabling mutual assistance amongst the national competent maritime authorities. Synergies among the main actors need to be build in order to implement the European Cybersecurity Strategy in all CIIs.

The "integrated Maritime Policy for the EU", COM (2007) 575, and the "Strategic goals and recommendations for the EU's maritime transport policy until 2018", COM (2009) they underline that ports play a key role in maritime transport and security, they will increase the efficiency of ports as nodes for maritime and intermodal transport, through the use of advanced ICT systems. However, emphasis need to be provided that these advanced ports' ICT systems need to be accompanied with appropriate security measures and security policies.

The "Regulation (Marucheck, Greis et al.) No 725/2004 and Directive 2005/65/EC on enhancing ship and port facility security" implied the adoption of the ISPS Code by every member state. However, these Directives need to be updated and enforce ships and port facilities to adopt not only the ISPS (which covers the physical security-safety-) but also cyber security standards (e.g. ISO27001 [14], ISO27005 [13]) in order to ensure the security of their digital systems.

The new European transport policies and recommendations (TEN-T Guidelines, 2011 White Paper on Transport) focus on the freight transport corridors and their effective security management including all regional stakeholders, by incentivizing the role of ports and logistic platforms towards the deployment of more efficient, safety and environmental solutions. These policies need to provide guidelines to the ports and to the logistic platforms on how they can enhance their security.

The SAFE Framework of Standards to Secure and Facilitate Global Trade needs to become more concrete in the provision of risk assessment and mitigation planning.

The environmental objectives of the Green Paper for an EU Maritime Policy to treat the oceans and seas in a holistic way in order to achieve a Sustainable Development through an integrated, intersectoral and multidisciplinary approach need to consider the environmental impact that security threats and incidents may have and enhance their objectives by enforcing methods to the ports and ships to assess and mitigate these threats. Supply Chain Services (SCS), as the ones considered in the MITIGATE project [15, 16], validated the existence of such threats and provided a risk assessment methodology and tool that the SCS providers may adopt.

The E.C. European Ports Policy COM (2007) 616 on the advancement of modernization measures of the European Ports in the future need to emphasise that such measures need to include physical and cyber security measures for the trustworthy advancement of the ports.

In 2011, the Commission published the White Paper "Roadmap to a single European Transport Area—White paper on competitive and sustainable transport", COM (2011) 144 without emphasising the role of security in this singe European Transport Area.

The Green Paper Towards a New Culture of Urban Mobility, the Action Plan on Urban Mobility, the White Paper on Transport and the CIVITAS initiatives, aim to minimise the consequences of pollution in the urban environment, and to increase living standards, economic competitiveness, and social cohesion. Such **initiatives need to encourage the security culture and awareness to the whole maritime sector**. Transnational cooperation is needed to enhance the security of the maritime sector.

7.3 MITIGATE Best Practices

The MITIGATE system is an innovative, scalable Risk Assessment solution that empowers Officers, IT personnel, technical operators and ICT experts of the Supply Chains to collaborate for the identification, assessment and mitigation of risks associated with cybersecurity assets and SC processes. In order to achieve its objectives, the Mitigate adopts and implements a bouquet of flexible and configurable self-driven functions and procedures [15, 16, 27] which constitute the conceptual pillars for building a solution that assists SC operators to improve their current cyber level of their SC Services.

This section documents a number of best practices [16] that can be successfully applied in the MITIGATE system and used by SC operators as guidelines to construct their incorporated risk management models, including heterogeneous CIs (such as airports, railway infrastructures) operating in a variety of business processes and different types of port infrastructures (including infrastructures of different sizes and different business activities). These best practices can be utilized to arm them with the tools and technologies required to efficiently identify, assess and treat their security problems and issues. The proposed guidelines will cover the following areas Assets Identification and IT infrastructure Representation as well as Supply Chain Services

Analysis and Representation. Various other best practices have also been thoroughly analysed, such as Vulnerabilities Management (Open Intelligence), Threats and Controls Management (Open Intelligence), Threat Scenarios Specification, Supply Chain Risk Analysis, Attack paths Generation and Representation, Supply Chain Risk Management and Social Engineering and Open Intelligence. Due to space limitations these best practices are not thorougly presented in this paper.

7.3.1 Best Practice 1: Assets Identification and IT Infrastructure Representation

MITIGATE adopts an integrated intra and inter-organization asset management approach [15, 16, 27] that allows the creation of an IT asset inventory of all computing and networking related devices owned, managed, or otherwise used by the SC operators (Fig. 7.1). These devices include all computers (desktops, notebooks, servers), network devices (switches, routers, etc.), printers, appliances (network attached storage, network capable cameras, etc.), applications and IT systems in general (e.g. vessels, global navigation systems, ports' Industrial Control Systems (ICSs) and cargo

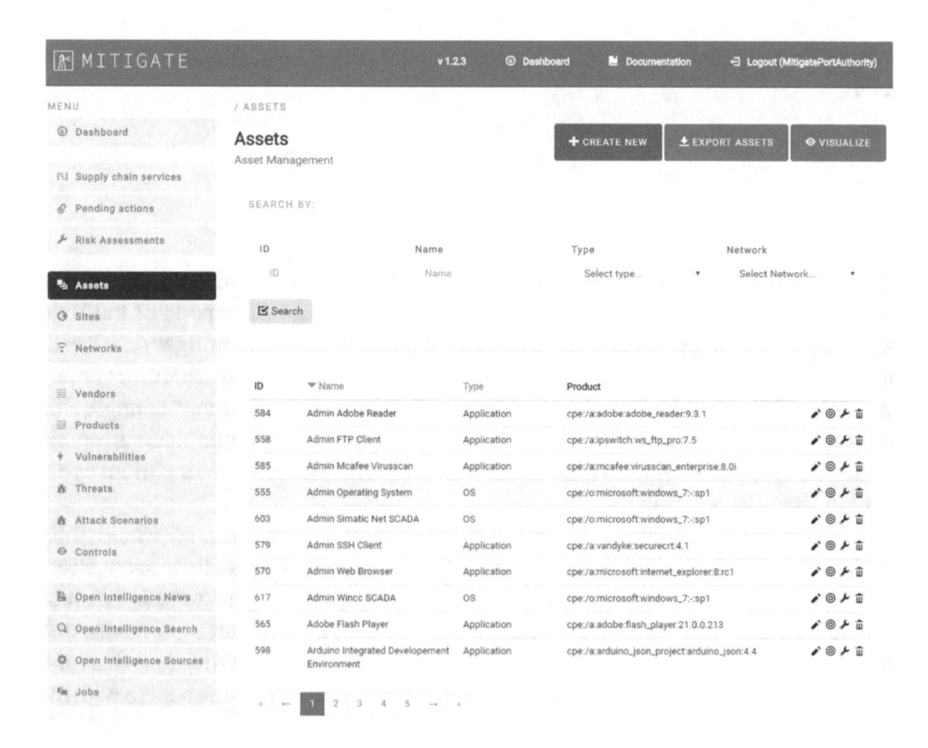

Fig. 7.1 Create a new asset

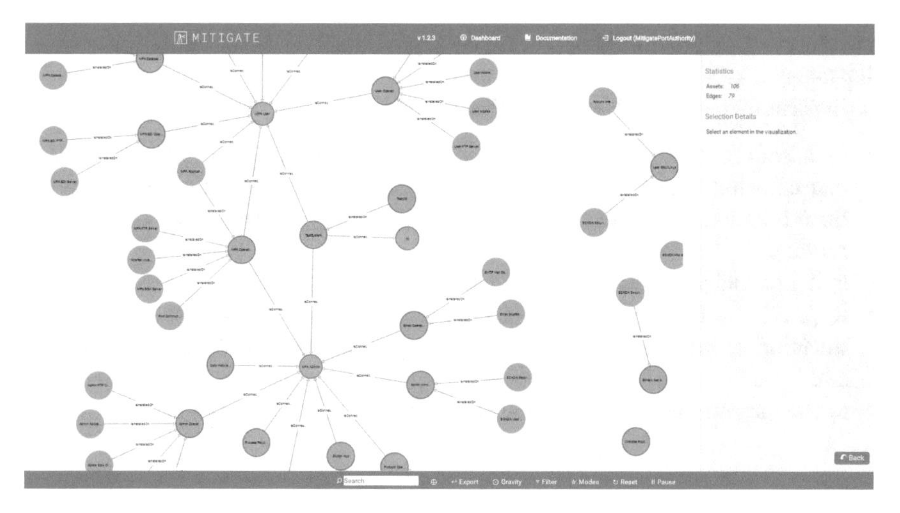

Fig. 7.2 Visualization of graph with all assets

management systems) that support the most critical operations and processes of the organizations.

A useful aspect of this approach is the specification of the main interrelations and interconnections that exist between the cyber-assets. This allows the SC operators to understand how these assets are used and interact with each other. Additionally, the MITIGATE system provides a visual representation of the entire infrastructure (Fig. 7.2) enabling the expansion of the existing knowledge on cyber-assets and the improvement of the data sharing of the spectrum [15].

7.3.2 Best Practice 2: Supply Chain Services Analysis and Representation

MITIGATE provides a collaborative, business-centric approach [15], which aims to facilitate knowledge sharing among interorganizational or extraorganizational business partnerships of the maritime and logistics industries. It applies business modelling practices to expand the organizational knowledge of the Supply Chain and creates new information to detect interruptions across the supply chain. The MITIGATE solution is considered collaborative because it promotes collaboration among the business partners of the Supply Chains and business-centric because it highlights their role over the supply chain's performance. MITIGATE arms the operators with a knowledge-based method for modelling various key components of the Supply Chain. The proposed knowledge-based method explores both process-based and asset-based views of knowledge within the Supply Chain. It focuses on both knowledge flows in the context of SCs and on knowledge content—its creation, storage and reuse and on providing support for the representation and retrieval of articulated, documented knowledge.

In particular, the MITIGATE's objectives [15] are:

- to introduce a technique for identifying and modelling the key components that constitute the SCs: the primary maritime stakeholders, the assets used by them functioning within the MLoSC, mapping ICT infrastructures with logistics services;
- to define and visualize prominent processes of the SCs;
- to provide a business analysis on the SC key-components enabling the identification of interactions among business partners, assets' cyber-dependencies; and
- to elicit implicit and explicit knowledge regarding the key-components of the SCs.

7.3.3 Best Practice 3: Vulnerabilities Management (Open Intelligence)

The organizations should be aware of the vulnerabilities that the assets comprising their IT infrastructure may have. The MITIGATE system makes use of open data sources where these vulnerabilities have been disclosed replicating all the vulnerabilities. In this way, the proposed system can act as a central repository for all custom and known vulnerabilities.

7.3.4 Best Practice 4: Threats and Controls Management (Open Intelligence)

The digital era puts the SC operators as well as all the organizations involved in the SCs under pressure to be aware of the threat landscape that their IT infrastructure is exposed to. Therefore, they should be armed with appropriate tools and solutions that will help them familiarize themselves with threats that may affect their organizations and the security controls that can be deployed or can be applied in order to mitigate the risks and deal with their defined threats and weaknesses.

In this context, the MITIGATE system can act as a comprehensive dictionary of known threats as well as the corresponding mitigation controls that can be used to advance SC operators understanding and enhance their defences.

7.3.5 *Best Practice 5: Threat Scenarios Specification*

Nowadays, all SC operators are facing an increasing trend where the threat landscape is evolving rapidly and becoming more complex with new aspects of threat emerging and new vulnerabilities discovered all the time. Since the nature of attacks of this sort are becoming even more targeted, sophisticate and ingenious; thus, the cyber criminals will continue to do the unexpected discovering new ways to break into ICT processes and operations.

Thus, the organizations should be able to identify and develop threat scenarios that includes all threats, vulnerabilities and assets information and illustrates a clear view of how each of these elements can be combined and used together. A threat scenario is a representation in which a threat can compromise an asset by exploiting vulnerabilities and weaknesses as well as taking advantage of the lack of adequate security controls.

MITIGATE provides the capability to declare statically a mapping between a Threat, a Vulnerability and an Asset (Fig. 7.3) contributing to increasing the intelligence of SC operators on cyber attacks.

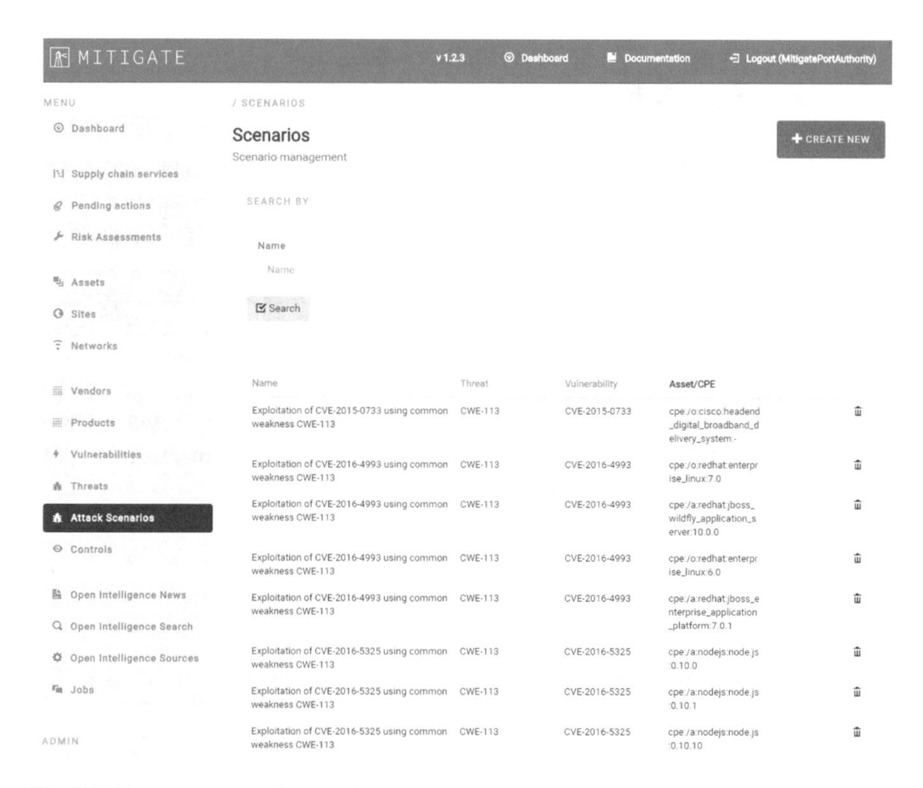

Fig. 7.3 Create a new attack scenario

7.3.6 Best Practice 6: Supply Chain Risk Analysis

The MITIGATE system aims to provide guidance to the supply chain operators on how to assess and organize the security issues associated with the SCSs in which they involved. In this context, the MITIGATE system encompasses and executes an evaluation process that implements the main steps of the proposed collaborative evidence-driven Maritime Supply Chain Risk Assessment.

The potential users of the proposed system are all business partners (port authorities, ministries, maritime companies, ship industry, customs agencies, maritime/insurance companies and other CIIs (e.g. airports)) involved in a Supply Chain Service. These organizations can use the system to:

- identify and measure all relevant cyber threats;
- evaluate the individual, cumulative and propagated vulnerabilities;
- assess the possible impacts; and
- derive and prioritize the corresponding risks.

7.3.7 Best Practice 7: Attack Paths Generation and Representation

The ICT-based supply chain arising from the ICT interconnections and interdependencies of a set of entities can be increasingly affected by multi-stage targeted cyberattacks (such as Stuxnet, Duqu, and Flame). These attacks aim at intruding multiple dependent organizations using them as stepping stones to reach the actual target. In general, various vulnerabilities, such as software vulnerabilities or inappropriate configuration settings, exist in information systems comprising the ICT-based maritime supply chain and can be exploited by attackers to gain access.

MITIGATE provides an attack path discovery method [29] that relies on unique characteristics, such as the attacker location, the attacker capability, assets interdependencies and which the entry and target points are in order to return all attack paths that exist in the examined supply chains (Fig. 7.4).

7.3.8 Best Practice 8: Supply Chain Risk Management

The MITIGATE system shows how an attacker can take advantage of the weaknesses and limitations that exist in the IT infrastructures involved in the SCs, conducting a sequence of attacks and exploiting multiple vulnerabilities in order to reach a specific target. These vulnerabilities trees produced expose the risks that are embedded in the IT systems as well as the risks that a combination of threat scenarios pose to the SC as a whole.

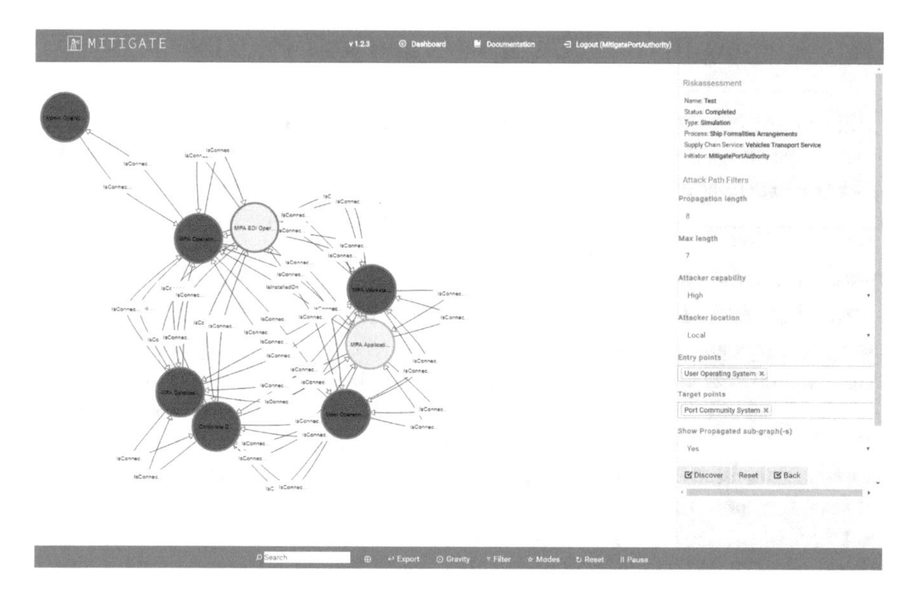

Fig. 7.4 The discovered attack graphs

In this context, the business partners need to receive guidelines on how to select the most appropriate security controls and optimize the actions they need to perform in order to minimize the expected damage.

In order to fulfil this goal, the MITIGATE aims to assure an acceptable risk level for all the business partners as well as the whole supply chain. In particular, the proposed system provide the necessary defensive capabilities and support [32] to determine which security controls must be implemented and which partners need to implement them in order to resolve the issues and the risks identified.

7.3.9 Best Practice 9: Social Engineering and Open Intelligence

World Wide Web is full of cyber-security related content. Social media like Twitter and Reddit, as well as security blogs, RSS feeds and general-purpose websites, contain invaluable information about disclosed vulnerabilities, cyber threats, exploitation methods and security controls. The Open Intelligence module collects information from various sources, analyzes their content, classifies their relevance to the cyber-security sector and stores the results for further browsing and analysis.

7.4 Mapping Between Challenges and Best Practices

In this section, we analyse the mapping between challenges identified in Sect. 7.2 and the best practices developed in Sect. 7.3 for successfully applying the MITIGATE system in various SC operators. The aim is to develop a better understanding of the best practices supported by MITIGATE to achieve effective security and risk management by overcoming specific maritime and SC related challenges. In the Table 7.1, we map the developed best practices with the identified challenges.

The strategy we followed for mapping is also mentioned in Table 7.1. Some interesting findings are presented regarding the results on this mapping. In Table 7.1, a map between challenges and best practices is shown: the best practices derived from the use of the MITIGATE system are presented vertically in the first blue column, while the SC key challenges addressed concerning cyber security are depicted horizontally in the blue sub rows. In order to specify the mapping, potential connections between the key challenges and best practices are identified in the white cells of the table. In particular, if a connection exists, this is marked with "+", otherwise it is marked with "o". It can be noticed that best practices are matched with almost all of the key challenges. Each best practice is strongly related ("+") with at least one challenge. In turn, each challenge is associated with at least one best practice. For example, the challenge SC1: High Interdependencies within the SCS can be addressed by two different best practices (Best Practice 1: Assets Identification and IT infrastructure Representation and Best Practice 2: Supply Chain Services Analysis and Representation).

The mapping of the SC security key challenges and the MITIGATE best practices proves that the system functionalities' of the MITIGATE approach are capable of addressing several aspects of the SC challenges and issues concerning cyber security facing all dynamic supply chains involving various interconnected transport infrastructures of various types (such as airports and railway infrastructures) and port infrastructures of different types and sizes (e.g. port logistics systems, vessel traffic systems, SCADA systems) to satisfy miscellaneous inter-organizational and cross-organizational processes, enabled by different sizes of enterprises (Small and medium-sized enterprises (SMEs) and Large Enterprises). According to the mapping of Table 7.1, we can conclude the following:

Challenge 1—High Interdependencies within the SCS: This challenge is associated with the first two Best Practices (BP1 and BP2) that help SC operators to identify, capture and represent the interdependencies of all transnational business partners involved within the SCS. In particular, The MITIGATE system addresses the cyber-security requirements of the business processes encompassing the SC performance through the modelling of the Supply Chain's services. This is achieved by adopting techniques, which analyses in-depth inter-organizational and cross-organizational key-concepts of SC critical services: key-management processes, business partners participating and IT Systems and assets. The aforementioned approach implements a combined technique of process-centric and asset-centric views. Moreover, the

Table 7.1 The mapping between the MITIGATE best practices and SC security key challenges

Best practice (BP)	SC security challenge (SCH)				
	SC1: high interdependencies within the SCS	SC2: low awareness of SCS security	SC3: lack of targeted standards and methodologies	SC4: economic crisis: a stumbling block to SC security	SC5: the policy dimension in enhancing maritime supply chain security
BP1: assets identification and IT infrastructure representation	+	o	+	o	o
BP2: supply chain services analysis and representation	+	o	+	o	o
BP3: vulnerabilities management (open intelligence)	o	+	+	o	o
BP4: threats and controls management (open intelligence)	o	+	+	o	o
BP5: threat scenarios specification	o	+	+	o	o
BP6: supply chain risk analysis	o	+	+	+	+
BP7: attack paths generation and representation	o	+	+	o	+
BP8: supply chain risk management	o	+	+	+	+
BP9: social engineering and open intelligence	o	+	o	o	+

process-centric views define the business processes and business partners' participation and collaboration for the provision of the supply chain service whereas the cyber-asset views, based on an asset-graph approach, identify the cyber-assets operating and their interrelation within the supply chain service.

Consequently, the MITIGATE system offers a cyber-asset inventory including all computing (e.g. desktops, notebooks, servers etc.) and networking-related devices (e.g. switches, routers, etc.) printers, appliances (e.g. network attached storage, network capable cameras, etc.), applications and IT systems in general owned, managed, or otherwise used by the SC operators. Such devices are vessel traffic monitoring systems, intermodal maritime-based logistics, SCADA components, such as Human Machine Interface (HMI), Master Terminal Unit (MTU), Programming Logic Controllers (PLCs), Supervisory stations, Remote Terminal Units (RTUs), sensor systems for controlling stevedoring equipment, such as gantry cranes, trailers and forklifts.

The modeling of supply chain services elicits information about the main cyberdependencies that exist among assets. This allows the SC operators to understand how these assets are used and interact with each other. Additionally, the MITIGATE system provides a visualization of the entire infrastructure, which expands the cyber-assets knowledge and improves the data sharing of the spectrum.

Challenge 2—Low awareness of SCS security: Nowadays, the SC operators have become increasingly dependent upon their IT infrastructures in supporting their SC services. However, it has been acknowledged that they do not adopt good IT security practices, i.e. they do not cope effectively with IT security (e.g. attacks, masquerading identities, network traffic monitoring) or privacy threats (e.g. theft/modification of personal data) causing tremendous damage and putting at risk (e.g. loss of reputation, loss of legal compliance, disruption of business operation) not only themselves but the whole value chain. Therefore, their IT security behaviour may become a problem as this may have a business impact on others. It should be noted that their non sufficient security management practices arise from, among other factors, their following characteristics:

- *Lack and ignorance of the security culture*: Risk-agnostic of their ICT security risks involved, as well as the resulting business risks (e.g. operational loss, breach of statutory obligations, customer loss, and damage to reputation) and the extended risk to their business as a whole, the business managers think of themselves as of no interest from a global perspective. There is a tendency to refute the benefits that may be achieved from implementing good IT security practices, and will only react once the business has suffered an attack (and sometimes only after repeated attacks).
- *Lack of attention* by their managers to address legislative or regulatory requirements either due to ignorance or to the cost required (e.g. expensive firewall, intrusion detection systems, database countermeasures) in order to become compliant (although a potential penalty for not doing so may be higher).

In this context, this challenge is associated with a number of Best Practices (BP3-9) that allows the SC operators to become security-aware, follow effective security and risk assessment practices and manage their IT security according to cyber-related information of up-to-dated security resources, that disrupt traditional security news feeds due to the MITIGATE "open intelligence" module. In particular, MITIGATE can help them develop security culture and consciousnesses within their enterprises and across the supply chain, contributing to:

- raise their IT security and privacy intelligence and culture;
- increase predictability and reduce uncertainty of SC business operations by lowering information security-related risks to definable and acceptable levels;
- continuously inform them on current and new threats and vulnerabilities as well as possible attack paths;
- provide information security awareness;
- guide them on selecting security countermeasures that fit to their needs;
- facilitate the communication and foster the collaboration both among the SC operators on issues of common concerns; and
- draw a more security-aware strategy that can be incorporated into the existing business operation and logic.

The provision of a more security-aware strategy gives to the enterprise a competitive advantage, which enhances its Risk Intelligence.
Challenge 3—Lack of targeted standards and methodologies: The existing maritime security standards, methodologies and tools are not appropriate for dealing with contemporary ICT-based ports CII and dynamic maritime supply chains (addressed in Sect. 7.5); they concentrate only on the physical security nature of the ports (safety). In particular, their main limitations are the following:

- They are overly focused on physical-security aspects and pay limited attention to CIIs. At the same time, they tend to ignore the complex nature of the ICT systems and assets used in the maritime sector (e.g. SCADA), along with their interrelationships. This is, for example, the case with several international standards and legislation (e.g. the International Ships and Port Facilities Security Code (ISPS), the International Safety Management Code and EC Regulation No 725/2004 on enhancing ship and port facility security, the EC Directive 2005/65 on enhancing port security), as well as with related risk assessment methodologies e.g. MSRAM (Maritime Security Risk Analysis Model) and MARISA (MAritime RISk Assessment).
- They do not adequately take into account security processes associated with international supply chains, which are nowadays ICT-enabled and therefore severely dependent on intentional and unintentional compromise of CIIs. This is reflected in the fact that up-to-date we have seen only limited/partial implementations of relevant standards, such as ISO 28000 [11] and ISO 28001 [12].

In addition, the various maritime standardization bodies (e.g. IMO, EMSA, EASA, TEN-T EA) do not refer to in their memorandum IT/cyber security. The ICT systems' security is only assessed in relation to port and ship safety. The relevant guidelines and standards (e.g. ISPS) do not describe security controls that need to be applied to the ICT systems of the port, in order for the port to gain compliance to regulations or standards. Finally, ports are not considered as Critical Information

Infrastructures and they do not use the appropriate standards and legislation for their protection (CCIP standards).

European ports and all the SC operators need to follow ICT security and directives as well as common security methodologies and practices in order to build a trusted, interoperable and secure E.U. maritime environment. The adoption of ad hoc, generic or even commercially-driven security practices provides an ambiguous level of the ports' ICT security management.

This challenge for effective management of ICT security is associated with Best Practices 1–8 that provide a continuous and systematic process of identifying, analyzing, mitigating, reporting and monitoring technical, operational and other types of security risks (risk management) as well as implementing appropriate security measures and controls. This process is based on the MITIGATE risk assessment methodology and takes into account the particularities, the needs and constraints of the ports and the SC ICT infrastructure.

Challenge 4—Economic crisis: a stumbling block to SC security: The World Economic Crisis prevents SC operators to invest in information security. In this context, despite the strong link between IT and information security, there is an increased pressure to reduce the financing of the information security functions. Thus, the minimal resources on budget or time prevent SC operators to evaluate and ensure security as a continuing activity. They are more concerned with the cost of information security than with the consequences. Although security management methodologies, standards and tools are available, the in-house expertise is insufficient for the appropriate parameterization and deployment of these|| generic and complicated methodologies and instruments.

A number of Best Practices (BP3-9) can be matched to this challenge providing strategic comprehensive business view of information security by helping the SC operators treat it as a factor that guarantees and enhances their viability in a competitive, turbulent and diverse globalized e-market. In this context, these operators will gain trust and confidence in the global digital markets so they can become antagonistic and key drivers in the local, national and European economy.

The main advantage of the MITIGATE approach is the adoption of new innovative and specialized security and risk management processes and tools that encapsulate several simplifications and incorporate steps and automation of the complex processes to better target the audience of the SC operators. In particular, MITIGATE offer an "easy-to use" risk management approach in intuitive and graphical way, taking into account the intricacies, problems and needs of the operators.

Challenge 5—The policy dimension in enhancing maritime supply chain security: Unable to comply with stricter security demands to the business value chains, SMEs and mEs find themselves losing business opportunities. Being the backbone of the economy and chief provider of jobs in many EU Member States, this may create severe damage to employment, innovativeness and competitiveness. Enhancing information security practices of SMEs and mEs has become an urgent need, especially these days that Europe needs to strengthen its economy. Although the involved ports are compliant and respect the IMO and E.C. legislation it seems that the various maritime authorities and organizations (e.g. IMO, EMSA, DG MARE, DG MOVE,

DG INFSO) do not impose upon the ports a holistic view of security measures in order to protect their CIs. In the current regulatory context for the maritime sector on global, regional and national levels, there is hardly any consideration given on cyber security elements. Most security related regulation only includes provisions relating to safety and physical security concepts, as can be found in the ISPS and other relevant maritime security and safety regulations, such as Regulation (EC) No 725/2004 on enhancing ship and port facility security. These regulations do not consider cyber-attacks as possible threats of unlawful acts. It seems that these authorities need to be more involved in extending their recommendations/directives in order to make the port operators aware in improving their security practices.

The MITIGATE Best Practices 6–9 address the challenge to develop policies and regulation that enhance the security culture of the entire Maritime Supply Chain as they are SC-tailored considering the particularities, needs and constraints of ports and ICT-based infrastructures. In particular, the MITIGATE approach provides: (i) an holistic risk model ranging from estimating the ports' cyber risks at the asset individual level (following a risk analysis formula of both asset-centric and process-centric views) to delivering reports about the expected risk impact and the risk exposure within the Supply Chain Service (following an attack path scenarios analysis that calculates risks concerning all the existing inter-organizational and cross-organizational cyberdependencies of the SC process), (ii) improves SC operators security awareness by giving them the opportunity to experiment on different defensive strategies (e.g. selecting alternative security controls of choice) and review and study the optimal results that mitigate the cyber-risks (e.g. highlight the responsible vulnerabilities and the applicable security controls).

Hence, the MITIGATE approach provides a first of a kind approach to limit the cyber security issues in Transportation environments, on the basis of their complex and diverse cyber nature. It can serve as a blueprint for managing cyber security risks and can therefore have a positive impact on the development of SC security aware policies through: (i) the harmonization of cybersecurity culture across EU stakeholders of Transportation; (ii) the establishment of secure and trusted CIIs across the global supply chain; (iii) the provision of effective SC risk management strategies that can be readjusted according to the SC requirements at national, regional and EU level.

Furthermore, the MITIGATE approach generates a cyber security framework, which can be considered to facilitate the application of EU-wide legislation on security assessment, taking into account the Commission's priority-set to meet the security requirements of the current and evolving threat landscape.

7.5 Conclusions

Supply Chains can be viewed as complex interconnected systems that play a vital role in the transportation and delivery of goods and services. SCs usually involve various Critical Infrastructures, mainly in the transportation sector, and exhibit intra-sector and cross-border dependencies. The primary goal of MITIGATE is to assess the risk of a IT-based supply chain, having in mind the cyber interconnections and inter-dependencies between the various entities within a SC. MITIGATE can be applied to assess the threats affecting all the business partners involved in a SC as well as the SC as a whole and can be used to project the expected individual, cumulative and propagated risks within it. The derived risk values are used in order to generate a baseline SC security strategy, identifying the least necessary security controls for each participant in the SC. This enables the SC participants to fine-tune their security strategies according to their business role as well as their dependencies.

It should be noted that in order to validate the MITIGATE methodology, we used case studies based on real-life maritime scenarios and data; however, in this paper, we showed that all the maritime and transport SCs have common characteristics and face similar challenges concerning cyber security. In this context, MITIGATE can fit to their needs and particularities.

To this end, a number of best practices in the form of guidelines for successfully applying of the MITIGATE system have been created that can be used by various transport infrastructures (such as airports and railway infrastructures) and different types of port infrastructures (including infrastructures of different sizes and different business activities.

Acknowledgements This work has been funded by the EU under the HORIZON2020 program's Mitigate and Sauron projects, Grant Agreements No. 653212 and No. 740477.

References

1. C. Alberts, J. Haller et al., Assessing DoD system acquisition supply chain risk management. CrossTalk **30**(3), 4–9 (2017)
2. S. Boyson, Cyber supply chain risk management: revolutionizing the strategic control of critical IT systems. Technovation **34**(7), 342–353 (2014)
3. K. Burgess, P.J. Singh et al., Supply chain management: a structured literature review and implications for future research. Int. J. Oper. Prod. Manag. **26**(7), 703–729 (2006)
4. P. Chen, L. Desmet et al., A study on advanced persistent threats. Lecture Notes in Computer Science (including subseries Lecture Notes in Artificial Intelligence and Lecture Notes in Bioinformatics). LNCS, vol. 8735 (2014), pp. 63–72
5. R. Cigolini, M. Pero et al., Reinforcing supply chain security through organizational and cultural tools within the intermodal rail and road industry. Int. J. Logist. Manag. **27**(3), 816–836 (2016)
6. D.J. Closs, E.F. McGarrell, *Enhancing security throughout the supply chain* (IBM Center for the Business of Government Washington, DC, 2004)
7. A. Couce-Vieira, S.H. Houmb, The role of the supply chain in cybersecurity incident handling for drilling rigs. Lecture Notes in Computer Science (including subseries Lecture Notes in Artificial Intelligence and Lecture Notes in Bioinformatics). LNCS, vol. 9923 (2016), pp. 246–255
8. J.K. Deane, C.T. Ragsdale et al., Managing supply chain risk and disruption from IT security incidents. Oper. Manag. Res. **2**(1), 4–12 (2009)
9. J.E. Gould, C. Macharis et al., Emergence of security in supply chain management literature. J. Transp. Secur. **3**(4), 287–302 (2010)

10. X. Gutiérrez, J. Hintsa et al., Voluntary supply chain security program impacts: an empirical study with basic member companies. World Cust. J. **1**(2), 31–48 (2007)
11. ISO 28000:2007, Specification for security management systems for the supply chain, Geneva, Switzerland: ISO/IEC
12. ISO 28001:2007, Security management systems for the supply chain—Best practices for implementing supply chain security, assessments and plans—Requirements and guidance, Geneva, Switzerland: ISO/IEC
13. ISO/IEC 27005:2008, Information technology—Security techniques—Information security risk management, ISO/IEC
14. ISO/IEC 27001:2013, Information technology—Security techniques—Information security management systems—Requirements, ISO/IEC
15. E.-M. Kalogeraki, D. Apostolou et al., Knowledge management methodology for identifying threats in maritime/logistics supply chains. Knowl. Manag. Res. Pract. **16**(4), 508–524 (2018)
16. E.-M. Kalogeraki, S. Papastergiou et al., A novel risk assessment methodology for SCADA maritime logistics environments. Appl. Sci. (Switzerland) **8**(9), 1477 (2018)
17. C. Keegan, Cyber security in the supply chain: a perspective from the insurance industry. Technovation **34**(7), 380–381 (2014)
18. S. Kenny, Strengthening the network security supply chain. Comput. Fraud Secur. **2017**(12), 11–14 (2017)
19. P. Kotzanikolaou, S. Papastergiou et al., Design and validation of the Medusa supply chain risk assessment methodology and system. Int. J. Crit. Infrastruct. **14**(1), 1–39 (2018)
20. P. Kotzanikolaou, M. Theoharidou et al., Interdependencies between critical infrastructures: analyzing the risk of cascading effects, in *International Workshop on Critical Information Infrastructures Security* (Springer, 2011)
21. P. Kotzanikolaou, M. Theoharidou et al., Assessing n-order dependencies between critical infrastructures. Int. J. Crit. Infrastruct. **9**(1–2), 93–110 (2013)
22. G. Lu, X.A. Koufteros, Adopting security practices for transport logistics: institutional effects and performance drivers. Int. J. Shipp. Transp. Logist. **6**(6), 611–631 (2014)
23. A. Marucheck, N. Greis et al., Product safety and security in the global supply chain: issues, challenges and research opportunities. J. Oper. Manag. **29**(7–8), 707–720 (2011)
24. J.T. Mentzer, W. DeWitt et al., Defining supply chain management. J. Bus. Logist. **22**(2), 1–25 (2001)
25. A. Nagurney, P. Daniele et al., A supply chain network game theory model of cybersecurity investments with nonlinear budget constraints. Ann. Oper. Res. **248**(1–2), 405–427 (2017)
26. M.S. Nikabadi, A. Jafarian et al., The effect of information security management on organizational processes integration in supply chain. Inf. Sci. Technol. **27**(2), (2012)
27. S. Papastergiou, N. Polemi, MITIGATE: a dynamic supply chain cyber risk assessment methodology, in *Smart Trends in Systems, Security and Sustainability*, ed. by X.S. Yang, A. Nagar, A. Joshi. Lecture Notes in Networks and Systems, vol. 18 (Springer, 2018), pp. 1–9
28. M. Pero, I. Sudy, Increasing security and efficiency in supply chains: a five-step approach. Int. J. Shipp. Transp. Logist. **6**(3), 257–279 (2014)
29. N. Polatidis, E. Pimenidis et al., From product recommendation to cyber-attack prediction: generating attack graphs and predicting future attacks, in *Evolving Systems* (Springer, 2018), pp. 1–12. ISSN: 1868-6478
30. N. Polemi, P. Kotzanikolaou, Medusa: a supply chain risk assessment methodology. Commun. Comput. Inf. Sci. **530**, 79–90 (2015)
31. D. Power, Supply chain management integration and implementation: a literature review. Supply Chain Manag. **10**(4), 252–263 (2005)
32. S. Rass, S. König et al., *Uncertainty in Games: Using Probability-Distributions as Payoffs*. Lecture Notes in Computer Science, vol. 9406 (2015), pp. 346–357
33. N.S. Safa, The information security landscape in the supply chain. Comput. Fraud Secur. **2017**(6), 16–20 (2017)
34. P.N. Sindhuja, Impact of information security initiatives on supply chain performance an empirical investigation. Inf. Manag. Comput. Secur. **22**(5), 450–473 (2014)

35. A.K. Sood, R.J. Enbody, Targeted cyberattacks: a superset of advanced persistent threats. IEEE Secur. Priv. **11**(1), 54–61 (2013)
36. C. Speier, J.M. Whipple et al., Global supply chain design considerations: mitigating product safety and security risks. J. Oper. Manag. **29**(7–8), 721–736 (2011)
37. G. Stergiopoulos, P. Kotzanikolaou et al., Risk mitigation strategies for critical Infrastructures based on graph centrality analysis. Int. J. Crit. Infrastruct. Prot. **10**, 34–44 (2015)
38. G. Stergiopoulos, P. Kotzanikolaou et al., Time-based critical infrastructure dependency analysis for large-scale and cross-sectoral failures. Int. J. Crit. Infrastruct. Prot. **12**, 46–60 (2016)
39. C. Tankard, Advanced persistent threats and how to monitor and deter them. Netw. Secur. **2011**(8), 16–19 (2011)
40. S. Véronneau, J. Roy, Security at the source: securing today's critical supply chain networks. J. Trans. Secur. **7**(4), 359–371 (2014)
41. K. Yoshifu, M. Itoh et al., Cybersecurity consulting services in the world of IoT. NEC Tech. J. **12**(2), 64–69 (2018)
42. S.H. Zailani, K.S. Subaramaniam et al., The impact of supply chain security practices on security operational performance among logistics service providers in an emerging economy: security culture as moderator. Int. J. Phys. Distrib. Logist. Manag **45**(7), 652–673 (2015)

Chapter 8
Risk Assessment for IoT-Enabled Cyber-Physical Systems

Ioannis Stellios, Panayiotis Kotzanikolaou, Mihalis Psarakis, and Cristina Alcaraz

Abstract Internet of Things (IoT) technologies have enabled Cyber-Physical Systems (CPS) to become fully interconnected. This connectivity however has radically changed their threat landscape. Existing risk assessment methodologies often fail to identify various attack paths that stem from the new connectivity/functionality features of IoT-enabled CPS. Even worse, due to their inherent characteristics, IoT systems are usually the weakest link in the security chain and thus many attacks utilize IoT technologies as their key enabler. In this paper we review risk assessment methodologies for IoT-enabled CPS. In addition, based on our previous work (Stellios et al. in IEEE Commun Surv Tutor 20:3453–3495, 2018, [47]) on modeling IoT-enabled cyberattacks, we present a high-level risk assessment approach, specifically suited for IoT-enabled CPS. The mail goal is to enable an assessor to identify and assess *non-obvious* (indirect or subliminal) attack paths introduced by IoT technologies, that usually target mission critical components of an CPS.

Keywords Internet of Things (IoT) · Cyber Physical Systems (CPS) · Risk assessment · Attack paths · Critical infrastructures

---

I. Stellios · P. Kotzanikolaou (✉)
SecLab, Department of Informatics, University of Piraeus, 85 Karaoli & Dimitriou, 18534 Pireas, Greece
e-mail: pkotzani@unipi.gr

I. Stellios
e-mail: jstellios@unipi.gr

M. Psarakis
ESLab, Department of Informatics, University of Piraeus, 85 Karaoli & Dimitriou, 18534 Pireas, Greece
e-mail: mpsarak@unipi.gr

C. Alcaraz
Computer Science Department, University of Malaga, Campus de Teatinos s/n, 29071 Málaga, Spain
e-mail: alcaraz@lcc.uma.es

G. A. Tsihrintzis and M. Virvou (eds.), *Advances in Core Computer Science-Based Technologies*, Learning and Analytics in Intelligent Systems 14,
https://doi.org/10.1007/978-3-030-41196-1_8

8.1 Introduction

Cyber physical systems consist of large-scale interconnected cyber and physical components, interacting with each other through various connectivity technologies. There are a multitude of devices and applications being deployed to serve critical functions as well as everyday operations, like smart grids, Supervisory Control and Data Acquisition (SCADA) systems, smart healthcare devices, wearables, Intelligent Transportation Systems (ITS) and vehicles, smart cities and many more.

A typical example of CPS are Industrial Control Systems (ICS), which are formed in hierarchical model. Field devices, such as Programmable Logical Controllers (PLC) and Remote Terminal Units (RTU) to Intelligent Electronics Devices (IED) which, in turn, they are managed through Human-Machine-Interfaces (HMI) from Command and Control (C&C) centers.

Although traditional ICS used to be, more or less, "closed" and isolated systems, the evolution of the IoT has also affected modern CPS. IoT technologies and protocols, used both in industrial and non-industrial environments (e.g. 6LoWPAN [45], CoAP [6]), allow even large-scale and mission critical industrial equipment to connect directly to the Internet (e.g. industrial robots, wind turbines, solar panel systems etc.). These new IoT technologies enable ICSs to become more flexible and interpolatable. They allow for remote monitoring and control (e.g. interconnected PLCs, Industrial robots), thus reducing management, surveillance and maintenance costs as well as increase the expected lifetime of old, yet very expensive ICSs, like those supporting Critical Infrastructures and services.

Traditional CPS used proprietary technologies, were isolated from the Internet and were built to be reliable and robust. Besides physical security concerns, no security mechanisms for older SCADA devices where present, since, they were physically and logically isolated, *air-gaped* systems. Their main line of defense was that were installed in highly secure areas with a much smaller attack surface than traditional IT infrastructure, and with dedicated *off-the-grid* communication channels.

While IoT enabling technologies have created new opportunities for the global economy, this unprecedented explosion in inter-connectivity and inter-dependency between billions of unsecured, energy constrained devices have raised a number of security issues and challenges.

Since modern CPS highly depend on computer functionality, network interconnectivity and machine-to-machine interaction in order to properly operate, an attack or disruption on a single component of a complex, large-scale CPS may concurrently affect the entire production line. The operating environment has evolved in a such a way that depends to a large extent on Internet connected/interconnected supply chains, networks and systems and thus reduces the ability to estimate inter/outer dependencies of such increased complexity.

To complicate things even more, latest policies in companies, like *Bring Your Own Devices* (BYOD), have enabled end-user devices such as smartphones, gadgets and laptops to connect and interact to corporate networks. On the other hand, concerns of privacy violations as well as a totally new set of attacks against mission critical IT

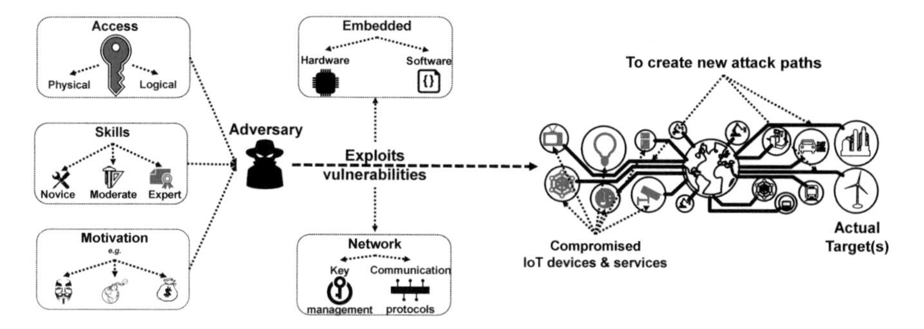

Fig. 8.1 High level representation of IoT-enabled attacks against CPS

systems and services, that can be launched from Internet using low-cost equipment and basic technical skills make the headlines more often. These kind of attacks not only are usually underestimated and in some cases hard to identify.

Assessing the risk of CPS systems and their related infrastructures has been the study of several research approaches [3, 7, 11, 29, 30, 35, 38] in the recent past. Moreover, several risk assessment methodologies [1, 4, 10, 12, 21, 27, 33] have been developed for the CPS systems focusing on IoT-enabled attack scenarios. In addition, security researchers [2, 16, 22, 51] proposed methodologies some of which incorporate Common Vulnerability Scoring System (CVSS) [13] in its latest version (3.0) for vulnerability assessment and complex access/attack path discovery.

Despite this extensive research detecting hidden/subliminal attack paths enabled by vulnerable IoT subsystems remains an hard task for an assessor, especially in facilities where IoT and mission critical systems coexist in close proximity. Our proposed a high-level risk assessment methodology focuses on identifying the risk that is introduced from subliminal access/attack paths by utilizing well established standards, such as ISO/IEC 7005:2011 [20], NIST SP800-30 [42] and SP800-39 [41] combined with newly introduced methodologies [2, 16, 22, 51]. Its main contribution is that, by modeling characteristics of the device/technology as well as the applicable in each case access and attack paths (see Fig. 8.1), it can be used as a guide to assess the risk that IoT devices may introduce in their related CPS. The discovered attack paths include both scenarios where the IoT enabling technology is the actual target and the ones where an adversary utilizes the IoT device in order to attack other mission-critical equipment.

IoT enabled attack scenarios. The FBI issued in 2012 an intelligence bulletin that reported an extensive fraud concerning thousands of smart meters in Puerto Rico [24]. The report states that even individuals with moderate technical skills, low cost tools and software that is available on the the Internet, may successfully alter the readings of the smart meters. According to the FBI the adversaries were company's employees that exploited the smart meters using an optical converter device which in turn enables the smart meter to communicate with a computer. Then using software that can be downloaded from the Internet, they managed to alter the settings for recording power consumption. The annual economical impact to the company is estimated to be over 400 million dollars. Clearly this case is a typical example of a direct attack path.

In 2017 security researchers [28, 37] were able to locate approximately 84.000 industrial robots, exposed to the Internet through FTP server (direct attack path) or through industrial routers (indirect path); 5.000 of these, did not even require any type of authentication. Among the vulnerabilities found were outdated software (application libraries, OS kernels), insecure web interfaces, publicly available firmware images, as well as wireless access to remote service facilities. In this attack scenario a "black hat" hacker may target a plethora of industrial robots so as to alter the robot state and force robots to produce defect products, install ransomware and/or injure the operators.

In December of 2015 and 2016 Ukrainian energy companies suffered from cyber attacks that targeted the smart grid. Utilizing spear-phishing techniques and sophisticated exploitation methods adversaries managed to take over interconnected field devices, such as circuit brakers, and inflict a massive blackouts that lasted for several hours and affected over 200,000 people (2015) [17, 26].

In another proof-of-concept attack scenario researchers proved that is possible to exfiltrate sensitive information which is stored in a air-gaped data center inside a highly secure facility. As described in [39, 40], an adversary manages to bypass proximity checks of a smart lighting system and by utilizing wardriving/warflying techniques (a drone equipped with off-the-shelf communication equipment), and take over smart lighting systems from a large distance (approx. 150 m). Then, by extending the functionality of the light bulbs, she manages to control light flickering in a way that the human eye cannot perceive thus creating a covert channel to extract the information.

Paper contribution. In this paper, we extend our previous work on modeling IoT-enabled attacks [47], and we present a targeted, high-level risk based approach that may be used to identify and assess IoT-enabled CPS. Such a methodology may assist a risk assessor to identify and assess *non-obvious* attack paths introduced by IoT technologies, such as indirect or subliminal attack paths against the critical components of a cyber-physical system.

Paper structure. In Sect. 8.2 we review the related work on risk assessment methodologies, from general purpose ones, to more targeted methodologies for IoT and CPS. In Sect. 8.3 we propose a high-level risk-based approach to model and assess IoT-enabled attacks. Finally, Sect. 8.4 concludes this paper.

8.2 Related Work

Most existing security risk and threat assessment methodologies examine a series of factors such as: (i) the assets that need to be protected, (ii) the threats and vulnerabilities that correspond to these assets, (iii) their value to the organization under assessment, (iv) the consequences (or impact) in case of security violations against the identified assets and (v) security controls that can reduce/eliminate the potential damage. The main goal of a risk assessment methodology is to provide guidance to an organization in order to minimize the risk and maximize the level of confidentiality, integrity and availability. The procedure of implementing the necessary security measures must be done in respect of the organization's needs in order to achieve the desired levels of confidentiality, integrity and availability and, at the same time, guarantee a satisfactory level of functionality.

IoT-specific risk assessment methodologies have been developed in the last few years in order to describe the ever growing risk that stems from the IoT systems and services. Atamli and Martin [4] present use cases of IoT enabled attack scenarios (power management, smart car and healthcare) so as to identify potential threats sources, classes of attack vectors and impact assessment applicable in devices such as Radio Frequency Identifiers (RFIDs), actuators, sensors as well as networking technologies. They also propose specific countermeasures that can reduce the risks evolved mainly in security and privacy.

In [12] Dorsemaine et al. access the risks introduced to a legacy Information System (IS) due to the integration of an IoT infrastructure. A practical example is then presented with the integration of a smart lighting system in a company's IT systems. The authors divide the IS into local environment, transportation, storage, mining and provision sectors. Then, they define security properties for the IoT systems of each IS sector, by focusing mainly on aspects such as confidentiality, integrity, availability, usability and auditability while also introduce additional properties for IoT components including energy, communication, functional attributes, local user interface and hardware/software resources. Finally they present the potential threats and the impact in all of the aforementioned attributes for an IS and IoT infrastructure.

In [27] Liu et al. propose a dynamical risk assessment method for complicated and constant changing IoT environments adopting features from an *Artificial Immune System* such as the distributed and parallel treatment, diversity, self-organization, self-adaptation, robustness etc. Through packet inspection from agents that are deployed in IoT systems, the proposed method locates abnormal behavior and responds by adapting appropriately the risk value.

A management framework for IoT devices, called Model-based Security Toolkit (SecKit), used to evaluate security policies that protect user's privacy, is presented in [33]. Seckit has been integrated in a framework, proposed by the *iCore* project, which enables usage control and protection of user data. Then a case study is presented in a smart home scenario.

Abie and Balasingham [1] propose a risk-based adaptive security framework for IoT enabled e-Health CP systems that estimates risk damages and future benefits using game theory and machine learning techniques. This enables the security mechanisms to adjust their security decisions accordingly.

A recent approach [10] about Medical Internet of Things (MIoT) points out difficulties that traditional risk assessment methodologies face when used in non-stable environments, such as the MIoT, where devices maybe added, removed or changed in their configuration. For assessing and managing threats the researchers adopt HMG IS1 and ISO/IEC 27033 standards and an existing threat analysis from the

Technology Integrated Health Management (TIHM) project. They taxonomize threats according to the severity level ranging from *very low* to *very high*, as well as the risk that emerges from IoT devices against other MIoT devices. In addition, for each MIoT device connected to the hub a multicheck process is proposed,

A survey [21] focusing mainly on cyber security management in industrial control systems depicts the current standards and future challenges that ICS face in the ever evolving threat landscape. Hot topics for future research, among others, considered to be the need for maintenance of security of ICS components throughout their lifetimes, interdependencies between large CPS, as well as real-time risk assessment.

In [22] Kott et al. describe Mission Impact Assessments (MIAs) in an effort to bridge the gap between operational decision makers and cyberdefenders. They managed to set a testbed (*Panoptesec*) that emulated cyber physical systems of an Italian water and energy distribution company as well as a prototype simulation platform named *Analyzing Mission Impacts of Cyber Actions (AMICA)* that simulate a military's air operations center they managed to discover high number of hidden network dependencies that weren't identified by human operators, unnecessary large volume communications between Human-Machine Interfaces (HMIs) and field devices and attacks against specific nodes of the network that, when used in a timely manner could lead to devastating results. The researchers proposed an abstractive threat modeling (e.g. [23]), for both adversaries and defenders, and emphasized on the challenges involved when modeling large scale, diverse and complex networks.

Agadakos et al. [2] proposed a methodology for modeling cyber-physical attack paths in IoT. In particular, they developed a framework that allowed the identification of IoT device types, interaction channels, as well as security and proximity features. Using the proposed framework they managed to simulate a home network that consisted of several home IoT devices. In particular, by using techniques such as passive sniffing for host discovery, they managed to discover attack scenarios that utilized hidden connectivity/interaction paths, security degradation (e.g. from authenticated to unauthenticated communication channels) and violations of transitions and states. According to the authors limitations of this work are considered to be the fact that the model may introduce false positives, since it does not filters unrealistic attacks and it is not easy to implement in large scale networks with mission critical systems.

Researchers in [5] propose a risk-based access control model for IoT technologies. Real-time data from IoT devices are utilized to dynamically estimate security risks through an risk estimation algorithm. The proposed model monitors and analyzes user behavior in order to detect abnormal action from authorized users

Recent methodologies that utilize the CVSS 3.0 have been also proposed by similar group of researchers [16, 51]. In [16], a framework for modeling and assessing the security of the IoT ecosystem based on previous work is proposed. The framework consists of five phases: (1) Data processing, (2) security model generation, (3) security visualization, (4) security analysis, and (5) model updates. In phase one system information and security metrics are introduced in order to construct the IoT network which is then used (phase two) to construct the extended Hierarchical Attack Representation Model (HARM) [19] and calculate all possible attack paths in the IoT network. In phase three attack graphs (in low, upper and middle layer) are utilized to visualize the IoT network whereas in four a security analysis, that takes into consideration e.g. nodes or vulnerabilities, is constructed and fed into an analytic modeling and evaluation tool (*Symbolic Hierarchical Automated Reliability and Performance Evaluator—SHARPE* [43]) for further security analysis. Finally in phase five proper defense strategies are decided. The researchers present scenarios such as a Sinkhole attack [32] in a smart home environment, wearable healthcare and environmental monitoring. According to the researcher the limitations in presented attack scenarios include the difficulty to depict all diverse connectivity paths, no-connectivity attack scenarios (e.g. Distributed-Denial-of-Service—DDoS) heterogeneity on communication protocols and static network topology.

8.3 Modeling Security Risks in IoT-Enabled CPS

In order to identify and assess the risks against CPS that derive from related IoT technologies, we will adopt and extend our modeling approach for IoT-enabled attacks, initially presented in [47]. From a high-level view, the adversary's access capabilities to the IoT device will be combined with the connectivity level of the IoT device with the target CPS, in order to identify and assess the attack paths against the target system that are enabled by the IoT device (see Fig. 8.2).

Note that such an approach can be combined with generic risk assessment methodologies, such as ISO27005 [20] and NIST800-30 [42]. The characteristics of the adversary will be used to assess the threat level of an attack whereas hardware, software and network vulnerabilities will be used to assess the vulnerability level. Finally, characteristics similar to ones of the aforementioned reviewed methodologies (e.g. [2, 16]) can also be incorporated.

Calculating the risk involved among complex cyber physical systems that operate in a environment with IoT devices and related technologies can be a quite a challenging task during a risk assessment. The proposed algorithm takes into consideration the available inputs/outputs, functions, network and software characteristics of each IoT device, the potential attack vectors (access paths) to the IoT device and the attack paths that originate from the IoT device. Then, by utilizing state-of-the-art methodologies, the vulnerability, threat and impact level are calculated. Finally, using the aforementioned calculated metrics the risk level for each attack path scenario is defined. Depending on the attack path, the IoT device may be used either as a target or as an amplifier of an attack.

8.3.1 Attack Vectors: Modeling the Adversary

To assess the access level, one must consider three key factors: Physical, proximity and network access. Listing the device's physical characteristics such as network and input/output interfaces as well as location is essential in order to determine all

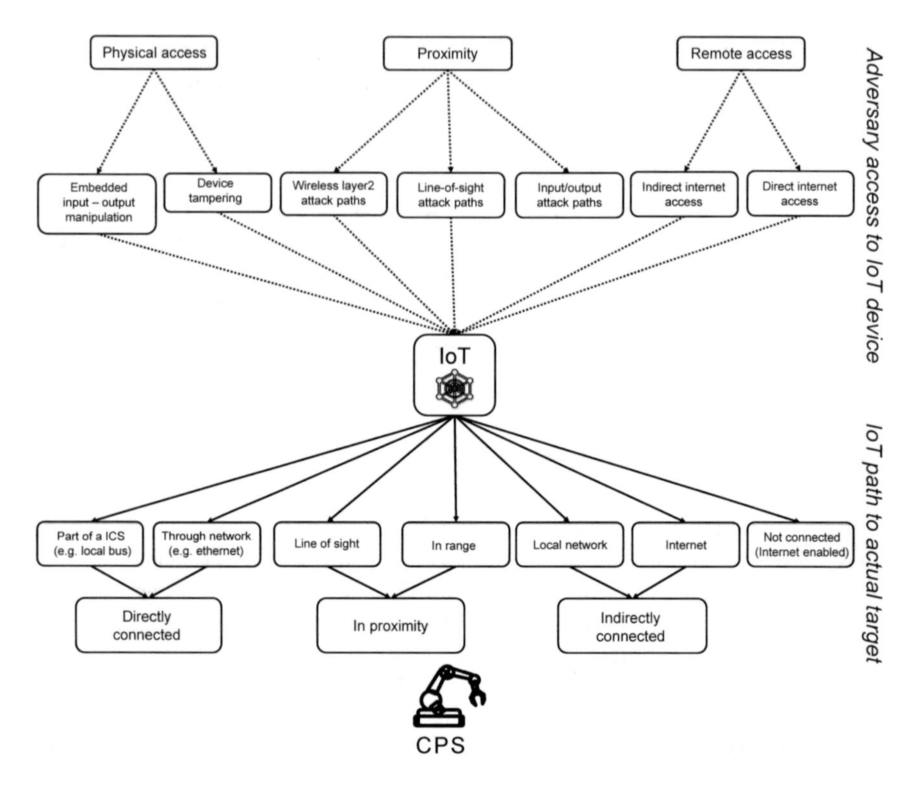

Fig. 8.2 An overview of the components of the proposed risk assessment methodology

possible direct machine-to-machine and human-to-device interactions that can take place. It is also a key factor in order to reduce the amount of the potential access paths per device thus eliminating impractical attack vectors (e.g. a layer-2-enabled temperature sensor cannot be directly accessed via layer-3 network).

8.3.1.1 Physical Access

Physical access mainly describes the ability of an actor (malicious or not) to access sensitive inputs/outputs and/or modify/replace the IoT device thus affecting the threat level. For example, if physical tampering of the IoT device is required for an attack, then this attack will be less likely to happen, in comparison with one than can be triggered remotely (e.g. through the Internet). On the other hand, a susceptible to tampering IoT device, placed in a public area with no physical access security controls enforced (e.g. a IP-enabled security camera outside a factory's premises), can be used as a point of entry to the company's internal mission critical systems, with severe consequences.

8.3.1.2 Proximity

Proximity attack vectors and paths can be hard to identify and may include bidirectional machine-to-machine and human-to-machine interactions. In [2] authors present an approach that can potentially depict all machine-to-machine interactions under the assumption that all activity may take place among devices in proximity through all available input/output channels and network interfaces in their predefined range. Furthermore, subliminal attack paths may also occur when the range for input/output and network interfaces can be extended and functionality characteristics can be manipulated in unpredictable ways, so as to server an adversary's needs, as various researched have recently been demonstrated [39, 40].

For example, several IoT enabling technologies utilize devices that include unsecured and vulnerable inputs/outputs (e.g. sensing systems, Light Emitting Diode (LED) lights). A typical example is the LIght Detection and Ranging (LIDAR), that researchers proved that can be manipulated from up to a certain distance [36, 50]. Wireless layer-2 network interfaces (e.g. Bluetooth, Zigbee) can also be used by adversaries to remotely adjust and even replace the legitimate system software.

8.3.1.3 Remote Access

If an attack can be triggered remotely by Internet adversaries, e.g. by abusing the connectivity features of the IoT, then such an attack has a high likelihood. The need for constant monitoring, remote management and control, cost reduction and increased productivity creates complex attack vectors that adversaries can use to their advantage. Since more and more IoT devices are Internet enabled attacks on such IoT systems are likely to increase in the near future. Recent real cyber attacks in Ukraine's smart grid [17, 26] considered to be the most prominent indirect connectivity attack scenarios: Through spear-phishing campaigns the adversaries manage to penetrate the corporate network and attack the remote controlled circuit brakers thus causing large-scale disruptions in the electricity network.

8.3.2 Impact Level: Identifying Attack Paths

In some cases, the IoT device itself is the actual target of an attack. Unfortunately, the manufacturers of IoT devices do not usually consider security as the top priority, at least in the case of consumer IoT products. Even for IoT devices that are used in sensitive cyber-physical operations, characteristics like the reliability of the device are usually favored instead of security, for example in the case of implantable medical devices or SCADA field IoT devices.

However, due to their increased connectivity features, IoT devices may also be used as a means to attack other CPS that are *indirectly* connected with the IoT device. For example, consider a car infotainment system that may be indirectly connected to critical control systems of the vehicle.

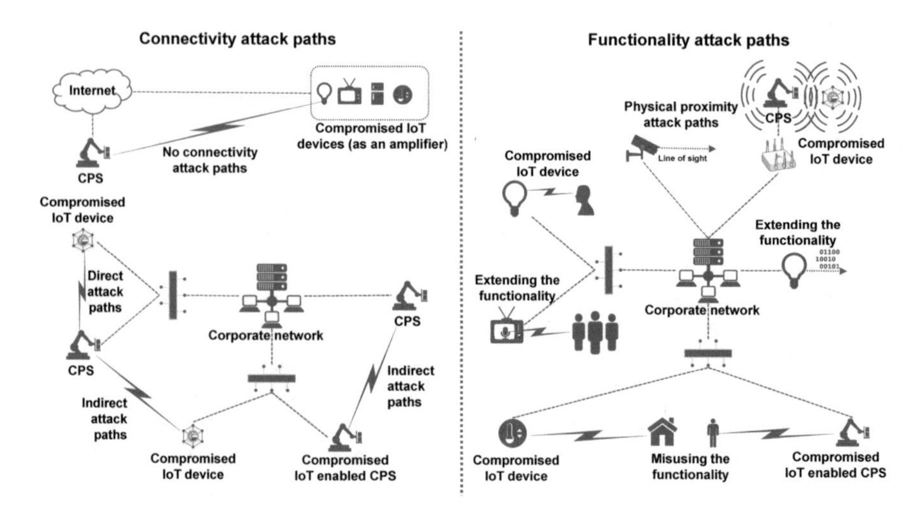

Fig. 8.3 On the left side of the figure are shown IoT-enabled attack paths that are based on direct, indirect and no connectivity features whereas on the right side are presented potential attacks paths that may occur in case of misuse/abuse/extend the functionality of the IoT device/service. In both connectivity/functionality attack paths the aforementioned characteristics induce risk that is not easy to identify and assess

Another category of subliminal attack paths, are those that involve vulnerable IoT technologies, whose functionality is *misused* or *extended* by an adversary, in an unpredictable way. For example, smart lights that are abused to create a covert channel for data exfiltration or even to attack patients and cause epileptic seizures [40]. IoT devices can be exploited so as to create, novel and hard-to-identify attack paths against other interconnected systems.

Figure 8.3 presents typical examples of such attack paths, which are explained bellow.

8.3.2.1 Connectivity Attack Paths

These may be realized due to the physical and/or logical connectivity of vulnerable IoT devices with other critical CPS components.

- *Direct connectivity attack paths* The IoT device is part of a critical system, or has direct connection to it. In this case, the IoT device is usually the actual target of the attack, since in most cases, it is considered a crucial component of the CPS [24]. In addition, if network segmentation is not well implemented, the IoT device may be used as an amplifier in order to disrupt other systems or processes, as described in [46].
- *Indirect connectivity attack paths*: Such attack paths usually involve indirect attacks that are not always obvious (as in the previous case). For example, by using zero-day (or even known) exploits against an intermediate system connected to an already compromized IoT device in order to extend the attack towards a third

system that is critical [31]. Vulnerable IoT devices may be used both as an amplifier [44, 48] or as the actual target of the attack [15, 25, 26]. Bring-Your-Own-Phone/Device (BYOP/D) policies may also cause indirect attack paths that may be hard to identify.

8.3.2.2 Attacks Paths Based on Massive Number of Compromised IoT Devices

Such attacks are usually based on the plethora of consumer IoT devices, such as smart home appliances and end-user devices. Although these devices are not connected (directly or indirectly) with critical CPS, they can still be used to create attack paths against critical systems (e.g. IoT-based botnets [9]). Compromised IoT devices in large numbers may cause Denial of Service (DoS attacks) against a critical system. A special case are concurrent Permanent DoS attacks that may target at the IoT devices themselves (e.g. ransomware attacks against consumer IoT devices). Although such devices are usually of low importance, a concurrent PDoS attack against thousands of devices may impose a high impact.

8.3.2.3 Functionality Attack Paths

Security researchers [40] and real incidents [49] have shown that it is possible to create subliminal attack paths by misusing or by extending the functionality of the IoT devices, to attack targets that are not connected (even indirectly) with the IoT device. Usually such attack are targeted against systems that are in some proximity with the vulnerable IoT device.

8.3.2.4 Physical Proximity Attack Paths

IoT devices installed near critical CPS components may be used to create subliminal attack paths against them [18]. Proximity may imply close distance, or even line of sight proximity. These attacks exploit common characteristics of wireless technologies (e.g. IEEE 802.15.4x wireless network adapters for ZigBee and WirelessHART protocols all use the same frequency for broadcasting). Such attacks are difficult to discover since they extend, alter or misuse the functionality of the IoT device in ways that cannot easily predicted [34]. Examples of such attacks include covert channels and data exfiltration attacks.

8.3.3 Calculating the Vulnerability Level

Since our methodology emphasizes on attacks against CPS that are IoT-enabled, and IoT technologies are usually the weakest link in the security chain due to their inherent limitations, identify the vulnerabilities of the IoT technologies involved. Vulnerability assessment should include at a minimum:

Embedded vulnerabilities on hardware (HW): IoT devices susceptible to physical tampering may allow an adversary to disable and/or extract sensitive hard-coded information, such as encryption keys and stored credentials, e.g. through the use of Correlation/Differential Power Analysis (CPA/DPA) techniques [39].

Embedded vulnerabilities on software (SW): Untested, outdated software and vulnerable update services may enable an attacker to entirely compromise the device from distance. This includes, but is not limited to, publicly available and unsigned firmware update files, the use of outdated vulnerable operating systems and Application Programming Interfaces (APIs) that have not been security tested. Lack of security techniques and practices such as firmware signing, secure OS parameterization and input sanitation may allow an adversary to remotely exploit interconnected CPS, with minimal effort.

Network vulnerabilities on communication protocols: Vulnerable encryption algorithms used at the network layer, may reveal sensitive information. With most of the IoT communication protocols based on inherently insecure wireless network protocols (e.g. WirelessHART, ZigBee, MiWi, WiFi) an adversary can remotely eavesdrop, inject and modify network messages in order to accomplish an attack. In addition, constraints of the IoT devices, such as energy and computational power, make them susceptible to inadequate key management schemes. The lack of support for Public Key Cryptography (PKC), the use of weak encryption algorithms (e.g. WEP encryption over WiFi), the use of a single embedded network key (such as in the ZigBee Light Link—ZLL protocol) or the absence of encryption mechanisms, may be exploited in order to disrupt highly sophisticated and critical systems.

8.3.4 Calculating the Risk

Based on the threat model described above, we describe a targeted, high-level risk assessment methodology, whose goal is to identify and assess *hidden risks* in CPS that stem from the IoT interaction. Following the RA standards, the calculation of the risks will be based on three basic phases: threat, vulnerability and impact assessment. Finally, the security risk of each identified attack path will be assessed. In the proposed methodology, one can use of typical Likert scales, to define the threat, vulnerability, impact and risk scales. This is common to most general purpose RA methodologies (e.g. [14, 42]), although each methodology may define a different scales for each risk factor.

By combining all the factors assessed in the previous phase, the risk of all possible IoT-enabled attack paths will be assessed in this phase. Essentially, during this process all the steps performed in the previous phases will be combined to methodologically output all the related risks. Also, it is crucial, when defining the IoT devices/technologies, to take into consideration mobile devices such as smartphones and laptops (BYOD), since, they are equipped with multiple inputs, outputs and wireless network interfaces, can be directly/indirectly (via the corporate network) connected to the Internet, and sometimes are in proximity of mission critical systems.The basic steps of this process are as follows:

1. Identify all IoT devices and enabling technologies.
2. Repeat for each of IoT device (say device i):

   2.1. **Access paths**: Identify all applicable access paths (physical, proximity, remote) to the IoT device/enabling technology:

   2.1.1 If the device can be physically accessed define all of embedded device's input, output and wired network interfaces (e.g. USB, Ethernet & Serial ports, sensors, speakers)

   2.1.2 Proximity access paths: Define all of input, output and wireless network interfaces characteristics (frequency active range etc.) (e.g. ZigBee, Bluetooth, WiFi, Z-Wave, microphone range and sensitivity etc.).

   2.1.3 Remote access paths: Define all enabled layer-3 network interfaces. Then for each network interface define all possible access paths (directly, indirectly), especially the ones that lead to Internet connectivity (e.g. Ethernet → Control Room → Corporate network → Web server).

   2.2 **Attack paths**: Identify all possible attack paths against any affected CPS:

   2.2.1 *Direct connectivity attack paths*: Identify all direct attack paths between the IoT device and any other system.

   2.2.2 *Indirect connectivity attack paths*: Identify all systems that are indirectly connected to the IoT using any network interfaces (wired or wireless).

   2.2.3 Identify attack paths against any affected CPS, related with IoT *extended/misused functionality*:

   2.2.3.1 *Physical proximity*: Identify systems that are in physical proximity in respect with the IoT device's wireless network interfaces (e.g. protocols that use the same bandwidth, devices that are in line of sight etc [34]).

   2.2.3.2 *Potential covert channels*: Examine devices for possible ways to create hidden covert channels (e.g. smart lamp systems have used as a covert exfiltration channels [40]; smart TVs/cameras for espionage [49]).

   2.2.3.3 *Other potential misuse*: Examine devices for any other possible misuse against other CPS. Examples of such misuse include abusing smart lamp systems installed in hospitals to cause epileptic seizures [40]; alter the functionality of IoT-enabled industrial robots to affect the production line [8, 28]; manipulate the functionality of thermostats to disrupt the operations of the data center [18]).

   2.3 **Calculate risk**: For each identified attack path k (with target system j):

   2.3.1 For each corresponding access path (attack vector):

2.3.1.1 Assess the threat level of the relative attack vector, denoted as T_{ijk}.

2.3.1.2 Assess the vulnerability of the IoT device, for the examined attack, denoted as V_{ijk}.

2.3.1.3 Assess the impact of the *actual target system* of the attack path, denoted as I_{ijk}.

2.3.1.4 Assess the risk of each examined attack path k that is triggered by IoT device i against the target system j as follows:

$$R_{ijk} = T_{ijk} V_{ijk} I_{ijk} \tag{8.1}$$

As a final step we propose the construction of a table with the calculated risk values of all IoT devices/enabling technologies in respect of the affected systems for all applicable paths. Metrics such as total risk ($R_{i_{total}}$) and Maximum Risk ($R_{i_{maxj}}$) per affected system, can be used in order to assess the criticality of each IoT device/technology i, and help prioritize the implementation of the appropriate security controls, so as to effectively reduce the organization's risk levels under a desirable threshold.

8.4 Conclusions

Insecure off-the-shelf IoT devices and relative technologies that may be connected to critical cyber-physical systems, or even nearby such systems, may enable an adversary to discover attack paths against CPS and cause severe damage to such systems. Assessing the risk introduced from IoT ecosystem in CPS is a very challenging task. In our work we present a high-level risk assessment approach that mainly focuses on the identification of subliminal access/attack paths. In order to do so, we examine all applicable access paths (physical, proximity, remote) to the IoT device/technology. Then we estimate all paths from the IoT device to other CPSs based on direct or indirect connectivity and dependency, or on the proximity of the IoT device to the target CPS. As a future work, we will extend the proposed high-level risk assessment approach to develop a detailed risk assessment methodology. We also plan to develop a tool that can be used to automate the identification and assessment of hidden and subliminal attack paths of IoT technologies against critical components of CPS.

References

1. H. Abie, I. Balasingham, Risk-based adaptive security for smart IoT in eHealth, in *Proceedings of the 7th International Conference on Body Area Networks* (ICST (Institute for Computer Sciences, Social-Informatics and Telecommunications Engineering), 2012), pp. 269–275
2. I. Agadakos, C.Y. Chen, M. Campanelli, P. Anantharaman, M. Hasan, B. Copos, T. Lepoint, M. Locasto, G.F. Ciocarlie, U. Lindqvist, Jumping the air gap: modeling cyber-physical attack paths in the internet-of-things, in *Proceedings of the 2017 Workshop on Cyber-Physical Systems Security and Privacy* (ACM, 2017), pp. 37–48
3. S. Amin, G.A. Schwartz, A. Hussain, In quest of benchmarking security risks to cyber-physical systems. IEEE Netw. **27**(1), 19–24 (2013)
4. A.W. Atamli, A. Martin, Threat-based security analysis for the internet of things, in *2014 International Workshop on Secure Internet of Things (SIoT)* (IEEE, 2014), pp. 35–43
5. H.F. Atlam, A. Alenezi, R.J. Walters, G.B. Wills, J. Daniel, Developing an adaptive risk-based access control model for the internet of things, in *2017 IEEE International Conference on Internet of Things (iThings) and IEEE Green Computing and Communications (GreenCom) and IEEE Cyber, Physical and Social Computing (CPSCom) and IEEE Smart Data (SmartData)* (2017), pp. 655–661
6. C. Bormann, A.P. Castellani, Z. Shelby, CoAP: an application protocol for billions of tiny internet nodes. IEEE Internet Comput. **16**(2), 62 (2012)
7. A.A. Cárdenas, S. Amin, Z.S. Lin, Y.L. Huang, C.Y. Huang, S. Sastry, Attacks against process control systems: risk assessment, detection, and response, in *Proceedings of the 6th ACM Symposium on Information, Computer and Communications Security* (ACM, 2011), pp. 355–366
8. C. Cesar, A. Lucas, Hacking robots before Skynet (IOActive) (2017), https://ioactive.com/pdfs/Hacking-Robots-Before-Skynet.pdf
9. S. Cobb, 10 things to know about the October 21 IoT DDoS attacks (2016), http://www.welivesecurity.com/2016/10/24/10-things-know-october-21-iot-ddos-attacks/
10. S. Darwish, I. Nouretdinov, S.D. Wolthusen, Towards composable threat assessment for medical IoT (MIoT). Procedia Comput. Sci. **113**, 627–632 (2017)
11. J. Depoy, J. Phelan, P. Sholander, B. Smith, G. Varnado, G. Wyss, Risk assessment for physical and cyber attacks on critical infrastructures, in *Military Communications Conference, 2005. MILCOM 2005* (IEEE, 2005), pp. 1961–1969
12. B. Dorsemaine, J.P. Gaulier, J.P. Wary, N. Kheir, P. Urien, A new threat assessment method for integrating an IoT infrastructure in an information system, in *2017 IEEE 37th International Conference on Distributed Computing Systems Workshops (ICDCSW)* (IEEE, 2017), pp. 105–112
13. P.M. Erdősi, The common vulnerability scoring system (CVSS) generations–usefulness and deficiencies
14. D. Evans, P. Bond, A. Bement, FIPS PUB 199 standards for security categorization of federal information and information systems. The National Institute of Standards and Technology (NIST) (2004)
15. N. Falliere, L.O. Murchu, E. Chien, W32. Stuxnet Dossier. White paper, Symantec Corporation. Secur. Response **5**(6) (2011)
16. M. Ge, J.B. Hong, W. Guttmann, D.S. Kim, A framework for automating security analysis of the internet of things. J. Netw. Comput. Appl. **83**, 12–27 (2017)
17. D. Goodin, Hackers trigger yet another power outage in Ukraine (2017), https://arstechnica.com/security/2017/01/the-new-normal-yet-another-hacker-caused-power-outage-hits-ukraine/
18. G. Hernandez, O. Arias, D. Buentello, Y. Jin, Smart nest thermostat: a smart spy in your home, in *Black Hat USA* (2014)
19. J. Hong, D.S. Kim, HARMs: hierarchical attack representation models for network security analysis (2012)

20. ISO: ISO/IEC 27005:2011 Information technology—security techniques—information security risk management. Technical report. International Standardization Organization (2011)
21. W. Knowles, D. Prince, D. Hutchison, J.F.P. Disso, K. Jones, A survey of cyber security management in industrial control systems. Int. J. Crit. Infrastruct. Prot. **9**, 52–80 (2015)
22. A. Kott, J. Ludwig, M. Lange, Assessing mission impact of cyberattacks: toward a model-driven paradigm. IEEE Secur. Priv. **5**, 65–74 (2017)
23. A. Kott, C. Wang, R.F. Erbacher, *Cyber Defense and Situational Awareness*, vol. 62 (Springer, 2015)
24. KrebsonSecurity, FBI: smart meter hacks likely to spread (2012), https://krebsonsecurity.com/2012/04/fbi-smart-meter-hacks-likely-to-spread/
25. D. Kushner, The real story of Stuxnet. IEEE Spectr. **50**(3), 48–53 (2013)
26. R.M. Lee, M.J. Assante, T. Conway, Analysis of the cyber attack on the Ukrainian power grid. SANS Industrial Control Systems (2016)
27. C. Liu, Y. Zhang, J. Zeng, L. Peng, R. Chen, Research on dynamical security risk assessment for the internet of things inspired by immunology, in *2012 Eighth International Conference on Natural Computation (ICNC)* (IEEE, 2012), pp. 874–878
28. F. Maggi, D. Quarta, M. Pogliani, M. Polino, A.M. Zanchettin, S. Zanero, Rogue robots: testing the limits of an industrial robots security. Technical report, Trend Micro, Politecnico di Milano (2017)
29. L. Maglaras, M.A. Ferrag, A. Derhab, M. Mukherjee, H. Janicke, S. Rallis, Threats, countermeasures and attribution of cyber attacks on critical infrastructures. Secur. Saf. **5**(16), 1–9 (2018). https://doi.org/10.4108/eai.15-10-2018.155856
30. L. Maglaras, M.A. Ferrag, A. Derhab, M. Mukherjee, H. Janicke, S. Rallis, Threats, protection and attribution of cyber attacks on critical infrastructures (2019), arXiv:1901.03899
31. E. Marin, D. Singelée, F.D. Garcia, T. Chothia, R. Willems, B. Preneel, On the (in)security of the latest generation implantable cardiac defibrillators and how to secure them, in *Proceedings of the 32nd Annual Conference on Computer Security Applications* (ACM, 2016), pp. 226–236
32. D. Martins, H. Guyennet, Wireless sensor network attacks and security mechanisms: a short survey, in *2010 13th International Conference on Network-Based Information Systems (NBiS)* (IEEE, 2010), pp. 313–320
33. R. Neisse, G. Steri, I.N. Fovino, G. Baldini, SecKit: a model-based security toolkit for the internet of things. Comput. Secur. **54**, 60–76 (2015)
34. C.P. O'Flynn, Message denial and alteration on IEEE 802.15.4 low-power radio networks, in *2011 4th IFIP International Conference on New Technologies, Mobility and Security (NTMS)* (IEEE, 2011), pp. 1–5
35. Y. Peng, T. Lu, J. Liu, Y. Gao, X. Guo, F. Xie, Cyber-physical system risk assessment, in *2013 Ninth International Conference on Intelligent Information Hiding and Multimedia Signal Processing* (IEEE, 2013), pp. 442–447
36. J. Petit, B. Stottelaar, M. Feiri, F. Kargl, Remote attacks on automated vehicles sensors: experiments on camera and Lidar, in *Black Hat Europe*, vol. 11 (2015), p. 2015
37. D. Quarta, M. Pogliani, M. Polino, F. Maggi, A.M. Zanchettin, S. Zanero, An experimental security analysis of an industrial robot controller, in *2017 IEEE Symposium on Security and Privacy (SP)* (IEEE, 2017), pp. 268–286
38. P.A. Ralston, J.H. Graham, J.L. Hieb, Cyber security risk assessment for SCADA and DCS networks. ISA Trans. **46**(4), 583–594 (2007)
39. E. Ronen, C. O'Flynn, A. Shamir, A.O. Weingarten, IoT goes nuclear: creating a zigbee chain reaction. IACR Cryptol. ePrint Arch. **2016**, 1047 (2016)
40. E. Ronen, A. Shamir, Extended functionality attacks on IoT devices: the case of smart lights, in *2016 IEEE European Symposium on Security and Privacy (EuroS&P)* (IEEE, 2016), pp. 3–12
41. R.S. Ross, NIST SP-800-39 Managing Information Security Risk–Organization, Mission, and Information System View. The National Institute of Standards and Technology (NIST), Gaithersburg (2011)
42. R.S. Ross, NIST SP-800-30rev1 Guide for conducting risk assessments. The National Institute of Standards and Technology (NIST), Gaithersburg (2012)

43. R.A. Sahner, K. Trivedi, A. Puliafito, *Performance and Reliability Analysis of Computer Systems: An Example-based Approach Using the SHARPE Software Package* (Springer Science & Business Media, 2012)
44. R. Santamarta, In flight hacking system (IOActive Research Labs) (2016), http://blog.ioactive.com/2016/12/in-flight-hacking-system.html
45. Z. Shelby, C. Bormann, *6LoWPAN: The Wireless Embedded Internet*, vol. 43 (Wiley, 2011)
46. R. Spenneberg, M. Brüggemann, H. Schwartke, PLC-blaster: a worm living solely in the PLC, in *Black Hat Asia*, Marina Bay Sands, Singapore (2016)
47. I. Stellios, P. Kotzanikolaou, M. Psarakis, C. Alcaraz, J. Lopez, A survey of IoT-enabled cyberattacks: assessing attack paths to critical infrastructures and services. IEEE Commun. Surv. Tutor. **20**(4), 3453–3495 (2018)
48. TrapX Research, Labs: Anatomy of Attack: MEDJACK.2—Hospitals Under Siege. TrapX Investigative Report (2016)
49. Wikileaks: Vault 7: CIA Hacking Tools Revealed—CIA malware targets iPhone, Android, smart TVs (2017), https://wikileaks.org/ciav7p1/
50. C. Yan, X. Wenyuan, J. Liu, Can you trust autonomous vehicles: contactless attacks against sensors of self-driving vehicle, in *DEF CON* (2016)
51. S.E. Yusuf, M. Ge, J.B. Hong, H.K. Kim, P. Kim, D.S. Kim, Security modelling and analysis of dynamic enterprise networks, in *2016 IEEE International Conference on Computer and Information Technology (CIT)* (IEEE, 2016), pp. 249–256

Part III
Computer Science-Based Technologies in IOT, Blockchains and Electronic Money

Chapter 9
Sustaining Social Cohesion in Information and Knowledge Society: The Priceless Value of Privacy

Stefanos Gritzalis, Maria Sideri, Angeliki Kitsiou, Eleni Tzortzaki, and Christos Kalloniatis

Abstract Within Information and Knowledge Society the concept of Privacy has been enriched including aspects related to digital life, while the right to online Privacy gains more and more attention daily due to several cases of privacy breaches. Privacy is associated with the control, access and use or misuse of personal information by others, including governments, companies and other users as well. Social Network Sites as a part of digital space have altered the way that people communicate and have contributed to the construction of online social networks. During online interaction, users disclose information about them or others, while at the same time they express their concerns about Privacy infringement that may come up due to self-disclosure practices, not restricting or reversing though their disclosure behavior. Thus the "Privacy paradox" phenomenon is recorded since users cannot balance between Privacy concerns and their need for disclosure. Privacy's circumvention destabilizes the trust between social actors, increases the feelings of insecurity and puts into risk social cohesion which is a prerequisite for the sustainability of our society. Legislation as well as technology may protect us, but sometimes they are not user friendly and sufficient. Users should protect themselves and other people in order to preserve their

S. Gritzalis (✉)
Laboratory of Systems Security, Department of Digital Systems, University of Piraeus, 18532 Piraeus, GR, Greece
e-mail: sgritz@unipi.gr

M. Sideri · A. Kitsiou · C. Kalloniatis
Privacy Engineering and Social Informatics Laboratory, Department of Cultural Technology and Communication, University of the Aegean, 81100 Lesvos, GR, Greece
e-mail: msid@aegean.gr

A. Kitsiou
e-mail: a.kitsiou@aegean.gr

C. Kalloniatis
e-mail: chkallon@aegean.gr

E. Tzortzaki
Laboratory of Information and Communication Systems Security, Department of Information and Communications Systems Engineering, University of the Aegean, 83200 Samos, GR, Greece
e-mail: etzortzaki@aegean.gr

G. A. Tsihrintzis and M. Virvou (eds.), *Advances in Core Computer Science-Based Technologies*, Learning and Analytics in Intelligent Systems 14,
https://doi.org/10.1007/978-3-030-41196-1_9

Privacy as a fundamental human right. In this paper, based on a literature review, we present the issue of Privacy in Social Network Sites focusing on factors that affect people's Privacy concerns and behavior while relating these to social cohesion.

Keywords Privacy · Privacy paradox · Privacy protective behavior · Social network sites · Human rights · Social cohesion

9.1 Introduction

In 1992, Beck [1] in his book titled "Risk Society: Towards a New Modernity" notes that technology and science within modern societies' development produce new forms of risks unknown in previous ages to which we are required to respond constantly. So, the Risk Society, as described by [1], raises itself the risks that threaten its existence. These risks are not limited to specific forms alone (e.g. environmental) but include a whole series of interrelated changes within contemporary social life (e.g. financial crisis, social inequalities, job insecurity, declining tradition influence, human rights jeopardy). In addition, risks are not restricted to one country only, but affect all countries and all social classes having global consequences. However, risks do not automatically lead to societies' destruction, since [1] incorporates in the concept of risk the ability to predict a future disaster in time, which can lead to the disaster's prevention. Nevertheless, even in this case, deterrence is not definitive, as the globalized post-modern society suffers from four systemic defections that amplify risk reproduction; exceeding limits, weakness in control, inadequate compensation for the damage caused and lack of knowledge and awareness regarding the risks [1]. In this way risks rebound.

In the frame above, the concept of sustainability has emerged in order to address the risks that society produces with reference to physical, social, economic and cultural level. The concept of sustainability has been broadened from its original frame, paying nowadays attention synthetically and simultaneously to all three pillars; economy, environment, society. Referring to the field of society, sustainability includes the proposal and the promise for social cohesion maintenance. Social cohesion expresses the extension and quality regarding relationship intensity between the members of a society, recording the degree of synergy between the social subjects. Synergy leads both to the establishment and strengthening of the social consciousness and its manifestation through expressions of social solidarity.

To better understand the concept of social cohesion, the study of social networks as multidimensional systems of communication and shaping of human practice and social identity [2] is required. Social networks are related to a person's social relationships, their characteristics and the way that people perceive and evaluate these relationships. Social networks are characterized by their extent, density, bonding, homogeneity, contact frequency between members, duration and reciprocity [3]. The emotional, psychological or financial support that people can acquire through their social networks constitutes the social support. This is linked to factors that affect the quality of life, such as life satisfaction and sense of well-being [4], while lack of social support and exclusion from networks are considered to reduce people's abilities to form their social identity, receive emotional support or material help and gain access to services and information [5]. In this respect, people's participation in social networks and human rights' respect within these are prerequisite for achieving and maintaining social cohesion in the frame of social sustainability and development both for the present and the future.

Several evolutions in the Information and Knowledge Society are inextricably linked to the processes of social development. Digital social networks for example coexist with offline social networks altering contemporary social life. In this frame, Social Networking Sites (SNSs), that have replaced many forms of offline social activities (e.g. communication, leisure activities, services provision) being tools of both private and public communication [6], are a place that intersects public and private practices. Promoting the social interconnection between users, facilitating and encouraging users' communication within and beyond the direct contacts of their networks [7], SNSs constitute, on the one hand, an appropriate field of social development, while on the other hand their usage raises multiple issues regarding human rights both at individual and social level.

The evolutions having taken place within Information and Knowledge Society highlight the necessity for human rights protection at digital level as well [8]. In this frame, the issue of privacy and its protection—not being a new social phenomenon though—takes a foreground place in the scientific community among IT, legal and social scientists, following a multidisciplinary approach. What is private and how private is intertwined with the public is an issue that is rooted in the very beginning of human presence. We should note that the distinction between private and public is related to the social context in which it occurs, underlining also that the social and cultural factors that determine the concept of privacy do not remain stable, altering thus the perceptions regarding private and public. Nevertheless, as noted by [9], after the technological evolution "*the classical concept of privacy has been greatly enriched*" (p. 508). Considering that within Information Society the relationships between the different information managers are complex, the distinction between private and public becomes even more obscure [10, 11].

The safeguarding of individual rights in the 18th and 19th centuries allowed the formulation of the right to privacy which is directly linked to the freedom of a person from all forms of control/surveillance and insult. At the same time, the legal introduction of the right to privacy has consistently led to the introduction of a constitutional protection obligation. So, theoretically, we live in a world where privacy is now legally enforceable and self-evident in every form of social practice, such as the use of SNSs. Is this real?

This paper addresses the right to privacy in online networks framed on Social Networking Sites. Section 9.2 refers to privacy in SNSs, focusing on the way SNSs have become a part of contemporary reality having effect on human experiences. This Section addresses also the Privacy Paradox phenomenon as the state of contradiction between privacy attitude and privacy behavior, includes subsections regarding the factors that affect users' disclosure practices and privacy concerns, while it also records measures that should be taken for privacy protection. Section 9.3 underlines that the right to privacy is one of the most endangered human rights in the context of globalization and emergence of the Information and Knowledge Society. Privacy violation affects social cohesion and puts thus into danger the whole society.

9.2 The Right to Privacy in Information Society. The Case of Social Networking Sites

Information holds a key role within Information Society. In this frame, control of information produces new conflicts that raise unique global risks [12] regarding individual rights, protection of personal data, and security of information. These risks come up because the social, economic and political functions of states directly depend on information circulating in information systems, while also depending on private sector [13–15] which due to the competition rules cannot provide security guarantees for the democratic orientation of the states. As a result, national governments challenge the control of information, establishing, limiting and applying laws that balance public and private interests [13, 16]. In this frame, the terms for privacy protection are being renegotiated globally. It is understood, thus, that the above-mentioned risks "*do not derive from external phenomena but from human decisions and actions*" [1] (p. 50) concerning the control and use of information according to the visible and latent interests of the social groups they serve.

A series of recent incidents, as that of the Snowden case in 2013 or of Cambridge Analytica in 2018, confirm that despite the constitutional requirements for privacy, governments or politicians in cooperation with companies use internet-based information and organize mass-tracking programs for citizens. Hundreds of millions of data are collected, while governments and private organizations/service providers refuse that they collect and distribute citizens' data.

Within a society where information is disseminated through every possible internet source becoming accessible to all and legislation has established general principles for privacy protection, while the states may have different starting points of legal culture, the interpretations of privacy become more and more obscure [16]. The regulatory framework for privacy protection is multidimensional concerning both the application of international law conventions, national regulations, decisions by independent authorities that manage information issues, and rules of private sector bodies through self-regulation [17]. Within this complex frame, keeping in mind that citizens are constantly expressing their anxiety about who has access to their data, it is particularly interesting to consider how citizens perceive themselves in online networks and take care to ensure their privacy—if they do so-, acknowledging it as an indefeasible right.

9.2.1 Users Privacy Experiences in Social Networking Sites

Social media are the outcome of the technological development during the last decades. Beyond a technological phenomenon, social media constitute a social phenomenon since their effect on human daily reality is catalytic. This is evidenced by the growing number of users, the new applications and the multiple environments to which social media have been exploited.

Social Networking Sites (SNSs)—one of the categories of social media- dominate in almost all human activities, facilitating the interaction between people, the online procurement of goods and services, business transactions, communication between the state and the citizens, the development of communities. In this context of ubiquitous presence of SNSs, social subjects adapt to a new "reality", the digital reality, that operates within the framework of social action, often shaping its own norms.

To participate in a SNS, the user has to build a profile that represents, in a way he/she chooses, the digital persona adopting specific methods to present and control his/her image. Users share a large amount of information in a variety of forms, such as personal data, photos, thoughts, experiences and preferences—sometimes true, sometimes false—leaving their digital footprint in every function. At the same time, this process raises users' anxiety regarding their privacy and the security of their data even though they voluntarily provide personal information and/or carelessly consent to its collection. SNSs provide users the facility to create new relationships or to preserve pre-existing, to self-present, to explore photos and profiles of other users, to activate post-communication forms such as commenting on messages posted or to have fun [18–22]. Pearson [23] has pointed out that the specific nature of SNSs creates intimacy feelings that encourage the information flow within them, allowing users to feel that they can maintain relationships not only at personal level but at professional also, as noted by [24].

SNSs are currently the most dynamically developing personal networking tool [6], as they contribute to the promotion of online interpersonal interactions based on the norms of daily interaction, allowing both the expression of personal identity and community building [25]. In this frame, it is clear that SNSs usage leads to the increase of users' material and symbolic resources reconstructing simultaneously the social status, since SNSs constitute the modern practice of participation in social networks in the Information Society. In this frame, the establishment of a collective digital culture, built on reciprocity and trust which are crucial for social development and sustainability is recommended. But what is the price?

Although users believe they can control the information they share, controlling thus their privacy, Conger et al. [26] point out that today information is not under the control of individuals, but of organizations that hold it. In the context above, users experience or learn about incidents ranging from personal data violation to online personalized advertisements. Violations of users' privacy may arise, in addition to those known as a result of the operation of governments and companies, by other users also, due to the multiple forms of unwanted or uncontrolled information disclosure, regardless the number of persons to whom it is disclosed, since information can be easily found and copied. Incidents regarding violation or misuse of personal information raise users' privacy concerns and anxiety about their visibility and vulnerability in digital environments. Users experience the feeling of intrusion into their personal lives, the concern that one knows their habits and preferences, controls their behavior and guides their daily practices. These anxieties will grow even more as the technological advances of mobile devices are moving fast forward [27]. Despite these, the number of SNSs users continues to grow steadily, because SNSs bear a form of glamor resulting from the combination of the possibilities they offer for self-presentation and social interconnection as [25] argues.

Privacy is a multidimensional concept and is perceived by people in different ways defined by a variety of parameters. Benson et al. [27] note that the literature provides five variables of privacy, including: "*perceived ability to control submitted information*", "*use of information*", "*notice*", "*perceived privacy*" and "*privacy protection behavior*" (p. 430).

9.2.1.1 Privacy Paradox

In order to understand the concept of privacy in digital environments, several studies [28–31] use the definitions of Westin and Altman. According to Westin, privacy is defined as one's right to determine what information is accessible, to whom and when, while in Altman's view privacy is determined as the selective control of individuals on others' access to their information, forming thus a social and dynamic process targeting the achievement of optimization in the relationship between information disclosure and withdrawal [28, 29, 31].

The relationship between privacy on SNSs and information disclosure is a multifacet issue [32–35]. Taddicken [28] has recorded the tense in the relation between users' desire and need to protect their privacy, and their desire to disclose personal information, which may lead them to underestimate the privacy risks resulting from personal information disclosure. As underlined by Buschel et al. [36] this relationship "*is characterized by a constant tension between secrecy and transparency. On the one hand, individuals are afraid of threats to their personal autonomy and freedoms stemming from a global data processing by governments and undertakings, while on the other hand they voluntarily proceed to the disclosure of personal data (eg by posting names, photographs, dates of birth, marital status....)*" (p. 642).

Referring to privacy in Web in general, Taddicken [28] points out that its ideal achievement is based on a balanced relationship between individuals' needs for social interaction and personal information disclosure, and their needs for privacy. So, as it happens with privacy in real life, SNSs' users need to balance their concerns regarding their visible content on a Web site to a variety of audiences with their desire to enjoy privileges because of their interactions in the website [30]. According to Petronio [37], the balance between privacy and self-disclosure is the core of human behavior and determines interpersonal relationships. The choice of more or less privacy changes according to wishes, social goals and specific context, influencing thus the ways in which interpersonal boundaries in relationships are being negotiated.

However, what has been observed in a number of researches is that SNSs users do not always manage to balance these needs. Several studies have dealt with the issue of privacy concerns' impact on users' behavior and have comparatively examined the stated attitude and the actual behavior, demonstrating that although users are interested in their privacy on SNSs and have concerns regarding the security of personal information [38–40] or feel vulnerable to privacy violations [30], these concerns are not followed by disclosing less information or changing privacy settings for example. Consequently, people fail, as Baek [41] explains, to turn their privacy concerns into privacy protection behaviors. In this way an inconsistency or discrepancy is revealed between views and attitudes on one hand and behavior on the other with reference to the informational privacy.

"Privacy paradox" [40, 42] finally emerges as the state of contradiction between privacy attitude and privacy behavior. In a recent literature review paper, Kokolakis [43] notes that although this is the dominant dichotomy when referring to privacy paradox, researchers have also compared privacy concerns with privacy behavior. Even though these two constructs are related, they are also fundamentally different, as "*privacy concerns could be quite generic and, in most cases, are not bound to any specific context, whilst privacy attitudes refer to the appraisal of specific privacy behaviours*" [43] (p. 123). Furthermore, [43] underlines that several studies investigate privacy intention instead of privacy behavior ignoring that often privacy intentions do not lead to protective behavior. Acquisti and Gross [40] attempting to interpret the "privacy paradox" explain that this discrepancy is likely to be based on users' trust towards service providers and other users if users consider providers to be honest with them [44] or if they recognize similarities between themselves and other users [45].

9.2.1.2 Factors Affecting Personal Information Disclosure

What impels people in online spaces such SNSs to reveal information about themselves and others, even though they really know or at least suspect that information is accessible? How could one interpret the fact that we often reveal more information during our online interactions with others than in our face to face communication? Many researches have attempted to uncover the factors that influence the decision-making process to disclose personal information on SNSs.

Benson et al. [27] have investigated the relationship between information disclosure and three important dimensions; control over personal information, user awareness, and security/privacy alerts. Information control is recognized as a key element in the perception of risk [46, 47] that derives from information disclosure. Brandimarte et al. [48] verified the hypothesis that the increased control individuals think they have regarding sharing and access to their information increases their willingness to disclose sensitive information, and if this increase is high, users will end up being more vulnerable, despite the fact that technologies are designed to protect them. So, [48] conclude that the perceived ability of people to control certain dangers shields their awareness or turns their attention to other dangers they cannot control.

Many researches have focused on users' general lack of awareness regarding the usage of their information by SNSs and third parties, including governments also [49, 50]. Christofides et al. [51] argue that awareness of the consequences resulting from privacy breaches predicts disclosure. The positive correlation between users awareness and information disclosure has been also supported by [27] who argue that when users have a better knowledge about the use of personal information, they are more likely to reveal more information. This finding is particularly interesting for user awareness programs.

The low level of knowledge has been shown to be related to the tendency or temptation to reveal personal information in order to gain small benefits [52–54]. Baek [41] predicts that in the future, in larger social environments, there will be a privacy protection gap "*given that knowledgeable users understand why their online privacy matters while less knowledgeable users may be easily persuaded to trade their privacy for transient benefits*" (p. 40). Digital literacy on the contrary seems to have positive effects on online privacy protection [55–57] being recorded as a prerequisite for the understanding of technical terms such as cookies and data mining [56, 57]. In this context, researches such as those of [58] and [57] have focused on users' lack of ability, knowledge and privacy protection skills, identifying this situation through the theory of cognitive deficiency. Debatin [59] referring to users' privacy literacy notes that it "*encompasses an informed concern for their privacy and effective strategies to protect it*" (p. 51), while [60] claim that "*online privacy literacy can be defined as a combination of factual or declarative knowledge ('knowing that') and the procedural ('knowing how') knowledge about privacy*" (p. 339). The first one refers to users' knowledge of the technical aspects regarding their data protection, the relevant laws and directives, while the second to their ability to use strategies in order to regulate their privacy and protect their data.

Self-disclosure has been also studied with reference to social influence [44, 61] and online trust [62, 63], revealing that both factors increase self-disclosure while, on the contrary, perceptions regarding the risks for privacy breach reduce it. In this frame, the influence by friends' practices regarding privacy settings or the social pressure that users receive from their social environment in order to participate in SNSs seem to affect disclosure behaviors. Tufekci [64] referring to the contribution of social factors, during a research addressed to students, has shown that students cannot avoid participating in Facebook, which acts as a social norm for their everyday life, although they have made progress regarding the personal information they reveal and share. As Ziegele and Quiring [31] point out "*(perceived) social norms seem to play an important role in determining personal and spatial access restriction to user profiles as well as the amount and the kind of information individuals provide within SNSs*" (p. 185). An important factor that also affects disclosure is people's need to adapt to the expectations of a group or community in order to avoid exclusion, indicating thus specific behavior that is determined mainly by their own representation of the group's expectations [65]. In this frame, one's need to feel being part of a group (sense of belonging) can limit privacy concerns.

Other approaches emphasize on the incentives that trigger users' disclosure behaviors. Social capital, social support, maintaining communication with others, starting new relationships, self-promotion and entertainment/fun have been recognized as such motivations for users' operation on SNSs [45, 66–71]. Steinfield et al. [72] report that Facebook users disclose personal information in order to acquire social capital benefits, while [73] underline that in order to achieve these benefits, disclosure needs to be permanent. In the frame above, self-disclosure is perceived as a privacy transaction, since users believe they will receive a reward if they reveal personal information and thus they behave in the opposite direction of protecting their privacy [74, 75]. The choice to disclose or conceal personal information constitutes consequently an act of balancing between the perceived benefits and the perceived costs [45].

Empirical researches in the field of psychology associate the control of information communicated and the information disclosure with personality treats, such as the need for popularity and self-esteem. For example, [76] have shown that Facebook users who disclose a large amount of information are possessed by a tension for self-promotion. They also demonstrated that those using Facebook for the creation of digital communities are the ones who reveal the most essential personal information and appear to be socially extroversial, while for both categories of users it was pointed out that the number of posts grew when they experienced periods of low self-esteem [76]. The reasons that lead to these behaviors, apart from the psychological factors, include reduced social cohesion and lack of satisfaction resulting from users' offline relationships as well.

The sensitivity level of information has also been reported as a factor for users' willingness to disclose information. The studies of [28, 77] have shown that information sensitivity raises the belief in risk while decreasing the desire for disclosure. So, users are more cautious when they reveal sensitive information in relation to less sensitive. A recent research however, regarding Greek Universities students' Facebook communities [78], showed that within these communities University students felt that they could share even the most inner information about their sexual life, thus limiting the concept of privacy.

The structure of SNSs has also a crucial role regarding users' information disclosure, as shown by [79] in the case of Facebook. In many cases, a user in order to use the services of a SNS, has to reveal information according to SNS's operating preconditions [38, 79, 80]. Ziegele and Quiring [31] note that providers, through technical features, try to maximize the amount of information they receive from users to make Websites more dynamic and attractive in order to make a profit. The high intensity usage of SNSs also leads to disclosure behaviors [68]. Finally, the state of anonymity has also been recognized as a factor influencing disclosure and reducing privacy concerns [74, 81].

Personal information disclosure, consciously or not, is ultimately a common practice among users involving heterogeneous audiences with different social relationships within users' networks. Although disclosure can be made either with full publicity and to unknown users or to specific individuals within users' network [28], the information is very easy to be found, copied, expanded and shared in both cases

[82]. As Taddicken [28] points out users who reveal personal information are often not sure who and how many people are included amongst the audiences at which the revelations have been made, due to the temporal and spatial segregation that exists in relationships developed between these audiences.

9.2.1.3 Privacy Concerns

Most researches regarding privacy concerns or related protective behaviors focus on individual level [53, 83, 84]. Alashoor et al. [85] have highlighted the impact of cultural values on users' privacy concerns and the way they may affect self-disclosure, noting that they also influence the assessment regarding the sensitivity of the personal information communicated. In this frame, researches [54, 86, 87] have focused at country level investigating how individuals from different cultural contexts evaluate privacy and respond to privacy concerns and privacy issues.

Baek [41] in his study regarding the factors of privacy concerns has included the dimension of perceived risk for other users, using the concept of comparative optimism as reported by [88]. Comparative optimism refers to the state of belief that the individual is more protected than others, mostly compared to more vulnerable others or groups. This situation may arise either from the underestimation of personal risk or from overestimating the vulnerability of others regarding online privacy violation, but in both cases it refers to knowledge about privacy protection [55].

Privacy, as already stated in the previous sub-section, is directly related to the control of personal information. Brandimarte et al. [48] underline that a distinction should be made between the act of voluntary information disclosure, the access and the use or misuse of information, emphasizing thus that the resulting cost depends on access and use/misuse of information, which people fail to conceive as they focus on the first level of control (release of information). In this frame, previous researches have shown that lower estimated control over personal information is associated with higher privacy concerns [89], while in other cases it has been pointed out that those who are indifferent to privacy feel that they have control over the information they reveal [40]. These findings verify [48] argument that "*perceived control over release or access of personal information can cause people to experience an illusory sense of security and, thus, release more information.* Vice versa, *lack of perceived control can generate paradoxically high privacy concerns and decrease willingness to disclose, even if the associated risks of disclosure may be lower*" (p. 342).

Privacy concerns also relate to the security of SNSs, as demonstrated by [40]. Acquisti et al. [90] explain that businesses trying to convince customers about the security of their personal data have introduced new techniques—self-regulatory transparency mechanisms—that provide alerts and include privacy statements and privacy seals. However, former researches have shown that privacy seals can increase the willingness to disclose information [91], thus putting aside privacy concerns. It seems, therefore, that privacy concerns are affected by users' confidence in privacy settings.

Age seems to be an important factor in privacy concerns also. As Steijn [92] explains, people belonging to different age groups vary in their perception of privacy and the way they can manage it. Boyd [93] reports that young people are willing to experiment with SNSs and this can lead them to behave inconsiderably or recklessly, while other researches have shown that young people have a higher level of privacy awareness [64, 94, 95]. Older people usually have more difficulty to understand and implement privacy settings and this turns them into potential high-risk users [96]. Van den Broeck et al. [97] investigated the use of Facebook, privacy concerns and the application of privacy settings in the three stages of adulthood (18–25, 25–40, 40–65) revealing differences between the three groups. Specifically, groups aged 25–40 and 40–65 years old were more vulnerable in terms of privacy protection than those of 18–25 years, who were recorded as conscious users with reference to privacy. Moreover, those aged 40–65 had greater privacy concerns than other age groups, although they admitted they were less likely to use privacy settings.

Gender constitutes a variable whose impact has been investigated in relation to privacy perceptions and privacy concerns. For example, Sheehan [98] and Acquisti and Gross [40] have shown that men are less concerned about online privacy, Jensen et al. [99] that women are generally more risk-averse, while other researchers [100–102] did not identify significant differences between genders in relation to privacy perception. With reference to teenagers, Feng and Xie [103] recorded no difference in privacy concerns between boys and girls, although the latter were more likely to have their profiles private and adopt privacy protection strategies to avoid victimization.

SNSs users' privacy concerns are related not only to the protection of their personal information disseminated to others who could exploit this, but are also related to the protection and management of their image in the frame of the relationships that they have developed within their network [30]. The extent to which these concerns are positive making users more cautious both in terms of quantity and quality of information they publish either for themselves or for others will ultimately determine the extent to which they will protect themselves from the potential problems that will arise from the exploitation of information.

9.2.2 Privacy Protection in Social Networking Sites

Several proposals have been made in order to ensure that personal information circulated on Internet and social media is kept safe, not accumulated and used by others -no matter who they are- without the explicit consent of the users. In this context, it is suggested that legislation should be strengthen in order to regulate the technological planning of data collection and the control of data acquired [36]. The European Commission in 2012 reformulated the European Union Data Protection Directive (1995) proposing the establishment of "the right to be forgotten", "privacy by default" and "privacy by design" in order to enhance privacy protection [104].

Referring to the providers, Cheung et al. [44] highlight the need to introduce "*more social features that foster users' interactions over the Social Networking Sites, such as person profile customization or news feed notification services*", while they also note that service providers "*can integrate intuitive privacy indices, showing users the level of privacy protection to alert them about the potential risks of self-disclosure in SNSs*" (p. 293). Nguyen and Mynatt [105] state that users could be helped to confront privacy issues if the configured information systems provide them with mechanisms and interfaces enabling them to understand their function and if these mechanisms become integrated into users' practices, values and sensitivities.

From a more technical point of view, software engineers consider privacy in a more technical sense mainly as a set of specific requirements that need to be fulfilled in order for a system or service to become privacy aware. In previous works [106–108] a method that assists software engineers in eliciting and modeling privacy requirements during system design is presented. The findings of these works show that for increasing users' privacy, it is of vital importance to understand the factors that overcome the close boundaries of an information system and its technical abilities (fulfilled requirements), as privacy is a multifaceted concept that is related to user's social and behavioral characteristics. Creating trustworthy systems and services that fulfill specific security and privacy requirements taken into consideration external non-technical factors is a solution towards this direction [109].

In addition to legislation's provisions and providers' obligations, it is important to activate users and enhance their awareness to use strategies in order to mitigate the risks resulting from disclosure to unwanted audiences [30]. These strategies relate both to personal information disclosure behaviors and the use of privacy control techniques provided by the Web sites. In the frame of the first dimension, users choose the type of information they record in their profile or the updates they share in their Status [30], control the network of their Friends [29, 110], retain different profiles, do not accept friend requests from strangers, delete comments or remove photos [111, 112]. Stutzman et al. [79] research, from 2005 to 2011, investigated Facebook users' behavior regarding personal information disclosure options, showing that the amount of information users choose to disclose to their friends has grown despite existing privacy concerns, while disclosure of information to profiles of strangers has decreased. With reference to the second dimension, that of technical control, users in order to protect their privacy use privacy settings [29, 30]. Johnson et al. [113] record that although the Facebook privacy control techniques allow users to successfully manage privacy threats from unknown external audience, they provide poor choices in relation to risk reduction arising from the existing Friends network. As pointed out by Stutzman et al. [79], Facebook, in recent years, in order to encourage the disclosure of personal information has changed the default settings when new users register on the Site.

Finally, educational programs aiming at raising users' awareness regarding potential risks deriving from self-disclosure on SNSs and adopting relevant protection behaviors are particularly important as shown in several researches [114–117]. Specifically, long term educational interventions are shown to have a significant impact on students' attitude, increasing both privacy awareness and concerns through acknowledging risks in SNSs and confronting them. Awareness increase leads users/students to adopt privacy protective behavior either by using personal strategies or employing technical mechanisms [118].

9.3 Ensuring Social Cohesion in Information and Knowledge Society

The digital revolution led to a new reality that essentially altered not only the way people perceive the social environment, but also social environment itself. Communication with friends, creation of new relationships, search for support from others, need to present oneself—sometimes even in the form of projection-, participation in communities of common interest, products and services' market, transactions with the state and other organizations are all fields mediated by the digital technology of SNSs. Indeed, social media and specifically SNSs have shaped new norms and practices in modern society, transforming among others the form of human interaction. The fundamental elements of users experiences on SNSs—in the sense that [119] refers to experiences as "relationships of power and forms of relationship between the Self and Others"—point out that within Information Society multiple aspects of social reality are being remodeled and redefined in particularly obscure and inconspicuous ways [1]. The flood of information [120] opens the path for more knowledge and individual and social rights on the one hand, while on the other it sets under negotiation concepts such as privacy and security.

In the context of SNSs, users "*when creating the personal information they want to share with others, they decide at the same time how they wish to be perceived by other members of the community*" [121] (p. 6), while providers with sophisticated techniques gather and process large amounts of information either by themselves or providing it to others (governments). Thus, as [120] points out, people's exposure to a "*flood of information*" raises conflicts about security, predictability, sense of belonging, stable personal identity, cohesion, unmediated experiences.

Within this flood, "*the distinction limits between personal data and personal data accessible to public are equally indistinguishable, which suggests that the possibilities of using and misusing personal data multiply*" [16] (p. 38). In this respect, the exercise of the power regarding personal data management runs throughout the social body, without being clear the conditions of enforcement and compulsion. The increase of control over individuals serves the purpose of safeguarding the well-being of social media large companies. As a result, besides the role of the state that changes [122], companies are increasingly involved in power, exercising ideological and political control [15]. This issue further reinforces [119] thesis on the development of "problem-making" when considering the protection of privacy and its effects on social cohesion, especially in the context of Information Society, by cultivating practices that pose problems on every political and social choice.

In this context, many researches on social media [123–126] use the "Panoptikon", a framework for monitoring prisoners developed by Jeremy Bentham in the late 18th century. "Panoptikon" extends into cyberspace. Potentially everyone can be seen by everybody. This reality alters the concept of privacy, while users often have the illusion of privacy which makes difficult to delimit the kind of information they should share [52]. This personalized exercise of power clearly illustrates the danger already identified by [127] regarding social systems of high differentiation, where the exercise of social control is pushed "*to the most intricate sphere of the meaning*" (p. 85) of the social actors, while simultaneously dominant established collective values are constructed. Through the operation of social media companies, specific interests are built up as values in relation to privacy and these are reproduced over the years, ending up in their encapsulation and integration by the community.

Although all may espouse these dominant values at the theoretical level, at empirical and experiential levels this may not happen [128]. Viégas [129] argues that there are differences between users' representations regarding how they feel about privacy and how they really react to its violation. So, although users have embraced or agreed to the general value of privacy protection, they may take up actions that contradict it accordingly to the effort to achieve the goals they have set in defending their individual interests. Underlining the phenomenon recorded as "Privacy Paradox" [52, 83, 130] which refers to the differentiation between the intention of social subjects to disclose personal information and the actual disclosure behavior, mediated by privacy concerns, it is important to note that the perception of privacy shows significant variations between socio-cultural systems [131, 132]. Users, according to the assessment of the situation they make through social media usage, show the extent to which they have incorporated the value of privacy protection and whether they are prepared to defend it through practices and actions in social media, in each case the value is specialized, goes beyond its abstract context and concerns specific purposes and interests. It should be noted that, even if there is complete consensus on the value of privacy protection, it is impossible to have full consensus on its evaluations. These evaluations result in the formulation of criteria of action directly linked to specific situations. However, the criteria cannot appeal to all social media users given the diversity that characterizes them.

The possibility of privacy violation is one of the greatest risks in the globalized environment of the Information Society [133], since it generates the lack of respect for the right to privacy as well as the control exercise over individuals, while at the same time creates significant opportunities and challenges for the delimitation of collective values and social behaviors in relation to privacy protection.

Thus, beyond the obvious responsibility of third parties, whoever those are that violate national and international conventions on the protection of human rights, we must think on the role of individuals/users in the process of revealing their information. Regardless the need to communicate with others, to join a team and gain benefits and despite the obvious and recorded privacy concerns, users themselves generate the risk, not only for themselves but for others too and potentially for the whole society, given their criteria of action and the collective representation that we all are somehow connected. Users' intention and need to interact with others even if they have to reveal personal information and their disclosure behavior constitute parts of a recurrent process of privacy risks generation. In this frame privacy risks re-occur as a result of the four systemic defections that contribute to the risk reproduction according to [1].

Based on [1] thesis that the concept of risk involves the possibility of timely forecasting that can lead to the prevention of future disaster, the role of users regarding their self-protection is of major importance, leading to a new form of social development in Information Society, within which one of the basic principles is that of personal responsibility. After all, as Foucault [119] records, from the moment when certain relations of power develop there are synchronically resistance possibilities too, which need to be equally resourceful, dynamic and productive.

The users' concerns and their anxieties with regard to privacy protection prove that they recognize the risk, while the strategies and techniques of protection they adopt show that they try to avert the risk. In this frame, social development ensure is based on a shift in intervention; from repressive to preventative intervention with an emphasis on users' awareness. As a matter of fact, this is a form of personal development that, as recorded by [134], collectively ensures social progress. On this basis, in order to make social development sustainable, its planning should be based on citizens' actions which will perpetuate and maintain it through participation and democracy, always taking into account the environmental constraints and the clear knowledge of their needs [135].

However, as [1] argues, prevention is not final as post-modern society suffers from the four systemic defects that contribute to risk reproduction. Thus, in cases where users exceeding the limits are unable to control the information they publish for themselves or others and to control who and how can access and use this information or in cases where users underestimate the harm that can come up for themselves or others while overestimating the perceived control over information, the risk for the members of society recurs. In this respect, the results of the [136] and [137] studies who argue that the co-responsibility of public and private organizations brings more effective measures for social development, need to be applied in terms of privacy protection also.

Nevertheless, according to the cognitive approach for social development planning, its production processes consist of overlapping interventions designed by experts and those benefited [138], requiring everyone's participation in the effort to improve the quality of life and social autonomy [139]. As [9] notes "*the concepts of society and privacy are completely interrelated, since without society there would be no need and demand for privacy*" (p. 507).

Privacy protection has been recognized as an important principle in all modern democracies [140] and its preservation has been identified as a major need [141]. In this context, social development principles based on privacy protection in social media can focus on users' personal development, their development as members of digital communities emphasizing on knowledge addressing to users' needs and goals for privacy protection, on practices assessment and on users' demand for social media providers adaption to their needs as well.

References

1. U. Beck, *Risk Society Towards a New Modernity* (Sage, London, 1992)
2. S. Chtouris, *Rational Symbolic Networks—Global States and National Hobbit* (Nisos Publ, Athens, 2004). (in Greek)
3. L.F. Berkman, T. Glass, Social integration, social networks, social support and health, in *Social Epidemiology*, ed. by L.F. Berkman, I. Kawachi (Oxford University Press, New York, 2000), pp. 158–162
4. E. Breeze, C. Grundy, A. Fletcher, *Inequalities in Quality of Life Among People Aged 75 Years and Over in Great Britain* (University of Sheffield, UK, 2001), https://www.growingolder.group.shef.ac.uk/GOProgSumms.pdf. Accessed Jan 2019
5. K. Walker, A. Macbride, M.L.S. Vachon, Social support networks and the crisis of bereavement. Soc. Sci. Med. **11**, 34–41 (1997)
6. K.Y. Lin, H.P. Lu, Why people use social networking sites: an empirical study integrating network externalities and motivation theory. Comput. Hum. Behav. **27**(3), 1152–1161 (2011)
7. P. Pai, D.C. Arnott, User adoption of social networking sites: eliciting uses and gratifications through a means–end approach. Comput. Hum. Behav. **29**(3), 1039–1053 (2013)
8. S. Gritzalis, Enhancing web privacy and anonymity in the digital era. Inf. Manag. Comput. Secur. **12**(3), 255–288 (2004)
9. L. Mitrou, Privacy protection in information and communication technology—the legal dimension, in *Privacy and Information and Communication Technologies—Technical and Legal Issues*, ed. by C. Lambrinoudakis, L. Mitrou, S. Gritzalis, S. Katsikas (Papasotiriou Publ., Athens, 2010), pp. 505–551. (in Greek)
10. T. Jones, T. Newburn, *Private Security and Public Policing* (Clarendon, Oxford, 1998)
11. G.T. Marx, Murky conceptual waters: the public and the private. Ethics Inf. Technol. **3**(3), 157–169 (2001)
12. J. Panousis, *Symbolic Constructions of Reality* (Entelecheia Publ, Athens, 2004). (in Greek)
13. A. Kitsiou, The social construction of crime in Information Society: the paradigm of free software movement. Ph.D. thesis. Department of Sociology, University of the Aegean, 2014 (in Greek)
14. J. Kallas, *Information Society and the Role of Social Sciences* (Nefeli Publ, Athens, 2006). (in Greek)
15. J.L. Cebrian, *The Network* (Stachi, Athens, 2000). (in Greek)
16. L. Mitrou, *Law in Information Society* (Sakkoulas Publ, Athens, 2002). (in Greek)
17. E. Simon, Introduction to the legal frame of information society, in *Information Society, Studies on Information Society. From Theory to Political Practice,* ed. by R. Pinter (Gondolat, Budapest, 2008)
18. K.T. Lee, M.J. Noh, D.M. Koo, Lonely people are no longer lonely on social networking sites: the mediating role of self-disclosure and social support. Cyberpsychol. Behav. Soc. Netw. **16**(6), 413–418 (2013)
19. E.M. Bryant, J. Marmo, A. Ramirez Jr., A functional approach to social networking sites, in *Computer-Mediated Communication in Personal Relationships*, ed. by K.B. Wright, L.M. Webb (Peter Lang, New York, 2011), pp. 3–20
20. J. Kim, J.R. Lee, The Facebook paths to happiness: effects of the number of Facebook friends and self-presentation on subjective wellbeing. Cyberpsychol. Behav. Soc. Netw. **14**, 359–364 (2011)
21. T.A. Pempek, Y.A. Yermolayeva, S.L. Yermolayeva, College students' social networking experiences on Facebook. J. Appl. Dev. Psychol. **30**, 227–238 (2009)
22. J.A. Bargh, K. McKenna, The Internet and social life. Annu. Rev. Psychol. **55**, 573–590 (2004)
23. E. Pearson, All the worldwide web's a stage: the performance of identity in online social networks. *First Monday***14**(3) (2009), https://firstmonday.org/article/view/2162/2127. Accessed Dec 2018
24. M. Trusov, R.E. Bucklin, K. Pauwels, Effects of word-of-mouth versus traditional marketing: findings from an internet social networking site. J. Mark. **73**, 90–102 (2009)

25. Z. Papacharissi (ed.), *A Networked Self: Identity, Community, and Culture on Social Network Sites* (Routledge, New York, 2011)
26. S. Conger, J.H. Pratt, K.D. Loch, Personal information privacy and emerging technologies. Inf. Syst. J. **23**(5), 401–417 (2013)
27. V. Benson, G. Saridakis, H. Tennakoon, Information disclosure of social media users. Does control over personal information, user awareness and security notices matter? *Inf. Technol. People***28**(3), 426–441 (2015)
28. M. Taddicken, The 'privacy paradox in the social web: the impact of privacy concerns, individual characteristics and the perceived social relevance on different forms of self-disclosure. J. Comput.-Mediat. Commun. **19**(2), 248–273 (2014)
29. F. Stutzman, J. Vitak, N.B. Ellison, R. Gray, C. Lampe, Privacy in interaction: exploring disclosure and social capital in Facebook, in *Proceedings of the Sixth International Conference on Weblogs and Social Media* (AAAI org, Ireland, Dublin, 2012), pp. 330–337
30. N. Ellison, J. Vitak, C. Steinfield, R. Gray, C. Lampe, Negotiating privacy concerns and social capital needs in a social media environment, in *Privacy Online: Perspectives on Privacy and Self-disclosure in the Social Web*, ed. by S. Trepte, L. Reinecke (Springer, Heidelberg, 2011), pp. 19–32
31. M. Ziegele, O. Quiring, Privacy in social network sites, in *Privacy Online: Perspectives on Privacy and Self-disclosure in the Social Web*, ed. by S. Trepte, L. Reinecke (Springer, Heidelberg, 2011), pp. 175–189
32. S.A. Rains, S.R. Brunner, The outcomes of broadcasting self-disclosure using new communication technologies responses to disclosure vary across one's social network. Commun. Res. **45**(5), 659–687 (2015)
33. N.N. Bazarova, Y.H. Choi, Self-disclosure in social media: extending the functional approach to self-disclosure motivations and characteristics on social network sites. J. Commun. **64**, 635–657 (2014)
34. T. Spiliotopoulos, I. Oakley, Understanding motivations for Facebook use: usage metrics, network structure, and privacy, in *Proceedings of the SIGCHI Conference on Human Factors in Computing Systems* (ACM, Paris, France, 2013), pp. 3287–3296
35. S.C. Walton, R.E. Rice, Mediated disclosure on Twitter: the roles of gender and identity in boundary impermeability, valence, disclosure, and stage. Comput. Hum. Behav. **29**, 1465–1474 (2013)
36. I. Buschel, R. Mehdi, A. Cammilleri, Y. Marzouki, B. Elger, Protecting human health and security in digital Europe: how to deal with the "privacy paradox"? Sci. Eng. Ethics **20**, 639–658 (2014)
37. S. Petronio, *Boundary of Privacy: Dialectics of Disclosure* (State University of New York Press, Albany, 2002)
38. M. Nguyen, Y.S. Bin, A. Campbell, Comparing online and offline self-disclosure: a systematic review. Cyberpsychol. Behav. Soc. Netw. **15**(2), 103–111 (2012)
39. D.M. Boyd, E. Hargittai, Facebook privacy settings: who cares? *First Monday***15**(8) (2010), https://firstmonday.org/article/view/3086/2589. Accessed Jan 2019
40. A. Acquisti, R. Gross, Imagined communities: awareness, information sharing, and privacy on the Facebook, in *Proceedings of the 6th Workshop on Privacy Enhancing Technologies (PET '06),* ed. by G. Danezis, P. Golle (Robinson College, Cambridge, UK, 2006), pp. 36–58
41. Y.M. Baek, Solving the privacy paradox: a counter-argument experimental approach. Comput. Hum. Behav. **38**, 33–42 (2014)
42. T. Dienlin, S. Trepte, Putting the social (psychology) into social media is the privacy paradox a relic of the past? An in-depth analysis of privacy attitudes and privacy behaviors. Eur. J. Soc. Psychol. **45**, 285–297 (2015)
43. Sp. Kokolakis, Privacy attitudes and privacy behaviour: a review of current research on the privacy paradox phenomenon. Comput. Secur. **64**, 122–134 (2017)
44. C. Cheung, Z. W. Y. Lee, T. K. H. Chan, Self-disclosure in social networking sites. Internet Res. **25**(2), 279–299 (2015)

45. H. Krasnova, S. Spiekermann, K. Koroleva, T. Hildebrand, Online social networks: why we disclose. J. Inf. Technol. **25**, 109–125 (2010)
46. W.M. Klein, Z. Kunda, Exaggerated self-assessments and the preference for controllable risks. Organ. Behav. Hum. Decis. Process. **59**, 410–427 (1994)
47. L.F. Nordgren, J. Van Der Pligt, F. Van Harreveld, Unpacking perceived control in risk perception: the mediating role of anticipated regret. J. Behav. Decis. Mak. **20**(5), 533–544 (2007)
48. L. Brandimarte, A. Acquisti, G. Loewenstein, Misplaced confidences: privacy and the control paradox. Soc. Psychol. Pers. Sci. **4**(3), 340–347 (2012)
49. J.C. Bertot, P.T. Jaeger, J.M. Grimes, Using ICTs to create a culture of transparency: e-government and social media as openness and anti-corruption tools for societies. Gov. Inf. Q. **27**(3), 264–271 (2010)
50. J.C. Bertot, P.T. Jaeger, D. Hansen, The impact of polices on government social media use: issues, challenges, and recommendations. Gov. Inf. Q. **29**(1), 30–40 (2012)
51. E. Christofides, A. Muise, S. Desmarais, Information disclosure and control on Facebook: are they two sides of the same coin or two different processes? CyberPsychol. Behav. **12**(3), 341–345 (2012)
52. S.B. Barnes, A privacy paradox: social networking in the United States. *First Monday***11**(9) (2006), https://firstmonday.org/article/view/1394/1312_2. Accessed Jan 2019
53. R. Gross, A. Acquisti, Information revelation and privacy in online social networks, in *Proceedings of the ACM Workshop on Privacy in the Electronic Society* (ACM, Virginia, USA, 2005), pp. 71–80
54. H.J. Smith, T. Dinev, H. Xu, Information privacy research: an interdisciplinary review. MIS Q. **35**(4), 989–1015 (2011)
55. Y.M. Baek, E. Kim, Y. Bae, My privacy is okay, but theirs is endangered: why comparative optimism matters in online privacy concerns. Comput. Hum. Behav. **31**, 48–56 (2014)
56. E. Hargittai, An update on survey measures of web-oriented digital literacy. Soc. Sci. Comput. Rev. **27**(1), 130–137 (2009)
57. Y.J. Park, Digital literacy and privacy behavior online. Commun. Res. **40**(2), 215–236 (2011)
58. B. Debatin, J.P. Lovejoy, A.K. Horn, B.N. Hughes, Facebook and online privacy: attitudes, behaviors, and unintended consequences. J. Comput. Med. Commun. **15**, 83–108 (2009)
59. B. Debatin, Ethics, privacy, and self-restraint in social networking, in *Privacy Online: Perspectives on Privacy and Self-disclosure in the Social Web*, ed. by S. Trepte, L. Reinecke (Springer, Heidelberg, 2011), pp. 47–60
60. S. Trepte, D. Teutsch, P.K. Masur, C. Eicher, M. Fischer, A. Hennhofer, F. Lind, Do people know about privacy and data protection strategies? Towards the "Online Privacy Literacy Scale" (OPLIS), in *Reforming European Data Protection Law*, ed. by S. Gutwirth, R. Leenes, P. Hert (Springer, Heidelberg, 2015), pp. 333–365
61. T. Zhou, Understanding online community user participation: a social influence perspective". Internet Res. **21**(1), 67–81 (2011)
62. C. Posey, P.B. Lowry, T.L. Roberts, T.S. Ellis, Proposing the online community self- disclosure model: the case of working professionals in France and the UK who use online communities. Eur. J. Inf. Syst. **19**(2), 181–195 (2010)
63. C. Dwyer, S. Hiltz, K. Passerini, Trust and privacy concern within social networking sites: a comparison of Facebook and MySpace, in*Proceedings of 13th Americas Conference on Information Systems, AMCIS 2007, paper 339,* https://aisel.aisnet.org/cgi/viewcontent.cgi?article=1849&context=amcis2007. Accessed Jan 2019
64. Z. Tufekci, Facebook, youth and privacy in networked publics, in *Proceedings of the Sixth International Conference on Weblogs and Social Media* (AAAI org, Dublin, Ireland, 2012), pp. 338–35
65. M. Ragnedda, Social control and surveillance in the society of consumers. Int. J. Sociol. Anthropol. **3**(6), 180–188 (2011)
66. N.B. Ellison, C. Steinfield, C. Lampe, The benefits of Facebook "friends": social capital and college students' use of online social network sites. J. Comput.-Mediat. Commun. **12**(4), 1143–1168 (2007)

67. M. Taddicken, C. Jers, The uses of privacy online: trading a loss of privacy for social web gratifications?, in *Privacy Online: Perspectives on Privacy and Self-Disclosure in the Social Web*, ed. by S. Trepte, L. Reinecke (Springer, Heidelberg, 2011), pp. 143–158
68. S. Trepte, L. Reinecke, The reciprocal effects of social network site use and the disposition for self-disclosure: a longitudinal study. Comput. Hum. Behav. **29**(3), 1102–1112 (2013)
69. C.M.K. Cheung, P.-Y. Chiu, M.K.O. Lee, Online social networks: why do students use Facebook? Comput. Hum. Behav. **27**(4), 1337–1343 (2011)
70. S. Utz, N. Kramer, The privacy paradox on social network sites revisited: the role of individual characteristics and group norms. Cyberpsychol. J. Psychosoc. Res. Cyberspace **3**(2) (2009), https://cyberpsychology.eu/article/view/4223/3265. Accessed Jan 2019
71. D. Boyd, J. Heer, Profiles as conversation: networked identity performance on Friendster, in *Proceedings of the Hawai'i International Conference on System Sciences (HICSS-39)* (IEEE, Kauai, 2006)
72. C. Steinfield, N. Ellison, C. Lampe, J. Vitak, Online social network sites and the concept of social capital, in *Frontiers in New Media Research*, ed. by F.L. Lee, L. Leung, J.S. Qiu, D. Chu (Routledge, New York, 2012), pp. 115–131
73. P.M. Valkenburg, J. Peter, The effect of instant messaging on the quality of adolescents' existing friendships: a longitudinal study. J. Commun. **59**, 79–97 (2009)
74. Z.J. Jiang, C.S. Heng, B.C. Choi, Research note-privacy concerns and privacy-protective behavior in synchronous online social interactions. Inform. Syst. Res. **24**(3), 579–595 (2013)
75. H. Xu, X.R. Luo, J.M. Carroll, M.B. Rosson, The personalization privacy paradox: an exploratory study of decision-making process for location-aware marketing. Decis. Support Syst. **51**(1), 42–52 (2011)
76. E.E. Hollenbaugh, A.L. Ferris, Facebook self-disclosure: examining the role of traits, social cohesion, and motives. Comput. Hum. Behav. **30**, 50–58 (2014)
77. N.K. Malhotra, S.S. Kim, J. Agarwal, Internet users' information privacy concerns (IUIPC): the construct, the scale, and a causal model. Inform. Syst. Res. **15**(4), 336–355 (2004)
78. M. Sideri, A. Kitsiou, C. Kalloniatis, S. Gritzalis, Privacy and Facebook universities students' communities for confessions and secrets: the greek case, in *e-Democracy 2015. CCIS*, vol. 570, ed. by S.K. Katsikas, A.B. Sideridis (Springer, Cham, 2015), pp. 77–94
79. F. Stutzman, R. Gross, A. Acquisti, Silent listeners: the evolution of privacy and disclosure on facebook. J. Priv. Confid. **4**(2), 7–41 (2013)
80. J.L. Gibbs, N.B. Ellison, C.-H. Lai, First comes love, then comes Google: an investigation of uncertainty reduction strategies and self-disclosure in online dating. Commun. Res. **38**(1), 70–100 (2011)
81. B.M. Okdie, Blogging and self-disclosure: the role of anonymity, self-awareness and audience size. Ph.D. dissertation. Department of Psychology in the Graduate School, University of Alabama, 2011
82. Z. Papacharissi, P.L. Gibson, Fifteen minutes of privacy: privacy, sociality, and publicity on social network sites, in *Privacy Online: Perspectives on Privacy and Self-disclosure in the Social Web*, ed. by S. Trepte, L. Reinecke (Springer, Heidelberg, 2011), pp. 75–89
83. P.A. Norberg, D. Horne, D. Horne, The privacy paradox: personal information disclosure intentions versus behaviors. J. Consum. Aff. **41**(1), 100–126 (2007)
84. Z. Tufekci, Can you see me now? Audience and disclosure regulation in online social network sites. Bull. Sci. Technol. Soc. **28**, 20–36 (2008)
85. T. Alashoor, M. Keil, L. Liu, H.J. Smith, How values shape concerns about privacy for self and others, in *Proceedings of the Thirty Sixth International Conference on Information Systems, ICIS 2015*, ed. by T. Carte, A. Heinzl, C. Urquhart (AIS Electronic Library, Atlanta, 2015)
86. H. Krasnova, N.F. Veltri, O. Günther, Self-disclosure and privacy calculus on social networking sites: the role of culture: intercultural dynamics of privacy calculus. Bus. Inf. Syst. Eng. **4**(3), 127–135 (2012)
87. P.B. Lowry, J. Cao, A. Everard, Privacy concerns versus desire for interpersonal awareness in driving the use of self-disclosure technologies: the case of instant messaging in two cultures. J. Manag. Inf. Syst. **27**(4), 163–200 (2011)

88. M. Helweg-Larsen, J.A. Shepperd, Do moderators of the optimistic bias affect personal or target risk estimates? A review of the literature. Pers. Soc. Psychol. Bull. **5**(1), 75–95 (2001)
89. C.M. Hoadley, H. Xu, J.J. Lee, M.B. Rosson, Privacy as information access and illusory control: the case of the Facebook news feed privacy outcry. Electron. Commer. Res. Appl. **9**(1), 50–60 (2010)
90. A. Acquisti, L.K. John, G. Loewenstein, What is privacy worth? J. Leg. Stud. **42**(2), 249–274 (2013)
91. K.-L. Hui, H.-H. Teo, S.-Y.T. Lee, The value of privacy assurance: an exploratory field experiment. MIS Q. **31**(1), 19–33 (2007)
92. W.M.P. Steijn, Developing a sense of privacy: an investigation into privacy appreciation among young and older individuals in the context of social network sites. Dissertation, Tilburg University, 2014, https://pure.uvt.nl/ws/portalfiles/portal/7737309/Steijn_Developing_05_09_2014_emb_tot_06_09_2015.pdf. Accessed Dec 2018
93. D. Boyd, What does the Facebook experiment teach us? *The Message* (2014), https://medium.com/message/what-does-the-facebook-experiment-teach-us-c858c08e287f. Accessed Jan 2019
94. S. Livingston, Taking risky opportunities in youthful content creation: teenagers' use of social networking sites for intimacy, privacy and self-expression. New Media Soc. **10**(3), 393–411 (2008)
95. K. Raynes-Goldie, Aliases, creeping, and wall cleaning: understanding privacy in the age of Facebook'. *First Monday***15**(1) (2010), https://firstmonday.org/article/view/2775/2432. Accessed Jan 2019
96. P.B. Brandtzæg, M. Lüders, J.H. Skjetne, Too many Facebook "friends"? Content sharing and sociability versus the need for privacy in social network sites. Int. J. Hum.-Comput. Interact. **26**, 1006–1030 (2010)
97. E. Van den Broeck, K. Poels, M. Walrave, Older and wiser? Facebook use, privacy concern, and privacy protection in the life stages of emerging, young, and middle adulthood. Soc. Media Soc. 1–11 (2015)
98. K. Sheehan, An investigation of gender differences in on-line privacy concerns and resultant behavior. J. Interact. Mark. **13**, 24–38 (1999)
99. C. Jensen, C. Potts, C. Jensen, Privacy practices of Internet users: Self-reports versus observed behavior. Int. J. Hum. Comput. Stud. **63**(1–2), 203–227 (2005)
100. M.S. Ackerman, L.F. Cranor, J. Reagle, Privacy in e-commerce: examining user scenarios and privacy preferences, in *Proceedings of the ACM Conference on Electronic Commerce* (ACM, Denver, U.S.A, 1999), pp. 1–8
101. K.B. Sheehan, Toward a typology of Internet users and online privacy concerns. Inf. Soc. **18**, 21–32 (2002)
102. M.Z. Yao, R.E. Rice, K. Wallis, Predicting user concerns about online privacy. J. Am. Soc. Inf. Sci. Technol. **58**, 710–722 (2007)
103. Y. Feng, W. Xie, Teens' concern for privacy when using social networking sites: an analysis of socialization agents and relationships with privacy-protecting behaviors. Comput. Hum. Behav. **33**, 153–162 (2014)
104. G. Cecere, F. Le Guel, N. Soulié, Perceived Internet privacy concerns on social networks in Europe. Technol. Forecast. Soc. Chang. **96**, 277–287 (2015)
105. D.H. Nguyen, E.D. Mynatt, *Privacy Mirrors: Understanding and Shaping Socio-Technical Ubiquitous Computing Systems* (Georgia Institute of Technology, USA, 2002)
106. E. Kavakli, C. Kalloniatis, P. Loucopoulos, S. Gritzalis, Incorporating privacy requirements into the system design process: the PriS conceptual framework. Internet Res. **16**(2), 140–158 (2006)
107. C. Kalloniatis, E. Kavakli, S. Gritzalis, PriS methodology: incorporating privacy requirements into the system design process, in *Proceedings of the SREIS 2005 13th IEEE International Requirements Engineering Conference—Symposium on Requirements Engineering for Information Security*, ed. by J. Mylopoulos, G. Spafford (IEEE, France, Paris, 2005), pp. 1–9

108. C. Kalloniatis, E. Kavakli, S. Gritzalis, Addressing privacy requirements in system design: the PriS method. Requir. Eng. J. **13**(3), 241–255 (2008)
109. M. Pavlidis, H. Mouratidis, C. Kalloniatis, S. Islam, S. Gritzalis, Trustworthy selection of cloud providers based in security and privacy requirements: justifying trust assumptions, in *Proceedings of 10th International Conference on Trust, Privacy and Security in Digital Business*, ed. by S. Furnell, C. Lambrinoudakis (Springer, Czech, Prague, 2013), pp. 185–198
110. N.C. Kramer, N. Haferkamp, Online self-presentation: balancing privacy concerns and impression construction on social networking sites, in *Privacy Online: Perspectives on Privacy and Self-disclosure in the Social web*, ed. by S. Trepte, L. Reinecke (Springer, Heidelberg, 2011), pp. 127–141
111. E. Litt, Understanding social network site users' privacy tool use. Comput. Hum. Behav. **29**, 1649–1656 (2013)
112. A.L. Young, A. Quan-Haase, Privacy protection strategies on Facebook. Information. Commun. Soc. **16**(4), 479–500 (2013)
113. M. Johnson, S. Egelman, S.M. Bellovin, Facebook and privacy: it's complicated, in *Proceedings of the Eighth Symposium on Usable Privacy and Security (SOUPS 2012)* (ACM, Washington, 2012), pp. 1–15
114. E. Vanderhoven, T. Schellens, M. Valcke, Exploring the usefulness of school education about risks on social network sites: a survey study. J. Media Liter. Educ. **5**(1), 285–294 (2013)
115. E. Vanderhoven, T. Schellens, M. Valcke, Educating teens about the risks on social network sites: an intervention study in secondary education. Communicar Sci. J. Media Educ. **43**(XXII), 123–131 (2014)
116. E. Vanderhoven, T. Schellens, R. Vanderlinde, M. Valcke, Developing educational materials about risks on social network sites: a design-based research approach. Educ. Tech. Res. Dev. **64**, 459–480 (2016)
117. R. Del Rey, J.A. Casas, R. Ortega, The ConRed program, an evidence-based practice. Communicar Sci. J. Media Educ. **39**(XX), 129–137 (2012)
118. M. Sideri, A. Kitsiou, E. Tzortzaki, C. Kalloniatis, S. Gritzalis, "I have learned that I must think twice before…". An educational intervention for enhancing students' privacy awareness in Facebook, in *Proceedings of the 7th International Conference, E-Democracy 2017,* ed. by S. Katsikas, V. Zorkadis (Springer, Cham, 2017), pp. 79–94
119. M. Foucault, *Power, Knowledge Ethics* (Ipsilon, Athens, 1987). (in Greek)
120. T.H. Eriksen, *Tyranny of the Moment Fast and Slow Time in the Information Age* (. Pluto Press, London, 2001)
121. M. Jakala, E. Berki, Communities, communication and online identities, in *Digital Identity and Social Media*, ed. by St. Warburton, St. Hatzipanagos (IGI Global, USA, 2013), pp. 1–13
122. E. Lampropoulou, *Internal Security and Control Society* (Kritiki Publ, Athens, 2001). (in Greek)
123. C. Norris, From personal to digital: CCTV, the panopticon and the technological mediation of suspicion and social control, in *Surveillance and Social Sorting: Privacy Risk and Automated Discrimination*, ed. by D. Lyon (Routledge, London, 2003), pp. 249–281
124. D. Lyon, *The Electronic Eye: The Rise of Surveillance Society-Computers and Social Control in Context* (Wiley, UK, 2013)
125. M. Kandias, L. Mitrou, V. Stavrou, D. Gritzalis, Which side are you on? A new panopticon versus privacy, in *Proceedings of 2013 International Conference on Security and Cryptography (SECRYPT)* (IEEE, Reykjavik, Iceland, 2013), pp. 1–13
126. L. Mitrou, M. Kandias, V. Stavrou, D. Gritzalis, Social media profiling: a panopticon or omniopticon tool? in *Proceedings of the 6th Biannual Surveillance and Society Conference* (Spain, Barcelona, 2014), pp. 1–15
127. A. Melucci, *Social Theory in the Information Era (*Trotta Editorial, 2002)
128. E. Daskalakis, *Criminology of Social Reaction: Traditions* (Sakkoulas Publ, Athens, 1985). (in Greek)
129. F.B. Viégas, Blogger's expectations of privacy and accountability: an initial survey. J. Comput.-Mediated Commun. **10**(3), 1–31 (2005)

130. G. D'Souza, J.E. Phelps, The privacy Paradox: the case of secondary disclosure. Rev. Mark. Sci. **7**, 1–29 (2009)
131. D.J. Solove, A taxonomy of privacy. Law Rev. **154**(3), 477–560 (2006)
132. M.B. Islam, J. Watson, R. Iannella, S. Geva, *What I Want for my Social Network Privacy* (NICTA, Australia, 2014)
133. A. Beldad, M. De Jong, M. Steehouder, I trust not therefore it must be risky: determinants of the perceived risks of disclosing personal data for e-government transactions. Comput. Hum. Behav. **27**(6), 2233–2242 (2011)
134. J. Midgley, *Social Development The Developmental Perspective in Social Welfare* (Sage, London, 1995)
135. J. Holmberg (ed.), *Making Development Sustainable* (Island Press, Washington D.C., 1992)
136. W. Streeck, Productive constraints: on the institutional conditions of diversified quality production, in *Social Institutions and Economic Performance*, ed. by W. Street (Sage, London, 1992), pp. 1–40
137. P. Cooke, K. Morgan, *The Associational Economy* (Oxford University Press, Oxford, 1998)
138. N. Uphoff, *Local Institutional Development: An Analytic Sourcebook with Cases* (Kumarian, West Hartford, 1986)
139. L.F. Salmen, *Listen to the People Participant-Observer Evaluation of Development Projects* (Oxford University Press, Washington, 1987)
140. S.E. Henderson, Expectations of privacy in social media. Mississippi Coll. Law Rev. **31**, 227–324 (2012)
141. J.E. Cohen, What privacy is for. Harvard Law Rev. **126**(7), 1904–1933 (2013)

Chapter 10
Can Blockchain Technology Enhance Security and Privacy in the Internet of Things?

Georgios Spathoulas, Lydia Negka, Pankaj Pandey, and Sokratis Katsikas

Abstract The Internet of Things (IoT) has changed the traditional computing models. While it has enabled multiple new computing applications, it has also raised significant issues regarding security and privacy. We are gradually shifting to using extended computing architectures, the nodes of which may be lightweight devices limited in hardware resources, scattered in terms of network topology and too diverse in terms of hardware and software to be efficiently administered and managed. Additionally, such nodes usually store, process and transmit sensitive private data of their users; thus, the risk of a security breach is significantly high. Blockchain technology, introduced through Bitcoin, enables the development of secure decentralized systems. It offers guarantees regarding data integrity, application logic integrity and service availability, while it lags behind in terms of privacy and efficiency. Because of the decentralized architecture of blockchain systems, there seems to be a good fit between blockchain and the IoT. Blockchain systems can be employed to develop solutions to some of the main security and privacy issues encountered in the IoT domain. In this chapter we discuss the convergence of the two technologies, we analyze possible use cases, where blockchain technology can enhance internet of things security and privacy, and we propose enhancements of blockchain technology to make it appropriate for application in the IoT domain.

Keywords Blockchain · Internet of things · Security · Privacy

---

G. Spathoulas · L. Negka
Department of Computer Science and Biomedical Informatics,
University of Thessaly, Lamia, Greece
e-mail: gspathoulas@uth.gr

L. Negka
e-mail: lnegka@uth.gr

P. Pandey · S. Katsikas
Center for Cyber and Information Security,
Norwegian University of Science and Technology, Gjøvik, Norway
e-mail: pankaj.pandey@ntnu.no

S. Katsikas (✉)
School of Pure and Applied Sciences, Open University of Cyprus, Nicosia, Cyprus
e-mail: sokratis.katsikas@ntnu.no; sokratis.katsikas@ouc.ac.cy

G. A. Tsihrintzis and M. Virvou (eds.), *Advances in Core Computer Science-Based Technologies*, Learning and Analytics in Intelligent Systems 14,
https://doi.org/10.1007/978-3-030-41196-1_10

10.1 Introduction

Blockchain technology, a shared Peer-to-Peer distributed ledger, is the underlying phenomenon of crypto-currencies that started with the proposal of a digital asset and payment system called "Bitcoin", which was introduced as an open source software in 2009 [39]. In a blockchain network there is no intermediary and the transactions are verified by a network of nodes before being recorded over a distributed ledger called blockchain [39]. Blockchains have begun to have a significant influence in the IoT domain and present a wide variety of opportunities for risk management. Blockchain technology and blockchain-based Smart Contracts have been well researched and some use cases have been proposed that adopt them in a range of applications in the IoT, including but not limited to the following: automated distribution and management of service-level information; automated management of warranty and maintenance information; management of ownership and transfer details; securing data records; protecting integrity of device software, and so on. Conoscenti et al. [14] provide a systematic literature review on the application of blockchain in IoT. Their work addresses the following six Research Questions (RQ):

- *RQ1)* What are the use cases of the blockchain beyond cryptocurrencies?
- *RQ2)* Are there any use cases applicable to the IoT?
- *RQ3)* What are the implementation differences with respect to the Bitcoin blockchain?
- *RQ3.1)* Which data are stored in the blockchain?
- *RQ3.2)* Which mining techniques are used?
- What is the degree of integrity (RQ4), anonymity (RQ5) and adaptability (RQ6) of the blockchain?

RQ1 and RQ2 aimed at reviewing the literature on the uses of blockchain technology beyond the initially developed Bitcoin and cryptocurrencies, and at identifying the blockchain applications that are applicable in the IoT context. RQ3 aimed at identifying the implementation choices that can be made in an IoT system, choices that are also different from the Bitcoin blockchain. For RQ4, ISO 25010 [2] was taken as the reference point for defining integrity, and characterizing the attacks that affect the blockchain and lead to undermining the integrity of the IoT. RQ5 addressed the need to protect the privacy of users by avoiding the linkage of IoT devices to their owners. RQ6 aimed at exploring whether the blockchain is adaptable to the number of transactions. For RQ6, the authors adopted the generic definition of adaptability from [2] and narrowed it down by defining the adaptability of the blockchain as its ability to scale with the number of transactions.

Conoscenti et al. [14] published their survey paper on the application of blockchain in the IoT about three years ago; since then there has been a tremendous growth in research and development in the domain, which has not, at the time of writing this chapter been fully. This chapter addresses this gap in the literature; it discusses the convergence of the two technologies, namely blockchain and IoT, it analyzes the potential use cases for leveraging blockchain in strengthening the security and privacy in the IoT, and proposes enhancements in blockchain that are required for its efficient and effective application in IoT. This chapter is divided into six sections. Section 10.1 presents an Introduction that highlights the context and background of the subject; Sect. 10.2 presents IoT Security and Privacy issues; Sect. 10.3 presents an overview of blockchain technology; the State of the Art is presented in Sect. 10.4; Sect. 10.5 presents an Analysis of the State of the Art; and Sect. 10.6 concludes the paper.

10.2 IoT Security and Privacy

The Internet-of-Things (IoT) has changed and continues to change the way we live in todays digitally connected society. IoT devices are increasingly finding applications in a range of contexts, from smart homes to smart cities, to smart grids, to smart farming, to the Internet-of-Medical-Things (IoMT), to the Internet-of-Military-Things (IoMiT), to the Internet-of-Vehicles (IoV), etc. At the same time, the pervasiveness of IoT devices raises concerns on security and privacy. For example, over 100,000 consumer devices were reported to have been compromised in 2014 to send over 750,000 phishing and spam emails [3]. In several application cases of IoT e.g. the Internet-of-Medical-Things and the Internet-of-Battlefield-Things, data confidentiality is crucial; thus, ensuring cyber-physical security of the data and devices, and privacy of data and computations becomes critical. A cyber-physical threat to the IoT system in any application area could be a result of ill-designed security. For instance, the entire IoT network is usually controlled by a team of Information Technology (IT) practitioners. In such a situation, it is not reasonable to expect that the IT team has detailed knowledge about the individual devices in the network, even though the team have been managing the network with full rights to install patches, remote access to devices, etc. Furthermore, centralized IoT networks have a higher risk of a single point of failure, thus hindering the scalability, and raising security and privacy concerns. In such a scenario, users depend upon third-party entities for data handling and for enforcing necessary security and privacy policies. Further, there is a risk of third-party entities indulging into mass surveillance, misuse of data etc. In this context, blockchain technology appears to be an important link in building a trusted, decentralized and secure environment for IoT applications.

10.3 Blockchain Technology

In 1991, Stuart Haber and W. Scott Stornetta were the first to present their work on a cryptographically secured chain of blocks [23]. Later, Bayer, Haber and Stornetta incorporated Merkle trees to the blockchain in 1992; this was to improve the efficiency so as to collect several documents into one block [7]. The concept of distributed blockchain was first introduced by an anonymous person or group known as Satoshi Nakamoto, in 2008, by publishing a whitepaper titled "Bitcoin: A Peer-to-Peer (P2P) Electronic Cash System" [39]. Blockchain (Bitcoin) was born when Satoshi Nakamoto solved a complex Game Theory conundrum called the Byzantine Generals Problem [34]; this ensured that, at a particular time, a block of assets could be transferred to only one other person, without the need for a third-party check. In 2009, the concept of distributed blockchain was implemented and released as an open-source software as a core component of the bitcoin digital currency. Bitcoin became the first digital currency (crypto currency) to solve the double spending problem through a blockchain, without requiring a trusted administrator [9]. The words *block* and *chain* were used separately in the original paper by Satoshi Nakamoto [39]; when the term moved into wider use it was still originally *block chain* [9] before becoming the single word "blockchain", by 2016.

Blockchain technology computationally answers the "Byzantine Generals Problem" [34]. In other words, blockchain answers the question of how individual users secure their data from non-trusted actors. The Byzantine Generals problem originates from a thought-experiment called the Two Generals Problem. The problem is illustrated by a scenario where two or more generals are to siege a city from opposite sides and they must coordinate their attack to be successful (to win). Let us assume that the General on one side, called *General A*, sends a message to the General on the other side, called *General B*, stating "attack at noon tomorrow", but the challenge is that the General *A* has no way to verify if General *B* has actually received the message; the risk is that if General *B* has not received the message, then if General *A* attacks in the afternoon, in absence of General *B*'s support he could potentially be marching towards defeat. On the other hand, if General *B* indeed receives the message sent by General *A*, General *B* has no way to verify if the message is authentic or a trap laid by the enemy. Let us assume that General *B* considers the message authentic and sends a response to confirm the planned attack. However, General *B* is now in the same situation as the one that General *A* was in before, i.e he has no way to verify if General *A* has received his response. Thus, there is a risk that General *A* will not be able to attack as planned, implying that General *B* will be the only one attacking as per the plan, risking his and his troops' lives. Nevertheless, General *A* could again send a message to General *B* to confirm the receipt of his acknowledgement message, but General *A* still has no way to verify if the message has reached General *B*, or even if the message was authentic in the first place. This puts General *A* in the same spot where General *B* was just in. This problem bounces back and forth into perpetuity like a never-ending loop, with neither of the Generals being ever confident about the authenticity and delivery of messages [27]. To relate the Byzantine Generals Problem to the blockchain we can illustrate it as follows: Person A can store almost anything of value into a 'digital lock box'. The content inside of the box can only be opened and changed with a unique private key. The information inside this box can then be shared on demand without the possibility of it being altered, changed, or replicated from its original form [22].

10.3.1 Blockchain Types

Buterin [10] categorized blockchains into three categories: Public Blockchain; Private Blockchain; and Hybrid Blockchain.

Public Blockchain

A fully open public ledger has no limitations with regards to reading- and writing permissions. Anyone can connect to the network, can access and add information. Anyone connected to the network has the right to participate in the consensus protocol, to verify the newly added blocks and ensure that they are not conflicting with previous blocks in the chain. The consensus protocol needs to be based on a cryptoeconomic mechanism, because of the open nature of the system and due to lack of trust between the nodes. A public ledger blockchain system operates without the requirement of trust between users; hence, it is considered to be fully decentralized.

Some of the state-of-the-art open source public blockchain platforms are Bitcoin, Ethereum, and Monero. The main characteristics of a public blockchain are as follows:

- Open to anyone for participation, without the need for any permission.
- Open to anyone to download the source code and run a public node on their local machine, validate the transactions in the network, and contribute to the consensus process. The consensus process is to determine the blocks that would be added to the (block)chain and their current state.
- Open to anyone to initiate transactions over the network and expect to see them added to the blockchain, after validation.
- Open to anyone to read transactions over a public block explorer. Transactions are transparent, but anonymous/pseudonymous.

Private Blockchain

A private blockchain enforces certain limitations on the reading- and writing permissions and is more tightly controlled than a public blockchain. Only a centralized group of participants, for example an organization, is granted the right to modify, add or read information. In a private blockchain system, a consensus protocol is usually not required, because of the trusted nodes. Private blockchain networks allow faster access to information, low cost transactions, and possibility to control the privacy level. Example applications of a private blockchain network include auditing, database management, etc. which are largely internal to a single organization; hence, public access to that information is not necessary in many cases. In other cases, public audit ability is desired. Private blockchains (such as MONAX, Multichain) exploit the blockchain technology by setting up internally verifiable groups and participants to approve the transactions. On the other hand, this has the risk of security breach like a traditional centralized system but has advantages in terms of scalability and compliance to data privacy rules and regulations.

Hybrid or Federated or Consortium Blockchains

As the name suggests, a hybrid blockchain, also called a consortium ledger, has some features of a public blockchain and some features of a private blockchain. In a consortium blockchain network, the consensus protocol is usually predetermined and managed by a predefined group of institutions [10]. A consortium blockchain system could e.g. have 25 participant institutions controlling one node, and every newly added block must be validated by at least 18 participant institutions before it can be added to the network. A hybrid blockchain system is thus partially decentralized. In a hybrid blockchain system, reading permissions could be granted to anyone or restricted to a group of participants. Furthermore, there is a hybrid solution to granting reading permissions as well, such that some parts of the information are open to the public while other parts are not.

Federated Blockchains (such as R3 (Banks), EWF (Energy), B3i (Insurance), Corda), operate under the leadership of a group, not allowing any other individual or institution to participate in the network transaction validation process. Federated Blockchains are much faster than public and private blockchains and provide much more privacy to transactions.

10.4 State of the Art

A lot of research on the convergence of blockchain and IoT technologies has been recently done. In this section research efforts relevant to applying blockchain solutions to enhance privacy and security for IoT ecosystems are reviewed and analyzed along three dimensions, namely (i) the security properties that the proposed systems aim to protect; (ii) the application domain in which these systems operate; and (iii) the technical maturity of the used blockchain infrastructure. Additionally, the main flaws or drawbacks of these proposals are identified, in order to define a clear path ahead for adapting blockchain technology to make it appropriate for solving IoT privacy and security issues.

10.4.1 Analysis Dimensions

10.4.1.1 Security Properties

We have identified five main security properties that blockchain solutions aim to protect; these are strongly coupled to fundamental security properties or combinations of these:

- **Confidentiality**: One of the main security issues in the IoT is the handling of sensitive personal data captured by IoT devices. On the other hand, the initial concept of blockchain is based on a publicly available ledger; therefore, any data

stored on blockchain networks is by default available to more than the strictly required users. Even though cryptography may be used to protect the data, this is not trivial in the context of the technical limitations imposed by blockchain technology.
- **Integrity**: The integrity of both data and procedures is crucial for any system, including the IoT ecosystem. Integrity is one of the main characteristics of blockchain technology as data that has been appended to the blockchain in the past cannot be removed or altered. Additionally, some blockchain solutions offer the ability to implement functionality; in this case, the integrity of the latter is also ensured.
- **Availability**: Systems need to be available to provide service to the end users. One of the main advantages of blockchain networks is that they are theoretically always available, or in other words their availability is not directly dependent on a single or a few points of failure. In this respect, blockchain technology has been extensively used to increase the availability of IoT systems and services.
- **Authentication**: Surprisingly, it is also common to use blockchain technology in order to implement authentication mechanisms in the IoT. Authentication is commonly achieved by means of challenging and proving the possession of a private key; given the restricted resources of IoT systems, this is not always easily forthcoming.
- **Non-repudiation**: Last, several blockchain/IoT research efforts provide non-repudiation mechanisms. Such mechanisms ensure that system actors are not able to argue on the content or the very existence of interactions with the system, to maliciously gain some benefit. This is also one of the straightforward applications of blockchain technology, as the integrity of past transactions is ensured by the protocol itself.

10.4.1.2 Application Domain

Research efforts that aim to improve IoT security through blockchain usually refer to a specific application domain. However, proposals addressing more than one domains are not uncommon. The main application domains we have identified are:

- **Smart home/city**: There are multiple proposals that address either smart home or smart city environments. These have been the first domains into which IoT applications were developed and the corresponding security and privacy issues were the first to be identified.
- **Supply chain**: Supply chain management is another domain where the application of IoT seems to offer added value. Integrating blockchain technology in such use cases can ensure the integrity of information collected along the whole supply chain, and thus increase trust in the process.
- **Data communication**: Data communication between IoT devices or between an IoT device and a central node is another common use case. Either through authentication schemes or access control mechanisms based on blockchain, several authors propose to enhance the security of such communications.

- **Data marketplace**: There are also some efforts to integrate blockchain technology with data marketplaces. IoT-produced data may be useful to other actors eager to pay in return for getting access to collected data. Blockchain technology is by design coupled with financial transactions, as its mechanics mainly work around a valuable token, the flow of which governs users' behavior.
- **Counterfeits**: Along with the growth in the usage of IoT devices, a new problem has emerged, related to counterfeit devices that either have low quality components or are maliciously designed to function differently than initially intended. Such devices may cause significant security or privacy issues. Blockchain technology has been proposed as a means of controlling IoT devices supply chains, to ensure that no counterfeit devices reach the end user.
- **Healthcare**: A critical domain to which IoT has started being applied to is healthcare. Due to the nature of this domain both availability of devices and data, and confidentiality of information are critical. There are multiple research efforts that relate to applying blockchain technology to healthcare IoT systems to make the latter more secure and safe.
- **Generic**: Finally, several proposals that are not domain-specific, and could theoretically fit any of the domains mentioned above, exist.

10.4.1.3 Technical Maturity

Due to the diversity in hardware, firmware and communication protocols and the limitations of computational resources in the IoT ecosystem, integrating any other technology with IoT systems is not trivial. Particularly when integrating a novel and relatively immature technology such as blockchain with IoT systems, technical validation of the outcome is of high importance. Merely proposing the use of blockchain in IoT systems is not enough, as the feasibility of the solution has to be demonstrated and validated.

For this reason, we have created a scale of 1 (less mature) to 5 (most mature) to rate the technical maturity of each one of the research efforts reviewed herein. This categorization is depicted in Fig. 10.3.

10.4.2 Literature

In this subsection the relevant literature sources are reviewed, categorized per the main security property or service they pertain to.

10.4.2.1 Generic (Several Properties/Services)

It was more common in the past, but is still happens to encounter research efforts that are too generic. Usually the authors reason about applying blockchain technology in IoT systems, in order to resolve any possible issue and without tackling any limitations.

The authors in [17] propose a lightweight, centrally managed architecture, based on that of the Bitcoin Blockchain, optimized for use in IoT ecosystems. Three tiers have been created (smart home, overlay network, cloud storage), with features that aim to eliminate the disadvantages of the blockchain (scalability problem, high resources, high delay) while maintaining its security level, and also improving availability and accountability in IoT ecosystems. Whilst the authors mix a lot of interesting ideas, their proposal is not mature enough to be applied to real world applications. The same authors propose a more detailed application of the same concept for a smart home case [19]. The description of the proposed system is more thorough and aims to ensure all three security properties for the IoT installation of the smart home, namely confidentiality, integrity and availability. The design is based on a single central node, called home miner, that facilitates the functioning of the system. Because of this approach the system is similar to a traditional centralized system rather than to a truly decentralized system. Using blockchain terminology, such as transactions or mining, is not enough to protect the IoT system. The same authors have further elaborated their approach [18] and presented the Lightweight Scalable Blockchain (LSB). To address scalability problems, computational costs and delays, they implemented a lightweight consensus algorithm, applied a distributed trust method and a distributed throughput management strategy, and separated the flow of the data from the transaction traffic. They also made a lot of progress in terms of evaluating their solution.

The work in [13] aims to enable device owners to manage the data they share in a community scenario where entities need to exchange private information generated by IoT devices. The proposed design has three main layers: A P2P network (e.g. network of IoT devices) for generating and storing private data; a blockchain layer used for certifying IoT devices and offering a way to check data integrity; and a set of access rules at the application layer for owners to set their desired privacy levels. Again, this is a relatively immature work without a technical implementation.

The authors in [8] propose a blockchain-based security architecture for smart cities, divided into four layers (Physical, Communication, Database, Interface).They provide a way to store and share IoT data from devices integrated into the smart city environment and to enable secure communication and data exchange between different smart cities. Their work is at early stages, without specifications or evaluation methodology.

In [35], the authors divide the IoT ecosystem in two layers (high level and edge level) and implement blockchain technology in both of them. This aims to facilitate blockchain adoption for IoT ecosystems and to lower the complexity and computation required for its use, without sacrificing the provided level of security. Eventually, the goal is to provide a secure wide-area network of Internet of Things. There is no implementation and no justification for the validity of the proposal.

The work in [20] presents a hybrid system comprising five layers, that aims to solve the majority of existing security and privacy issues of the IoT, especially in the healthcare domain. A patient-centric approach is assumed, to give patients control over their EMRs (Electronic Medical Records). Blockchain technology and several other cryptographic techniques are employed to that end. The first layer is the Overlay Network, that recognizes certified IoT devices as nodes and groups them into clusters, to increase scalability. Each of the clusters is associated to a Cluster Head, responsible for key management of devices, patients, and healthcare providers. Next, a Cloud Storage layer is used for storing patient data. It is connected to and cooperates with the Overlay Network for verification purposes. Smart Contracts are another layer, charged with alerting responsible parties when abnormal data is obtained from a patient. The remaining two levels are the main actors of the platform. These include healthcare providers, patients, or wearable IoT devices. The authors provide a very detailed analysis of their approach, without however providing information on how this can be practically applied to the healthcare domain.

A framework to enable secure data transmission between connected nodes, in this case IoMT devices, is presented in [16]. The project also aims at reducing the gigantic volumes of storage required by IoMT devices to process medical records in real time, as well as the replacement of cloud services that are currently being used for storage, since the cloud is a low security and privacy solution. The solution has hashes of all data, obtained either through real time or remote observation of patients, uploaded to the blockchain. The actual data is stored off-chain, since the blockchain cannot accommodate its size. Physicians and health practitioners, as well as care givers have access to the EMRs (Electronic Medical Records).

10.4.2.2 Authentication

An interesting technical implementation is presented in [31], where the authors discuss the risks of creating a single SSH key that is copied to every device a user needs to have access to. They address the key management problem by proposing a custom, private blockchain that will have a block added to it when an SSH public key is added, rotated or revoked. The approach combines collective signing and a custom blockchain to create a secure and easy-to-use, decentralized SSH-key management system.

For the purposes of identity management, the authors of [61] present the Blockchain-based Identity Framework for IoT (BI-FIT). This framework stores device-owner identities on the blockchain and correlates them with the device identities via a signature that has been created with the owner's private key. The signatures are used for authentication and device identification purposes. The whole scheme is user-centric and aims to facilitate the application of security mechanisms and real time monitoring.

The authors in [47] explore the scenario of an information distribution system that is blockchain-assisted, but all blockchain related operations are performed by a gateway which then provides IoT devices with an API. The scheme is presented as a secure way to identify and locate IoT devices. The authors point out security prerequisites for such a system, and explore whether these can be met using blockchain and smart contracts and how current security schemes can be empowered through blockchain.

The goal of [21] is to present a protocol that fulfills both authentication and authorization purposes, and is also sufficiently scalable for widespread use in the IoT domain. The proposed scheme implements blockchain technology and allows seamless integration of new devices (no physical intervention required), while it can adapt to existing authentication techniques. Additionally, it enables continuous identity verification and authorization of devices at the gateway level, even when these are moving.

A blockchain-based, multi layered ID-management framework is proposed by the authors of [48]. In this architecture, all IoT devices are considered to be nodes on the blockchain, but may belong to different categories (lightweight, full, communication), based on the intensity of the computations they can handle and on their connection life. Easy identity verification is achieved though the generation of a unique ID for each IoT device, that is also coupled with the blockchain wallet ID. Attackers are discouraged from repeatedly creating fake IDs because of the cost of such a practice. No proof-of-concept implementation is discussed.

The authors in [49] propose a blockchain-based method for identity and credibility verification for IoT that makes use of self-organizing Blockchain Structures (BCS) to counter the problems that Blockchain-IoT integration presents, such as computational requirements or network throughput. Devices are assigned an identification id and a private key to be used for credibility verification, generated by a Manage Server (MS). Manage Servers also have ids and private keys and are responsible for providing calculation and storage. BCS are small blockchain networks, each managed by a MS. All actions, such as adding or deleting a device, are recorded on the BCS the device was part of. Different BCSs may have a hierarchical relationship to each other. This flexible structure enables IoT devices to form blockchain networks that may not overwhelm their functioning. On the other hand the security guarantees of the approach need to be furthered researched.

In [59], the concept of a physical-logical link through physical chip identification is presented with the purpose of preventing illegal spoofing of physical addresses. The authors advocate replacing the SSD controllers' cash memories with Identification RAMs (IDRAMs), and using them to generate a secret key to pair with the IoT devices public key. The authors confirm that blockchain technology can be utilized to protect data between logical addresses, and by extension, thanks to the link, physical addresses as well.

10.4.2.3 Privacy

Enigma [63] offers the novel opportunity to execute data computation while keeping it private. It is designed to connect to an existing blockchain and load private and intensive computation to an off-chain network. The blockchain stores proof of correct computations for verification purposes, while the off-chain storage and computations are also linked to the blockchain. Enigma offers a management overlay for multi-party computation to enhance its integrity and efficiency.

Another interesting approach is the ChainAnchor [25] architecture which is a blockchain-based, privacy-preserving platform for the commissioning of IoT devices into a cloud ecosystem. It supports device owners who sell their devices' data to service providers, and incentivizes both parties to use the framework. ChainAnchor builds on EPID (Enhanced Privacy ID), and utilizes the blockchain as a means to anonymously register devices for commissioning and decommissioning. It also enables devices to prove that they are genuine without requiring the involvement of a trusted third party. While the concept seems promising, the level of technical implementation is too low.

The work in [15] utilizes a smart contract-based access control architecture previously proposed for the IoT, enhanced in terms of privacy. The authors present an ecosystem comprising service providers, devices (or a cluster of), and storage devices for storing the collected data. The smart home that all the IoT devices are connected to, and the user-owner of the smart home are also identified as main actors of the system. Moving the storage location of the data to trusted nodes and utilizing blockchain to manage access, offers increased privacy.

The authors of [51] propose a network model that aims to preserve privacy and to enable access control by combining attribute-based encryption and blockchain technology. Cluster devices have an important role in the framework. They are defined as devices capable of handling intense computations that are responsible for the processing of data that other IoT devices transmit to them. Miners are necessary for the verification of transactions and are rewarded with tokens that enable them to access data. Attribute Authorities exist to provide Attribute Based Encryption (ABE) and as a way to specify access rights. The approach is interesting, but not much in terms of implementation is provided.

Beekeeper 2.0 [60] was created to mitigate the risk of leaking sensitive information in the context of blockchain-enabled IoT systems. It is a novel approach that simultaneously enables devices to trade data with each other, servers to perform homomorphic multiplications of any degree, as well as additions of encrypted data without ever having access to plaintext data of the devices. The main actors of this framework are the IoT devices, the servers, and the blockchain validators. Servers come to use for devices by processing encrypted data, when being requested to do so. They communicate with the devices through blockchain transactions. The validators of the blockchain, in addition to the usual general verification duties, are also responsible for verifying commitments. Any dishonest behaviour by a server is detected by the data owner and by the blockchain validators.

The blockchain based framework presented in [6] is broken down in three tiers. The first tier includes the Devices, as constrained or unconstrained nodes, as well as the Patient, which functions either as a gateway or as an aggregator. A private blockchain per patient is employed, and this tier is responsible for the creation of new Electronic Health Records (EHR). Data from IoT devices is used for completion of block attributes and the registration, after which the private key of the patient is generated. Before a device can send data, it needs to be authenticated by the patient system. The second tier, made up from Authorities (Hospitals, Labs etc.), is responsible for both accessing existing EHRs and the generation of new blocks. It is implemented in a public blockchain. A block is added to the authority's chain after it has been visited by a patient, and the same block is sent to the cloud. Authorities can access patient data, but are unable to tamper with it, since it, is recorded in an established block. To achieve perfect privacy, Pseudonym Based Encryption (PBE) is brought into service. The third tier is described as a public blockchain to ensure the compliance of various cloud servers, but is not explored further in this work.

A novel architecture with built-in privacy and adaptability, called modular consortium blockchain architecture for IoT and blockchains, is what the authors of [4] have proposed and implemented. This scheme aims to secure IoT devices communication and data exchange on top of a software stack of blockchains on the IPFS (Inter-Planetary File System). To address scalability problems, the authors have divided the workload on many smaller private blockchains called sidechains, that join up to form a consortium network, which they can be added to or removed from at any given time. The sidechains log hashes for all activity related to sensor data from sidechain members and the data itself is stored on the IPFS, while a public blockchain run by the whole network keeps records of all access requests between consortium members, and their outcome, to enable accountability. This distribution helps overcome privacy issues. Each IoT network associated with a sidechain has IoT devices and a single validator as its members. Devices send encrypted data to the validator, who in turn logs the data and its hash to the IPFS and to the blockchain respectively. A smart contract is activated to enforce access control in the sidechain; to ensure that only data by authorized origins reaches the validator; and to store the public keys of requesters with access rights along with the public keys of the data they have access to. A similar, Access Control performing smart contract, runs on the consortium blockchain and also stores the devices that are entitled to submit access requests. To become a requester, one must first join the network, then sign a request transaction with their private key, and if they end up receiving the IPFS file hash, decrypt it with their private key to access the information.

10.4.2.4 Access Control

The authors of [62] have designed a platform that grants users of mobile phones access over data provided by service-providing entities. For that purpose, a blockchain is being used as an access control manager, storing a pointer to the data and sending the actual information to a distributed, private, key-value data store. The proposed approach is interesting, but the authors do not provide information on the practical specifications of the proposed blockchain implementation, such as the number of nodes or the security of the consensus mechanism.

The authors of [24] introduce Bubbles of Trust, a decentralized system based on blockchain, which aims to facilitate the authentication of devices against each other. The proposed system creates virtual secure zones (bubbles) around master IoT devices. Follower devices are identified by signing object IDs with the associated private keys and are given authentication tickets. Devices with tickets can request to be associated with a bubble, in order to be considered trusted by other bubble members. A bubble is protected and non-member devices cannot access it. Member IoT devices should only communicate with other members of their bubble, as those are the only verified trusted devices. All communications are practically associated with blockchain transactions and must therefore be validated according to bubbles' membership. The proposal is technically thorough and sufficiently tested.

The authors in [30] present a framework for access control based on smart contracts. The application comprises three different kinds of smart contracts. Several Access Control Contracts (ACCs) are implemented, each offering a different access control method for a subject-object pair. ACCs provide two kinds of access control verification, namely static, based on predefined rules; and dynamic, based on the object's behavior. Information about misbehavior is sent to a single Judge Contract (JC), which is responsible for imposing penalties. Finally, registering and updating of allowed interactions is achieved through the Register Contract (RC). The proposed framework has been implemented by using the Ethereum platform and Rpi3 as IoT devices, in order to showcase its validity.

The authors of [46] have developed the ControlChain system, which relies heavily on blockchain technology. It is a scalable, user friendly and compatible to existing access control mechanisms model for authentication and authorization optimization. It utilizes an off-chain side channel, for the propagation of time-sensitive data, as well as four different blockchains. The Relationships Blockchain is responsible for storing the public data and relationships of all entities. The Context Blockchain stores contextual info on users and devices that are then taken into account in authorization decisions, and the Accountability Blockchain holds all information about entity behavior, actions performed and access control permissions. Finally, the Rules Blockchain stores authorization guidelines by owners to objects, or objects to themselves. While the authors describe general blockchain implementations, they present E-ControlChain, a proof of concept implementation to be deployed on the Ethereum network.

EdgeChain [45] is an edge-IoT framework that aims to connect the account of every IoT device to edge cloud resources. The main idea is to regulate how light devices may access and utilize offered resources in a secure way. Blockchain technology is used to record all transactions and activities, and smart contracts are brought in as means to enforce rules and regulate device behavior. The authors aim to construct a framework that will drive efficient utilization of resources from IoT devices by controlling their activity through behavioral economics. While testing has been done on a private blockchain, the scheme has been built on the Ethereum network.

In [41], a framework that utilizes blockchain technology to enforce access control policies in IoT networks is presented. It is a lightweight, mobile, scalable design that does not directly integrate the blockchain functionality into the devices, thus ensuring that even those with the most limited resources can be part of the network. Entities named managers interact with the contract to add and update access control rules. They do not need to be continuously connected to the network and have no requirements to meet in terms of computational strength or memory. Management control hubs are a special type of node in wireless sensor networks that, while not part of the blockchain, are continuously connected to a blockchain node. They need to have high performance capability, as they handle all the requests for information on access control policies on behalf of devices with limited resources. Each IoT device needs to be registered under a hub, and to use a public key as its unique identity. The whole scheme, while tested on a private Ethereum network, is meant to be carried by a public blockchain.

The authors in [50] secure information exchange in the healthcare ecosystem, and hand information access control over to the patients, through a blockchain-dependent framework. They propose the use of mobile edge computing (MEC) for securing in-home therapy management. They combine this with a blockchain infrastructure to offer low-latency, secure, anonymous, and always-available therapeutic data communication. They propose trustless nodes of two kinds; edge nodes, that analyze and share with the cloud the data that IoT devices forward to them, and cloudlet server nodes, that reinforce the data processing, storing and analysis. Trusted nodes have the duty of verifying the therapy transactions within the blocks. While the authors discuss the use of a blockchain system and a tor layer that can enhance data exchange in such scenarios, they do not analyze how this blockchain system practically functions. They provide some use case implementations but with limited technical documentation.

A prototype for tracking the supply chain and detecting counterfeits by bringing blockchain technology and Physical Unclonable Functions (PUFs) into service is being proposed by the authors in [28]. They make use of a custom private blockchain to store PUF data and info for each Integrated Circuit (IC) to enable authentication. On each block, verified transactions are stored to record the transfer of ownership for an IC between owners. The protocol enhances security by enabling only legitimate IP address owners to profess themselves initial owners of an IC, and only current owners to fire a transferring transaction.

In their first two works on the FairAccess framework [42, 43], the authors describe its earliest and most simplified version. According to their design, any subject identified with a requester address wishing to access protected data will be able to submit a request through its wallet, which is acting as a Policy Enforcement Point (PEP) and is charged with regulating the protected resources. The PEP expresses the request through a GetAccess transaction and shares it with the miners, who will evaluate it and rule whether the request is to be accessed or denied. The evaluation is done by comparing the transactions unlocking script to the GetAccess locking script. Transactions deemed valid will be recorded on the blockchain. The authors extend their approach in [44], but the justification of their approach suffers from insufficient analysis.

The authors of [52] introduce the concept of the Internet of Smart Things, where Smart Things are defined to be devices that have been provisioned with Artificial Intelligence (AI) features that enable them to be autonomous. A permission-based blockchain protocol (Multichain), that offers secure communication at low cost, is employed to create a network of such devices. While the contribution of the work is unclear, the authors present an implementation of their approach.

The work in [54] describes an auditable, resilient, integrity preserving, blockchain-based framework for sharing and storing IoT data. Resources are organized in streams, per which ownership and sharing rights are defined. They are stored off-chain (on-premise storage/cloud/distributed P2P network), while a corresponding identifier is stored on the blockchain, making it tamper-proof. End-to-end encryption of the data takes place before data is stored. Access control is enforced through blockchain, on which the access permissions are stored for each data stream. A blockchain transaction contains information on the ownership of the stream and its corresponding access permissions. Stealth addresses are used to preserve privacy. Storage nodes, in the case of an access request, consult the blockchain to determine whether to grant access or not.

10.4.2.5 Integrity

A blockchain-based framework is proposed as a means to provide a Data Integrity Service for transactions of IoT data for both data owners and consumers [37]. Both of them act as blockchain nodes. The proposed system is built on the blockchain and implemented in a smart contract, while cloud storage services are used as general purpose data storage. The proposed service framework is reliable and able to handle an increasing number of clients that trade for IoT data. The implementation is based on a custom blockchain network, while a significant part of the functionality has been left as future work.

In [40], through an Ethereum Smart Contract, an application that enables the detection of Counterfeit IoT devices is realized. This is achieved by tracking the ICs that IoT devices will consist of down the supply chain, logging their owners until they are ultimately part of a device. Furthermore, authentication is made possible, since all ICs and IoT devices are linked to their own PUF-derived unique ID. The authors provide a proof of concept implementation for the blockchain functionality of their approach.

The authors of [38] present a three-level blockchain solution for ensuring data integrity and identity verification. The first blockchain layer, named IoT, includes devices like sensors and gateways. It is secure, time predictable, and achieves energy-efficient communication, through a Proof of Trust protocol, which is based on the Trustful Space-Time Protocol (TSTP). Fog is the second blockchain level of this architecture, and its based on the Proof of Luck consensus algorithm. It enables secure communication between gateways and the cloud storage and the blockchain, and protects against data loss. The final blockchain level is the cloud. It provides semi-trusted data storage and identity verification and can function with any agreement algorithm. The authors have systematically approached the problem of securing the whole IoT ecosystem stack, but the implementation they provide does not prove the concept.

An architecture meant to be integrated into a Smart Factory environment in order to increase data privacy and ensure its integrity is proposed in [58]. It is broken down in 5 layers with specific responsibilities. The sensor layer includes the IoT sensor devices, along with a microcomputer with sufficient computing power, that collects information from factory equipment. The entities of the Management Hub layer are the blockchain nodes and are record blocks. They are also responsible for parsing, encrypting and packaging collected data, and for including it in the blocks they generate. The Storage layer is where blockchain technology comes in, in the form of a private network. The Application layer is responsible for providing services to users, and last but not least, the Firmware layer encompasses the technologies required to guarantee the smooth cooperation of all layers. To ensure data integrity and information privacy, the SHA256 hashing algorithm and Elliptic Curve Cryptography are employed. A use case scenario is provided, without however any actual implementation.

The authors in [26] present an attribute-based blockchain model for the management of IoT devices. In this framework there are two types of nodes, each with their respective set of attributes. Primary nodes are responsible for block generation and for storing transactions in them, while backup nodes handle the creation of new transactions, and read transaction information from a block, provided they have the right attributes that enable them to decrypt the data. Blocks are classified in different categories, such as a New Transaction Block (NTB) which is generated for each new IoT device to be traded, carrying a unique identifier for it, as well as all device and transaction related info. Device Maintenance Blocks (DMtB) are created when maintenance is required for a device; they come with maintenance related info, and the device's identifier. Access Control and other security policies are stored in Device Management Blocks (DMgB). The main actors in this case are manufacturers, sellers, purchasers and administrators. Manufacturers produce and trade IoT devices, and may need to maintain equipment. Thus, they have the right to interact with NTBs and DMtBs. Sellers also trade IoT devices and can send messages to NTB blocks. The Purchasers are the ones that receive ownership of a device and can write into NTBs. An Administrator is an entity that is responsible for regulating IoT devices, and managing Access control policies; they interact with DMgBs. An Authority Agency is the point where users (manufacturers, sellers, purchasers) apply to join the network. This entity has to verify the applicants' identities, and then distribute public parameters and a master key to each of the accepted ones. All of this information is then included in an Authority Agency-signed certificate.

The authors of [11] focus on laying out the requirements, principles and design for a blockchain-based Sensor Data Protection System (SDPS). Furthermore, they have implemented a SDPS system using Ethereum, that satisfies the aforementioned criteria. Their scheme combats odometer fraud on mileage data gathered from cars, and makes the collection, processing and exchange processes of that data inviolable, in a scalable, economically feasible manner. The architecture is called CertifiCar and through it data is collected from sensors, cross-validated and transmitted via blockchain transactions. Raw data is stored in secure mass storage, and hashes derived from it are stored in the blockchain. Through the comparison of stored hashes to hashes calculated from raw information, data consistency can be verified. The authors went through three possible implementations before settling on the final prototype. The end product can detect continuous odometer fraud, has a smartphone app that gives data owners the options of sharing with clients only the current odometer value, thus protecting any detailed car usage information, and also lets them share historic data to establish trust. The product was evaluated in a field test with 100 cars and by a focus group of 16, as well as through various interviews and workshops.

DroneChain [36] is a blockchain-based architecture used for ensuring data integrity in a scenario where drones are collecting data from IoT devices. The main entities of this system are the drones, which are enrolled as data collecting nodes and are identified by a unique ID. Their functions are to collect sensor data and send it to a Control System (CS), from which they also receive commands. Control Systems in this framework are also identified by an ID. After receiving data from the drones, they are charged with hashing it, and sharing hashed data and with the blockchain network and original data with the cloud server. Blockchain can then be used to ensure the integrity of exchanged data or commands.

The authors in [5] present a design that replaces the single server of the ACE authorization framework [53] with a blockchain, and utilizes the OSCAR security model [57] to build a more secure access control architecture for the IoT. The components of the scheme are: Recourse Servers (RS) that generate and store protected resources; Proxy Servers that store encrypted data if necessary; and Recourse Owners (RO) that legally claim the RS and the data they generate. Blockchain is used to manage the authorization requests and grant access in a secure way. The framework was tested and evaluated using a private Ethereum network.

The difficulties of merging blockchain technology with the IoT are analyzed, and popular techniques for achieving the integration are showcased in [55]. The authors present a Hyperledger Fabric-based system, structured in five distinct layers. The first layer consists of sensors, connected to Raspberry Pi devices, which make up the second layer, namely the Edge Device layer. Raspberry Pi devices function as peer nodes and structure data acquired from sensors into transaction format. After those transactions have been peer validated, they are sent to the Orderer nodes, that exist in the Cloud layer and unburden Raspberry Pi devices of block creation. Valid data is encompassed in blocks and committed blocks are broadcast back to peers.

The utilization of Ethereum smart contracts and PUFs is being proposed in [29] for the purpose of preventing attackers from impersonating IoT devices or tampering with the data sent by legitimate devices to spread malicious software. IoT devices can register to the BlockPro network, through smart contracts. Two smart contracts have been designed for this system. One is charged with ensuring safe communication between registered IoT devices, by acting as an intermediary between the devices and the servers of the network, by checking the legitimacy of participating devices. The second smart contract can only be invoked by the first, and is responsible for manipulating information, uploading to and receiving from the blockchain already verified data. Servers/miners are the nodes that deploy the smart contracts and are registered as trusted hosts. They are also responsible for maintaining optimal blockchain operation.

IoT data integrity is protected through the use of blockchain technology in a series of works [12, 32, 33, 56] related to the GHOST research project [1] which deals with the security of smart home installations. Three different use cases are described that are related to forms of consent, software integrity and IP blacklisting. In each installation a smart home gateway functions as a blockchain node, equipped with an account (pair of keys) coupled to the home owner. In the first use case, traditional forms of consent have been replaced by a blockchain mechanism. Through this smart home users accept the terms of use for their installation set up by their service provider. The users are not able to use the system, without having accepted the terms of use. The second use case is related to ensuring software integrity for the system itself. The hash of the software installed on the device is periodically calculated and it is compared to the valid software's hash stored in the blockchain. Finally, the reputation of each external IP is built upon reports submitted by all installed gateways. Specifically, each gateway is equipped with a risk engine component that calculates risks associated with each connection and thus each IP. If this risk is high, the gateway reports the specific IP as malicious to the blockchain. The reputation score for each IP is calculated upon multiple factors, such as time, number of different gateways reporting the IP as malicious or requests to remove the reports. The platform of choice is a private Ethereum network built by using smart home gateways as nodes.

10.5 State of the Art Analysis

10.5.1 Trends in State of the Art

An extensive literature review was conducted to identify papers that have been published over the last four years and are relevant to the employment of blockchain technology to resolve privacy or security issues in the IoT domain. Searched databases were the ACM Digital Library, IEEE Xplore, and Springer. Google Scholar was also employed for retrieving additional research results published elsewhere. Combinations of the following keywords, were used to identify works relevant to this study: *blockchain, IoT, security, privacy, integrity, authentication, availability* and *non-repudiation*. Papers related to IoT, blockchain, and at least one of security, privacy, integrity, authentication, availability were studied. The references of these papers were also analyzed to identify additional relevant publications. This exhaustive search resulted in the identification of 73 papers relevant to this subject. A more detailed analysis of those led to discarding several papers, either because of low relevance or of very low quality. This procedure reduced the number of the papers to be analyzed to 49. For all these, the following have been analyzed:

- Security goals
- Main security goal
- Relevant application domains
- Technical maturity.

From Fig. 10.1 it is obvious that there is an increasing trend in the number of publications, starting in the early 2017. Before then one or two papers per semester appear. Starting from the first semester of 2017 the number of published papers is continuously increasing. The lower number in the first semester of 2019 is due to

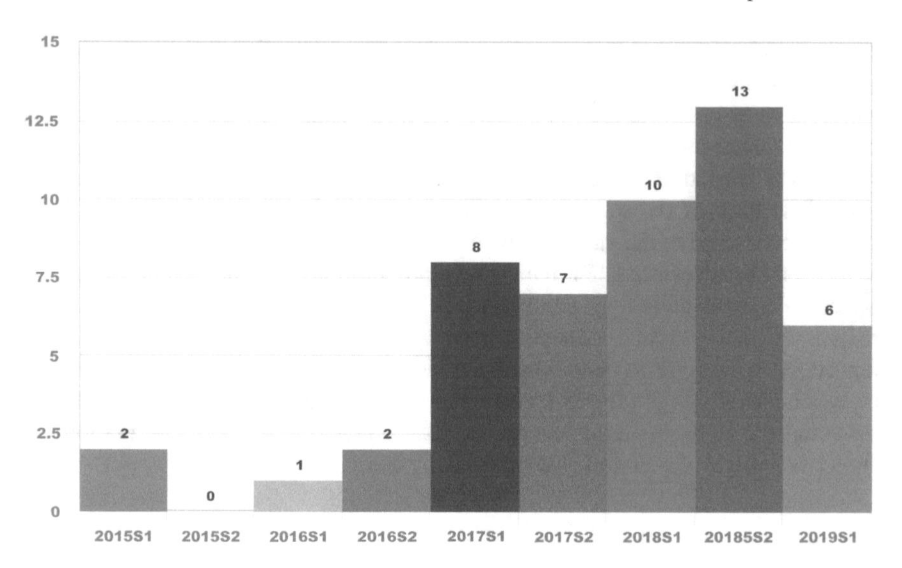

Fig. 10.1 Papers per semester

the fact that we have been able to analyze publications only in the first three months of the semester, at the time of conducting our survey. It is, however, evident that applying blockchain to IoT systems in order to enhance privacy and security is a hot topic and we expect to see more publications in the coming years.

The distribution of the application domains to which examined papers apply is depicted in the pie chart of Fig. 10.2. It has to be mentioned that multiple papers proposed systems or methodologies applicable to more than one application domains. A large portion of the papers stated that the proposed methods are domain agnostic and can function in every scenario. Apart from that, there were two dominant application domains, namely smart home/cities and data communication. Our view is that smart home/cities is the most common use case for IoT systems, so it is normal to have more blockchain integration efforts in that domain. Regarding the data communication it seems that there is a good fit that enables to provide enhanced communication schemes between IoT devices by using blockchain technology. Integrity of data and immutable access control mechanisms are the main benefits of this approach. The remaining papers are split among other domains. The only domain that is statistically more significant than others is healthcare, presumably due to the criticality of its applications.

A good overview of the results we found is depicted in Fig. 10.3. Most of the papers had more than one security goals but in order to classify those we have identified the main security goal for each one. Out of 49 papers, 14 were mainly related to integrity, 13 to access control, 7 to privacy, 7 to authentication, while 8 of the papers equally aimed to multiple security goals so they were classified as generic. The majority of the papers mainly deal with integrity, as blockchain can secure data and offers an immutable storage resource, although there are significant limitations

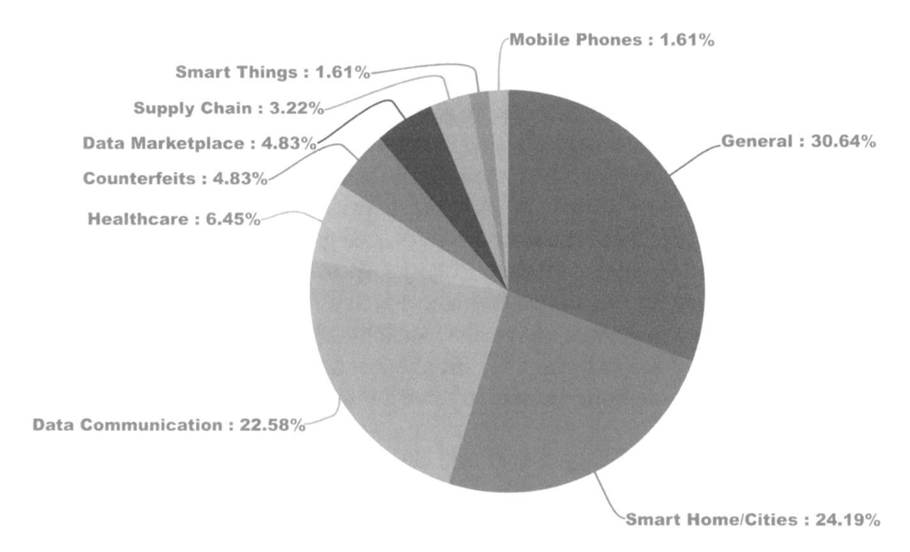

Fig. 10.2 Application domains distribution

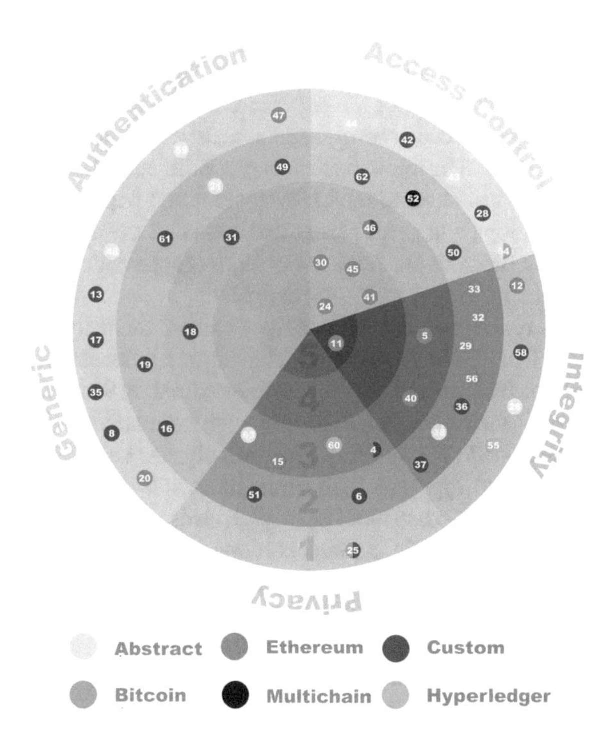

Fig. 10.3 Technology radar for blockchain application to IoT security and privacy

in terms of size and efficiency. Equally frequent are papers that set up access control mechanisms over blockchain technology. Theoretically, blockchain offers a trusted decentralized architecture that can apply access control rules, without any actor in the system being able to maliciously alter or cease this functionality. There are some efforts that aim to protect user privacy, mainly by combining blockchain mechanisms with cryptographic workflows, in order to create private workflows the integrity of which is also assured. There is another medium-sized group of papers that deals with authentication. Authentication schemes usually rely on a central node against which other nodes authenticate. This is problematic and inefficient for heterogeneous networks of numerous IoT devices, thus some blockchain related approaches have been proposed throughout the last years. Finally there is an important portion of papers that propose holistic solutions that aim to solve more than one security problems for IoT ecosystems.

Figure 10.3 depicts all papers analyzed through a technology radar approach. Specifically, the circle is split into five different triangular subareas (each one differently colored) which correspond to the five different taxonomies identified according to the main security goal of each paper: integrity, access control, privacy, authentication and generic. The five rings in the circle represent the five different levels of technical maturity (1–5) of the papers. Each paper is denoted on the figure with a small circle placed at the proper triangular subarea and the proper ring. Additionally, the color of the paper circle corresponds to the blockchain implementation employed in each one of the papers, to depict any correlations. The different options for blockchain implementation are Ethereum, Bitcoin, Hyperledger, Multichain or custom implementations. There are some papers that only present abstract implementations and some others that state that the approaches presented can be applied to any blockchain. Both cases seem problematic, as papers that belong to those do not technically justify their claims.

It is obvious that the average technical maturity of the proposed solutions is very low. Out of 49 papers only 5 are in the two inner rings (technical maturity 4 or 5). This means that most of the papers, approximately 90%, propose unjustified schemes that even if theoretically sound, come with a relatively high risk of not being applicable to real world IoT systems. Additionally, it is evident that the more technically mature a proposed approach is, the more likely it is that this approach is implemented in one of the ready to use blockchain platforms (mainly Ethereum). Papers that discuss the custom blockchain implementations tend to be located in the outer rings of the circle. Building a custom secure blockchain system is a difficult procedure that requires a lot of effort and should be avoided in the first place. It is also evident that the main security goal of the paper is correlated to some extend to the technical maturity of the paper. Papers aiming to access control seem to be of a higher technical level, papers aiming to generic solutions seem to be of a lower technical level and papers in the other taxonomies seem to be of a medium technical level.

It is evident that trying to apply blockchain technology to appropriate specific problems is more successful in general than trying to apply it to any generic problem. Additionally, trying to employ already built and tested platforms is also beneficial for the proposed approach than trying to build everything from scratch.

10.5.2 Common Problems

Through conducting an extensive analysis of papers regarding the incorporation of blockchain technology into the Internet of Things, we have identified patterns of disadvantageous approaches to the problem. Those precarious practices are presented and discussed in this subsection.

10.5.2.1 Blockchain Limitations

While some would argue that the most appropriate approach for a secure, decentralized application would be to use an established blockchain network, under no circumstances can it be said that this comes without drawbacks, and those unavoidably affect any design that chooses to adopt them. Ethereum is the most popular platform for implementations that require anything more than simple transactions, but the security it provides comes at a cost. Fees for single transactions may sum up to high amounts that make the proposed applications too expensive to be deployed.

Currently, Ethereum can process 15 transactions per second. Keeping in mind that there is a significant number of applications running on it, many of which require an equally large amount of transactions to go through each day, and factoring in the individuals that are also using the blockchain, we end up with a considerable scalability problem. Thankfully, there is a number of both proposed and recently activated solutions for this predicament. It must be noted that such problems are not exclusive to the Ethereum network. The Bitcoin network has a maximum block size that severely limits its throughput and results in an average of 3–7 transactions per second, meaning it is also suffering from scalability issues.

Another platform that is relatively frequently opted for is Hyperledger Fabric. While it is more scalable than other major blockchain networks, its consensus algorithm is not viewed as particularly secure, it does not offer a miner incentive and does not provide complete transparency or immutability. This makes it suitable for quite a limited variety of use cases that only allow trusted parties to participate.

When proposing the use of any of these platforms for deploying a blockchain application, the aforementioned limitations should be taken into account. This is not the case in most of the analyzed papers, as the authors neglect these issues during the design of the proposed methodologies and end up with schemes that are inefficient or even inapplicable to the IoT.

10.5.2.2 Custom Blockchain Implementations

In an effort to escape scalability limitations, delays, and most importantly, costs, many opt for creating a custom blockchain instead of using an existing one. This is a choice that almost unavoidably means sacrificing security. While such a route could function acceptably in a very targeted or privatized framework, it is unlikely to be beneficial in any other context. A blockchain network requires numerous nodes in order to operate the way it is supposed to. It has to be a vigorous network, with nodes that are widely dispersed, otherwise it cannot reach its full potential. Custom blockchains can rarely claim to have the required number of nodes to guarantee the level of security they are supposed to be providing. Most rely on achieving it as the network grows, which still leaves them quite vulnerable during their early life, and possibly longer, since there is no guarantee that the implementation will be adopted widely enough to achieve the number of desired participants.

10.5.2.3 Partially Decentralized Schemes

The integration of blockchain technology with the IoT is a tricky endeavor, since most commonly used IoT devices are very limited in terms of computational power. Most observed formats opt for sacrificing decentralization to overcome that obstacle. They choose to burden few, but more capable devices with the weight of the calculations required to participate in a blockchain network. Constrained devices are therefore dependent upon those gateways for handling their communication with the blockchain, resulting in a hierarchy in the system.

Additionally, in an attempt to organize authorization and verification processes, many frameworks have resorted to grouping devices and appointing managers that have to coordinate actions and communication, again creating some sort of central management. It is also a common practice to give certain entities the responsibility of granting and revoking permission to participate in such groups or even the whole network itself. Even though it is clear that this is being done for security reasons, it still differentiates one device from another, and creates inequality in an environment that is supposed to be functioning without central authorities.

Such designs are plagued by similar risks as fully centralized systems do, especially if the more powerful nodes are very few. They pose enticing targets for attackers, and the damage caused by a malicious device that has more freedom, power, and is responsible for the regulation of participating devices, is quite difficult to control. A denial of service (DOS) attack against those elevated nodes could render the entire system unusable.

Furthermore, to deal with the fact that storing large amounts of information in the blockchain would come at an extraordinary cost, some choose to store protected, sensitive data the traditional way, using on-site storage, or the cloud. Storing resources in that manner provides malicious entities with a single target that would give them access to the entirety of the sensitive, private information.

10.5.2.4 Unsecured Edge Devices

It comes without saying that an attacker will always go for the weakest component of a whole. In the cases being studied, those are the edge devices of each network. Most of the architectures presented up till now take few to no precautions towards securing the edge IoT devices, thus undermining most of their work, since the network will always be only as secure as those devices are. At best, means for detecting and isolating a compromised device are provided, and ways to limit the damage it can cause are established, but the designs that actually guard against attackers in the first place are few to nonexistent. While such "tamper-evident" implementations might be able to correct the damage or revert the alterations on the data, they cannot effectively prohibit third parties from accessing it.

On top of that, the vast majority of proposals attempt to ensure data integrity or validity only after data has left the device it was procured from. We have seen an abundance of approaches when it comes to ensuring information is safe, private and remains unchanged after being collected by an IoT device, but a notable lack of measures to verify data and ensure it is tamper proof before it is transferred out of the devices.

10.5.2.5 Lack of implementation

Since almost all of the works are proposing frameworks and architectures to be implemented, some sort of evaluation of the suggestion is expected to identify whether it would be operational and beneficial to the ecosystem. Even through most works include such a section, the vast majority only explores the blockchain side of their scheme, and few to no tests are conducted with regards to the IoT devices. It is hard to simulate an implementation of that scale, and yet necessary, to verify it will continue to work as intended when many entities are using the system. Lack of such testing results in proposals that may very well turn out to be not advantageous at all, or even not applicable in real-world scenarios.

10.5.3 Proposed Approach

While there has been a lot of research with respect to applying blockchain technology to solve IoT security and privacy problems, it seems that it has not been as effective as required. In our view, the main axes around which such research efforts should focus in the near future are :

- **Technical validation**: The IoT ecosystem is characterized by high diversity in terms of hardware devices, firmware and software installed on those, communication protocols and middleware devices used to connect IoT devices to wide area networks. Apart from that, such systems are characterized by limited resources in terms of computational power, storage capacity or energy consumption as many devices operate on batteries. Proposing a blockchain integration scheme for IoT systems may be promising, but until it is implemented and technically validated it can be regarded as non-functional in real world environments. New approaches have to be at least tested against a general IoT setup, to check they are applicable under IoT limitations. Even if such a test is successful, diversity in IoT environments has also to be taken into account, in order to propose a methodology applicable in the IoT in the long run.

- **Research on the device level**: Another common pattern is methodologies that secure the network communication or the data storage for IoT devices, but in the same time neglect to secure the devices themselves. A system is as secure as its weakest link. It does not make sense to design immutable blockchain systems for managing IoT devices output, if we have not first secured the devices themselves in order to be sure about the integrity of the data those output to our system. There are some interesting research efforts that propose the use of hardware security techniques such as trusted execution environment or the integrity checking of hardware through PUF technology, to ensure that devices function legitimately. Such features have to be integrated into blockchain-based systems in order to establish trust upon the complete workflow.
- **Build upon existing blockchain technology**: Many researchers propose custom blockchain implementations that fit with IoT limitations, at least in the context of specific applications. This approach is problematic as building a secure custom blockchain platform, requires significant effort, that is not usually committed or even available. Additionally, most of the security properties of blockchains stem from the assumption that there are multiple nodes connecting to each other, and maliciously altering information on the system requires the collaboration of too many nodes, which cannot happen. Custom blockchain implementations usually drive to blockchain networks consisting of too few nodes, so this assumption does not hold.
- **Genuinely decentralized design**: Current blockchain technology comes with significant drawbacks that makes its application problematic. The most obvious of those are scalability issues, speed and privacy. It is common to try to overcome these drawbacks by proposing the use of either private blockchain networks or permissioned ones in which some nodes hold more power than others. While these approaches partially resolve blockchain inherent problems, they also come with reduced security guarantees. The right direction is to improve blockchain technology, in order to resolve the existing problems while avoiding to create insecure blockchain networks.

10.6 Conclusions

Blockchain technology, a shared Peer-to-Peer distributed ledger, which is the underlying artefact of crypto-currencies that started with the proposal of a digital asset and payment system, has disrupted the technological development in various domains. Blockchains have begun to have a significant influence in the IoT domain and present a wide variety of opportunities for enhanced security and privacy in IoT. The Internet of Things has changed traditional computing models. While it has enabled multiple new computing applications, it has also created significant issues related to security and privacy. The world is now gradually moving towards using extended computing architectures, the nodes of which may be lightweight devices limited in hardware resources, scattered in terms of network topology and too diverse in terms of hardware and software, to be efficiently administered and managed. Additionally, such nodes usually store, process and transmit sensitive private data of their users, so the risk of any security event is significantly high. Blockchain technology enables the development of secure decentralized systems. It offers guarantees regarding data integrity, application logic integrity and service availability, while it lacks in terms of privacy and efficiency. In this chapter, we have presented a case for convergence of blockchain technology and the IoT, we have analyzed potential use cases where blockchain technology can be harnessed to enhance security and privacy in the IoT, and we have proposed a set of required features in blockchain technology for it to be effectively applied in the IoT.

References

1. https://www.ghost-iot.eu/
2. ISO/IEC 25010:2011(en) Systems and software engineering Systems and software Quality Requirements and Evaluation (SQuaRE) System and software quality models
3. Proofpoint Uncovers Internet of Things (IoT) Cyberattack (Jan 2014), https://www.proofpoint.com/us/proofpoint-uncovers-internet-things-iot-cyberattack
4. M.S. Ali, K. Dolui, F. Antonelli, IoT data privacy via blockchains and IPFS, in *Proceedings of the Seventh International Conference on the Internet of Things* (ACM, 2017), p. 14
5. O. Alphand, M. Amoretti, T. Claeys, S. Dall'Asta, A. Duda, G. Ferrari, F. Rousseau, B. Tourancheau, L. Veltri, F. Zanichelli, IoTChain: a blockchain security architecture for the Internet of Things. IEEE Wirel. Commun. Netw. Conf. WCNC **2018-April**(October), 1–6 (2018). https://doi.org/10.1109/WCNC.2018.8377385
6. S. Badr, I. Gomaa, E. Abd-Elrahman, Multi-tier blockchain framework for iot-ehrs systems. Proc. Comput. Sci. **141**, 159–166 (2018)
7. D. Bayer, S. Haber, W.S. Stornetta, Improving the efficiency and reliability of digital time-stamping, in *Sequences Ii* (Springer, 1993), pp. 329–334
8. K. Biswas, V. Muthukkumarasamy, *Securing Smart Cities Using Blockchain Technology* (2017). https://doi.org/10.1109/HPCC-SmartCity-DSS.2016.0198, https://www.researchgate.net/publication/311716550
9. J. Brito, A. Castillo, *Bitcoin: A Primer for Policymakers* (Mercatus Center at George Mason University, 2013)
10. V. Buterin, *On Public and Private Blockchains* (2015). https://ethereum.github.io/blog/2015/08/07/on-public-and-private-blockchains/
11. M. Chanson, A. Bogner, D. Bilgeri, E. Fleisch, F. Wortmann, Privacy-preserving data certification in the internet of things: leveraging blockchain technology to protect sensor data. J. Assoc. Inf. Syst. (2019)
12. A. Collen, N. Nijdam, J. Augusto-Gonzalez, S. Katsikas, K. Giannoutakis, G. Spathoulas, E. Gelenbe, K. Votis, D. Tzovaras, N. Ghavami et al., Ghost-safe-guarding home IoT environments with personalised real-time risk control, in *International ISCIS Security Workshop* (Springer, Cham, 2018), pp. 68–78
13. M. Conoscenti, A. Vetr, J.C. De Martin, Peer to peer for privacy and decentralization in the internet of things, in *2017 IEEE/ACM 39th International Conference on Software Engineering Companion (ICSE-C)* (2017), pp. 288–290. https://doi.org/10.1109/ICSE-C.2017.60

14. M. Conoscenti, A. Vetro, J.C. De Martin, Blockchain for the internet of things: a systematic literature review, in *2016 IEEE/ACS 13th International Conference of Computer Systems and Applications (AICCSA)* (IEEE, 2016), pp. 1–6
15. T.L.N. Dang, M.S. Nguyen, An approach to data privacy in smart home using blockchain technology, in *2018 International Conference on Advanced Computing and Applications (ACOMP)* (IEEE, 2018), pp. 58–64
16. N. Dilawar, M. Rizwan, F. Ahmad, S. Akram, Blockchain: securing internet of medical things (iomt). Int. J. Adv. Comput. Sci. Appl. **10**(1), 82–89 (2019)
17. A. Dorri, S.S. Kanhere, R. Jurdak, *Towards an optimized blockchain for IoT* (October), 173–178 (2017). https://doi.org/10.1145/3054977.3055003
18. A. Dorri, S.S. Kanhere, R. Jurdak, P. Gauravaram, LSB: a lightweight scalable blockchain for IoT security and privacy. Tech. rep. https://arxiv.org/pdf/1712.02969.pdf
19. A. Dorri, S.S. Kanhere, R. Jurdak, P. Gauravaram, Blockchain for IoT security and privacy: the case study of a smart home, in *2017 IEEE International Conference on Pervasive Computing and Communications Workshops (PerCom Workshops)* (IEEE, 2017), pp. 618–623
20. A.D. Dwivedi, G. Srivastava, S. Dhar, R. Singh, A decentralized privacy-preserving healthcare blockchain for IoT. Sensors **19**, 326 (2019)
21. A. Fayad, B. Hammi, R. Khatoun, An adaptive authentication and authorization scheme for IoTs gateways: a blockchain based approach, in *2018 Third International Conference on Security of Smart Cities, Industrial Control System and Communications (SSIC)* (IEEE, 2018), pp. 1–7
22. P. Francis, *Blockchain, The Byzantine Generals Problem, and The Future of Identity Management* (2016). https://medium.com/@philfrancis77/blockchain-the-byzantine-generalproblem-and-the-future-of-identity-management-6b50a2eb815d
23. S. Haber, W.S. Stornetta, How to time-stamp a digital document, in *Conference on the Theory and Application of Cryptography* (Springer, 1990), pp. 437–455
24. M.T. Hammi, B. Hammi, P. Bellot, A. Serhrouchni, Bubbles of trust: a decentralized blockchain-based authentication system for IoT. Comput. Secur. **78**, 126–142 (2018). https://doi.org/10.1016/j.cose.2018.06.004, http://www.sciencedirect.com/science/article/pii/S0167404818300890
25. T. Hardjono, N. Smith, Cloud-based commissioning of constrained devices using permissioned blockchains, in *Proceedings of the 2nd ACM International Workshop on IoT Privacy, Trust, and Security* (ACM, 2016), pp. 29–36
26. Q. He, Y. Xu, Z. Liu, J. He, Y. Sun, R. Zhang, A privacy-preserving internet of things device management scheme based on blockchain. Int. J. Distrib. Sens. Netw. **14**(11), 1550147718808750 (2018)
27. A. Heikkila, *The Blockchain and The Byzantine Generals Problem* (2017). http://techblog.cosmobc.com/2017/03/16/blockchain-byzantine-generals-problem/
28. M.N. Islam, V.C. Patii, S. Kundu, On IC traceability via blockchain, in *2018 International Symposium on VLSI Design, Automation and Test (VLSI-DAT)* (IEEE, 2018), pp. 1–4
29. U. Javaid, M.N. Aman, B. Sikdar, Blockpro: blockchain based data provenance and integrity for secure IoT environments, in *Proceedings of the 1st Workshop on Blockchain-enabled Networked Sensor Systems* (ACM, 2018), pp. 13–18
30. X. Jiang, Y. Shen, Y. Zhang, J. Wan, S. Kasahara, Smart contract-based access control for the internet of things. IEEE Internet of Things J. **PP**(c), 1–1 (2018). https://doi.org/10.1109/jiot.2018.2847705
31. L. Kokoris-Kogias, L. Gasser, I. Khoffi, P. Jovanovic, N. Gailly, B. Ford, Managing identities using blockchains and CoSi, in *HotPETs 2016—9th Workshop on Hot Topics in Privacy Enhancing Technologies (EPFL-TALK-220210)* (2016). https://infoscience.epfl.ch/record/220210/files/1_Managing_identities_bryan_ford_etc.pdf
32. C.S. Kouzinopoulos, K.M. Giannoutakis, K. Votis, D. Tzovaras, A. Collen, N.A. Nijdam, D. Konstantas, G. Spathoulas, P. Pandey, S. Katsikas, Implementing a forms of consent smart contract on an IoT-based blockchain to promote user trust, in *2018 Innovations in Intelligent Systems and Applications (INISTA)* (IEEE, 2018), pp. 1–6

33. C.S. Kouzinopoulos, G. Spathoulas, K.M. Giannoutakis, K. Votis, P. Pandey, D. Tzovaras, S.K. Katsikas, A. Collen, N.A. Nijdam, Using blockchains to strengthen the security of internet of things, in *International ISCIS Security Workshop* (Springer, Cham, 2018), pp. 90–100
34. L. Lamport, R. Shostak, M. Pease, The byzantine generals problem. ACM Trans. Program. Lang. Syst. (TOPLAS) **4**(3), 382–401 (1982)
35. C. Li, L.J. Zhang, A blockchain based new secure multi-layer network model for internet of things, in *2017 IEEE International Congress on Internet of Things (ICIOT)* (IEEE, 2017), pp. 33–41
36. X. Liang, J. Zhao, S. Shetty, D. Li, Towards data assurance and resilience in IoT using blockchain, in *MILCOM 2017-2017 IEEE Military Communications Conference (MILCOM)* (IEEE, 2017), pp. 261–266
37. B. Liu, X.L. Yu, S. Chen, X. Xu, L. Zhu, Blockchain based data integrity service framework for IoT data, in *Proceedings—2017 IEEE 24th International Conference on Web Services, ICWS 2017* (2017), pp. 468–475. https://doi.org/10.1109/ICWS.2017.54
38. C. Machado, A.A.M. Fröhlich, IoT data integrity verification for cyber-physical systems using blockchain, in *2018 IEEE 21st International Symposium on Real-Time Distributed Computing (ISORC)* (IEEE, 2018), pp. 83–90
39. S. Nakamoto, *Bitcoin: A Peer-to-peer Electronic Cash System* (2008)
40. L. Negka, G. Gketsios, N.A. Anagnostopoulos, G. Spathoulas, A. Kakarountas, S. Katzenbeisser, Employing blockchain and physical unclonable functions for counterfeit IoT devices detection, in *Proceedings of the International Conference on Omni-Layer Intelligent Systems* (ACM, 2019), pp. 172–178
41. O. Novo, Blockchain meets IoT: an architecture for scalable access management in IoT. IEEE Internet of Things J. **5**(2), 1184–1195 (2018)
42. A. Ouaddah, A. Abou Elkalam, A. Ait Ouahman, Fairaccess: a new blockchain-based access control framework for the internet of things. Secur. Commun. Netw. **9**(18), 5943–5964 (2016)
43. A. Ouaddah, A.A. Elkalam, A.A. Ouahman, Towards a novel privacy-preserving access control model based on blockchain technology in IoT, in *Europe and MENA Cooperation Advances in Information and Communication Technologies* (Springer, 2017), pp. 523–533
44. A. Ouaddah, A.A. Elkalam, A.A. Ouahman, Harnessing the power of blockchain technology to solve IoT security and privacy issues, pp. 1–10, 2018 (2017). https://doi.org/10.1145/3018896.3018901
45. J. Pan, J. Wang, A. Hester, I. AlQerm, Y. Liu, Y. Zhao, Edgechain: an edge-IoT framework and prototype based on blockchain and smart contracts. IEEE Internet of Things J. (2018)
46. O.J.A. Pinno, A.R.A. Grégio, L.C. De Bona, Controlchain: a new stage on the IoT access control authorization. *Concurrency and Computation: Practice and Experience*, p. e5238
47. G.C. Polyzos, N. Fotiou, Blockchain-assisted information distribution for the internet of things, in *2017 IEEE International Conference on Information Reuse and Integration (IRI)* (IEEE, 2017), pp. 75–78
48. H. Qiu, M. Qiu, G. Memmi, Z. Ming, M. Liu, A dynamic scalable blockchain based communication architecture for IoT, in *International Conference on Smart Blockchain* (Springer, 2018), pp. 159–166
49. C. Qu, M. Tao, J. Zhang, X. Hong, R. Yuan, Blockchain based credibility verification method for IoT entities. Secur. Commun. Netw. **2018** (2018)
50. M.A. Rahman, M.S. Hossain, G. Loukas, E. Hassanain, S.S. Rahman, M.F. Alhamid, M. Guizani, Blockchain-based mobile edge computing framework for secure therapy applications. IEEE Access **6**, 72469–72478 (2018)
51. Y. Rahulamathavan, R.C.W. Phan, M. Rajarajan, S. Misra, A. Kondoz, Privacy-preserving blockchain based IoT ecosystem using attribute-based encryption, in *2017 IEEE International Conference on Advanced Networks and Telecommunications Systems (ANTS)* (IEEE, 2017), pp. 1–6
52. M. Samaniego, R. Deters, Internet of smart things-iost: using blockchain and clips to make things autonomous, in *2017 IEEE International Conference on Cognitive Computing (ICCC)* (IEEE, 2017), pp. 9–16

53. L. Seitz, G. Selander, E. Wahlstroem, S. Erdtman, H. Tschofenig, Authentication and authorization for constrained environments (ace). Internet Engineering Task Force, Internet-Draft draft-ietf-aceoauth-authz-07 (2017)
54. H. Shafagh, L. Burkhalter, A. Hithnawi, S. Duquennoy, Towards blockchain-based auditable storage and sharing of IoT data, in *Proceedings of the 2017 on Cloud Computing Security Workshop* (ACM, 2017), pp. 45–50
55. J.C. Song, M.A. Demir, J.J. Prevost, P. Rad, Blockchain design for trusted decentralized IoT networks, in *2018 13th Annual Conference on System of Systems Engineering (SoSE)* (IEEE, 2018), pp. 169–174
56. G. Spathoulas, A. Collen, P. Pandey, N.A. Nijdam, S. Katsikas, C.S. Kouzinopoulos, M.B. Moussa, K.M. Giannoutakis, K. Votis, D. Tzovaras, Towards reliable integrity in blacklisting: facing malicious IPS in ghost smart contracts, in *2018 Innovations in Intelligent Systems and Applications (INISTA)* (IEEE, 2018), pp. 1–8
57. M. Vučinić, B. Tourancheau, F. Rousseau, A. Duda, L. Damon, R. Guizzetti, Oscar: object security architecture for the internet of things. Ad Hoc Netw. **32**, 3–16 (2015)
58. J. Wan, J. Li, M. Imran, D. Li et al., A blockchain-based solution for enhancing security and privacy in smart factory. IEEE Trans. Ind. Inform. (2019)
59. H. Watanabe, H. Fan, A novel chip-level blockchain security solution for the internet of things networks. Technologies **7**(1), 28 (2019). https://doi.org/10.3390/technologies7010028, https://www.mdpi.com/2227-7080/7/1/28
60. L. Zhou, L. Wang, T. Ai, Y. Sun, Beekeeper 2.0: confidential blockchain-enabled IoT system with fully homomorphic computation. Sensors **18**(11), 3785 (2018)
61. X. Zhu, Y. Badr, J. Pacheco, S. Hariri, Autonomic identity framework for the internet of things, in *Proceedings—2017 IEEE International Conference on Cloud and Autonomic Computing, ICCAC 2017* (2017), pp. 69–79. https://doi.org/10.1109/ICCAC.2017.14
62. G. Zyskind, O. Nathan, A.S. Pentland, Decentralizing privacy: using blockchain to protect personal data, in *Proceedings—2015 IEEE Security and Privacy Workshops, SPW 2015* (2015), pp. 180–184. https://doi.org/10.1109/SPW.2015.27
63. G. Zyskind, N. Oz, A.S. Pentland, *Enigma: Decentralized Computation Platform with Guaranteed Privacy*. Tech. rep. (2015). https://arxiv.org/pdf/1506.03471.pdf

Chapter 11
The Future of Money: Central Bank Issued Electronic Money

Pankaj Pandey and Sokratis Katsikas

Abstract The modern day society driven by a variety of electronic devices and high-speed internet is changing its perception and practice of paper currency, mode of economic and financial transactions, and so on. The usage of cash is increasingly reducing because of the ease of payments facilitated by cards, mobile phone apps and contact-less chips, online payment systems, etc. Furthermore, cryptotokens (cryptocurrencies), such as bitcoin, have fueled the interest of society and policymakers in investigating the usefulness and limitations of a central bank-backed electronic fiat currency. Also, blockchain/distributed ledger technology, the technology enabling the cryptotokens, has gained a lot of attention from almost all the sections of the society for its ability in providing a decentralized transaction verification process while maintaining the features similar to the traditional cash currency. This chapter presents an overview of the concepts and potential features, potential primary models of issuing electronic fiat currency, as well as the key design principles from a technical perspective and a high-level architecture of a central bank-backed electronic fiat currency. We present the key aspects of currency in modern society, the advancements in the field of financial technologies and how could these be harnessed to launch a central bank-backed electronic fiat currency. We also discuss the position of various central banks and governments on cryptocurrencies, blockchain technology, and the initiatives related to issuing electronic fiat currency.

Keywords Central bank · Digital currency · Electronic currency · e-Currency · Fiat currency

P. Pandey · S. Katsikas (✉)
Norwegian University of Science and Technology, Gjøvik, Norway
e-mail: sokratis.katsikas@ntnu.no; sokratis.katsikas@ouc.ac.cy

P. Pandey
e-mail: pankaj.pandey@ntnu.no

S. Katsikas
Open University of Cyprus, Latsia, Cyprus

G. A. Tsihrintzis and M. Virvou (eds.), *Advances in Core Computer Science-Based Technologies*, Learning and Analytics in Intelligent Systems 14,
https://doi.org/10.1007/978-3-030-41196-1_11

11.1 Introduction

Heraclitus of Ephesus, an ancient Greek philosopher, said that "*Change is the only constant*" [64]; and the world has seen a phenomenal change in the design, attributes, and usage of currency since ancient times. The island of Yap in Micronesia is known for *Rai Stones* as its currency [37, 38, 78, 80]. Rai stones are large doughnut-shaped limestones quarried on several of the Micronesian islands, such as Palau and Guam which are then transported to Yap to be used as currency. Rai stone currency is typically between 1.4 in. and 12 feet in diameter. The monetary value of a specific stone would depend on several factors, such as the size and craftsmanship, if many people or no one died transporting the stone, if the stone was brought in by a famous sailor, etc. [38]. Rai stones are still used in transactions of social importance, such as marriage, political deals, inheritance, etc. [78]. Equally interesting is the use of *Tea Bricks* as currency in many countries of Central Asia, China, Mongolia, and Tibet [12, 27, 47, 80]. On the other end of the globe, *Salt Bricks* were used as currency in Africa [8, 25, 80]. Mongolian and Siberian nomads preferred tea bricks as currency over metallic coins [27, 47]. Siberians used the tea bricks as an edible currency until the end of World War II [12, 27]. In the parts of Western Africa, long *T-shaped rod* called *Kissi penny* was used as currency [80]. These T-shaped rods were in size range of 6–16 in. Kissi pennies were later discontinued with the introduction of colonial money, but they continued to be in use for various social rituals in the societies of Poro and Sande, in marriages, and to be placed on graves and tombs [30].

In today's world, currency in a traditional sense means an inconvertible piece of paper, which a government has decreed as legal tender, known as a *fiat currency* in economics [39, 63]. Fiat currency is not backed by any physical commodity but the value of the currency is derived from the mechanics of demand and supply [39, 63]. Hence, unlike the historic currencies which were based on commodities such as Silver or Gold [63], fiat currency is solely based on the faith and credit of the economy [39]. Thus, an economy based on fiat currency works on the confidence and faith of the public in the "currency's" ability to serve the purpose of storage of value (purchasing power) [39, 63].

Rohan Grey and Jonathan Dharmapalan presented an interesting analogy between *Currency* and *Grease* on the one hand and *Economy* and *Car*, on the other hand [41]. Authors say that the role of currency in an economy is similar to the role of grease in car wheels, referring to the phrase *greasing the wheels of the economy*. Once the purpose of *grease* in the economy is served, the old currency notes are removed and destroyed in time [41]. The process of greasing the economy is a cumbersome, inefficient and expensive process, which can be addressed to a large extent, if not completely, by introducing an Electronic Fiat Currency (EFC).

A fiat (physical) currency costs heavily in designing, printing, shipping, collection, destruction and replacement, and is limited to in-person transactions only [41]. Furthermore, Central banks across the globe are struggling to address the anaemic economic growth and an increase in public demand for strong regulation and realignment of budgetary priorities [41]. On the other hand, an exponential growth in innovation and adoption of digital financial technologies offers new opportunities and challenges for the digital economy. One potential opportunity for central banks to harness the innovation and evolution in digital financial technologies is to introduce a central bank backed Electronic Fiat Currency (EFC).

The objectives of this chapter are to present an overview of the concepts and features relevant to a central bank-backed electronic fiat currency. We attempt to answer a series of questions related to EFC, namely: (i) What is the position of international community on the adoption of blockchain technology, cryptocurrency, and issuing a central bank-backed EFC?; (ii) What are the key initiatives of various governments and central banks on piloting and researching on EFC?; (iii) What are the concepts, and potential and preferred attributes of an EFC?; (iv) How could the identified attributes of EFC be translated into models of issuing EFC?; (v) What are the key technical design principles relevant to issuing an EFC?; and (vi) What could be the potential technical architecture of an EFC system?

To ensure a common understanding of the terminology used in this chapter and in the larger context of prevailing popular discussion on EFC, a list of key terms with their respective definition is presented in Table 11.1.

Table 11.1 Relevant terminologies and their respective definitions

Terminology	Definition
Blockchain	"Blockchain is nothing else but a DLT with a specific set of features. It is also a shared database-a log of records-but in this case shared by means of blocks that, as the name indicates, form a chain. The blocks are closed by a type of cryptographic signature called a hash; the next block begins with that same hash, a kind of wax seal. That is how it is verified that the encrypted information has not been manipulated, and that it can't be manipulated" [9]
Cryptocurrency	"A digital or a virtual currency that relies on secure cryptographic algorithms and technology for its creation and transactional operations (includes Bitcoin, Litecoin, Ethereum, etc)" [88]
Cryptotoken	Another name for cryptocurrency. Originates from a computational and cryptographic perspective
Digital currency	"A digital payment mechanism that is denominated in fiat currency (for example a central bank digital currency)" [88]
Distributed Ledger Technology (DLT)	"A DLT is simply a decentralized database that is managed by various participants. There is no central authority that acts as arbitrator or monitor. As a distributed log of records, there is greater transparency making fraud and manipulation more difficult and it is more complicated to hack the system" [9]
Electronic (or Digital) Fiat Currency (EFC)	A central bank-backed electronic payment mechanism which may or may not be based on Blockchain/DLT and is a legal tender to the same extent as cash
Virtual currency	"A digital representation of value, issued by private developers, and denominated in its own unit of account. In other words, it is not legal tender in fiat currency or central bank money. In terms of this definition, Bitcoin is a virtual currency" [88]

This chapter is divided into ten Sections where Sect. 11.1 presents an Introduction to the progression of currency from ancient times to now, objectives of this chapter and key terminologies relevant to this chapter are defined; motivation for this chapter is discussed in Sect. 11.2; Sect. 11.3 highlights the position of various governments and central banks on blockchain, cryptocurrency, and central bank-backed EFC; Sect. 11.4 briefly discusses the initiatives related to EFC by various central banks; Sect. 11.5 explains the context of money and EFC; Concepts and Attributes of EFC are discussed in Sect. 11.6; Sect. 11.7 presents the potential primary models of issuing EFC; the design principles for an EFC from a technical perspective are discussed in Sect. 11.8; a high-level technical architecture for an EFC is presented in Sect. 11.9; and Sect. 11.10 concludes the chapter.

11.2 Motivation

In 2008, the first cryptotoken named *Bitcoin* was launched [72], and since then there has been an exponential growth in the number and types of cryptotokens [23, 102], often termed as a cryptocurrency or digital currency or e-currency, etc [96]. The interest in launching a central bank-backed electronic fiat currency is primarily because of the blockchain technology, the underlying technology behind Bitcoin and other cryptotokens.

Blockchain technology involves the creation of digital tokens for digital files, such as documents or transactions [42]. These digital tokens can be considered as digital fingerprints of the files. These digital fingerprints are saved in groups called *block*. The individual blocks are then linked in a chain of blocks, and each subsequent block has a digital token from the previous block. Thus, it becomes impossible to modify the information in an old block in the chain without modifying the subsequent blocks. The ability of blockchain to secure the data and history of transaction lead it to be called as *The Trust Machine* by the Economist [94]. The World Economic Forum conducted an expert survey in 2015 and reported that the majority (57%) of respondents estimated that by the year 2025 the 10% of the world's GDP will be registered in a blockchain [104]. In December 2015, Goldman Sachs, an influential international investment bank, stated that "*Silicon Valley and Wall Street are betting that the underlying technology, the Blockchain, can change well everything*" [103].

A section of society sees these cryptotoken as an alternate form of payment and has led to explosive growth in the price (value) of these cryptotokens against the standard currencies such as United States Dollar (USD) and other fiat currencies as well as against other cryptotokens [81]. Many economists have compared this speculation-driven rise in the price of cryptotokens with the famous *Tulip Mania* of 1630s in the Netherlands [69, 100]. Nonetheless, the buzz surrounding the cryptotokens has aroused interest in the question: *whether a government or central bank should lawfully issue its own electronic (crypto) fiat currency?*

By design, an EFC should be fully regulated by the respective central banks and governments. Unlike most cryptocurrencies, EFCs would not be decentralised but would represent a fiat currency albeit in an electronic form. Each unit of EFC should act as a secure electronic equivalent of a traditional paper currency and would normally be powered by blockchain or DLT infrastructure. Hence, if a central bank issues an EFC, the central bank should not only be its regulator but the holder of clients' accounts as well. On the other hand, similar to the distribution of physical cash a central bank can issue an EFC in a decentralised form. The goal of an EFC should be to have the best of the two worlds, i.e., the best of cryptocurrencies and characteristics of the conventional banking system. An EFC's aim would be to take the features of convenience and security from cryptocurrencies and combine them with the time-tested functionalities of the traditional banking system where the circulation of money is regulated and backed with reserves.

The discussion on launching an EFC is not new. The proponents of EFC, advocate for a staged abolition of physical fiat currency, starting with high denomination currency notes such as EUR 500, CHF 1000, INR 1000, INR 500, and USD 100 [48, 73, 82, 83, 89, 95]. The general perception is that high currency notes provide low-cost anonymity to money hoarders, tax evaders and criminals. Hence, if money hoarders are forced to hoard money in several bags filled with small denomination notes or criminals have to carry trade with several bags filled with small denomination notes then the cost of anonymity increases and the risk of detection and apprehension goes up. In 2016, the Government of India demonetised all the notes of Rupees 500 and 1000 denomination [79]. Also, the European Central Bank (ECB) announced that the EUR 500 notes would be phased out starting in 2018 [33]. However, the complete production (issuance) of EUR 500 became effective on 26-Apr-2019 when Germany and Austria stopped to issue any new EUR 500 notes [61]. All the other European countries, block of national central banks, had terminated the production of EUR 500 notes in Jan 2019 but Germany and Austria had asked for additional three months to stop issuing EUR 500 notes. EUR 500 notes were taken off the production lines because of increasing suspicion that the notes are being used for payments related to illegal activities [33, 61]. EUR 500 notes in circulation continue to be legal tender and can be used for payment and store of value.

The American Economist Dr Kenneth Saul Rogoff has been vocal about abolishing cash and has published numerous articles on the subject. In 2016, Dr Rogoff published a book titled *The Curse of Cash* in which he urged the Government of the United States of America and Federal Bank to phase out USD 100, 50 and 20 denomination notes, leaving only small denomination notes in circulation [82]. Among the various suggestions made by Rogoff, he made a key suggestion to provide free debt account to low-income families and free smartphones, if necessary to make the electronic currency a viable option for everyone [82]. Rogoff proposed that these debit accounts and required smartphones can be offered in two ways: either through the private banks by offering subsidies by the government or at the expense of bank; the second method is that the government does the job [82]. Accounts supported through government initiatives are the preferred mode of offering a debit account as this would cover the people that the private banking sector is less likely to cover even with subsidies. Rogoff says that these debit accounts may even offer the anonymity feature, thereby compensating for the loss of currency notes [82]. Rogoff suggests restricting the level of information that a government can access and monitor payments data. Furthermore, to curb black money in the economy, this type of restricted monitoring would only be available to small individual accounts below a threshold; thus, it prevents the potential misuse of anonymous currency in criminal transactions [82].

Dr Rogoff's idea of a debit account sets the stage for government-backed electronic fiat currency. Dr Rogoff coins the idea of *Bencoin* named after Benjamin Franklin [82]. Inspiration for a central bank issued electronic currency such as Bencoin emerges from the popularity and acceptance of cryptotokens such as Bitcoin. Bencoin aims to harness the technology behind Bitcoin to create a superior transaction clearing mechanism. Dr Rogoff says that [82]:

> For the moment, there are just too many uncertainties, but over a long enough time frame, it is not hard to imagining that this kind of idea, or perhaps a later generation approach to digital currencies, will make a case for a digital government currency competing.

A key difference in the existing private cryptotokens such as Bitcoin, commonly known as cryptocurrency, and a central bank issued electronic fiat currency would be that the fiat electronic currency would serve the purpose of a legal tender which private cryptotokens (cryptocurrencies) cannot. Hence, it is critical for government or central bank to back an EFC. Furthermore, as an EFC would not be linked to any physical commodity, there is a high risk of EFC becoming worthless if people lose faith in the ability of the currency to serve as a medium of storing purchasing power. The purchasing power of a currency is directly linked to the way a country governs itself and the state of the economy of the country [39, 63].

EFC represents the future of innovation in the payment system, efficient implementation of monetary policy, and regulation of the black economy. An EFC would leverage the strengths of existing banking infrastructure while addressing the limitations of existing money markets, and coordination between fiscal and monetary policy [41]. On the one hand, the interest in EFC is fuelled by cryptotokens used to run a Distributed Ledger Technology (DLT) infrastructure, and on the other hand, there has been an exponential growth in adoption of DLT in various domains. Hence, an EFC based on the usage of DLT infrastructure would have a wide range of impact on the economy and financial markets. The potential impacts include disruption in business models and systems, particularly on the retail payment services possibly making them faster and less expensive [41]. However, this may raise potential risks, several policy issues for central banks and governments, and issues of financial scalability and monetary policy are likely to become more prominent [41].

11.3 Global Position on Blockchain, Cryptocurrency, and Central Bank-backed Electronic Fiat Currency

This section presents an overview of the position of different countries on Blockchain technology, Cryptocurrency, and Central Bank-backed Electronic Fiat Currency.

Table 11.2 Central banks' position towards blockchain technology [99]

Position	Countries or central banks
Blockchain-friendly	France, Hong Kong, Kazakhstan, New Zealand, South Korea
Recognise potential	Canada, Bahamas, Eastern Caribbean, Kenya, Russia, Turkey
Technology limitations seen as a barrier to adoption	Canada, China, European Central Bank, Germany, Hong Kong, Japan, Netherlands, Norway, South Africa, South Korea, USA

11.3.1 Central Banks' Position Towards Blockchain Technology

The Blockchain technology is generally considered to have the potential to improve a range of public service functions, particularly reducing the cost and third-party dependency for verification of certain processes. These processes may or not be financial in nature. Hence, a range of countries are open to blockchain technology and its adoption in different sectors. In the financial sector, the potential of blockchain technology is primarily seen in cross-border payments, collateral in electronic format, and settlement of financial securities (instruments). However, several countries believe the technology is not yet mature for mass adoption and poses several risks. Furthermore, in several cases, the performance of blockchain-based systems does not seem to be better or comparable to the settlements systems of the advanced economies. Thus, these nations intend to avoid the unknown risks attached to the blockchain technology. A summary of the position of various countries on the blockchain technology [99] is presented in Table 11.2, valid as of January 2019.

11.3.2 Central Banks' Position on Cryptocurrencies

Central banks are generally not positive towards cryptocurrencies, and the arguments are as follows:

- Many central banks do not accept them as some form of currencies because they do not meet the basic functional definitions of money.
- Cryptocurrencies are a medium to speculate, highly volatile, and they distort the economy by fuelling bubbles in investments.
- Cryptocurrencies are vulnerable to be used as an instrument of money laundering, criminal proceeds, cybercrime and high risks of loss.

The European Central Bank called the cryptocurrencies as "the evil spawn of the financial crisis". The position of various central banks on cryptocurrencies [99] is summarised in Table 11.3, valid as of January 2019.

Table 11.3 Central banks' position towards cryptocurrencies [99]

Position	Countries or central banks
Generally negative	Azerbaijan, Australia, Bank for International Settlements, Canada, China, European Central Bank, Finland, France, Hong Kong, India, Indonesia, Japan, Malaysia, Netherlands, New Zealand, Norway, Russia, South Africa, South Korea, Switzerland, United Kingdom, USA
Policy: Ban	China, India, Indonesia, Morocco, South Korea, Thailand
Policy: Monitor	– No real risk to financial stability: Australia, India, United Kingdom, USA – Ready if action required: Brazil, Malaysia, Russia, Singapore, Switzerland, United Kingdom – Issued investor warnings: Canada, China, France, European Central Bank, Germany, Hong Kong, United Kingdom
Policy: Regulate	– Integrate into existing financial regulatory framework: Canada, Hong Kong, Indonesia, Japan, Mauritius, Singapore, USA – New regulatory framework: France (ICOs), Gibraltar, Japan, Mauritius, Thailand – Consulting on regulations: Israel, Russia, South Africa
Policy: Support	Brazil, Bermuda, Germany, Israel, Japan, Malta, Sweden, Switzerland, Ukraine, Venezuela

11.3.3 Central Banks' Position on Electronic Fiat Currency

The lack of standardised definition and specification of a Central Bank-backed Electronic Fiat Currency affects the exploration of opportunities and assessment of risks. A summary of the position of various Central Banks on Electronic Fiat Currency [99] is presented in Table 11.4, valid as of January 2019.

11.4 Initiatives of Several Central Banks on EFC

Bank for International Settlements (BIS) published a report this year in which it claimed that as much as 70% of financial authorities across the globe are engaged in research on issuing a central bank backed electronic fiat currency [7]. However, the motivation of these authorities and their plans on implementation vary significantly from one country to the other. BIS report is based on a survey conducted with 63 central banks in different countries; out of 63 central banks surveyed by BIS, 22 are from advanced economies while 41 are from emerging market economies [7]. The banks surveyed by BIS represent about 80% of the global population and more than 90% of economic output.

Furthermore, the global stance on a central bank-backed electronic fiat currency seems to be tilting in its favour. For example, the International Monetary Fund (IMF), which has once denounced [49] the plan of issuing a central bank-backed EFC by the

Table 11.4 Central banks' position on electronic fiat currency [99]

Position	Countries or central banks
Recognise potential value of EFC	– Maintain public access to Central Bank liability in event of declining use of cash: Norway, Sweden – Facilitate de-cashing: Curaçao and Sint Maarten, Israel – Improve cross border transaction systems: Canada, Hong Kong, Saudi Arabia, Singapore, United Kingdom – Modernise interbank settlement systems: Bank for International Settlement, Singapore, Thailand – Address underserved markets: Bahamas
Against EFC implementation	Azerbaijan, Australia, Denmark, ECB, Estonia, Germany, Hong Kong, India, Israel, Japan, New Zealand, Norway, South Korea, Switzerland, USA
Studying long-term potential of EFC	Brazil, Canada, China, Indonesia, Israel, Norway, Singapore, UAE, United Kingdom, USA, European Central Bank and Japan
Actively developing pilots	Bahamas, Eastern Caribbean, Kazakhstan, Philippines, Russia, South Africa, Sweden, Thailand, Ukraine, UAE, Uruguay
Launched EFC	Iran, Marshall Islands, Senegal, Tunisia, Venezuela
Abandoned EFC project	Ecuador

Republic of the Marshall Islands, has now urged [59] the international community to positively "consider" the idea. IMF's Managing Director Christine Lagarde said that although she is "not entirely convinced" with the concept "I believe we should consider the possibility to issue digital currency. There may be a role for the state to supply money to the digital economy" [59].

Some of the initiatives on developing a central bank backed EFC are listed below:

- In 2015, Tunisia launched a blockchain-based fiat electronic currency called eDinar [87]; thus becoming the first country in the world to issue government-backed electronic fiat currency. eDinar is also known as BitDinar and Digicash. The issuance and distribution of eDinar are entirely under the ambit of a government body called La Poste or La Poste Tunisian. La Poste is not a bank but an authorised financial institution. The architecture of blockchain-based DigiCash platform is shown in Fig. 11.1.
  The DigiCash architecture incurs transaction fees, although the transaction fees are not significant as the maximum transaction amount is capped at one dinar. eDinar was created with the technical support of a Switzerland-based blockchain company called Monetas [87]. Now, the central bank of Tunisia, Banque Centrale

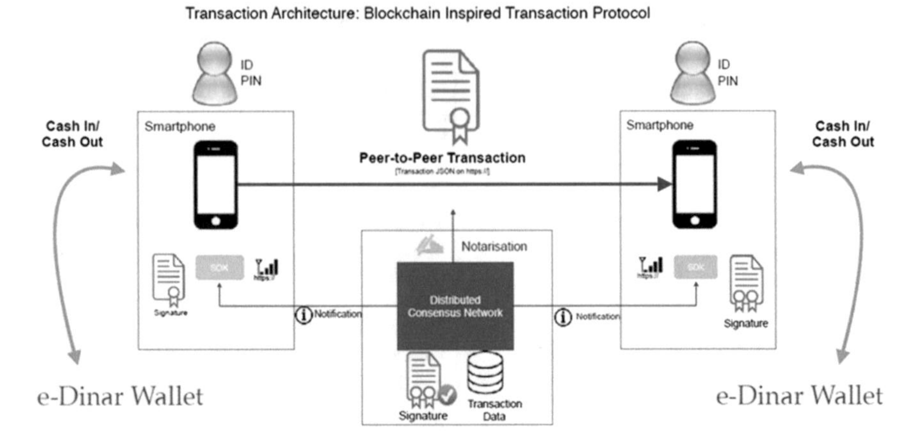

Fig. 11.1 Technical architecture: DigiCash platform [21]

de Tunisie, has announced that it has now created a working group to study the development and issuance of a sovereign cryptocurrency bond [85]. El Abassi, the governor of the central bank, said that the blockchain technology offers a mechanism to central banks to combat money laundering, fight cross-border terrorism, manage remittances, and limit informal economies [85].

– Senegal is another country which is an early adopter of a sovereign electronic fiat currency. In December 2016, Senegal issued its national currency over a blockchain and called it eCFA [18]. eCFA is the electronic version of Senegal's paper-based fiat currency called CFA franc. eCFA was developed jointly by the local bank called Banque Regionale de Marches (BRM) and an Ireland-based technology company called eCurrency Mint Limited [18]. eCFA can only be issued by an authorised financial institution and is entirely under the control of the central bank. eCFA has been designed as a legal tender to work alongside the paper-based fiat currency. If eCFA is found efficient, then this could be extended to other countries which are a member of the West African Economic and Monetary Union (WAEMU) namely Benin, Burkina Faso, Cote d'Ivoire, Guinea-Bissau, Mali, Niger, and Togo [18].
– The Central Bank of Uruguay, in November 2017, issued a plan to carry out the pilot on the issuance and usage of a digital version of Uruguay's fiat currency Peso [5]. The Central Bank stressed that "*it is not a new currency, it is the same Uruguayan peso that, instead of having a physical support, has a technological support*" [5]. The central bank's plan proposed to let a total of 10,000 mobile service subscribers of Antel, the state-owned telecommunications company, to download a mobile application (app) with an integrated digital wallet. The first tranche of digital tokens to be issued was planned to consist of 20 million Uruguayan pesos [5]. Apart from the Central Bank of Uruguay and Antel, RGC, IBM, IN Switch, and RedPagos were part of the pilot project. RGC has the role of system provider; IBM to provide storage support, circulation and control; IN Switch to provide user management and transfers; and RedPagos for ticketing. The head of the Central

Bank of Uruguay said that Uruguay "*is very much in the vanguard*" of developing a digital currency [2].

- In 2017, Sweden's Central Bank, Riksbank, started a project called e-Krona to investigate the potential of issuing a blockchain-based electronic fiat currency which could complement cash [90, 92]. Riksbank published its first report on e-Krona [90] and an action plan in 2017 [91] followed by the publication of its second report on e-Krona in 2018 [92]. These documents highlighted that the Riksbank "*has not yet taken a decision on whether to issue an e-Krona and the aim is not for an e-Krona to replace cash*" [90, 92]. Furthermore, Riksbank reported that the motivation to study the concept of a central bank-backed electronic fiat currency emerges from the fact that the country is witnessing a decline in the popularity of cash [90, 92]. For instance, in a survey conducted in 2018, it was found that only 13% of people living in Sweden paid in cash for their most recent purchase against the corresponding figure of 39% in 2010 [92].
- The Monetary Authority of Singapore (MAS) released a report in June 2017 in which it disclosed about its project on a blockchain-based "*tokenized form of the Singapore Dollar*" [26]. The project is called *Project Ubin*. The project is a collaborative effort of the Monetary Authority of Singapore and blockchain consortium R3, which is a consortium of financial services firms and is focussed on developing a blockchain-based pilot to carry out cross-border payments [26]. Interestingly, in January 2018, in an interview with the Financial Times, Ravi Menon, the Managing Director of the Monetary Authority of Singapore, criticised the idea of a central bank-backed electronic fiat currency and questioned the reasoning behind the central banks issuing electronic currency to the (non-bank) public [34].
- The President of Venezuela, Nicolas Maduro has announced in December 2017 that the government is working on launching an electronic currency backed by Venezuela's gold, minerals, and oil reserves [97]. In January 2018, the President announced that the government would soon issue 100 Million Petros backed by an equivalent number of barrels of oil [17]. The President further announced that the Petro could be freely converted to and from several fiat currencies such as the Russian Rubble, Chinese Yuan, Turkish Lira, and Euro. Finally, in February 2018, the government of Venezuela launched its sovereign electronic currency called Petro (PTR) [58].
- In March 2018, the Republic of Marshall Islands launched a blockchain-based electronic fiat currency called Sovereign (SOV) [60]. SOV, the digital legal tender, was first introduced in late February 2018, when the government of the Republic of Marshall Islands passed a law called the Declaration and Issuance of the Sovereign Currency Act [66]. SOV is a result of a collaboration between the government of the Republic of Marshall Islands and Neema, an Israel based Financial Technologies (FinTech) company. Neema's CEO, Barak Ben-Ezer, said that SOV "*is completely decentralized and the government cannot control the money supply*" after the Initial Coin Offering [22].
- In January 2019, four local banks in Iran developed a blockchain-based electronic currency which is pegged to gold and called PayMon [105]. Interestingly, the PayMon was launched in less than a week after the Central Bank of Iran's publication

on regulating the future cryptocurrencies [70]. It appears that Iran is likely to launch a state-backed electronic fiat currency [46] to circumvent the Western Sanctions [106]. PayMon was developed in cooperation of three private banks, namely Parsian Bank, the Bank of Pasargad, and Bank Mellat in addition to the cooperation of state-owned financial institution, Bank Melli Iran [105]. The technical support to the project was provided by a company called Kuknos, specializing in blockchain technology. PayMon's objective is to tokenize and thereby monetize the assets and excess bank properties.

- Saudi Arabia and the United Arab Emirates in January 2019 announced an agreement to jointly work on developing a blockchain-based electronic currency to facilitate cross-border trading [98]. According to the Emirate News Agency, the official news agency of UAE, the currency thus designed "*will be strictly targeted for banks at an experimental phase with the aim of better understanding the implications of blockchain technology and facilitating cross-border payments*" [98]. The primary objective of this joint project between the two countries is to determine the impact of centralized currencies, and also look at protecting customer interests, developing standards for the underlying technology and consider cybersecurity risks [98]. Saudi Arabia's financial news portal Argaam reported in February 2019 that six commercial banks from Saudi Arabia and the UAE have joined hands for the proposed electronic currency project called Aber [4].
- The Bank of Thailand has been studying the prospect of releasing its own electronic fiat currency. The details of the project surfaced in Aug 2018; and according to the Thai Central Bank, having its own electronic fiat currency would reduce the transaction cost and validation time [6]. The project is named as *Inthanon* and is being developed jointly by R3 blockchain consortium and a technology company called Wipro Limited [19]. The developed prototype is a blockchain-based system for interbank settlement in electronic fiat currency between the Bank of Thailand and eight commercial banks [6, 19].

11.5 Context of Money and Electronic Fiat Currency

Julius Paulus Prudentissimus, the chief legal adviser of Emperor Severus Alexander in ancient Rome, described the fundamental reasoning behind a government-backed currency as [15]: (i) *a unit of account* to price goods and services; (ii) *a method to store value*; and (iii) *a medium of exchange* to ease economic and financial transactions. Furthermore, Julius Paulus Prudentissimus understood the utility of currency and that it does not depend on the substance of its material but the nominal quantity of the currency. In other words, the power of the currency is attached to the confidence of the public in the institutions managing the monetary system of the state [84].

The pursuit of *a stable unit of account* was followed by leaders like Jevons [52], Marshall [65], Wicksell [101], Fisher [35], Buchanan [16], Hayek [43], Bordo [14], Black [13], Cagan [20], Dorn and Schwartz [29], Patinkin [77], and Dorn [28]. Friedman pursued the aspect of currency's usage as *an efficient medium of exchange* [36].

Friedman argued that a government-backed currency should fetch the same rate of return as other risk-free assets. However, the two goals of a currency, i.e., a stable unit of account and an efficient medium of exchange, appeared impractical for want of interest payment on paper currency, and therefore, Friedman proposed a mechanism of steady deflation instead of price stability.

Today, after nearly two millennia from the work of Julius Paulus Prudentissimus in the Roman empire, the modern day society consisting of electronic devices and high-speed internet is changing the society's perception and practice of paper currency, mode of economic and financial transactions, and so on. The usage of cash is increasingly reducing because of the ease of payments facilitated by cards, mobile phone apps and contact-less chips, online payment systems, etc.

The cash in circulation as a percentage of nominal Gross Domestic Product (GDP) in advanced economies in the period of 2000–2017 [86] is shown in Fig. 11.2. Advanced economies consist of the Euro zone, Japan, Sweden, and the United States of America. Interestingly, since 2008, Sweden has witnessed a sharp decline in the ratio of cash to nominal GDP, thus progressing towards becoming the most cashless country in the world.

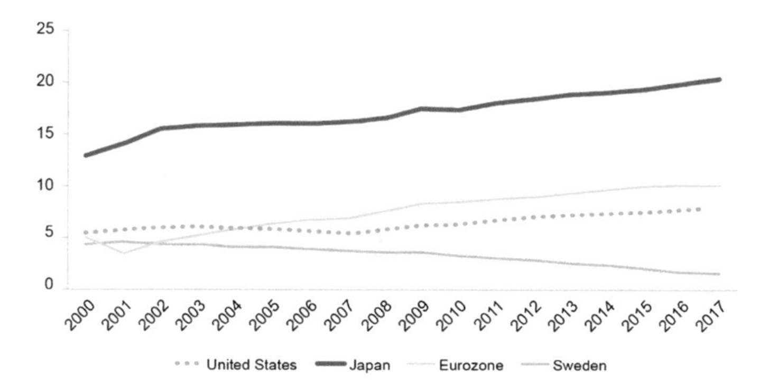

Fig. 11.2 Cash in circulation in advanced economies (% of GDP) [86]

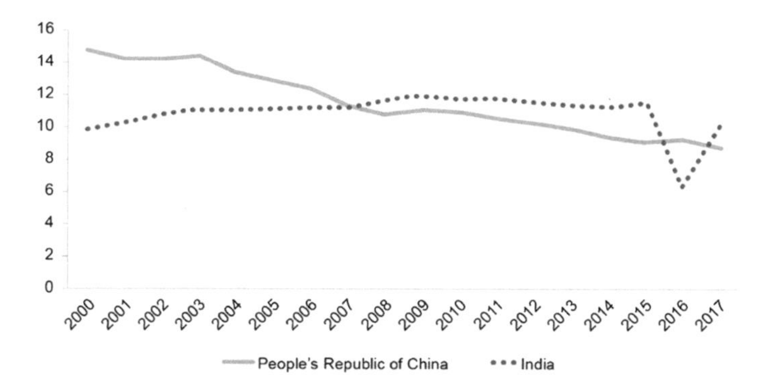

Fig. 11.3 Cash in circulation in People's Republic of China and India (% of GDP) [86]

The circulation of cash as a percentage of nominal GDP in emerging economies of the People's Republic of China and India in the period of 2000–2017 [86] is shown in Fig. 11.3. The sudden dip in cash to GDP ratio of India is because of the demonetisation program of Government of India through which Rs 1000 and Rs 500 notes were pulled out from the circulation. The trend of declining cash to GDP ratio is more evident in the case of the People's Republic of China. This declining trend appears to reflect the wider adoption of cashless payment tools such as Alipay and WeChat pay.

Table 11.5 indicates that the circulation of currency notes issued by central banks has dropped to around 1.5% in Sweden and Norway, while Australia, Denmark, Sweden, and Norway have exhibited a decline in terms of GDP in selected advanced economies [86].

The central bank of Sweden, Riksbank, is concerned about the decline in usage of cash by the citizens of Sweden [92]. The Norges Bank, the central bank of Norway, shares the same concern on the decrease in usage of cash by Norwegians [76]. Both central banks believe that the existing payments systems would not be able to deliver several important services if the cash is to disappear, as these services are critically dependent on cash [76, 92]. For instance, the provision of a risk-free alternative to deposits in a private bank and an independent back-up solution required to keep the ordinary electronic payments systems in service during the time of crisis is dependent on the cash system [54]. An electronic fiat currency system would be able to provide continuity of services mentioned above [54]. Furthermore, the achievement of both the goals of currency, i.e. a stable unit of account and an efficient medium of exchange, can now be achieved by harnessing the modern day financial technologies to design and launch a central bank-backed electronic fiat currency.

On the other hand, the central bank of Denmark, Danmarks Nationalbank, is sceptical about electronic fiat currency [74]. Danmarks Nationalbank believes that the current cash system can provide a holding of safe assets in the form of government bonds or government-insured bank deposits. Hence, no major benefit is seen for EFC. On the issue of robustness, the Danmarks Nationalbank is concerned about the functioning of an EFC as an independent system because of its dependence on the electricity and communication network.

The central bank of the Czech Republic, Czech National Bank, says that there are at least two major monetary policy actions providing enough incentives and motivation to investigate electronic fiat currency [68]. Electronic fiat currency would offer an opportunity to set negative interest rates as a tool to implement a drop of *helicopter money*. Helicopter money is a tool of monetary policy which involves printing and distribution of a large sum of money to the public to stimulate the economy during a deflationary period [44]. Agarwal and Kimball argue that by not implementing the one-to-one peg between currency notes and central bank deposits, a central bank can constrain the circulation of currency notes with a negative interest rate [1]. This approach can be useful in introducing an electronic fiat currency by limiting the circulation of currency notes. Another motivation to investigate the usefulness of an electronic fiat currency is the decline in revenues from seigniorage [54]. This is because the source of income for central banks is the income earned on the assets

Table 11.5 Cash in circulation in selected economies [86]

Country	Cash to nominal GDP ratio (%)				
	2000	2005	2010	2015	2017
Australia	3.9	3.6	3.5	4.1	4.1
Canada	3.2	3.2	3.4	3.7	3.9
Denmark	2.8	3.0	2.9	2.9	2.9
Eurozone	4.8	6.2	8.3	9.9	9.9
India	9.8	11.1	11.7	11.4	10.1
Japan	12.8	16.0	17.4	19.4	20.4
Norway	3.1	2.6	2.1	1.7	1.5
People's Republic of China	14.6	12.7	10.9	9.0	8.7
Republic of Korea	3.4	2.8	3.4	5.5	6.2
Singapore	6.8	6.9	6.9	8.1	9.5
Sweden	4.1	3.8	3.0	1.7	1.3
United Kingdom	3.2	3.3	3.8	4.0	4.1
United States of America	5.7	6.0	6.5	7.8	8.2
Country	Cash (in billions of local currency)				
	2000	2005	2010	2015	2017
Australia	27	35	48	67	74
Canada	35	45	57	75	84
Denmark	37	47	53	60	62
Eurozone	338	521	795	1,038	1,112
India	2,129	4,082	9,070	15,699	16,974
Japan	67,620	83,773	86,856	103,120	111,508
Norway	47	52	54	53	48
People's Republic of China	1,465	2,403	4,463	6,322	7,065
Republic of Korea	21,425	26,136	43,307	86,757	107,908
Singapore	11	15	22	34	42
Sweden	98	111	105	73	58
United Kingdom	34	46	60	76	84
United States of America	584	785	980	1,416	1,607

Source CEIC, US Federal Reserve of St. Louis, IMF

that back the issuance of currency notes plus there is no interest to be paid on the cash. Therefore, a decline in the use of currency notes would lead to a decline in seigniorage. In such a scenario, issuance of electronic fiat currency could be one of the potential mechanisms to arrest the fall of seigniorage. The World Bank believes that an electronic fiat currency could be a strong tool for financial inclusion in developing and economically weaker countries [54].

The research on EFC is growing, and so is the interest of private players and central banks in EFC. One such idea is *Fedcoin* [57]. This idea of a central bank backed electronic currency began in 2013 on several discussion boards over the internet [3, 55, 56, 71]. In 2016, the Bank of England discussed the idea of an EFC over a blockchain containing deposit reserves [11]. Ben Broadbent, Deputy Governor at the Bank of England, said that "*it seems likely that a distributed ledger would make that process easier, opening up the balance sheet to a wider variety of financial firms. One might go further, giving access to non-financial firms, or perhaps even individual households. In the limit, a distributed ledger might mean that we could all of us hold such balances*" [11]. Many other central banks such as Uruguay's Central Bank [5], Eastern Caribbean Central Bank (ECCB) [31], Monetary Authority of Singapore [26], Bank of Canada [32], the People's Bank of China [53], Riksbank of Sweden [92], and the Dutch Central Bank [75] have been working on issuance and piloting of an electronic fiat currency. In June 2019, the International Telecommunication Union (ITU) announced a partnership with Stanford University, USA to support the initiatives of different Central Banks in implementing the pilot of their electronic fiat currency [50]. The partnership is intended to offer technical support to Central Banks piloting the launch of an EFC and an open forum to share the lessons learnt among the Central Banks, technology platform providers, payment system organizations, academia, and telecommunication companies.

11.6 Concept and Attributes of EFC

The Bank for International Settlements (BIS) reported that there is no proper definition of a central bank-backed digital (electronic) currency [24]. BIS says that a central bank-backed digital currency can refer to several different concepts. However, it is envisioned as a new form of a central bank-issued currency. Thus implying that a central bank-backed digital currency is a liability for the central bank, denominated in the existing unit of account, thereby serving as a store of value as well as a medium of exchange. For a better understanding of the concept of a central bank-backed digital currency, it should be viewed in the context of other forms of currency. Bech and Garratt presented a taxonomy of money in the form of a Venn-diagram [10]. Bech and Garratt's money taxonomy is called as '*The Money Flower*' and is depicted in Fig. 11.4.

The money flower consists of four key properties of a currency, namely an issuer of currency (central bank or the government); form of currency (physical or digital); accessibility (widely or restricted); and underlying technology (token- or account-

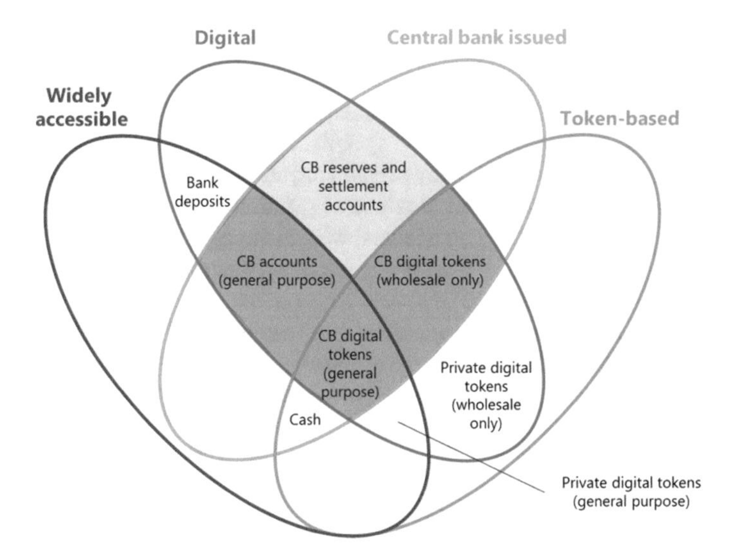

Fig. 11.4 The money flower: a taxonomy of money [10]

based). Green [40] and Mersch [67] classified the underlying technologies of a currency into two forms: token of stored value or accounts. Cash and several privately issued cryptocurrencies are token-based, while the balances in reserve accounts and commercial bank paper-money are forms of account-based currency. The dark grey region in the 'Money Flower' represents a central bank-backed electronic fiat currency.

We envision (define) a central bank-backed electronic fiat currency (EFC) as *a financial technologies driven monetary instrument that serves as legal tender and means of settlement across digital payment networks and can be easily converted to bank deposits and other forms of government-backed liabilities which includes central bank reserves and interest-earning government securities. An EFC would offer (nearly) real-time transactions thereby addressing the liquidity distribution problem imposed by physical cash movement.*

A central bank-backed electronic fiat currency can be conceptualised in at least two forms. EFC can be made available to the public (retail) or could be limited to usage for large and time-critical payment between banks (wholesale) [41, 54]. EFC available to the public would imply that like cash payments EFC would offer opportunities to make anonymous payments. On the other hand, if anonymity is not a critical feature of EFC, then the public can be allowed to obtain cash in digital form from the accounts held with the central bank. Technically, this has been possible for quite some time; however, no central bank has experimented with this yet. On the other hand, for an EFC with its usage limited to banks, the new (blockchain) technology has the potential to increase efficiency and reduction in the cost of the settlement. Some central banks have launched field trials, but several technical challenges have been reported that must be addressed before the technology could be adopted at an operational level.

For the success of EFC, it is important to ensure that the attributes of EFC are sufficiently attractive to stakeholders viz public, banks, non-banking financial institutions, and others. To this end, many private players have unsuccessfully attempted to design a customer-friendly digital payment solution [51]. However, it is difficult to predict which solution (design attributes) will find acceptance in public and which will not. Yet, it is possible to identify some attributes that may have an impact on the demand for EFC.

The attributes of a potentially successful EFC can be identified from the recorded objective of various governments concerning digital currency technology and the payment sector in general. Some of the recorded objectives of governments are [45]:

- To offer clarity and certainty on the applicability of existing rules and regulations for the parties dealing in electronic money;
- To offer an appropriate technical, functional and legal environment for EFC to flourish including the support to provisions of banking, financial and other professional services;
- To support the research and development of new technology, and technical infrastructure to promote competition and innovation in financial services, payment systems and associated sectors;
- To provide a conducive environment for monetary and financial stability by continuously monitoring the adoption of EFC and ascertaining the risks.

Norges Bank has identified some factors that may influence the demand for an EFC. These factors are [76]:

- What are the needs that an EFC can fulfil?
- How well does the EFC function, and whether it is perceived as an attractive and secure solution or not?
- What payment situation can the proposed EFC be used in? Could it be used for online shopping, physical user locations, bill payments and payments to friends?
- How efficient and cost-effective is it to convert EFC into bank deposits and vice-versa?
- What players (stakeholders) are granted access to the infrastructure?
- Are there any limitations on the amount of funds that can be stored in EFC?
- How attractive is the relative return on the stored fund?
- How secure is the storage system credit risk, operational risk, cyber risk, etc.?
- Which other government-backed securities and assets are available that can be easily used for value storage, conversion to EFC and vice versa?

A study by the International Monetary Fund (IMF) has pointed out that the central bank-backed EFCs have the potential to execute fast and secure cross-border financial transactions [62]. Thus, an EFC has the potential to significantly enhance the payment systems' efficiency and offer a nearly cost less medium of exchange. EFCs would particularly be useful for lower-income groups and small businesses. Lower-income groups would benefit from EFC by mitigating their heavy reliance on cash while the small businesses would benefit by avoiding the substantial cost incurred on handling cash as well as the transaction cost associated with payments made via credit and debit cards. Bank of England's researchers believe that the estimated productivity gains from an EFC adoption would be similar to the gains made from substantial reduction in distortionary taxes [45]. Furthermore, the issuance of an EFC and its wide acceptance leading to a gradual withdrawal of paper currency would be helpful in arresting tax evasion, money laundering, and other illegal transactions [45]. This benefit would be much more pertinent for developing countries than for developed countries, yet the developed countries would benefit immensely from EFCs. The benefits for developing countries are more than the developed countries because of their economies' heavy dependency on cash, thereby leading to higher level of tax evasion, money laundering, and other illegal activities.

11.7 Potential Primary Models of Issuing EFC

The central bank of Norway, Norges Bank, has formed a working group to assess the potential models for an electronic fiat currency system. Norges Bank's working group has identified two primary models for the working of an electronic fiat currency system, namely *Account-based Model* and *Value-based Model* [76]. Sweden's central bank Riksbank in its detailed report on developing and launching an electronic fiat currency in Sweden called *e-krona* has meticulously assessed both the models, account-based model and value-based model, for issuance of an electronic fiat currency [90, 92]. Norges Bank has identified a third type of model called hybrid model, which is based on the blockchain (distributed ledger) technology [76]. Each of these models has its own properties and functionalities that need to be carefully chosen while issuing an electronic fiat currency to have a significant impact. Generally, the properties and functionalities of a value-based model are quite similar to the properties and functions of currency notes [76, 90, 92]. On the other hand, the properties and functions of an account-based model are in close resemblance to an ordinary bank deposit [76, 90, 92].

A key difference between an account-based model and a value-based model is that the storage of value and processing of transactions is centralised in an account-based model while in a value-based model, both the storage and processing are decentralised [76, 90, 92]. However, both models imply that the public would be able to hold money with the central bank and an underlying registry would be required, thus implying that both models can be implemented over the same technical system. Also, several different connected systems and participants would be required to implement the functioning of electronic fiat money in practice.

Both the account-based model and the value-based model, assume that there is a central registry recording all the transactions, thus implying that the transactions in electronic fiat currency can all be traced [76, 90, 92]. In an account-based model, the payment system is based on an owner registry to establish the identity of the account owner [76, 90, 92]. This is critical to ensure that the payment meant for a person is received by the right person. Hence, anonymous transactions are impossible in an account-based model for electronic fiat currency [76, 90, 92]. Furthermore, a transaction registry is required to avoid double-spending. The registry provides information on the balance EFC, and the bearer of value on which EFC is stored is also registered [76, 90, 92]. Digital wallets and payment cards can be used as a bearer of value.

In an account-based model, the central bank holds the EFC account like a traditional bank account [76, 90, 92]. An EFC account held with a central bank offers real-time transaction between the parties as the transactions are executed in a closed system without the need of involving any external agents [76, 90, 92]. An account-based EFC system must also be able to execute transactions with other systems, thus allowing the flow of money between the systems [76, 90, 92]. The underlying structure of an account-based EFC system would consist of EFC holding accounts (person, business, others), execution of the transaction to and from external systems (person-to-person, person-to-government authority, consumer-to-business, business-to-business) and between EFC accounts held at the central bank [76, 90, 92]. There are two possible alternate mechanisms in which a central bank can offer account-based EFC [76, 90, 92]. One mechanism is to offer a solution like the one provided by the banks traditionally. Under this mechanism, services additional to the basic services such as payment cards, mobile apps, etc. can be provided. Such a mechanism can provide full control and supervision over the transactions [76, 90, 92]. However, this mechanism could be expensive to set up, run, and maintain. Furthermore, this mechanism would require additional functions to be integrated with the external system of the present and future. The alternate mechanism is that the account-based services for EFC are restricted to the basic functions of EFC account holding, deposit, withdrawal, and transactions between the accounts held at the central bank [76, 90, 92]. A central bank can offer limited accounts-based services by providing a standardised interface to connect to the external payment service providers [76, 90, 92]. In this mechanism, a relatively small operational commitment would be needed from the central bank. Hence the cost of development, running, and maintenance of this mechanism would be lower than the first mechanism [76, 90, 92].

A value-based EFC system is like the existing system of currency notes, and hence would primarily be suitable for small value transactions in the presence of the transacting parties [76, 90, 92]. A value-based EFC can be stored on a transaction medium such as a payment card or an app installed on a phone belonging to the user [76, 90, 92]. A card-based EFC solution can be imagined as a traditional gift card which is preloaded with a certain value. As the transactions would take place through card-readers, an offline payment mechanism is thus available. An app-based payment solution would facilitate an online transaction between the parties using the said app. A value-based EFC model would normally allow anonymous transaction like the transactions made with currency notes [76, 90, 92]. However, technically, it is possible to design the system such that a transaction valued above a predefined threshold is not entirely anonymous [76, 90, 92].

Table 11.6 Comparison between Cash, Account-based EFC, and Value-based EFC [76, 90, 92]

Attributes	Cash	Account-based EFC	Value-based EFC
Real-time Payments	Yes	Yes	Yes
Underlying Register	No	Yes	Yes
Store of Value	Yes	Yes	Yes
Physical Presence Required	Yes	No	Yes for prepaid cards; No for applications
Anonymous Payments	Yes	No	Yes, below a threshold and for card-based payments
Legal Structure	Yes	Yes	Yes
Interest Payment	Yes, while in bank deposits	Yes	Not as a rule
Credit Risk	No	No	No
Traceability	No	Yes	Yes; Not if the prepaid card changes hand
Offline Payments	Yes	No	Yes
Usability	No additional technical tools required	Via mobile apps and website	Additional technical tools such as card readers or special smart phone technology required

A comparison between traditional cash currency and the two forms of EFC (account-based and value-based)[76, 90, 92] is presented in Table 11.6.

The hybrid model of issuing EFC is based on a blockchain/DLT and can be categorised into two categories, namely DLT-based model for the retail level launch of EFC and a DLT-based model for the wholesale level launch of EFC. The DLT-based model for retail level EFC is popular among the central banks in emerging economies. The motivation behind this is to join the technological developments in the field of financial technologies, to promote financial inclusion, and to reduce the cost of printing and handling cash currency. Some countries, such as Ecuador, India, Israel, Uruguay, Republic of Marshall Islands, and the Peoples Republic of China, have either been theoretically investigating the model or have run pilots. However, with the developments in 2019, it appears that the enthusiasm of the Reserve Bank of India is fading away [93]. The DLT-based model for the wholesale level launch of EFC is the most popular model which has seen many pilot runs across the globe. Some of these pilots are briefly discussed in Sect. 11.4 of this chapter. The main objective of these pilot runs was to encourage the central bank's understanding of the blockchain/DLT and its usefulness in the wholesale financial services market. The wholesale financial services include services like Real-Time Gross Settlement Systems (RTGS), cross-border interbank payments and settlements, securities (financial instruments) clearing and settlement systems, and so on. Table 11.7 presents a matrix of models of issuing an EFC and features of various central bank-backed EFC initiatives.

Table 11.7 Matrix of EFC issuing models and selected EFC initiatives

EFC issuing model	Underlying technology	Target user	Country initiatives
Account-based EFC	Non-DLT/Blockchain	Retail	Sweden
Value-based EFC	Non-DLT/Blockchain	Retail	Sweden
Digital token	DLT/Blockchain	Retail	Uruguay, Venezuela
Digital token	DLT/Blockchain	Wholesale	Singapore, Thailand

11.8 Desired Design Principles

In this section we present a set of desired design principles for an EFC. These design principles are identified from a technical perspective rather than an economic perspective.

- *Cyber Security*: An important characteristic required in an EFC is the security of EFC or the risk of loss. As is evident from several security breaches in the wallet or system of crypto tokens, which has fuelled the research on EFC, user' concerns on the security of EFC is one of the primary characteristics. A security breach may negatively influence the confidence and trust of a user, and it may also affect the acceptance of EFC by other stakeholders. In such a scenario, central banks and governments may be required to provide a level of insurance against the loss for the EFC held in a bank account and authorised wallet, but the insurance should be limited to a threshold to be determined by the authorities. Furthermore, there would be concerns about the chances of an unknown system glitch or the possibility of a well-known 51% attack if the services are provided over a distributed ledger. 51% attack in a distributed ledger system implies that if the majority of miners could gain enough computing power, then there is a theoretical possibility of overtaking the system by the attackers.
- *Privacy*: Several cryptotokens facilitate transactions over a distributed ledger without the need to provide any sensitive personal information or payment details. This attractiveness of anonymity is often driven by the desire to bypass laws and regulations of the land. Bitcoin (crypto token) is one of the preferred media of the transaction on the Dark Web. The top six Dark Markets, in 2014, recorded around USD 650,000 worth of transactions in bitcoin. In this respect, an EFC is potentially vulnerable to illicit use. However, the illicit use of currency notes is also possible. Hence, with an appropriate regulation and governance model, the privacy in transactions with EFC may be restricted to a threshold level, and beyond that, the transacting parties would be required to provide some information. The disclosure of user information would, in this case, be aligned to the principles of Know Your Customer.
- *Technical Issues*: The idea of an EFC finds its roots in the launch of Bitcoin (cryptotoken), and exponential growth in price, acceptance and number of cryptotokens is likely to influence the design and infrastructure of an EFC to be designed over a

distributed ledger. The EFC distributed ledger would play a critical role in building consensus among the participants to ensure the integrity of recorded transactions. Malicious agents may attempt to influence the transaction record by introducing the participants to verify fraudulent transactions. Hence, if the integrity of the ledger cannot be guaranteed and two differing versions may coexist, then that would negatively affect the trust on EFC.

- *Processing Speed*: Bitcoin and other crypto tokens running on a distributed ledger are known for faster clearance and settlement of transactions than the traditional systems. However, the processing speed of these ledgers varies according to the technical architecture of the system. In this regard, an EFC running over a distributed ledger system would need to be carefully designed to offer a faster clearing and settlement of EFC transactions.
- *Usability*: The adoption of payment methods and mechanisms is critically dependent on the 'ease of use' of the application and system. The ease of use can be determined based on factors such as the number of steps needed to complete the payment process, whether the process is intuitive and easy to integrate with other processes. Hence, the potential use/wide acceptance of EFC would depend on the usability advantages offered by EFC/distributed ledger technology over the existing methods.
- *Cost*: From the discussion on distributed ledger technology/blockchain and cryptocurrencies, it is evident that the distributed ledger may offer lower transaction costs against the costs incurred in other payment methods. In some distributed ledger mechanisms, the processing of payments is rewarded with newly issued units of the currency (token) which could be designed as a mechanism to earn "capital gains" translated in government-backed currency units. Hence, an EFC may be designed in such a manner that it works as an attractive alternative to some customers, particularly to those involved in cross-border payments involving high transaction fees. Furthermore, transactions in EFC over a distributed ledger would not require an intermediary to process the transactions, thus reducing the processing cost. However, there is a limitation in processing transactions over a distributed ledger which is that the transaction costs mechanism in such schemes is not always transparent, and several other costs such as conversion fees for currency conversion, etc. may exist.
- *Volatility*: If an EFC is not controlled and regulated by a central bank, then the users of the currency may witness substantial volatility and liquidity risk. If a digital currency is not backed and controlled by a central bank or the government of the state, then the currency in question would not be different from privately issued cryptocurrencies. In such a scenario, the users of the currency holding it in a currency account, let us say after having received a payment for some goods or services, would be exposed to substantial risk of volatility similar to the kind of volatility we have seen in privately issued cryptocurrencies in the last few months and years. On the one hand, volatility could be an opportunity for speculators to gain from it, but on the other hand, the volatility may pose an obstacle in its wider adoption. Hence, adequate design elements from an economic perspective and regulatory frameworks must be considered while designing an EFC.

- *Irrevocability*: Looking at the latest developments and initiatives (discussed in Sect. 11.8), an EFC would be based on a distributed ledgers technology. Distributed ledger Technologies do not offer a dispute resolution mechanism, and the transactions are irrevocable, thus, reducing the risk of a payment being reversed because of fraud or chargebacks. However, this feature is like a double-edged sword. Payees, such as merchants, would find this feature attractive but the payers, such as consumers and customers, may see it as a limitation in the system, thereby deterring the adoption.

11.9 High Level Technical Architecture

After having discussed the potential models of issuing an EFC and features of an EFC from a technical perspective, this section presents a potential architecture to issue an EFC. The EFC issuing central bank would be expected to provide the required technical infrastructure (platform) consisting of an account structure for an account-based EFC and a register enabling the issuance and redemption of a value-based EFC. The EFC platform would be needed to be able to communicate with other relevant and required systems and applications in the banking and financial sector. Such applications and systems normally include user applications, external payment systems, support systems, settlement systems, and so on. The proposed architecture is depicted in Fig. 11.5.

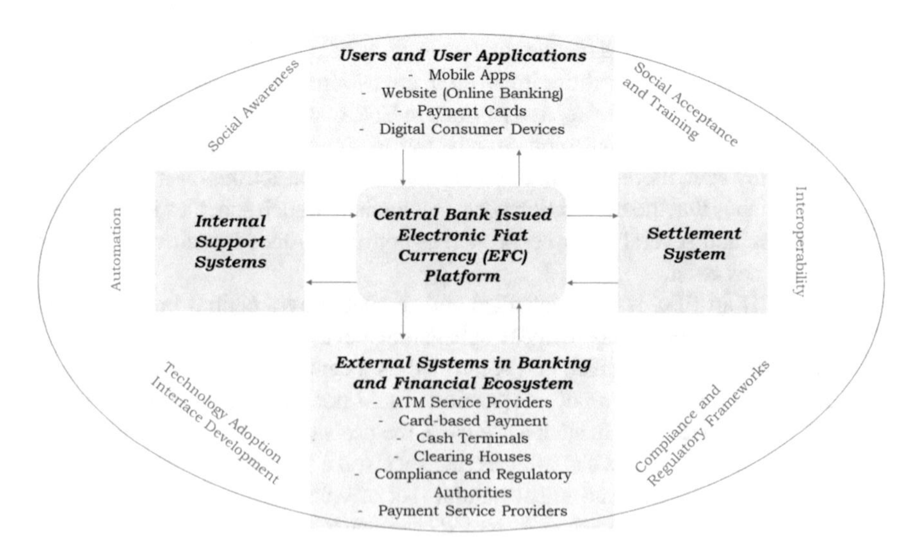

Fig. 11.5 Possible architecture of an EFC system

The various components of the proposed architecture, depicted in Fig. 11.5, are discussed below.

- *EFC Platform*: The EFC platform would consist of a central register for the holders of EFC, regulatory framework and the transaction and settlement conditions to be applied. The EFC platform also consists of the necessary logic to process and implement a variety of payment. The EFC platform would be the central/core part of the EFC infrastructure, and the platform would manage all the communication with other systems and participants in the banking and financial infrastructure. The EFC platform would settle the payments between the parties transacting in EFC, and the regulatory framework would be in full control of the central bank.
- *Users and Applications*: The central bank would need to provide a user application to the potential users of EFC to let them transact in EFC. Hence, there must be one or more value-bearing applications facilitating transactions in EFC. The central bank may launch a mobile application, website for online banking, and a payment card to allow transactions in EFC. Furthermore, given the advancements in digital technologies and the world moving towards a digitally connected society, consumer products such as smart watches, smart refrigerator, etc. can be equipped with a digital wallet for EFC to pay and receive payments in EFC. In such a scenario, the central bank would be expected to provide open source infrastructure and tools to let the relevant parties develop novel payment services in and around the EFC payment infrastructure.
- *External Systems*: The overall system architecture of EFC infrastructure would have a bearing on the way EFC systems connect and communicate with external systems, such as ATMs, payment applications, and so on. For an EFC to reach a grassroots level, the central banks would need to connect their EFC system with the entities involved in providing services related to ATMs. Furthermore, various payment apps providing services for e-commerce businesses and otherwise would need to be connected to the EFC system to facilitate payments in EFC. Also, if the companies want to use EFC for various business payments such as for employee salaries, they would need to have a connection to the EFC system. If a retail user (households) wants to use EFC for various bill payments, including to government bodies and private organisations, they would need to have a connection with EFC systems. The users and account holders of EFC would often need a combination of the above services to be connected to the EFC system.
- *Internal Support Systems*: The EFC platform would need to be connected to a range of internal systems of the central banks issuing and managing an EFC. A connection between the EFC platform and internal systems is needed to implement relevant control functions and administer the entire ecosystem of the banking and financial infrastructure. Some of the internal support systems of the central banks that would need to be connected to the EFC platform include the systems related to managing debt, statistics, and reporting systems in addition to the regulatory compliance systems such as money laundering checks and financing of terrorist activities.

- *Settlement Systems*: A central bank issuing an EFC would need to connect the EFC system with its settlement system. The central bank must be able to transfer EFCs from and to the EFC system smoothly and securely, thereby allowing it to control the flow of EFC in the monetary system. Hence, the EFC platform would need to connect to the settlement system of the central bank. Payments in EFC that would be transacted through the EFC platform and with a bank account or vice-versa would thus be instantly settled through the settlement system of the bank.

11.10 Conclusion

In this chapter, we looked at the key aspects of currency in modern society, the advancements in the field of financial technologies and how could these be harnessed to launch a central bank-backed electronic fiat currency. In particular, we discussed the motivation for such a study, the context of money and electronic fiat currency, concept and attributes of an electronic fiat currency, potential primary models of issuing an electronic fiat currency, the key design principles from a technical perspective. We presented a high-level architecture for an electronic fiat currency system, the position of various central banks and governments on cryptocurrency, blockchain technology, and the initiatives related to issuing an electronic fiat currency. From this discussion, it is evident that the concept of a central bank-backed electronic fiat currency has been steadily drawing the attention of various stakeholders across the globe. On the other hand, we also found that several countries or central banks are against the idea of issuing a central bank-backed electronic fiat currency. The reasons given by the opponents of such an initiative are that the society is not yet ready, the heavy cost of infrastructure development, 'moral hazard' and so on. Hence, the overall actual economic and policy implications of issuing a central bank-backed electronic fiat currency are largely unknown. However, in the light of exponential growth in the adoption of cryptocurrency and innovation in payment technologies, it would not be prudent for central banks and governments to not act on bringing in necessary technological and legal advancements in terms of launching an Electronic Fiat Currency for the digitally advancing society of today and future. Inaction on the part of the central bank with regards to modern financial technologies and not accepting or launching an electronic fiat currency may lead to loss of monetary control and severe economic downturns.

References

1. R. Agarwal, M. Kimball, Breaking through the zero lower bound. Technical Report (International Monetary Fund, 2015)
2. Agencia EFE, Uruguayan Central Bank to test digital currency (2017). https://www.efe.com/efe/english/business/uruguayan-central-bank-to-test-digital-currency/50000265-3385232

3. D. Andolfatto, Fedcoin: on the desirability of a government cryptocurrency (2015). http://andolfatto.blogspot.com/2015/02/fedcoin-on-desirability-of-government.html
4. Argaam, Six Saudi, UAE Banks to participate in digital currency project (2019). https://www.argaam.com/en/article/articledetail/id/593036
5. Banco Central del Uruguay, El BCU present un plan piloto para la emisin de billetes digitales (2017). English Translation https://www.bcu.gub.uy/Comunicaciones/Paginas/Billete_Digital_Piloto.aspx
6. Bank of Thailand (BOT) Press Release, Announcement of project Inthanon collaborative partnership (Wholesale Central Bank Digital Currency) (2018). No. 54/2018. https://www.bot.or.th/Thai/PressandSpeeches/Press/News2561/n5461e.pdf
7. C. Barontini, H. Holden, Proceeding with caution a survey on central bank digital currency. Technical Report. BIS Papers No. 101 (Monetary and Economic Department, Bank for International Settlements, 2019)
8. BBC, Salt blocks used as currency (2014), Contributed by Powell Cotton Museum Archives http://www.bbc.co.uk/ahistoryoftheworld/objects/vDn91YroQr-CC4OpxxEtDw
9. BBVA, What is the difference between DLT and blockchain? (2018). https://www.bbva.com/en/difference-dlt-blockchain/
10. M.L. Bech, R. Garratt, Central bank cryptocurrencies. BIS Quarterly Review. Bank for International Settlements (2017)
11. Ben Broadbent, Deputy Governor for monetary policy of the Bank of England, Central banks and digital currencies (2016). https://www.bankofengland.co.uk/speech/2016/central-banks-and-digital-currencies
12. W. Bertsch, The use of tea bricks as currency among the Tibetans. Tibet J. **34**(2), 35–80 (2009)
13. F. Black, A gold standard with double feedback and near zero reserves, in *Business Cycles and Equilibrium* (Basil Blackwell, Oxford, 1987)
14. M. Bordo, The gold standard: the traditional approach, in *A Retrospective on the Classical Gold Standard, 1821–1931* (University of Chicago Press, USA, 1984)
15. M.D. Bordo, A.T. Levin, Central bank digital currency and the future of monetary policy. Technical Report (National Bureau of Economic Research, 2017)
16. J. Buchanan, Predictability: the criterion of monetary constitutions, in *Search of a Monetary Constitution* (Harvard University Press, USA, 1962), pp. 155–183
17. J. Buck, President Maduro: Venezuela to issue first 100 million petros (2018). https://cointelegraph.com/news/president-maduro-venezuela-to-issue-first-100-million-petros
18. BUSINESS WIRE, eCurrency Mint Limited and Banque Rgionale De Marchs launch new digital currency in Senegal (2016). https://www.businesswire.com/news/home/20161103006949/en/eCurrency-Mint-Limited-Banque-R%C3%A9gionale-De-March
19. BUSINESS WIRE, Wipro, R3 build blockchain-based solution prototype to power digital currency in Thailand (2019),. https://www.businesswire.com/news/home/20190507005443/en/Wipro-R3-Build-Blockchain-Based-Solution-Prototype-Power
20. P. Cagan, A compensated dollar: better or more likely than gold?, in *The Search for Stable Money: Essays on Monetary Reform* (University of Chicago Pres, USA, 1987)
21. M. Chakchouk, Blockchain in Tunisia: from experimentations to a challenging commercial launch, in *ITU Workshop on Security Aspects of Blockchain Geneva, Switzerland*, 21 March 2017 (2017)
22. G. Chavez-Dreyfuss, Marshall Islands to issue own sovereign cryptocurrency (2018). https://www.reuters.com/article/us-crypto-currencies-marshall-islands/marshall-islands-to-issue-own-sovereign-cryptocurrency-idUSKCN1GC2UD
23. CoinMarketCap, All cryptocurrencies (2019). https://coinmarketcap.com/all/views/all/
24. B. Cur, J. Loh, Central bank digital currencies. resreport, Committee on payments and market infrastructures, markets committee, bank for international settlements (2018). https://www.bis.org/cpmi/publ/d174.pdf
25. Dave, Salt: Ethiopias white gold (2016). http://davehoggan.com/white-gold/
26. Deloitte and Monetary Authority of Singapore, The future is here—project Ubin: SGD on distributed ledger. techreport, Monetary Authority of Singapore (2017)

27. A.E. Dien, *Six Dynasties Civilization* (Yale University Press, USA, 2007)
28. J. Dorn, *Monetary Alternatives: Rethinking Government Fiat Money* (Cato Press, Washington DC, USA, 2017)
29. J. Dorn, A. Schwartz, *The Search for Stable Money: Essays on Monetary Reform* (University of Chicago Press, USA, 1987)
30. E.D. Earthy, The social structure of a Gbande Town, Liberia. Man, 203–205 (1936)
31. ECCB, ECCB to issue worlds first blockchain-based digital currency (2019). https://www.eccb-centralbank.org/news/view/eccb-to-issue-worldas-first-blockchain-based-digital-currency
32. W. Engert, B.S.C. Fung, Central bank digital currency: motivations and implications. resreport, Bank of Canada. Staff Discussion Paper/Document danalyse du personnel 2017–16 (2017)
33. European Central Bank, Directorate General Communications, ECB ends production and issuance of 500 banknote (2016). https://www.ecb.europa.eu/press/pr/date/2016/html/pr160504.en.html
34. Financial Times, Singapore sounds cautious note on cryptocurrencies (2018). https://www.ft.com/content/2d433cda-f54e-11e7-8715-e94187b3017e
35. I. Fisher, A compensated dollar. Q. J. Econ. **27**(2), 213–235 (1913)
36. M. Friedman, *A Program for Monetary Stability* (Fordham Press, New York, USA, 1960)
37. M. Friedman, The island of stone money. resreport E-91-3, The Hoover Institution, Stanford University. Working Paper in Economics (1991)
38. C.L.C. Gillilland, The stone money of yap: a numismatic survey. resreport 23, Smithsonian Studies in History and Technology (1975)
39. D. Goldberg, Famous myths of "Fiat Money". J. Money Credit. Bank., 957–967 (2005)
40. E. Green, Some challenges for research in payment systems, in *The Future of Payment Systems* (Routhledge Publishers, Milton Park, 2008)
41. R. Grey, J. Dharmapalan, The macroeconomic policy implications of digital fiat currency. Technical Report. The Case for Digital Legal Tender' Paper Series (eCurrency Mint, Ltd.) (2017). ISBN: 978-0692849651
42. M. Gupta, *Blockchain for Dummies*, IBM Limited edn. (Wiley, 2017)
43. F.A. Hayek, *Denationalisation of Money—The Argument Refined: An Analysis of the Theory and Practice of Concurrent Currencies*, 3rd edn. (The Institute of Economic Affairs, London, UK, 1990)
44. T. Hirst, What is helicopter money? (2015). https://www.weforum.org/agenda/2015/08/what-is-helicopter-money/
45. HM Treasury, *Digital currencies: response to the call for information*. Technical Report (HM Treasury, 2015)
46. M. Huillet, Iran: model of state-issued digital currency now ready, says govt minister (2018). https://cointelegraph.com/news/iran-model-of-state-issued-digital-currency-now-ready-says-govt-minister
47. Imperial Tea Garden, Tea money (2018). https://www.imperialteagarden.com/blogs/tea/tea-money
48. Indo-Asian News Service, Abolish Rs 500, 1,000 notes to curb black money: Chandrababu Naidu (2016). https://www.indiatoday.in/india/story/abolish-500-1000-notes-black-money-chandrababu-naidu-346215-2016-10-12
49. International Monetary Fund, Republic of the Marshall Islands. Technical Report. IMF Country Report No. 18/270 (International Monetary Fund, 2018)
50. ITU News, ITU and Stanford University to launch new partnership supporting pilots of digital fiat currency (2019). https://news.itu.int/itu-stanford-university-launch-new-partnership-supporting-pilots-digital-fiat-currency/
51. W. Jack, T. Suri, Mobile money: the economics of m-pesa. Working paper 16721, National Bureau of Economic Research (2011). https://doi.org/10.3386/w16721, http://www.nber.org/papers/w16721
52. W. Jevons, A tabular standard of value, in *Money and the Mechanism of Exchange* (Appleton, New York, 1875)

53. C. Jia, Central bank unveils plan on digital currency (2019). http://www.chinadaily.com.cn/a/201907/09/WS5d239217a3105895c2e7c56f.html
54. J.P. Koning, Approaches to a central bank digital currency in Brazil. resreport, R3 LLC (2018)
55. J. Koning, Why the fed is more likely to adopt bitcoin technology than kill it off (2013). http://jpkoning.blogspot.com/2013/04/why-fed-is-more-likely-to-adopt-bitcoin.html
56. J. Koning, Fedcoin (2014). http://jpkoning.blogspot.com/2014/10/fedcoin.html
57. J. Koning, Fedcoin: a central bank-issued cryptocurrency. resreport (2016)
58. R. Krygier, Venezuela launches the petro, its cryptocurrency (2018). https://www.washingtonpost.com/news/worldviews/wp/2018/02/20/venezuela-launches-the-petro-its-cryptocurrency
59. C. Lagarde, Winds of change: the case for new digital currency (2018). IMF Managing Director, Speech at Singapore Fintech Festival. https://www.imf.org/en/News/Articles/2018/11/13/sp111418-winds-of-change-the-case-for-new-digital-currency
60. S. Liao, The Marshall Islands replaces the US dollar with its own cryptocurrency (2018). https://www.theverge.com/2018/5/23/17384608/marshall-islands-cryptocurrency-us-dollar-usd-currency
61. C. Look, European authorities stop printing high-value 500-euro notes (2019). https://www.bloomberg.com/news/articles/2019-04-26/european-authorities-stop-printing-high-value-500-euro-notes
62. T. Mancini Griffoli, M.S. Martinez Peria, I. Agur, A. Ari, J. Kiff, A. Popescu, C. Rochon, Casting light on central bank digital currencies. Technical Report (International Monetary Fund, 2018)
63. N.G. Mankiw, *Brief Principles of Macroeconomics* (Cengage Learning, 2014)
64. J.J. Mark, Heraclitus of Ephesus (2010). https://www.ancient.eu/Heraclitus_of_Ephesos/
65. A. Marshall, Remedies for fluctuations of general prices. Contemp. Rev. **51**, 355–375 (1887)
66. Republic of the Marshall Islands, Declaration and Issuance of the Sovereign Currency Act 2018 (2018). https://rmiparliament.org/cms/images/LEGISLATION/PRINCIPAL/2018/2018-0053/DeclarationandIssuanceoftheSovereignCurrencyAct2018_1.pdf
67. Y. Mersch, Digital base money: an assessment from the ecb's perspective. Speech at the Bank of Finland (2017)
68. Mojmr Hampl, Vice-Governor, Czech National Bank: a digital currency useful for central banks?, in *7th BBVA Seminar for Public Sector Investors and Issuers, Bilbao* (2018). https://www.cnb.cz/en/public/media-service/speeches-conferences-seminars/speeches/a-digital-currency-useful-for-central-banks/
69. S. Mossavar-Rahmani, B. Nelson, M. Weir, M. Minov, A. Ubid, F. Asl, M. Dib, M.C. Rich, (Un)Steady as she goes. Outlook, Investment Management Division, Goldman Sachs (2018). https://www.goldmansachs.com/what-we-do/investment-management/private-wealth-management/intellectual-capital/isg-outlook-2018.pdf
70. M. Motamedi, Iran's Central Bank issues draft rules on cryptocurrency (2019). https://www.aljazeera.com/news/2019/01/iran-central-bank-issues-draft-rules-cryptocurrency-190129051653656.html
71. S. Motamedi, Will bitcoins ever become money? A path to decentralized central banking (2014). https://tannutuva.org/2014/will-bitcoins-ever-become-money-a-path-to-decentralized-central-banking/
72. S. Nakamoto et al., Bitcoin: a peer-to-peer electronic cash system (2008)
73. N. Nathan, Time to withdraw Rs 500 and Rs 1,000 currency notes (2016). https://economictimes.indiatimes.com/blogs/counter-point/time-to-withdraw-rs-500-and-rs-1000-currency-notes/
74. D. Nationalbank, Central bank digital currency in Denmark? techreport 28, Danmarks Nationalbank (2017)
75. M. Nikolova, Dutch fin regulator retains critical stance regarding central bank digital currency (2018). https://financefeeds.com/dutch-fin-regulator-retains-critical-stance-regarding-central-bank-digital-currency/

76. Norges Bank, Central bank digital currencies. resreport, Norges Bank (2018), Norges Bank Papers NO1 | 2018
77. D. Patinkin, Irving fisher and his compensated dollar plan. Fed. Reserv. Bank Richmond Econ. Q. **79**(3), 1 (1993)
78. R.M. Poole, The tiny island with human-sized money (2018). http://www.bbc.com/travel/story/20180502-the-tiny-island-with-human-sized-money
79. Press Information Bureau, Government of India, Ministry of Finance, With a view to curb financing of terrorism through the proceeds of Fake Indian Currency Notes (FICN) and use of such funds for subversive activities, and for eliminating Black Money, Government decides to cancel the legal tender character of the High Denomination bank notes of Rs.500 and Rs.1000 from the expiry of the 8th November, 2016 (2016). http://pib.nic.in/newsite/PrintRelease.aspx?relid=153406
80. A.H. Quiggin, *A Survey of Primitive Money: The Beginning of Currency*, 1st edn. (Routledge, 2018)
81. O. Rinaldi, E. Lam, Bitcoin is approaching its highest price of 2019 (2019). https://www.bloomberg.com/news/articles/2019-07-10/bitcoin-renaissance-gains-momentum-as-year-s-high-back-in-sight
82. K.S. Rogoff, *The Curse of Cash* (Princeton University Press, 2016)
83. P. Sands, Making it harder for the bad guys: the case for eliminating high denomination notes. Technical Report 52, Mossavar-Rahmani Center for Business and Government, Harvard Kennedy School, USA. M-RCBG Associate Working Paper Series (2016)
84. J.A. Schumpeter, *History of Economic Analysis* (Allen & Unwin (Publishers) Ltd, 1954)
85. P. Semler, Kabul, Tunis in sovereign crypto bond race (2019). https://www.asiatimes.com/2019/04/article/kabul-tunis-in-sovereign-crypto-bond-race/
86. S. Shirai, Money and central bank digital currency. ADBI Working Paper Series, Asian Development Bank Institute, Tokyo (922) (2019)
87. E. Smart, Tunisia becomes first nation to put nations currency on a blockchain (2015). https://dcebrief.com/tunisia-becomes-first-nation-to-put-nations-currency-on-a-blockchain/
88. South African Reserve Bank, Procurement Division, Expression of interest, request for expression of interest from propsective solution providers in anticipation of a feasibiity project for the issuance of electronic legal tender—a central bank digital currency issued and backed by the South African Reserve Bank (2019). EOI Number: MR01/2019-0
89. L.H. Summers, Its time to kill the 100 bill (2016). https://www.washingtonpost.com/news/wonk/wp/2016/02/16/its-time-to-kill-the-100-bill
90. Sveriges Riksbank, The Riksbanks e-Krona project. resreport Report 1, Sveriges Riksbank (2017)
91. Sveriges Riksbank, The Riksbanks e-Krona Project Action Plan for 2018. techreport, Sveriges Riksbank (2017)
92. Sveriges Riksbank, The Riksbanks e-Krona Project. resreport Report No. 2, Sveriges Riksbank (2018)
93. The Economic Times, Draft law proposes 10-year jail term for dealing in cryptocurrency (2019). https://economictimes.indiatimes.com/news/economy/finance/draft-law-proposes-10-year-jail-term-for-dealing-in-cryptocurrency/articleshow/69693984.cms
94. The Economist, The promise of the blockchain: the trust machine (2015). https://www.economist.com/leaders/2015/10/31/the-trust-machine
95. The Editorial Board, The New York Times, Getting rid of big currency notes could help fight crime (2016). https://www.nytimes.com/2016/02/22/opinion/getting-rid-of-big-currency-notes-could-help-fight-crime.html
96. The Law Library of Congress, Global Legal Research Center, Regulation of cryptocurrency around the world. resreport, The Law Library of Congress, Global Legal Research Center, USA (2018)
97. A. Ulmer, D. Buitrago, Enter the 'petro': Venezuela to launch oil-backed cryptocurrency (2017). https://in.reuters.com/article/venezuela-economy/enter-the-petro-venezuela-to-launch-oil-backed-cryptocurrency-idINKBN1DY081

98. WAM/Rasha Abubaker, Saudi-Emirati Powerhouse Announces 7 Joint Initiatives in Vital Sectors (2019). http://wam.ae/en/details/1395302733616
99. O. Ward, S. Rochemont, An addendum to "A Cashless Society—Benefits, Risks and Issues (Interim paper)" Understanding Central Bank Digital Currencies (CBDC). resreport, Institute and Faculty of Actuaries, UK (2019)
100. M. Wendorf, What do bitcoin and Tulip Mania have in common? (2019). https://interestingengineering.com/what-do-bitcoin-and-tulip-mania-have-in-common
101. K. Wicksell, *Interest and Prices: A Study of the Causes Regulating the Value of Money* (Gustav Fischer Press, Sweden, 1898)
102. Wikipediam, List of cryptocurrencies (2019). https://en.wikipedia.org/wiki/List_of_cryptocurrencies
103. O. Williams-Grut, GOLDMAN SACHS: 'The Blockchain can change... well everything' (2015). https://www.businessinsider.com/goldman-sachs-the-blockchain-can-change-well-everything-2015-12
104. World Economic Forum, Deep shift: technology tipping points and societal impact. techreport REF 31081, Global Agenda Council on the Future of Software & Society, World Economic Forum (2015)
105. A. Zmudzinski, Four Iranian banks support gold-backed cryptocurrency (2019). urlhttps://cointelegraph.com/news/four-iranian-banks-support-gold-backed-cryptocurrency
106. M.J. Zuckerman, Iran and Russia discuss transacting in crypto to avoid international sanctions (2018). https://cointelegraph.com/news/iran-and-russia-discuss-transacting-in-crypto-to-avoid-international-sanctions

Part IV
Computer Science-Based Technologies in Mobile Computing

Chapter 12
Lightweight Stream Authentication for Mobile Objects

Mike Burmester and Jorge Munilla

Abstract Conventional authentication is a temporal action that takes place at a specific point in time. During the period between this action and when the associated task(s) is (are) executed several events may occur that impact on the task(s), e.g., an authenticated user may take a short break without logging out. This is a vulnerability that may lead to exploits. For applications where such exploits are a concern, authentication should be dynamic with a continuous monitoring loop, where trust is updated while the tasks associated with the authentication are executed. Continuous user authentication addresses this issue by using biometric user traits to monitor user behavior. In this paper we extend this notion for applications where monitoring *mobile objects* has to be a continuous process, e.g., for liveness probing of unmanned aerial vehicles (UAVs), or to protect UAVs (with WiFi based UAVs an attacker may use a WiFi de-authentication attack to disconnect an authorized operator and then take control of the vehicle while the operator is trying to re-establish connectivity). We propose a lightweight stream authentication scheme for mobile objects that approximates continuous authentication. This only requires the user and object to share a loosely synchronized pseudo-random number generator, and is provably secure.

Keywords Stream authentication · Continuous authentication · Pseudo-random number generators · Forward and backward security

---

This material is partly based upon work supported in part by the National Science Foundation under Grants DUE 1241525, DGE 1565215, and by the NSA/DoD under Grants H98230-17-1-0419, H98230-17-1-0322.

M. Burmester (✉)
Department of Computer Science, Florida State University, Tallahassee, FL 32306-4530, USA
e-mail: burmester@cs.fsu.edu

J. Munilla
Department of Communication Engineering, Universidad de Málaga, 29071 Málaga, Spain
e-mail: munilla@ic.uma.es

G. A. Tsihrintzis and M. Virvou (eds.), *Advances in Core Computer Science-Based Technologies*, Learning and Analytics in Intelligent Systems 14,
https://doi.org/10.1007/978-3-030-41196-1_12

12.1 Introduction

The deployment of emerging technologies has made it possible to connect smart systems and smart objects via heterogeneous networks so that users can access and share resources in practical and efficient ways. This smart network, known as the Internet of Things (IoT), can only be effective if the behavior of its components is trusted, which in turn requires efficient authentication enforcers.

The IoT extends client-server Internet applications to capture a broad class of client-object applications where a client monitors and controls mobile objects, such as underwater, overwater, land and aerial vehicles. For such applications, one-time authentication is not an effective way of managing trusted behavior, since trust degrades over time. There are several reasons for this, the main one being that by extending the model to capture physical events, the impact of time and space on trust becomes more prominent, since physical systems inexorably progress towards disorder (characterized by high entropy). For security applications the degradation of trust is exacerbated by malicious behavior that can target specific system vulnerabilities (e.g., the vehicle can be highjacked).

Continuous object authentication maintains the trust between the user and the associated task over time, thus guaranteeing that any attempt to compromise the task will be frustrated. For one-time authentication, the corresponding computation latencies are not typically a major concern. However for continuous authentication this concern is more substantial, and we require solutions for which the verification does not interfere with the authentication process (the next authenticator can only be sent after the previous one is verified).

Our main contribution in this paper is to propose a stream authentication system for mobile objects that can be tailored to address the degradation of trust over time and space in a practical and efficient way. This system only requires that the user and mobile object share a loosely synchronized pseudo-random number generator. It is based on a lightweight one-time Radio Frequency Identification authentication protocol that is proven secure in the Universal Composability security framework [3].

Paper outline. In Sect. 12.2 we overview the literature related to continuous user authentication. Then in Sect. 12.3 we motivate our application for mobile object stream authentication and discuss the threat model. In Sect. 12.4 we present a lightweight stream authentication protocol that is resilient to de-synchronization attacks and that addresses man-in-the-middle relay attacks, and then extend this protocol to capture robustness against traffic analysis attacks. In Sect. 12.5 we show that this is secure in the standard model. We summarize our results in Sect. 12.6.

12.2 Related Work

Continuous biometric authentication systems are built around two major forms of biometrics: those based on behavioral characteristics and those based on physiological attributes (stable body traits) [16]. Behavioral characteristics include learned movements such as handwritten signatures, keyboard dynamics, mouse movements, gait and speech. Biometric systems based on behavioral characteristics are not intrusive, but are not very accurate. Physiological attributes include fingerprints, face, iris, and hand. Biometric systems that use such attributes are considered to be more robust; however they are also intrusive.

In one of the earliest works on continuous biometric authentication, Shephard [14] proposed using keyboard dynamics for the continuous monitoring of user behavior. Gascon et al. [5] investigated typing motion behavior on mobile devices, and Sitová et al. [15] investigated hand movement, orientation and grasp on mobile devices. Patel et al. [11] investigated touch dynamics, face recognition, gait dynamics and keystroke dynamics on mobile devices, and Liu et al. [7] discussed leveraging breathing for continuous authentication. Frank et al. [4] proposed using touch features that can be extracted from the touchscreen of a phone. Murmuria et al. [8] proposed using power consumption, touch gestures and physical movement for continuous authentication on mobile devices. Al Solami et al. [16] give an overview of the main issues and limitations of continuous authentication based on behavioral characteristics.

Klosterman [6] explored the use of biometric traits that capture unforgeable features of users and discussed the issues involved in the integration of enhanced biometric authentication. Niinuma et al. [9] proposed using soft biometric traits such as face features, gender, ethnicity etc. Continuous authentication can also be based on contextual information. For an overview of continuous authentication systems that use physiological, cognitive and behavioral biometrics, the reader is referred to Traore [17].

These works focus on continuous user authentication rather than object authentication. Saadeh et al. [13] proposed an architecture for mobile object authentication in the context of IoT applications, that is based on elliptic curve identity based digital signatures. Several authors investigated multicast stream authentication for Internet applications (e.g. TV Internet broadcasts). In particular, Perrig et al. [12] proposed several efficient solutions, including TESLA (Timed Efficient Stream Loss-tolerant Authentication). Ueda et al. [18] investigated the video coding standard H.264/AVC and proposed a stream authentication scheme in which the authentication is carried out at the network abstraction layer. These systems also use digital signatures (for non-repudiation).

We conclude by observing that biometrics are not secrets and therefore their primary use is to enhance security rather than prove security.

12.3 A Motivating Example and the Threat Model

Consider an unmanned aerial vehicle (UAV), i.e., a drone, which is remotely piloted by a user. Communication is long-range Wi-Fi (IEEE 802.11), that is essentially line-of-sight, so subject to interference and sudden dropouts (a lossy channel), as well as eavesdropping, rogue access, and hacking attacks (an insecure channel). The drone has a GPS autopilot system, a camera, and can fly using a programmed route, e.g., return to home. Our goal is to secure such applications. Although this is a special

IoT application involving the remote control of an object using a wireless channel, it is a typical scenario for applications where stream authentication can be used to support trusted behavior over a period of time in a semi-continuous way.

12.3.1 Threat Model

Some of the most common vulnerabilities of UAV's are,

- Spoofing

  External references, such as GPS satellite signals: these are very weak and easy to overlay with a spoofed signal originating from a local transmitter with a strong signal (or an adversarial drone that is nearby); the spoofed signal simulates a GPS signal that leads to a falsified position.
  Internal references, such as the inertia navigation system measurements: these may cause value drifts resulting in de-synchronization.

- Exposure of the ground control station to malware: the attacker may gain access to the control station and inject malware.
- The communication channel: the attacker may succeed in disabling the protection of the communication channel.
- For Web-based UAV implementations: the attacker may use SQL injections for backend database manipulation to access private information, or Cross-Site Scripting (XSS) attacks that target web pages using a browser API, or more generally malware that targets the Internet [10].
- DoS or DDoS attacks: the attacker may flood the bandwidth or resources of a system to disrupt the normal traffic of a target (a jamming attack), or use falsified or malicious data.
- Relay attacks and distance fraud: these are man-in-the-middle attacks in which the attacker interposes between the user and the UAV and relays messages, to falsify their distance or impersonate the UAV.

Fault-handling mechanisms include the redundancy used to compensate for missing or faulty references, and fail-safe states. We get a fail-safe state when there are conflicting reference signals or the communication link is lost. In such cases the functionality of the UAV is restricted to enable the continuation of the mission (e.g., by hovering or landing until the conflict is resolved, or the link is restored). There are several fault-handling strategies for mission critical applications that depend on the fault. These include: hover, land, return-to-base and self-destruct.

For security analysis our threat model assumes a Byzantine adversary that is modeled by a probabilistic polynomial-time Turing machine (PPT). The adversary controls the delivery schedule of all communication channels, and may eavesdrop on, or modify, their contents. The adversary may also instantiate new communication channels and directly interact with honest parties. In particular, this implies that the adversary can attempt to perform impersonation, reflection, man-in-the-middle, and any other passive or active attacks that involve user-to-drone communication.

12.4 Lightweight Stream Authentication for Mobile Objects

We first present a basic stream authentication protocol for UAV. This builds on the one-time authentication protocol by Burmester et al. [3] for Radio Frequency Identification applications, and inherits its lightweight computation and communication features. In this protocol, the user and drone share a loosely synchronized pseudo-random number generator (same algorithm and seed), say $g = g(state)$, and are mutually authenticated by exchanging consecutive numbers drawn from g.

Figure 12.1 shows the flows of the protocol in the 'honest' case when no flows are substituted or dropped. The user U starts by sending the first pseudo-random number r_0 drawn from g to the drone D. Then D draws two numbers r_0, r_1 from its pseudo-random number generator (PRNG) and checks if the received number r_0' (we assume that an authenticator sent as r_i is received as r_i') is correct. If $r_0' \neq r_0$ it aborts, otherwise it sends r_1 to U. U draws the next two pseudo-random numbers r_1, r_2 from its PRNG and checks if the received number $r_1' = r_1$. If $r_1' \neq r_1$ it aborts, otherwise it sends r_2 to U. This process is repeated over and over again at regular intervals to get a stream,

$$r_0, r_1, r_2, \ldots, r_{2i}, r_{2i+1}, \ldots ,$$

of authenticators that authenticate U, D (mutual authentication): r_{2i} authenticates U while r_{2i+1} authenticates D.

This first approach is not resilient: flows may be dropped (availability), or substituted (integrity), resulting in the de-synchronization of the PRNGs of the user and

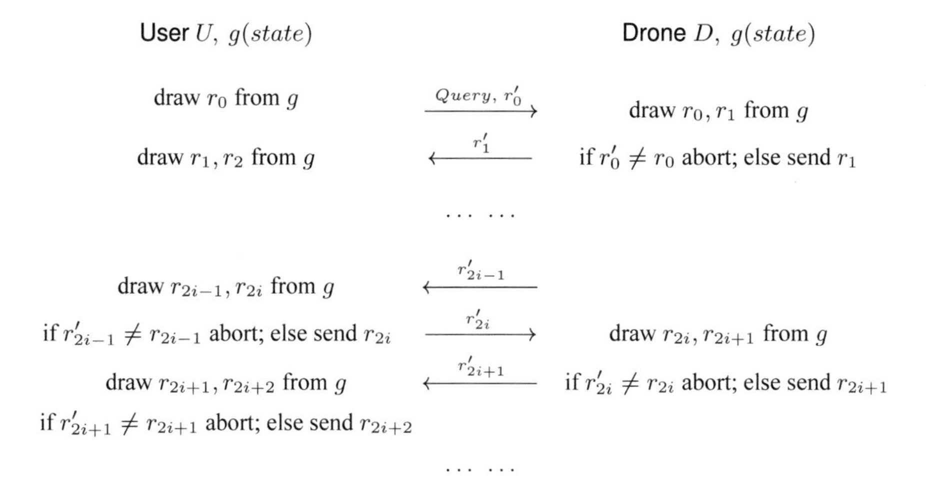

Fig. 12.1 The flows of the basic lightweight stream authentication protocol

the drone. To address this, and other security related issues, we now show to modify this protocol.

12.4.1 A Lightweight Stream Authentication Protocol

As in the basic protocol, U, D share a (loosely) synchronized PRNG $g = g(state)$. For resiliency, U, D store in non-volatile memory a list containing consecutive numbers drawn from g, the current state $g(state)$ of g, counters $flow$, acc (accept), and a countdown timer tmr:

$$(list_x;\, g(state_x), flow_x, acc_x, tmr),\;\; x \in \{u, d\}.$$

The list of U has three numbers: $[r_i, r_{i+1}, r_{i+2}]$ while the list of D has four numbers: $[r_{i-2}, r_{i-1}, r_i, r_{i+1}]$. Initially the shared PRNG is synchronized, and U, D store the lists $[r_0, r_1, r_2]$, $[\bot, \bot, r_0, r_1]$. To update a list, the functions

$$\begin{aligned} update_u\colon\; [r_{i+2}, r_{i+3}, r_{i+4}] &\leftarrow [r_i, r_{i+1}, r_{i+2}], \\ update_d\colon\; [r_i, r_{i+1}, r_{i+2}, r_{i+3}] &\leftarrow [r_{i-2}, r_{i-1}, r_i, r_{i+1}], \end{aligned}$$

are used, where r_{i+3}, r_{i+4} are the next two pseudo-random numbers drawn from g by U after r_{i+2}, and r_{i+2}, r_{i+3} are the next two numbers drawn from g by D after r_{i+1}. The notation used for incrementing a counter ctr is $ctr^{\uparrow}$.

Figure 12.2 shows the flows of two rounds of the lightweight stream authentication protocol for $j = 0, 1, \ldots$. Timers are set in the first round and checked when a response is received. If a timer has not expired then the corresponding counter acc is incremented ($acc^{\uparrow}$). Note that if the pseudo-random numbers sent have the correct values then the responses are in the stored lists, and do not need to be computed. This is an essential requirement to address man-in-the-middle relay attacks.

The timing of the flows is controlled by U. A regular sequence of discrete points in time $T = t_0, t_1, \ldots,$ is used to transmit authenticators. We assume that the time period $\Delta t = (t_i - t_{i-1})$ is significantly greater than the transmit-receive turnaround time δt.

In the protocol we distinguish two types of flows: flows of U that correspond to repeated actions (transmissions) until a certain event occurs (a certain message is received) indicated by double arrows ($\Rightarrow$), and flows of D that correspond to single events (one transmission) indicated by single arrows ($\rightarrow$). We now describe the protocol in more detail.

Protocol. Repeat for $j = 0, 1, \ldots$

Step $4j$ $\quad U$ $[list_u\colon [r_{4j}, r_{4j+1}, r_{4j+2}]]$
If $j > 0$ and no authenticator r'_{4j-1} was received in response to authen-

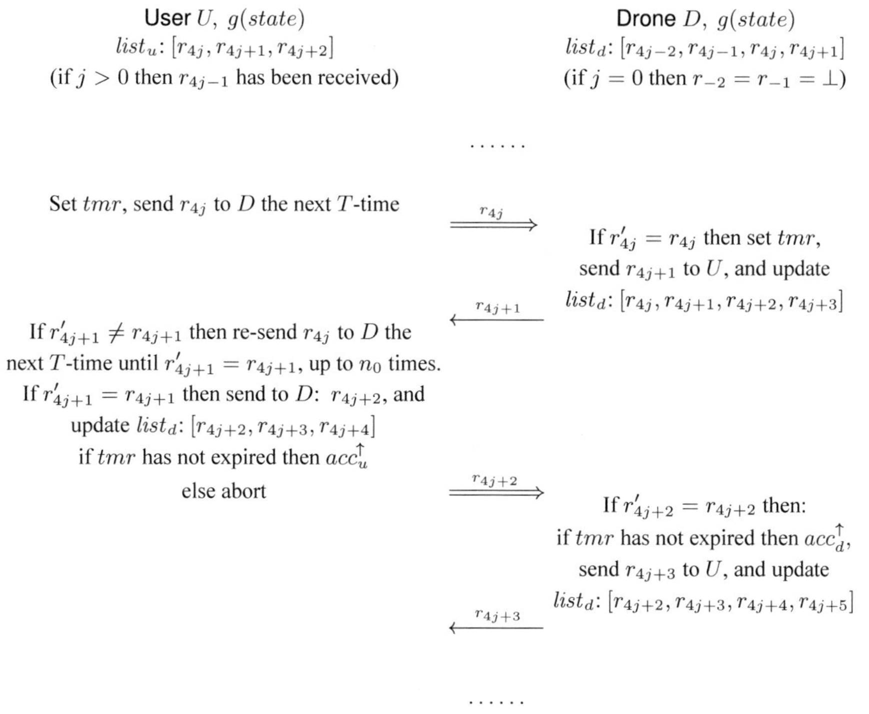

User U, $g(state)$		Drone D, $g(state)$
$list_u$: $[r_{4j}, r_{4j+1}, r_{4j+2}]$		$list_d$: $[r_{4j-2}, r_{4j-1}, r_{4j}, r_{4j+1}]$
(if $j > 0$ then r_{4j-1} has been received)		(if $j = 0$ then $r_{-2} = r_{-1} = \perp$)
	......	
Set tmr, send r_{4j} to D the next T-time	$\xrightarrow{r_{4j}}$	
		If $r'_{4j} = r_{4j}$ then set tmr, send r_{4j+1} to U, and update $list_d$: $[r_{4j}, r_{4j+1}, r_{4j+2}, r_{4j+3}]$
	$\xleftarrow{r_{4j+1}}$	
If $r'_{4j+1} \neq r_{4j+1}$ then re-send r_{4j} to D the next T-time until $r'_{4j+1} = r_{4j+1}$, up to n_0 times. If $r'_{4j+1} = r_{4j+1}$ then send to D: r_{4j+2}, and update $list_d$: $[r_{4j+2}, r_{4j+3}, r_{4j+4}]$ if tmr has not expired then $acc^{\uparrow}_u$ else abort		
	$\xrightarrow{r_{4j+2}}$	
		If $r'_{4j+2} = r_{4j+2}$ then: if tmr has not expired then $acc^{\uparrow}_d$, send r_{4j+3} to U, and update $list_d$: $[r_{4j+2}, r_{4j+3}, r_{4j+4}, r_{4j+5}]$
	$\xleftarrow{r_{4j+3}}$	
	......	

Fig. 12.2 Flows of the lightweight stream authentication protocol for $j = 0, 1, \ldots,$ showing the authenticators for two rounds, the updates of stored lists, and when authenticators are accepted

ticator r_{4j-2} sent previously, or if $r'_{4j-1} \neq r_{4j-1}$, then *flow*$^{\uparrow}_u$ and re-send to D: r_{4j-2} the next T-time, until $r'_{4j-1} = r_{4j-1}$, up to n_0 times.
If $j > 0$ and $r'_{4j-1} = r_{4j-1}$ then update $list_u$:
$[r_{4j}, r_{4j+1}, r_{4j+2}] \leftarrow [r_{4j-2}, r_{4j-1}, r_{4j}]$.
If $j = 0$, or $j > 0$ and $r'_{4j-1} = r_{4j-1}$, then $flow^{\uparrow}_u$, set tmr and, send at the next T-time to D: r_{4j} ;
else abort.

Step $4j+1$ D $[list_d$: $[r_{4j-2}, r_{4j-1}, r_{4j}, r_{4j+1}]$; when $j = 0$ then $r_{4j-2} = r_{4j-1} = \perp]$
If $r'_{4j} = r_{4j}$ then *flow*$^{\uparrow}_d$, set tmr, send to U: r_{4j+1}
and update $list_d$: $[r_{4j}, r_{4j+1}, r_{4j+2}, r_{4j+3}] \leftarrow [r_{4j-2}, r_{4j-1}, r_{4j}, r_{4j+1}]$.

Step $4j+2$ U $[list_u$: $[r_{4j}, r_{4j+1}, r_{4j+2}]]$
If no authenticator was received or $r'_{4j+1} \neq r_{4j+1}$ then *flow*$^{\uparrow}_u$ and re-send to D: r_{4j} the next T-time, until $r'_{4j+1} = r_{4j+1}$, up to n_0 times.
If $r'_{4j+1} = r_{4j+1}$ then *flow*$^{\uparrow}_u$, send to D: r_{4j+2} ,
update $list_u$: $[r_{4j+2}, r_{4j+3}, r_{4j+4}] \leftarrow [r_{4j}, r_{4j+1}, r_{4j+2}]$, and
if $timer$ has not expired then $acc^{\uparrow}_u$
else abort.

Step $4j+3$ $\quad D \quad [list_d\colon [r_{4j}, r_{4j+1}, r_{4j+2}]]$
If $r'_{4j+2} = r_{4j+2}$ then $flow_d^{\uparrow}$, send to U: $\quad r_{4j+3}$
update *list*: $[r_{4j+2}, r_{4j+3}, r_{4j+4}, r_{4j+5}] \leftarrow [r_{4j}, r_{4j+1}, r_{4j+2}, r_{4j+3}]$
if tmr has not expired then $acc_d^{\uparrow}$.

This protocol involves a series of rounds in which authenticators r_j drawn from loosely synchronized PRNGs are used to stream authenticate U, D. Counters $flow_u$, $flow_d$ keep track of the number of authenticators exchanged, while counters acc_u, acc_d keep track of the number of times U, D accept. There is an upper bound n_0 on the number of times U may attempt to get a response from D: if this is exceeded then we have protocol failure. Normally, if no flows are compromised or dropped then in each 2-round exchange, the values of $flow_u$, $flow_d$ increase by two and the values of acc_u, acc_d increase by one. Finally timers are used in Step $4j$ and Step $4j+1$ to thwart relay attacks (distance fraud): the countdown time of a timer should be accurate and reflect the sender-receiver turnaround time.

12.4.2 A Robust Lightweight Stream Authentication Protocol

We next extend the stream authentication protocol by refreshing the shared PRNGs. PRNGs are refreshed to ensure resilience against traffic analysis attacks that exploit the correlation between the successive numbers drawn from a PRNG (state entropy leakage). In particular, that the adversary cannot predict with probability better than a certain threshold,

- the next number drawn, and/or
- the internal state of the PRNG,

until the PRNG is next refreshed. Refreshing a PRNG involves updating its internal state with fresh (high entropy) randomness. For our application this randomness is provided by user U when needed, using authenticated encryption.

To explain how this is done we first briefly discuss the requirements for robustness of PRNGs. Let n be the security parameter and ε a number that is negligible (as a function of n).

Definition 1 [2, 3] The pair of functions $g(\cdot)$ and $refresh(\cdot, \cdot, \cdot)$ with:

- $(r, s') \leftarrow g(s)$: r a pseudo-random n-bit string; s, s' the current and next state of g, both of length at least n,
- $s' \leftarrow refresh(k, x, s)$: k the refresh key; x a refresh bitstring of length at least n,

is a *robust* PRNG if there is a threshold n_0 such that for any probabilistic polynomial-time (PPT) observer $\mathcal{O}$:

- *Resilience*: given $m < n_0$ successive numbers $r_1, \ldots, r_m$ drawn from g, $\mathcal{O}$ cannot predict the next number with probability better than $1/2^n + \varepsilon$.

User $U,\ g(s),\ k$		Drone $D,\ g(s),\ k$
$list$: $[r_{4j}, r_{4j+1}, r_{4j+2}]$		$list$:$[r_{4j-2}, r_{4j-1}, r_{4j}, r_{4j+1}]$
r_{4j-1} has been received		
Compute & store $x_{4j} = enc_k(s^{ref} \| r_{4j})$		
s^{ref} true random	$\xrightarrow{(r_{4j}, x_{4j})}$	If $r'_{4j} = r_{4j}$ then decrypt x'_{4j};
		if x'_{4j} is valid then
	$\xleftarrow{r_{4j+2}}$	$update^{ref,d}\ [r_{4j}, r_{4j+2}, r_0, r_1]$
If $r'_{4j+2} \neq r_{4j+2}$ then re-send (r_{4j}, x_{4j})		
until $r'_{4j+2} = r_{4j+2}$, up to n_0 times.		
If $r'_{4j+2} = r_{4j+2}$ then $update^{ref,u}[r_0, r_1, r_2]$		
else abort.		

Fig. 12.3 Refreshing the PRNG of the lightweight stream authentication protocol

- *Forward security*: given $m < n_0$ successive numbers $r_1, \ldots, r_m$ drawn from g prior to refreshment, $\mathcal{O}$ cannot distinguish these from truly random numbers with probability better than $1/2^n + \varepsilon$, even if $\mathcal{O}$ learns the state of g after it has been refreshed.
- *Backward security*: given $m < n_0$ successive numbers $r_1, \ldots, r_m$ drawn from g after refreshment, $\mathcal{O}$ cannot distinguish these from random with probability better than $1/2^n + \varepsilon$, even if $\mathcal{O}$ learns the state of g before it has been refreshed.

For our application we take the length of the state s of g to be n (the security parameter), and the refresh bitstring to be: $x = enc_k(s' \| r)$, where: $s' = s^{ref}$ is a truly random bitstring of length n, r is a pseudo-random number of length n and enc_k is a blockcipher with secret key k (e.g., AES).

We next present the modifications that are needed so that the lighweight stream authentication protocol in Sect. 12.4.1 is robust. To refresh the generator g, user U and drone D share, additionally, a refresh key k, and use the following refresh update functions:

- $update^{ref,u}$: $[r_0, r_1, r_2] \leftarrow [r_i, r_{i+1}, r_{i+2}]$
- $update^{ref,d}$: $[r_i, r_{i+2}, r_0, r_1] \leftarrow [r_{i-2}, r_{i-1}, r_i, r_{i+1}]$

where $r_{i-2}, r_{i-1}, r_i, r_{i+1}, r_{i+2}$ are successive numbers drawn from the current generator $g(s)$, and r_0, r_1, r_2 are the 1st, 2nd and 3rd number drawn from the refreshed generator $g(s^{ref})$ (to simplify our notation we re-use the old labels).

Figure 12.3 shows the two flows needed to refresh the generator g of the lightweight stream authentication protocol. Refreshing g should take place at regular intervals to maintain robustness. Below we describe in more detail this protocol.

Protocol (Refreshing generator g of the lightweight stream authentication protocol).

Step $4j$ $\quad U\ \ [j > 0$; *list*: $[r_{4j}, r_{4j+1}, r_{4j+2}]$; r_{4j+1} has been received; s^{ref} is a uniformly random n-bit string.]

Compute the encryption $x_{4j} = enc_k(s^{ref} \| r_{4j})$, store this value, $flow_u^{\uparrow}$ and, send to D the next T-time: (r_{4j}, x_{4j}) .

Step $4j+1$ D $[r_{4j-2}, r_{4j-1}, r_{4j}, r_{4j+1}]$
If $r'_{4j} = r_{4j}$ then decrypt x'_{4j}, check that its length is $2n$-bits and that, the last n-bits are those of r_{4j}; if so, then take the first n-bits to be s^{ref}, $update^{ref,d}$: $[r_{4j}, r_{4j+2}, r_0, r_1] \leftarrow [r_{4j-2}, r_{4j-1}, r_{4j}, r_{4j+1}]$, $flow_u^{\uparrow}$ and send to U: r_{4j+2} .

Step $4j+2$ U $[[r_4 j, r_{4j+1}, r_{4j+2}], x_{4j}]$
If $r'_{4j+2} \neq r_{4j+2}$ then $flow_u^{\uparrow}$ and re-send to D (r_{4j}, x_{4j}) the next T-time until $r'_{4j+2} = r_{4j+2}$, up to n_0 times.
If $r'_{4j+2} = r_{4j+2}$ then $update^{ref,u}$: $[r_0, r_1, r_2] \leftarrow [r_{4j}, r_{4j+1}, r_{4j+2}]$ and discard x_{4j}; else abort.

This protocol has one round in which U sends D an authenticator r_{4j} together with an authenticated encryption x_{4j} of the randomness s^{ref} needed to refresh generator g and r_{4j}. Since r_{4j} authenticates U, and x_{4j} encrypts $(s^{ref} \| r_{4j})$, s^{ref} is authenticated by U. D responds with the authenticator r_{4j+2}, not r_{4j+1} as in the stream authentication protocol, to show that the message previously sent includes an authenticated encryption (otherwise an adversary can substitute (r_{4j}, s^{ref}) by r_{4j}, resulting in U and D using different updating functions and getting desynchronized). D decrypts x'_{4j} and then parses it to get the randomness s^{ref} and r'_{4j}: if r'_{4j} is the same as the stored value r_{4j} then s^{ref} is used to refresh generator g. After refreshing the generator, both U and D continue with the stream authentication protocol as in Sect. 12.4.1 (for index $(j+1)$).

In this protocol the numbers that U, D store are updated in a different way than with the stream authentication protocol. This is because, to keep synchronized, U, D, must store some numbers drawn from the generator g before it is refreshed, and some numbers after it is refreshed. For example D must store the numbers r_{4j}, r_{4j+2}, that will be needed if U does not receive r'_{4j+2}: in this case U will re-send (r_{4j}, x_{4j}), and D must check r'_{4j} and respond with r_{4j+2} (again). However, if D does receive $r'_{4j+2} = r_{4j+2}$, then U can update its stored numbers using the refreshed generator $g(s^{ref})$. If the refreshed numbers are $r_0, r_1, \ldots$, then U will send r_0 to D (as in the stream authentication protocol). To check the correctness of the received r'_0, using stored values, r'_0 must be the third number stored by D.

12.4.3 Features of the Robust Lightweight Stream Authentication Protocol

Synchronization. At all times U and D share at least one pseudo-random number. In the stream authentication protocol (Sect. 12.4.1) the first number that U stores (say r_j) is either the first number or the last number that D stores. A similar argument applies for the refreshing protocol (Sect. 12.4.2): either the first number that U stores (say r_{4j}), or the last number (r_{4j+2}), is also stored by D.

Efficiency. When the adversary is passive only one pseudorandom-number needs to be exchanged to authenticate a party. The accept rate of U, D, as measured by the counters $accept_u$, $accept_d$ and $flow_u$, $flow_d$, is one half (it is possible to increase the rate to get close to 3/4, but this will increase the cost when dealing with disrupted/corrupted flows).

Relay attacks and distance fraud attacks. These attacks are online man-in-the-middle attacks in which the adversary tries to impersonate U or D by interposing between them and exchanging their (possibly modified) messages. Distance bounding protocols based on round-trip delay measurements are the main defense against such attacks. These estimate the propagation time as accurately as possible so as to determine the distance between the controller and drone. Determining the processing time is essential in order to isolate the propagation component from the overall measured time, and therefore variable processing times constitute a major problem for distance bounding. The processing time must be as short as possible, since the adversary may be able to overclock the drone to absorb the delay introduced by its own devices. The lightweight stream authentication protocol with its fixed nearly-zero processing delay is particularly suitable for protection against such attacks. This is because in this protocol U, D pre-compute their response. For a detailed discussion on how the challenge-response round trip is computed, and then used to thwart distance fraud attacks we refer the reader to the one-time Radio Frequency Identification authentication protocol by Burmester et al. [3] that addresses such attacks, and on which the stream authentication protocol is based.

12.5 Security Proof

We briefly sketch a proof of security using the real-or-random (ROR) formal model [1]. This model is based on indistinguishability, and assumes a probabilistic polynomial-time (in the security parameter n) adversary $\mathcal{A}$, as defined in Sect. 12.3.1.

In the ROR model *semantic* security (privacy) is defined by an experiment Test^{str}, in which a passive (eavesdropping) adversary $\mathcal{A}$ must distinguish the messages exchanged from random. At the beginning a bit b is chosen uniformly at random: if $b = 1$ then the real message is returned while if $b = 0$ a random string of equal length is returned. $\mathcal{A}$ has to guess b. The advantage of $\mathcal{A}$ is: $\mathsf{Adv}^{str-sec}(\mathcal{A}) = 2Pr[b' = b] - 1$. For semantic security this must be negligible. For the stream authentication protocol (Sect. 12.4.1) the numbers $r_0, r_1, \dots$, exchanged by U and D are drawn from a PRNG, and therefore cannot be distinguished from random in probabilistic polynomial time. Hence $\mathsf{Adv}^{str-sec}(\mathcal{A}) = \varepsilon$ (negligible).

For authentication, $\mathcal{A}$ is active and can substitute or drop messages, and use impersonation, reflection or other active attacks. In this case an experiment $\mathsf{TestPair}^{str-aut}$ is used in which $\mathcal{A}$ has "oracle" access to instances Π_u^j, Π_d^{j+1}, Π_u^{j+2} of U, D, that generate the flows r_j, r_{j+1}, r_{j+2} of an authentication session. If $\mathcal{A}$ succeeds in getting either U or D to accept (increment $accept_u$ or $accept_d$) for one or more substituted flows $\hat{r}_a \neq r_a, a \in \{j, j+1, j+2\}$, then the experiment returns $b = 1$; otherwise it returns $b = 0$. The advantage of $\mathcal{A}$ in this experiment is: $\mathsf{Adv}^{str\text{-}aut}(\mathcal{A}) = Pr[b = 1]$, that must be negligible for mutual stream authentication. Again, for the stream authentication protocol the numbers r_j, r_{j+1}, r_{j+2} are drawn from a PRNG: if any of these is (are) substituted for different numbers, then the receiving party will detect this and not accept them. It follows that: $\mathsf{Adv}^{str\text{-}aut}(\mathcal{A}) = \varepsilon$.

A similar argument holds for the refreshing protocol (Sect. 12.4.2). In this case the encryption x_j must authenticate the refresh randomness s^{ref}. We require that the experiment $\mathsf{TestPair}^{str\text{-}ref}$ returns $b = 1$ when $\mathcal{A}$ succeeds in getting either U or D to refresh their generator with different randomness $\hat{s}^{ref} \neq s^{ref}$ when one or more of the flows r_j, x_j, r_{j+2} are substituted by different ones, and otherwise return $b = 0$. As with the advantage for mutual stream authentication, the advantage for refresh authentication: $\mathsf{Adv}^{str\text{-}ref}(\mathcal{A})$ is negligible because if $x_d = enc_k(s^{ref} \| r_j)$ is substituted by a different encryption, then a different value for r_j must have been used. This will be detected by the receiver drone D (D will not refresh).

12.6 Conclusion

Secret sharing and threshold cryptography are powerful cryptographic tools for fault-tolerant multiparty communication and computation. Similarly shared clocks, even if only loosely synchronized, will check replay attacks. In this chapter we show how loosely synchronized pseudo-random number generators can be used to realize light-weight stream authentication for mobile objects in a formal security framework. For this realization we prove: (a) availability (the stream authentication cannot be de-synchronized), (b) privacy (semantic security), (c) mutual stream authentication, (d) robustness with forward and backward security, and (e) man-in-the-middle relay attack protection.

References

1. M. Abdalla, P.-A. Fouque, D. Pointcheval, Password-based authenticated key exchange in the three-party setting, in *Public Key Cryptography—PKC 2005, 8th International Workshop on Theory and Practice in Public Key Cryptography, Proceedings* (2005), pp. 65–84
2. B. Barak, S. Halevi, A model and architecture for pseudo-random generation with applications to/dev/random, in *Proceedings of the 12th ACM Conference on Computer and Communications Security* (ACM, 2005), pp. 203–212
3. M. Burmester, J. Munilla, Lightweight rfid authentication with forward and backward security. ACM Trans. Inf. Syst. Secur. (TISSEC) **14**(1), 11 (2011)
4. M. Frank, R. Biedert, E. Ma, I. Martinovic, D. Song, Touchalytics: on the applicability of touchscreen input as a behavioral biometric for continuous authentication. IEEE Trans. Inf. Forensics Secur. **8**(1), 136–148 (2013)
5. H. Gascon, S. Uellenbeck, C. Wolf, K. Rieck, Continuous authentication on mobile devices by analysis of typing motion behavior. Sicherheit **2014**, 1–12 (2014)

6. A.J. Klosterman, G.R Ganger, Secure continuous biometric-enhanced authentication. Technical Report, (Carnegie-Mellon University Pittsburgh, PA, Department of Computer Science, 2000)
7. J. Liu, Y. Dong, Y. Chen, Y. Wang, T. Zhao, Poster: leveraging breathing for continuous user authentication, in *Proceedings of the 24th Annual International Conference on Mobile Computing and Networking* (ACM, 2018), pp. 786–788
8. R. Murmuria, A. Stavrou, D. Barbará, D. Fleck, Continuous authentication on mobile devices using power consumption, touch gestures and physical movement of users, in *International Workshop on Recent Advances in Intrusion Detection* (Springer, Cham, 2015), pp. 405–424
9. K. Niinuma, U. Park, A.K. Jain, Soft biometric traits for continuous user authentication. IEEE Trans. Inf. Forensics Secur. **5**(4), 771–780 (2010)
10. Top OWASP, Top 10–2013: the ten most critical web application security risks. *The Open Web Application Security Project* (2010)
11. V.M. Patel, R. Chellappa, D. Chandra, B. Barbello, Continuous user authentication on mobile devices: recent progress and remaining challenges. IEEE Signal Process. Mag. **33**(4), 49–61 (2016)
12. A. Perrig, R. Canetti, J.D. Tygar, D. Song, Efficient authentication and signing of multicast streams over lossy channels, in *Proceedings 2000 IEEE Symposium on Security and Privacy, 2000. S&P 2000* (IEEE, 2000), pp. 56–73
13. M. Saadeh, A. Sleit, K.E. Sabri, W. Almobaideen, Hierarchical architecture and protocol for mobile object authentication in the context of iot smart cities. J. Netw. Comput. Appl. **121**, 1–19 (2018)
14. S.J. Shepherd, Continuous authentication by analysis of keyboard typing characteristics, in *Proceedings, European Convention on Security and Detection, 1995* (IET, 1995), pp. 111–114
15. Z. Sitová, J. Šeděnka, Q. Yang, G. Peng, G. Zhou, P. Gasti, K.S. Balagani, Hmog: new behavioral biometric features for continuous authentication of smartphone users. IEEE Trans. Inf. Forensics Secur. **11**(5), 877–892 (2016)
16. E. Al Solami, C. Boyd, A.J. Clark, A.K. Islam, Continuous biometric authentication: can it be more practical?, in *2010 IEEE 12th International Conference on High Performance Computing and Communications (HPCC)* (2010), pp. 647–652
17. I. Traore, *Continuous Authentication Using Biometrics: Data, Models, and Metrics: Data, Models, and Metrics* (IGI Global, 2011)
18. S. Ueda, Y. Shinzaki, H. Shigeno, K.-I. Okada, H. 264/avc stream authentication at the network abstraction layer, in *Information Assurance and Security Workshop, 2007. IAW'07. IEEE SMC* (IEEE, 2007), pp. 302–308

Dr. Mike Burmester is a Professor of Computer Science at Florida State University and Director of the Center for Security and Assurance in IT (C-SAIT). He is editor of four Journals in Information Security, has organized several Workshops and Conferences in this area, and published five Books, 20 Book Chapters, and over 200 journal and refereed conference publications covering a wide range of security topics including: privacy/anonymity, pervasive/ubiquitous network systems, lightweight cryptographic applications, RFID and sensor applications and trust management.

Dr. Jorge Munilla is an Associate Professor in the Engineering Department of the University of Málaga (Spain). He has been guest researcher in the IAIK Krypto Group of the University of Graz, Austria, in 2006, and visiting faculty member in the Center for Security and Assurance in IT (C-SAIT) of the Florida State University, USA, in 2009, 2011 and 2015, and in the Center for Computer and Information Security Research of the University of Wollongong, Australia, in 2012 and 2014. His fields of interest include resilient cyber-physical systems, security in pervasive/ubiquitous systems (RFID and IoT) and the application of big data and machine learning techniques. He is co-author of 2 book chapters, more than 20 papers in international journals and 30 conference participations.

Chapter 13
This is Just Metadata: From No Communication Content to User Profiling, Surveillance and Exploitation

Constantinos Kapetanios, Theodoros Polyzos, Efthimios Alepis, and Constantinos Patsakis

Abstract Mobile devices have become an indispensable part of our daily lives. Practically, most of our everyday communication is performed through mobile devices which host third party apps and provide for various means of interaction with diverse levels of security. Android is by far the most widely used mobile operating system, with a user base in the scale of billions. However, while Android Open Source Project (AOSP) is paving the way for all manufacturers, Android market is so fragmented that those who are using the latest version are only a small minority. Moreover, Android comes in several flavours as manufacturers tailor it to their needs. However, this tailoring often prevents users from getting the latest updates. In fact, as we show, manufacturers may not follow the security and privacy guidelines of AOSP, exposing their users to unexpected threats. In this work we study a yet unpatched vulnerability by most major manufacturers, and partially fixed in AOSP, which allows for an adversary to extract important information from the victim's device. To this end, we showcase that unprivileged apps, without actually using any permissions, can harvest a considerable amount of valuable user information. This is achieved by monitoring and exploiting the file and folder metadata of the most well-known messaging apps in Android, which have been hitherto considered secure, deriving thereby usage statistics in order to elicit user profiles, social connections, credentials or other sensitive information.

Keywords Android · Access control · Privacy · Smartphones · Metadata

---

C. Kapetanios · E. Alepis · C. Patsakis (✉)
Department of Informatics, University of Piraeus, Piraeus, Greece
e-mail: kpatsak@unipi.gr

T. Polyzos
Department of Informatics, University of Athens, Athens, Greece

G. A. Tsihrintzis and M. Virvou (eds.), *Advances in Core Computer Science-Based Technologies*, Learning and Analytics in Intelligent Systems 14,
https://doi.org/10.1007/978-3-030-41196-1_13

13.1 Introduction

Back in 2013 when Edward Snowden started unravelling the story behind many secret surveillance projects such as the PRISM,[1] the agencies defended themselves by responding that they were just collecting metadata and not actual content. Therefore, many government agents were arguing that these surveillance actions could not be considered as privacy invasive. For instance, Dianne Feinstein while she was defending NSA phone records program declared [21]:

> As you know, this is just metadata. There is no content involved. In other words, no content of a communication. That can only be, these records, I'm not talking about content, the records can only be accessed under heightened standards.

Consequently, other revelations that followed were far more indicative of the intrusive methods employed in these projects; therefore, the "battle" for the "metadata" argument was soon forgotten. Nevertheless, in this paper we argue that there is far more knowledge in such metadata information than one can even anticipate. Unfortunately, users do not have to protect themselves only from the prying eyes of government surveillance projects, since software companies have already shown similar rogue behaviours under the pretext of user profiling designed towards offering better user experience and personalised recommendations. In fact, the latest methods used for user profiling are so privacy invasive that cannot be considered lightheartedly benign. In the most sinister scenario, usage metadata can be exploited by rogue applications to trick users into providing sensitive information like credentials.

13.1.1 Motivation

Android, having long ago surpassed the boundary of one billion users, is currently by far the most widely used mobile platform [26]. However, Android is fragmented in several aspects. Firstly, while there are many versions (API levels) of Android, notably only a handful of users have access to its latest version. As illustrated in Fig. 13.1, almost half a year after its debut only 21.5% have installed the latest Android version and 7.5% use the latest update (version 8.1). While this may indicate that the vast majority of the users do not enjoy the new features of the platform, it also suggests that not all of its users have the same security updates installed.

Apart from the versioning fragmentation, we also have the variety of Android "flavours", also known as the different Android versions that manufacturers ship their devices with. The reason for these diverse flavours is the fact that manufacturers, despite following AOSP, they actually tailor the OS to their needs, most commonly by adding/removing apps and features to bind them with the firmware or by changing UI elements. Furthermore, it is also debatable whether all of the vendors follow the AOSP

[1] https://www.washingtonpost.com/news/wonk/wp/2013/06/12/heres-everything-we-know-about-prism-to-date/?utm/term=.2e201efd7097.

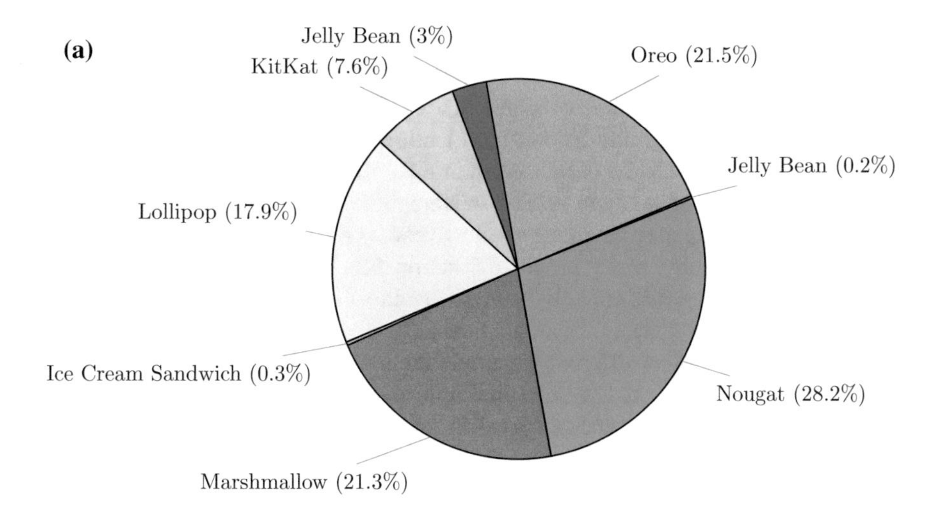

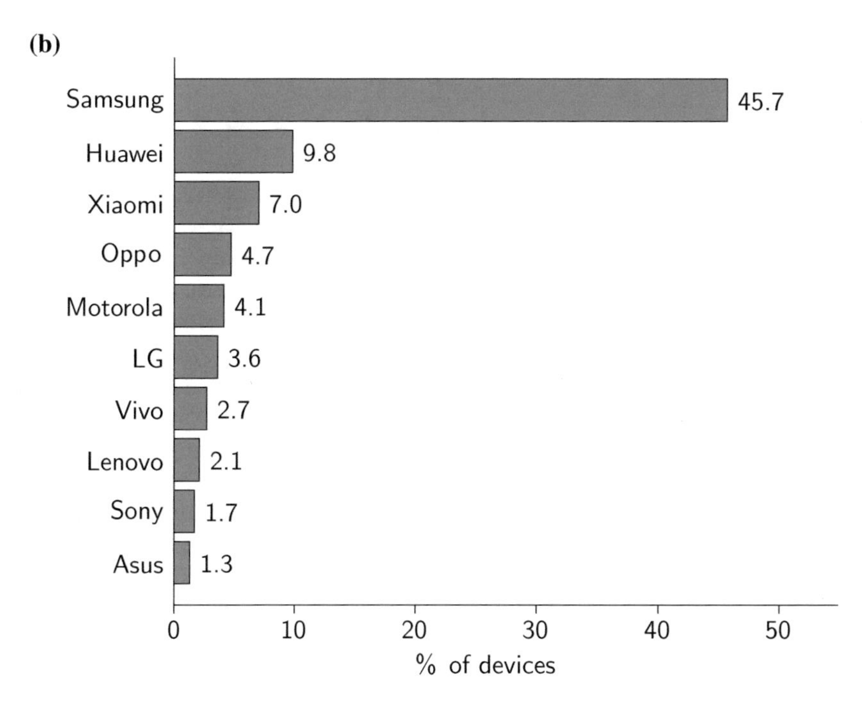

Fig. 13.1 **a** Android version statistics. *Source* Android Developer, **b** market share of Android manufacturers. *Source* Appbrain

guidelines regarding security and privacy requirements when they modify the original source code according to their needs. Google notifies them of already reported and fixed vulnerabilities in several versions. Since vulnerabilities are accompanied by Common Vulnerabilities and Exposures Identifiers (CVEs), it is easy to monitor whether a specific patch has been pushed into the device. Although major vendors usually report fixes of publicly known vulnerabilities,[2] updates not falling in line with the CVE system may be disregarded by vendors or even deliberately omitted if they happen to coincide with a provided functionality. Recently, using SnoopSnitch [20] it was discovered that not only Android vendors fail to provide security updates to their users or their updates are incomplete [17]. Finally, apps may target different API levels which implies different targetted functionality, but also different security standards. Note that the red vertical line indicates the API level 24 below which apps are vulnerable to our attacks, regardless of manufacturer and installed Android version.

13.1.2 Main Contributions

The main contribution of our paper is threefold. Firstly, we illustrate that a PRISM-like surveillance project collecting only communication metadata could be easily implemented through the distribution of seemingly benign apps that do not use any dangerous permissions. These apps could exploit privacy leakages partially fixed in AOSP, but not yet handled by the majority of smartphone vendors. More precisely, in this work, we demonstrate that by simply monitoring metadata available from the most well-known messaging apps in Android, "useful" and personal information about social interactions can be extracted by non-privileged apps. Table 13.1 contains the communication apps we examined. As a result, we show that the use of metadata bypasses the Android security mechanisms, allowing an adversary to extract a lot of sensitive information about individuals without requesting any dangerous permissions from the users. For instance, by simply exploiting inherent Android mechanisms, and without interacting with service providers or backdooring any communication app, one can determine with overwhelming probability not only when users interact with messaging apps, but also whether two users communicate with each other. The disclosure of this vulnerability aligns perfectly with the emerging belief that preventing privacy leakages in mobile environments is far more complex than one would expect [25].

Secondly, we discuss how the above approach can be used by unprivileged apps to derive usage statistics to build valuable user profiles, a case that would otherwise have required "system level" permissions. Beyond user profiling, these "monitoring" events can be further exploited by malware to timely interfere with user interaction resulting in a number of "unpleasant" situations, both for the users and also for the companies involved. Notably, Google is well aware of such threats; therefore with

[2]For instance see Samsung: https://security.samsungmobile.com/securityUpdate.smsb.

Table 13.1 Apps that are investigated and the reported installations according to Google Play

Application	Installations
BBM	100,000,000–500,000,000
Facebook messenger	1,000,000,000–5,000,000,000
ICQ	10,000,000–50,000,000
imo	100,000,000–500,000,000
KakaoTalk	100,000,000–500,000,000
LINE	100,000,000–500,000,000
QQ international	5,000,000–10,000,000
Signal	5,000,000–10,000,000
Skype	1,000,000,000–5,000,000,000
Snapchat	500,000,000–1,000,000,000
Telegram	100,000,000–500,000,000
Viber	500,000,000–1,000,000,000
WeChat	100,000,000–500,000,000
Wire	1,000,000–5,000,000
WhatsApp	1,000,000,000–5,000,000,000

the introduction of Nougat, apart from pushing many changes into the Android, they also tried to fix such leakages by further locking the contents of the corresponding `/data/data` directory. More precisely, as stated in Android 7.0 Behavior Changes:

> In order to improve the security of private files, the private directory of apps **targeting** Android 7.0 or higher has restricted access (0700). This setting prevents leakage of metadata of private files, such as their size or existence. [6]

Finally, our results indicate that these changes have not been implemented at all by some major manufacturers or they have been implemented partially by others, creating thereby a non-uniform and inconsistent landscape of different implementations with varying functionality, which exposes users' security and privacy.

13.1.3 Vulnerable Audience

Quantifying the vulnerable audience is rather complicated due to the difficulties in quantifying the constraints that have to be met to exploit the vulnerability. Therefore, it is easier to identify who is definitely secure. In this regard, users who are running AOSP since Nougat and whose applications are all running on API level above 23 are considered secure. In practice, this initially means that 70% of the devices are vulnerable as they are not running Nougat, hence the proper permissions have not been implemented. From the remaining 30% of the market we have no precise numbers, but qualitative data. First, most manufacturers have not implemented the "0700 policy", consequently their users are exposed. Second, even in the cases where some of the manufacturers have implemented this policy, the vast majority of apps does not target API levels above 23. Therefore, users of such applications are also exposed. Using Tacyt [11], we managed to identify developer trends since the beginning of the year, based on the target API levels of the updates they pushed to Google Play. We filtered the results to applications which have a tangible amount users, so our results refer to apps having more than 100 K downloads. For finance apps, developers pushed 38 (58%) of the 66 updates for secure API levels and 28 (42%) for insecure. Similarly, for communication apps 41 (57%) of the 71 updates was made for secure API levels and 30 (43%) for insecure. The above indicate that above 42% of the updates that are currently pushed for apps which by definition handle sensitive data are targeting insecure API levels. Notably, from all apps in Google Play having more than 100 K downloads which updated their APKs since the beginning of the year, 791 (39%) targeted insecure API levels, and 1229 (61%) targeted secure ones.

13.1.4 Organization of This Work

The rest of this work is organized as follows. In the next section, we provide an overview of Android internals related to our work. More precisely, we discuss Android app permissions and methods to derive the foreground app and usage statistics. Then, in Sect. 13.3 we discuss our metadata collection methodology which we use in Sect. 13.4 to showcase two different threat scenarios. In the first one, we illustrate how an unprivileged app can monitor the usage of the most well-known communication apps to derive social connections. In the second use case, we present a UI replication attack having Paypal as our reference. The article concludes in Sect. 13.5 where we summarize our contributions and findings and we propose remedies that could be applied in all the aforementioned attacks to mitigate user profiling and exploitation.

13.2 Related Work

13.2.1 Android Permissions

Although mobile devices may not compare with desktop computers in terms of computational capacities, they can provide an augmented user experience as they can quickly adapt to their context and interact with the user through various means due to the plethora of embedded sensors they are equipped with. However, many of these sensors such as microphone, camera and GPS can leak really sensitive information and therefore access to these resources is only permitted upon user consent. As of Android Marshmallow, the management of permissions has become more fine-grained, thus allowing users to grant and revoke consent whenever deemed necessary.

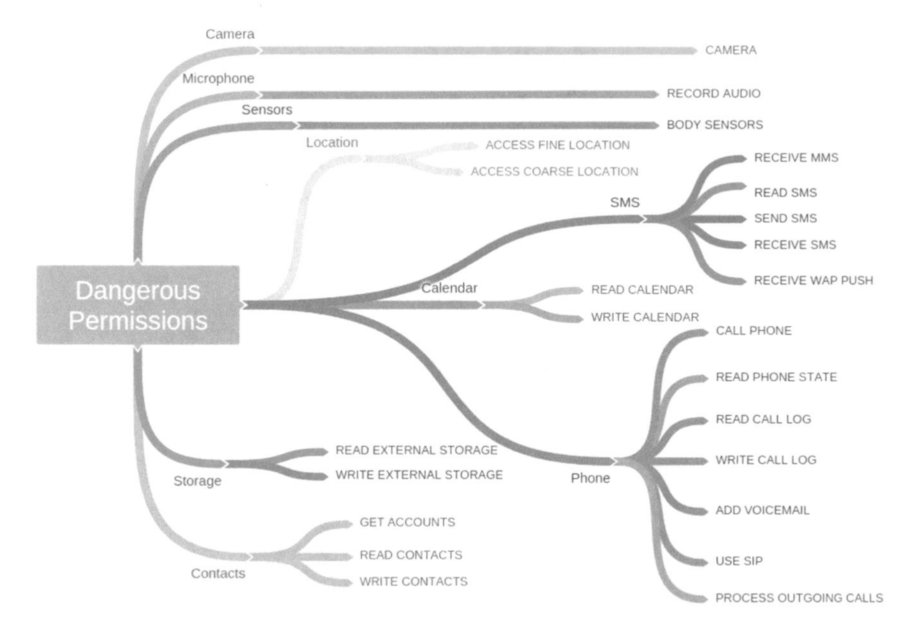

Fig. 13.2 Dangerous permissions in Android

Depending on the risk the users are exposed to, the permissions are mainly categorised as normal and dangerous. In this regard, normal permissions include, among others, access to the Internet, accelerometers and vibration while dangerous permissions include access to the microphone, camera, GPS, phone calls etc. The full list of dangerous permissions in Android is illustrated in Fig. 13.2. Apart from the level of the risk exposure involved, another significant difference between these two groups of permissions is in their management. While dangerous permissions can be revoked or granted whenever the user wants, normal ones are automatically granted once the app is installed and they cannot be revoked at all. Notably, if an app requires only normal permissions to be granted, the latest versions of Android prevent the user from even reading which these permissions are.

Apart from these two categories of permissions, Android has two more: Signature and SignatureOrSystem. The signature permission is designed for interoperability and enables applications which are signed with the same certificate to access the same resources even though only one of them is granted this access. SignatureOrSystem is a special permission designed for manufacturers to enable installing their applications in the Android system image and pertains to many elevated permissions such as rebooting the device or clearing caches. For a more detailed overview of Android permissions the interested reader may refer to [3].

To further protect the Android ecosystem, Google requires the user to explicitly grant some app permissions through completely different permission management screens (see Fig. 13.4b–d), which differ significantly from the traditional permission screen supplied for handling dangerous permissions (seen in Fig. 13.4a).

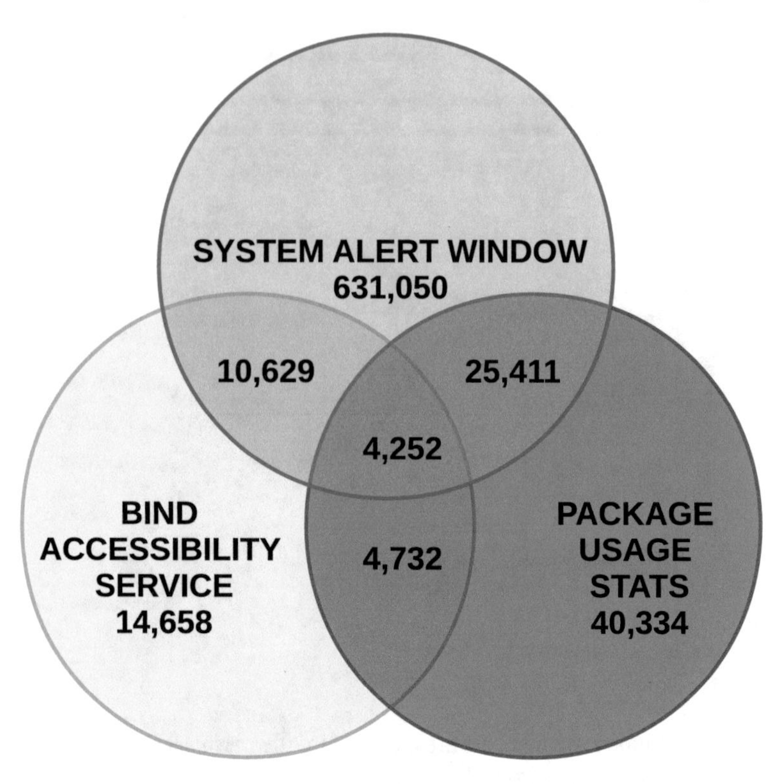

Fig. 13.3 Usage of some system permissions according to Tacyt. Numbers are reported in app versions

In the former category, we have permissions such as SYSTEM_ALERT_WINDOW, BIND_ACCESSIBILITY_SERVICE, WRITE_SETTINGS and PACKAGE_USAGE_STATS. To prevent users from carelessly granting these permissions, Android provides a completely different interface, and in principle, the corresponding settings are well-hidden in the menus so that users will grant access only when deemed necessary (see Fig. 13.4 for comparison). Nevertheless, these permissions are used by thousands of apps, as reported by Tacyt and seen in Fig. 13.3. To understand the extent of the risk that these permissions expose their users to, one has to consider that the SYSTEM_ALERT_WINDOW allows an application to overlay every Android activity and therefore can utterly deceive the user. The BIND_ACCESSIBILITY_SERVICE permission allows an application to imitate user tapping on the screen. Therefore, once granted to an app, it can perform any action on the user's device. Finally, the PACKAGE_USAGE_STATS will be discussed in detail in the following paragraphs.

Beyond the aforementioned categorisation of permissions, Android has the same Linux-based mechanism for UID/GID based access control. All users and groups are assigned with an ID (see Listing 1). As implied by this code excerpt, apps are

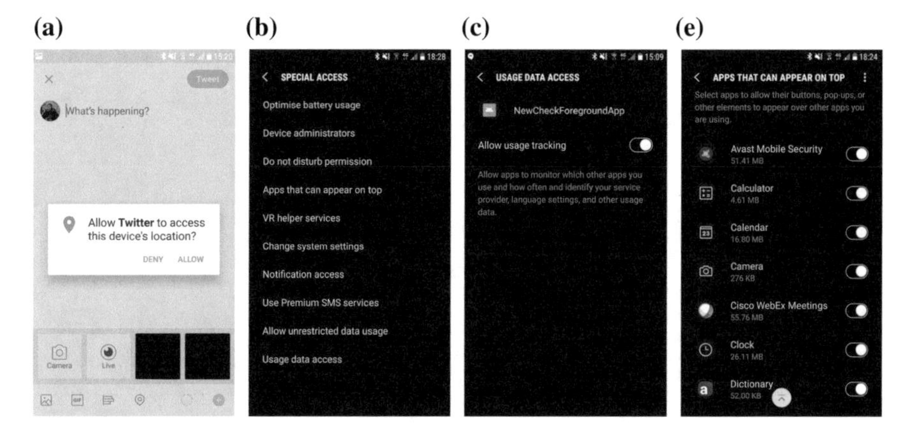

Fig. 13.4 Different interfaces for managing permissions in Android. **a** Granting a dangerous permission, **b** managing some system permissions, **c** allowing an app to track usage of other installed apps, **d** managing the permission to overlay other apps

assigned an ID above 10000, referred to as AID. Once a user grants a permission to an app, the app is added to the corresponding group, and it can access the particular resource. Therefore, if an app belongs to groups 1006 and 1021, it can access camera and GPS.

13.2.2 Android Foreground App

Android has an inherent problem regarding its UI: the lack of actual proofs of the identity of the application running in the foreground. This stems from two reasons, the size constraints of the devices which imply further constraints to UI, and also user permissions. According to the first one, UI components in Android, and mobile devices as a whole are stacked one on top of the other to fit into the small monitor of these devices. As a result, this size constraint prevents users from being able to determine the "actual" foreground app successfully. As for the second, users do not have many permissions, or even "high level" permissions, to determine themselves, or even by installing additional apps, the list of open apps and services in AOSP. Therefore, Android users blindly trust the Android UI.

In previous API levels, applications could monitor open apps, e.g. using the `getRunningTasks` method of `ActivityManager` as of API level 1. Nevertheless, as of API level 21, this method is no longer available to third-party applications. Google stated that:

> …the introduction of document-centric recents means it can leak person information to the caller.

```
#define AID_ROOT 0 /*traditional unix root user*/
/*The following are for LTP and should only be used for testing*/
#define AID_DAEMON 1 /*traditional unix daemon owner*/
#define AID_BIN 2  /*traditional unix binaries owner*/
#define AID_SYSTEM 1000 /*system server*/
#define AID_RADIO 1001 /*telephony subsystem, RIL*/
#define AID_BLUETOOTH 1002 /*bluetooth subsystem*/
#define AID_GRAPHICS 1003 /*graphics devices*/
#define AID_INPUT 1004 /*input devices*/
#define AID_AUDIO 1005 /*audio devices*/
#define AID_CAMERA 1006 /*camera devices*/
#define AID_LOG 1007 /*log devices*/
#define AID_COMPASS 1008 /*compass device*/
#define AID_MOUNT 1009 /*mountd socket*/
#define AID_WIFI 1010 /*wifi subsystem*/
#define AID_ADB 1011 /*android debug bridge (adbd)*/
#define AID_INSTALL 1012 /*group for installing packages*/
#define AID_MEDIA 1013 /*mediaserver process*/
#define AID_DHCP 1014 /*dhcp client*/
#define AID_SDCARD_RW 1015 /*external storage write access*/
#define AID_VPN 1016 /*vpn system*/
#define AID_KEYSTORE 1017 /*keystore subsystem*/
#define AID_USB 1018 /*USB devices*/
#define AID_DRM 1019 /*DRM server*/
#define AID_MDNSR 1020 /*MulticastDNSResponder (service discovery)*/
#define AID_GPS 1021 /*GPS daemon*/
#define AID_UNUSED1 1022 /*deprecated, DO NOT USE*/
...
#define AID_APP 10000 /*TODO: switch users over to AID_APP_START*/
#define AID_APP_START 10000 /*first app user*/
#define AID_APP_END 19999 /*last app user*/
    ...
```

Listing 13.1: Excerpt from available UIDs/GIDs in Android as defined in AOSP source code [15].

However, security researchers managed to derive the foreground apps through leaks from the `procfs`, as Android, like all Linux-based operating systems, uses it to store information of the processes that are executed by the OS.

Towards this end, Chen et al. [10] monitored offline the memory consumption of each activity in an application by tracking the memory allocation of the corresponding `/proc/[pid]/statm` file. They hypothesize that there is a specific footprint when shifting from one activity to another which can be used to identify app and activities. The latter can be further improved by monitoring network traffic through `/proc/net/tcp6`.

Bianchi et al. in [8] also identify the foreground application by using `procfs`. In this case, the leakage is from the file `/proc/[pid]/cgroups` whose contents change from `/apps/bg_non_interactive` to `/apps` when an app is sent to the foreground.

Finally, Alepis and Patsakis [2] exploited the `oom_adj_score` file in `procfs`, a file used by Android to monitor resource allocation and release. Depending on app usage, Android modifies this file which is stored under the directory `/proc/[pid]/` of each app. By pruning all the system applications, the least likely process to be killed is the foreground app.

As of Android Nougat, access to /proc/[pid]/ is prohibited to other apps; therefore all the aforementioned attacks are not applicable to the two latest Android versions. However, as of Android Lollipop, developers may use two additional methods to accomplish foreground app detection. Namely, either through the utilization of the UsageStatsManager API which requires the PACKAGE_USAGE_STATS permission and allows an app to collect statistics about the usage of the installed apps, or through the AccessibilityService API which requires the BIND_ACCESSIBILITY_SERVICE permission. Regarding the first case, this kind of information is presumably vital for the Android ecosystem, considering that it requires system permission to be collected. Yet, Android does not intend to allow an app to derive anything else apart from aggregated statistics about the usage of the installed apps. Therefore, the app can get statistics but they are not very fine-grained: the app can collect aggregated usage data for up to 7 days for daily intervals, up to 4 weeks for weekly intervals, up to 6 months for monthly intervals, and finally up to 2 years for yearly intervals, always depending on the chosen interval. Nevertheless, this service gives developers the ability to build a list of usage statistics per app, List<UsageStats>, and consequently query each of its items through the built-in getLastTimeUsed() method, to derive the foreground app. The second option, utilizing the AccessibilityService, includes handling the onAccessibilityEvent() callback and checking whether the TYPE_WINDOW_STATE_CHANGED event type is present, to determine when the current window changes. Finally, the target windows is further checked to determine whether it is the case of an activity by the method PackageManager.getActivityInfo() and the foreground app is revealed. It is important to note that Google has warned developers about this permission, that she will remove apps from the Play Store if they use accessibility services for "non-accessibility purposes" [24].

In the following paragraphs, we discuss in detail why app usage statistic data and also foreground application detection are actually considered such sensitive user information, using concrete examples of their malicious usage.

13.3 Collecting App Metadata

13.3.1 Assumptions and Desiderata

In our threat model, we assume that the victim has been tricked into installing a malicious app in his Android device. This assumption is considered standard in most Android related attacks [12, 13, 27]. Therefore it is aligned with the current literature. To relax possible constraints, we further assume that the device is not rooted; therefore, the attack could be launched in every stock Android installation. Finally, we assume that the application does not request any dangerous permissions from the user. The latter is rather crucial as dangerous permissions require further run-time user interaction in OS versions following Android Marshmallow and they can also be revoked later, as already discussed. Moreover, dangerous permissions can also deter users from installing an app.

Regarding code and library dependencies, we assume that the malicious app, apart from listing the corresponding app files, it does not request any additional shared library and does not make any suspicious API calls. To hide the actual filenames from possible static code analysis, we consider that the adversary collects them on runtime through a remote server. To this end, the app can use the Firebase or Azure cloud infrastructure that would be utilized for all regular interaction tunnel all of its traffic through legitimate servers, as used in [18] by social botnets, and hence to hide its communication with the C&C server.

Having a "zero permission" app, that is an app requesting only normal permissions, which communicates only with the Google servers not only makes the app look benign for the user, but it also bypasses many static security controls such as permissions, API calls, and network connections, that many consider as key indicators for identifying malicious apps [1, 7, 16, 23, 28, 30].

Finally, we make two weak assumptions to monitor whether two or more users are communicating via a specific app. The first one is that the users are simultaneously online, and thus there is no significant delay in message delivery between apps. This is a common case when users are interacting with "instant" messaging apps since by default these apps are used in real-time mode. The second weak assumption is that both apps are synchronised with a remote clock for precise timing. The latter is required for preventing errors due to time lags and for detecting whether user A contacted user B or the other way round. This can also be achieved by sending instant reports to the server once an activity has been recorded.

13.3.2 Basic Concept

While Android apps are isolated from each other, as they belong to different users, specific metadata can be extracted from their corresponding files. As the underlying filesystem of Android is `ext4` [19], as in many Linux installations, the file permissions are also similar to other Linux systems. Practically, depending on the user permissions, a user is allowed to (r)ead, (w)rite or e(x)ecute a file. Therefore, listing the contents of a directory or a file depends on the already assigned user permissions. All user installed apps in Android are installed in a directory `/data/data/<package_name>`, where `package_name` stands for the name of the installed app's package, and it is of the form `com.xyz`.

As in any system, misconfiguration of file permissions can leak much sensitive information which may lead to full compromise. More specifically, apps store their different "types" of stored data in corresponding subdirectories:

- databases/: Storage for the app's databases
- lib/: Storage for libraries and app helpers

- files/: Storage for app related files
- shared_prefs/: Storage for shared preferences and usually app settings
- cache/: Storage for caches

If an app does not want to share specific information with other apps, the contents of the underlying files and directories under folder `/data/data/com.xyz` are by default inaccessible by other apps, hence users. Nevertheless, if the permissions of the corresponding files and folders are not properly set and "someone" knows the absolute location of a file, even though he may not be able to access the contents of the file, he might be able to derive evidence of its existence, or even metadata about it. Typically, in Linux-based systems, this can be performed via various commands like:

```
ls -l /data/data/com.xyz/secret_file
stat /data/data/com.xyz/secret_file
```

Both commands, among other data, contain the last modification time and size of the file. Clearly, this seems only a small piece of information which can be collected by any installed app in Android without requesting any permission from the user, not even the most profound one: "Storage". Correspondingly, the research question in stake is whether this small leak of information can be used to derive sensitive information about the user. As already discussed, Google as of Android Nougat decided to remove all possible access from these files by setting the file permissions to 0700. Nevertheless, this change was not assigned to any CVE and, to the best of our knowledge, app permissions in file level do not follow a specific "default" pattern. In fact, throughout our research, we found different permission patterns in app files, discussed in more detail in the next section. It is worthy to notice that even if a CVE had been assigned and a patch had been made available, as recently shown by Lell and Nohl [17], manufacturers may often lie about their integration. Moreover, applications do not follow a specific pattern in the way they store their data. Even though, as already discussed, there are directories regarding libraries, caches, shared preferences, databases and file storage, inside these "default" directories, there is no specific rule to be followed. As a result, in many occasions related to data storage, apps have their data spread among different local databases, most often SQLite. Therefore, different user interactions correspond to changing different files which can then be traced to reveal the interaction.

The obvious drawback of the aforementioned issues is that many vendors could leak which applications are currently running or even the kind of interaction they are performing. Although running applications can be considered as trivial information, it is a considerably critical piece of information for the Android ecosystem since, apparently, by monitoring running applications one can derive usage statistics, vital information for user profiling and mobile targeted advertisement. In the most sinister scenario, if one can derive which is the foreground app, he can easily proceed in tricking the users into disclosing their credentials by pushing forward a forged Android activity that imitates the UI of the user-initiated app, running in the foreground [2, 8, 10, 22].

According to the official source [5], in Android Nougat, API level 24 and above, the private directory of apps has restricted access to improve the security of private files. As a result, as of API level 24, apps cannot access the `/proc/PID` directory for other PIDs, rendering all relevant to the "proc" directory approaches to detect foreground apps useless. Nevertheless, as we are illustrating in this paper, other directories and specific file metadata enable us to achieve the same result in the vast majority of Android smartphones to date.

13.3.3 Methodology

We opted for a black box methodology to see the problem in its full extent. Therefore, our goal was to determine how much information can be collected from an adversary who does not have any access to the apps' internals.

To bypass the restrictions of file access permissions in AOSP, we started our experiments with two devices, a rooted smartphone and one using AOSP. First, we installed the same apps and versions in both devices to have a common reference point, and we created fake user profiles for each application when deemed necessary. Then, we went through each application independently and started interacting with it, keeping track in the rooted smartphone of which files were changing upon every action. In this regard, the rooted phone can be considered as the "file change sensor" module, while the non-rooted phone provided the necessary "triggers" to initiate specific user actions.

Finally, after having noted all the involved files for the metadata exploitation, we implemented an Android app which acts as a background service. The service silently monitors in the background the metadata of specific files of the installed apps and subsequently transmits any detected changes to a remote database. Apart from this passive role, the app is also capable of presenting an Android activity whose UI is rendered from data that it fetches from the database when the proper trigger is sent.

The basic concept is illustrated in Fig. 13.5. Once the user installs the malicious app X, it retrieves the specs of the device and the Android OS version and a list of all installed apps from Android's Package Manager system service. X communicates this list to its server to obtain a list of files that it should monitor. Once received, X running as a background service, monitors the aforementioned files at predetermined intervals to derive the last modification time and size, which are then sent to the C&C server (see Fig. 13.5a).

To efficiently monitor the filesystem changes in every directory recursively, we used the `inotifywait` utility. The events that we monitored were: *create*, *delete*, *attrib* and *modify*. Since `inotify` cannot be directly installed in Android, we exploited the features of a well-known terminal emulator for Android devices, Termux.[3] The feature of Termux that we aimed is the `apt` integration that allows the user to install Linux packages, in our case `inotify-tools`.

[3]https://termux.com/.

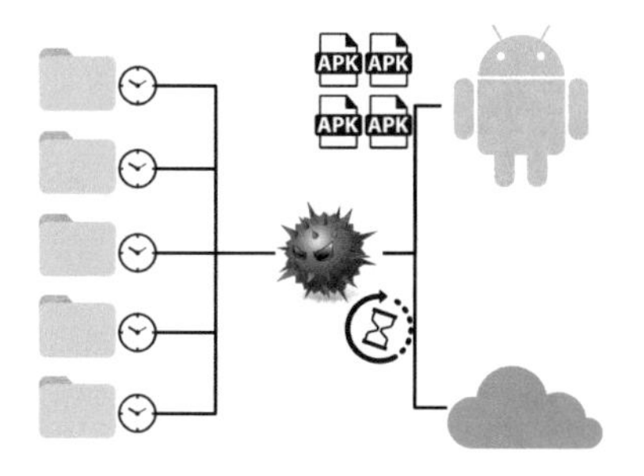

(a) Collecting local information. The malicious app periodically checks the modification time of specific files.

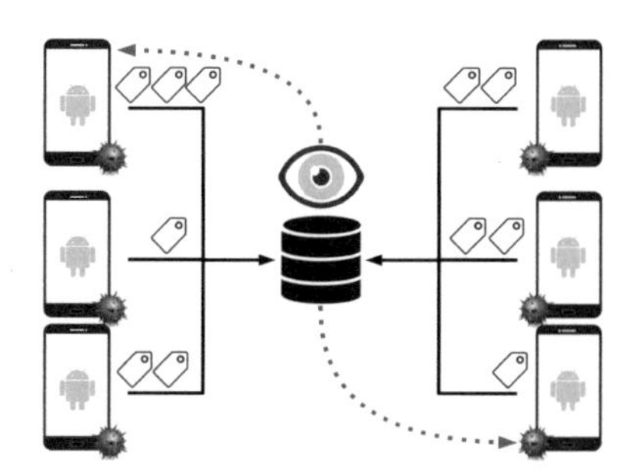

(b) Correlating information from many devices in the C&C server and pushing commands to specific devices.

Fig. 13.5 Basic concept

13.4 Experimental Results

As discussed, Google introduced in AOSP the 0700 permissions in Nougat, which at the moment of writing means that officially 70% of all Android devices are vulnerable. However, the number of the actually affected devices is, alarmingly, far bigger. In our experiments we tested several devices from top major Android vendors running Android API level$\geq$ 24, that is devices running Nougat and Oreo. More precisely, we experimented with devices from Samsung, LG, Xiaomi, Huawei and HTC. The "pure" Android devices (Nexus and Pixel) were the only ones that did not present any metadata leakages. On the contrary, devices which did not run pure Android, even if they were from the same manufacturer, leaked file metadata. To leak this information, we used the native `os.stat` method for each file we identified.

In what follows we present three different threat scenarios that stem from the described metadata leakage. In the first two cases, we focus on surveillance, while in the latter we focus on harvesting credentials.

13.4.1 Monitoring the Android Filesystem

In this scenario we try to derive usage statistics from the OS regarding the following actions:

- Add/Remove a user to contacts.
- Send/receive an SMS.
- Make/receive a call.
- Enable/disable GPS
- Shoot a photo.

Table 13.2 Identified Android actions

	Version	Call		Message		GPS		Contact		Camera
		Make	Receive	Send	Receive	On	Off	Add	Delete	
HMD Global TA-1024 (Nokia 5)	8.1.0			✓	✓	✓	✓			
Huawei Nexus 6P	8.1.0	✓	✓	✓	✓	✓	✓	✓	✓	✓
LGE LG-H870	8.0.0			✓	✓	✓	✓			
Xiaomi Mi A1	8.0.0			✓	✓	✓	✓			
Xiaomi MI 6	8.0.0	✓	✓	✓	✓	✓	✓			
HUAWEI VTR-L09	8.0.0			✓	✓	✓	✓			
Xiaomi Redmi 4A	7.1.2	✓	✓	✓	✓	✓	✓	✓	✓	
Xiaomi Redmi Note 5A Prime	7.1.2		✓	✓	✓	✓	✓			
Samsung SM-J510FN	7.1.1	✓	✓	✓	✓	✓	✓	✓	✓	
Xiaomi Redmi Note 4	7.0			✓	✓	✓	✓	✓		
Samsung SM-G930F	7.0			✓	✓	✓	✓	✓	✓	
Samsung SM-G935F	7.0			✓	✓	✓	✓	✓	✓	
Samsung SM-J710F	7.0			✓	✓	✓	✓	✓	✓	
HUAWEI PRA-LX1	7.0	✓	✓	✓	✓					
Samsung S6	7.0			✓	✓	✓	✓	✓		
Samsung SM-A500FU	6.0.1	✓	✓	✓	✓	✓	✓	✓	✓	
Xiaomi Redmi Note 3	6.0.1			✓	✓			✓	✓	✓
Xiaomi Redmi Note 4	6.0			✓	✓	✓	✓	✓	✓	
Plaisio Computers SA Turbo-X_A2	6.0			✓	✓	✓	✓	✓	✓	✓
Huawei P9 lite	6.0	✓	✓	✓	✓	✓	✓			
HUAWEI ALE-L21	6.0			✓	✓	✓	✓			✓
Sony D2303	5.1.1			✓	✓	✓	✓	✓	✓	
Motorola XT1032	5.1			✓	✓	✓	✓	✓	✓	
Samsung SM-J320FN	5.1.1			✓	✓	✓	✓			✓
Samsung SM-J320F	5.1.1			✓	✓	✓	✓	✓	✓	
HTC One	5.0.2			✓	✓	✓		✓	✓	✓

without requesting any permission from the user, just by monitoring filesystem changes. To this end, we monitored a series of files of Android, as seen in Table 13.4, to determine the aforementioned actions. While several variations are depending on the underlying Android flavour, it is evident that an adversary can easily determine a wide set of actions as illustrated in Table 13.2.

13.4.2 Communication Apps

In this threat scenario, we try to derive usage statistics from communication apps. Since locality plays a crucial role in their usage (see Fig. 13.6), we studied the most well-known, secure and widely used communication apps in Google Play. The list (see Table 13.1) contains 15 apps, all of which have at least one million installations while their vast majority has more than 100 million downloads.

In our experiments, we aimed to detect more than the fact of using the apps. To this end, we wanted to determine whether we could identify users communicating with each other, when and how. Using the methodology detailed in the previous section, we performed the following actions in each app and device:

Fig. 13.6 Most popular apps per country. *Source* SimilarWeb

- Add a user to contacts.
- Remove/Ban a user from contacts.
- Send a text message.
- Initiate a call.
- Initiate a video call.

The above actions were performed multiple times from each device to determine the files affected each time, how fast these changes were made, and to prune any coincidental file changes. It is worthwhile to note that most of these apps would periodically change their files as they communicate with the corresponding service and receive some status messages. While some of these changes created noise in our samples, they were easy to be pruned based on simultaneous similar changes made to other installed apps. Apparently, the changes were initially monitored only in the rooted device. Then, knowing the locations of the altered files, we validated our results with the non-rooted device too.

The results of our experiments are illustrated in Table 13.3. From the 15 apps, only 6 were secure (their row is highlighted in gray), meaning that no metadata could be extracted from them. Then, for three more we could detect that some app interaction occurred, yet it could not be precisely identified ("Unidentified Action" column in Table 13.3). The reason for the latter is that some apps used a small number of databases, one or two. Thus, every interaction was mapped to the same files, not giving us the opportunity to track "specific" actions. Yet, the most interesting results are for the rest 6 apps, most of which have users on the scale of billions.

More precisely, for Facebook messenger, we can determine when the user blocks someone. Moreover, we can ascertain when the user initiates a communication with another user. While we cannot determine whether this occurs via a message or a call, we can trace the communication and determine from the respective timestamps who is sending and who is receiving if both devices have the monitoring app installed. In the case of Line, we can only detect when the user receives a message. In Skype we can tell that a user has performed one of the following actions: add/block a contact, send a message or receive a call. In the case of Snapchat, we can determine when a user receives or sends messages and, if both communicating users have installed the monitoring app, we can determine from the timestamps which one is the sender and the receiver. Additionally, we are able to trace an interaction which is either the addition/removal of a contact or the sending/receipt of a snap. We argue that if both parties have installed our monitoring app, the success rate of our approach is further increased. For Wire we can determine when the user adds a new contact and when the user interacts with others, but without being able to determine the mean (distinguish between messages and calls). Finally, in the case of WhatsApp, we can determine when a user initiates a phone call. However, we cannot determine exact message receipt/sending, or call received in the same device. Again, if both parties have installed our monitoring app, these results can be further fine-tuned.

The full list of the affected files for each application is illustrated in Table 13.4. Figure 13.7 illustrates part of the problem in terms of file permissions. Files and folders of WhatsApp have different file permissions, allowing an adversary to extract a lot of metadata about them.

Fig. 13.7 Partial structure and permissions of the folder that Whatsapp is installed as recorded in the rooted phone

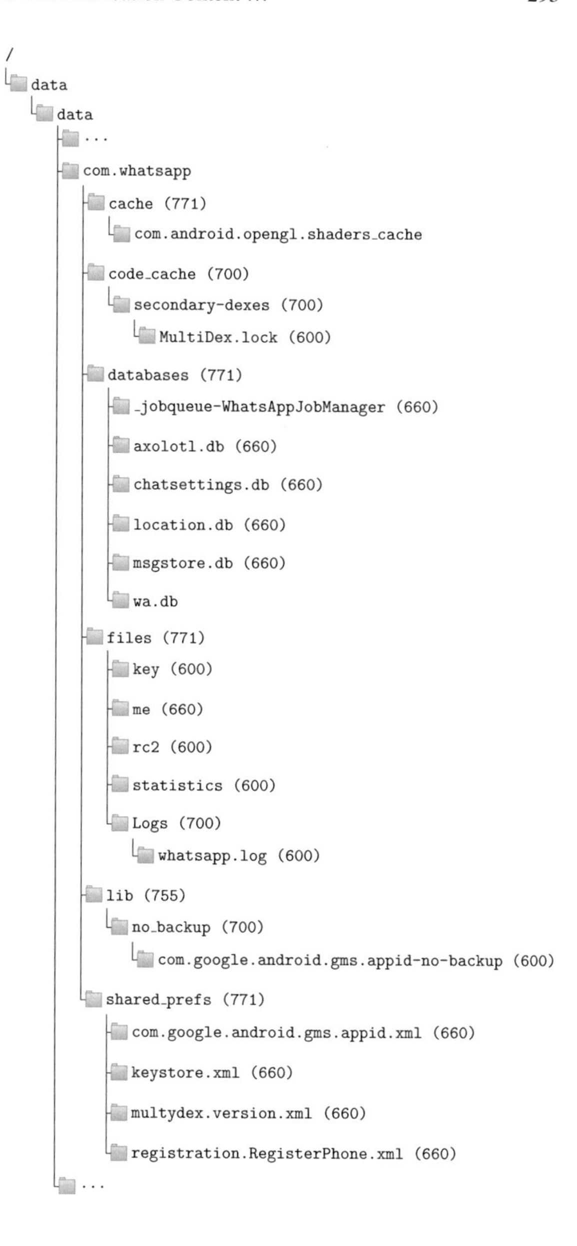

Based on the above, it is apparent that unauthorized apps can derive a lot of valuable information from communication apps such as when they have been used and, should the monitoring app has been installed in both devices, to whom is interacting with whom, how and when.

Table 13.3 Actions that can be deduced from metadata per app. Cells marked with + in the same row denote that an action has been identified and can be narrowed down to one of them. Same applies for *. Snapchat does not enable calls, but "snaps"

	Contact			Message		Call		Unidentified
	Add	Delete	Block	Send	Receive	Make	Receive	Action
BBM								
Facebook messenger			✓	✓+	✓+	✓+	✓+	
ICQ								✓
imo								✓
Line					✓			
KakaoTalk								✓
QQ International								
Signal								
Skype	✓+		✓+	✓+			✓+	
Snapchat	✓+	✓+		✓*	✓*	✓+	✓+	
Telegram								
Viber								
WeChat								
Wire	✓			✓+	✓+	✓+	✓+	
WhatsApp				✓+	✓+	✓	✓+	

13.4.3 Forged UI

Currently, there exist malware targeting bank applications which use overlays and UI replication of numerous banking applications.[4] Malware of this kind include families such as Bankbot, Bankun, Koler and SlemBunk, while there are other malware families like ransomware, e.g. Lockdroid, which also exploit these capacities. MazarBot and Svpeng Malware are two malware families which are even worse as they manage, after obtaining administrator privileges on the infected device, to trick users to give away their banking credentials. Moreover, new variants of Svpeng have keylogger capacities thanks to Accessibility Services. To quantify the problem, it must be highlighted that, according to CheckPoint [9], 74% of ransomware, 57% of adware, and 14% of banker malware abuse the notorious `SYSTEM_ALERT_WINDOW` permission.

While there are plenty of ways for the aforementioned malware to overlay the actual UI such as those described in [2, 4, 14, 29], the key ingredient of all is to timely present the overlay to the user. As discussed, these methods cannot be applied for API levels >23; therefore, other methods could be used to trick the user into disclosing the foreground app, e.g. forged shortcuts and notifications [2, 22]. However, since apps can retrieve the list of installed apps, they could try to use the metadata to infer when a targeted app is used.

To showcase the issue, we used Paypal as our reference. Using the same methodology, we monitored which files of Paypal are changed once the user logs in or logs out. The exact list of `data/data/com.paypal.android.p2pmobile/` direc-

[4]https://blog.avast.com/mobile-banking-trojan-sneaks-into-google-play-targeting-wells-fargo-chase-and-citibank-customers.

tory is illustrated in Listing 2. Therefore, an attacker can easily determine when the user signs in to Paypal.

The attack is now straightforward: an app that requests only normal permissions monitors the Paypal's target file. When the user logs in, the malicious app comes to the foreground, replicating the UI of the Paypal app. To prevent detection, the app can collect its data regarding the fake UI construction during runtime from the Internet and replicate the UI through a web form that is displayed within a web view. Since the user was already using the app, he considers that there was an error in typing his credentials and types them again, without knowing that they are disclosed to the malicious app. Once collected, the credentials are sent to the C&C server, and the app launches Paypal to allow the user to interact with the right app. The concept is illustrated in Fig. 13.8. Note that in this scenario the malicious app does not need any special permission to overlay Paypal as it uses UI replication and the forged UI completely covers the genuine one. Its only requirement is to come to the foreground, something that all apps in Android have the permission to accomplish.

```
shared_prefs/PresentationAccount.RememberedUserState.xml
shared_prefs/FoundationAccount.AccountState.xml
shared_prefs/FoundationCore.DeviceInfoState.xml
shared_prefs/version.6.shared.keys.xml
files/com.paypal.android.AccountInfo.secure
files/CoreStateData
files/AdjustIoActivityState
```

Listing 13.2: Files changed when a user logs in to Paypal.

13.4.4 User Metadata Experiment

In our independent research, to determine the extent of the problem we have also conducted a supplementary experiment involving distinctly statistical user data. More precisely, we created an app that collects both hardware and software specs from users' devices, accompanied by a test about the metadata leakage in question. The resulting data bundle from each user was sent to our server for further processing. This limited yet enlightening experiment justified our claims, about Android fragmentation that results in vendors failing to conform to AOSP security guidelines, while the same is true about app developers too. More precisely, our short-term experiment involved 120 random users. The analysis of the results showed that almost half of them, namely 56 users (46%) owned an Android smartphone with an OS API level beyond 23 (Marshmallow). This already means that 54% of the users in our evaluation were already vulnerable to the metadata leakage. In the remaining 46%,

Table 13.4 Files affected by each action per application

	Contact			
	Add	Delete	Block	Send
Facebook messenger /data/data/com.facebook.orca/databases/			prefs_db graphql_cache threads_db2	prefs_db threads_db2
imo /data/data/com.imo.android.imoim/				databases/imofriends.db
kakaoTalk /data/data/com.kakao.talk/databases/				KakaoTalk.db
Line /data/data/jp.naver.line.android/databases/				
Snapchat /data/data/com.snapchat.android/databases,	com.snapchat.android.analytics.framework com.snapchat.android.analytics.frameworkshadow tcspahn.db unlockable_encrypted_database.db	com.snapchat.android.analytics.framework com.snapchat.android.analytics.frameworkshadow tcspahn.db unlockable_encrypted_database.db		com.snapchat.android.analytics.framework com.snapchat.android.analytics.frameworkshadow tcspahn.db
WhatsApp /data/data/com.whatsapp/databases/				msgstore.db-wal msgstore.db-shm axolotl.db-shm axolotl.db-wal
Wire /data/data/com.wire/databases/				mixpanel ZGlobal.db-shm ZGlobal.db-wal evernote_jobs.db

	Message	Call	
	Receive	Make	Receive
Facebook messenger /data/data/com.facebook.orca/databases/	prefs_db threads_db2	prefs_db threads_db2	prefs_db threads_db2
imo /data/data/com.imo.android.imoim/	databases/imofriends.db	databases/imofriends.db	databases/imofriends.db
kakaoTalk /data/data/com.kakao.talk/databases/	KakaoTalk.db	KakaoTalk.db	KakaoTalk.db
Line /data/data/jp.naver.line.android/databases/	naver_line naver_line_push_history		
Snapchat /data/data/com.snapchat.android/databases,	com.snapchat.android.analytics.framework com.snapchat.android.analytics.frameworkshadow tcspahn.db	com.snapchat.android.analytics.framework com.snapchat.android.analytics.frameworkshadow tcspahn.db unlockable_encrypted_database.db	com.snapchat.android.analytics.framework com.snapchat.android.analytics.frameworkshadow tcspahn.db unlockable_encrypted_database.db
WhatsApp /data/data/com.whatsapp/databases/	msgstore.db-wal msgstore.db-shm axolotl.db-shm axolotl.db-wal		msgstore.db-wal msgstore.db-shm axolotl.db-shm axolotl.db-wal
Wire /data/data/com.wire/databases/	mixpanel ZGlobal.db-shm ZGlobal.db-wal evernote_jobs.db	mixpanel ZGlobal.db-shm ZGlobal.db-wal evernote_jobs.db	mixpanel ZGlobal.db-shm ZGlobal.db-wal evernote_jobs.db

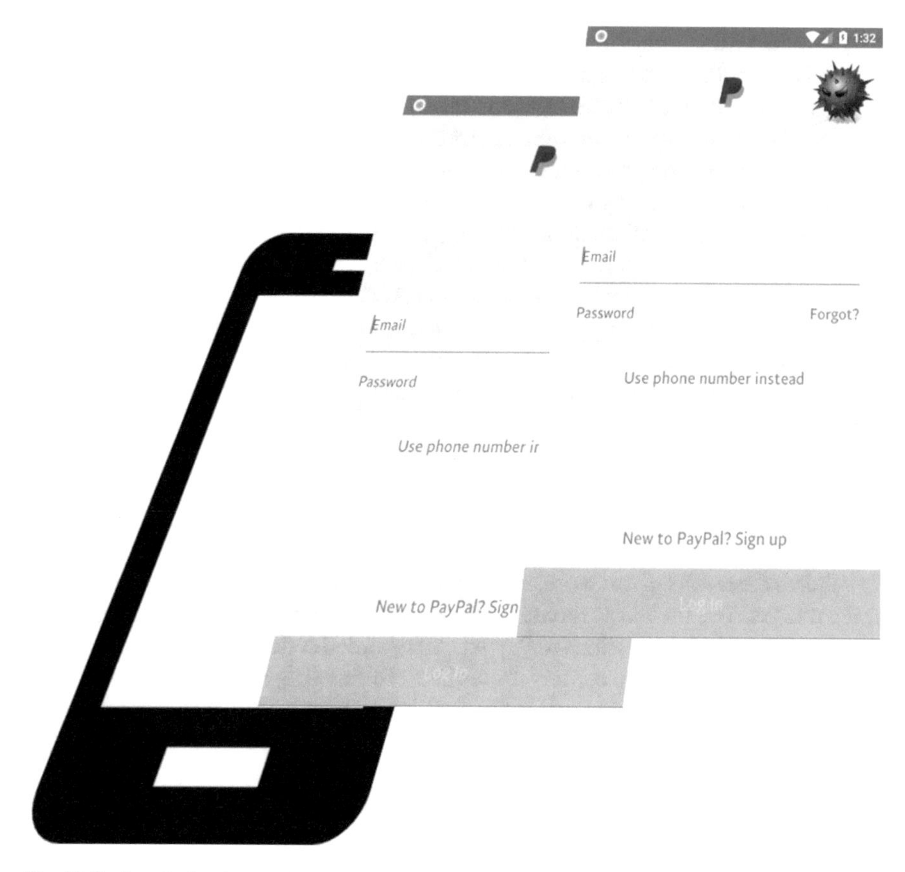

Fig. 13.8 Overlaying Paypal interface with forged interface

we further examined whether the vendors of the devices had incorporated the file permission security update which was introduced in Nougat.

The results showed that two major vendors, (see Fig. 13.1), namely Samsung and Xiaomi, do not meet the criteria, as well as OnePlus. For these three well-known manufacturers, we have discovered metadata leakage in popular messaging applications, which, most importantly, have a target API level (app specific targetSdkVersion) above 23. Regarding the case of Samsung, our experiment identified five of the most popular device models that are vulnerable, while we have found two popular models from Xiaomi and one from for OnePlus. Of course, these findings indicate that the problem is generic regarding the vendors and not specific devices. However, this argument is subject to the extent of the experiment.

Going one step further we have examined the number of vulnerable devices where the API level of Android is beyond 23 (Nougat and Oreo), yet the target app for the metadata leakage has a targetSdkVersion below 24. The results are quite alarming since we did not find any protected device in this case. Namely, the vendors that are affected by this vulnerability in our study are Samsung, Huawei, Xiaomi, LGE Nexus, LG, OnePlus, Sony and HMD Global (NOKIA). It is important to note that for these vendors, Google's models (Nexus/Pixel) are also included, while a significant number of these devices were running Android Oreo.

Concluding, our reported metadata leakage issue affects end-users in an overwhelming majority of cases since all parties involved can be found to be defective in the vulnerability in question. Summarizing, the OS is vulnerable in all versions prior to Nougat. Even in the cases after Nougat, device manufacturers may provide users with "insecure" devices by not following the AOSP guidelines. Even after Nougat and with conformance to the AOSP guidelines, app developers, including top companies (e.g. Facebook apps like Facebook, Facebook Messenger, and Instagram) may provide users with "insecure" apps, by not targeting their app API beyond 23.

13.4.5 Proofs of Concept

To validate the results of our work, we provide two deliberately stub implementations of the attacks. The first APK monitors the aforementioned communication apps and the second one the login actions to Paypal. All recorded actions are displayed as a log at https://monitor1webapp.azurewebsites.net/. To this end, we record a timestamp, the UserID (in this case AndroidID), the application which was being monitored and the action which was detected.

13.5 Conclusions

In this paper we have provided evidence that metadata leakage in not only existent, but in can easily lead to user data harvesting, user profiling and even user surveillance, impinging on users' lives. After analyzing the results of this study, it seems quite straightforward that the manufacturers need to fully embrace AOSP to implement security requirements effectively into their devices. However, at the time of writing this scenario is not the case, neither seems realistic or feasible given the huge deal of effort that would admittedly require. Therefore, the solution probably lies in the developers' hands. A potential patch that could be easily deployed and would help towards a solution of the "metadata leakage" problem for each app would be to utilize app subfolders with random names. To do this, the app upon installation or update should not be assigned to any local database. Instead, upon its first execution, it should create a folder using a nonce for its name, and it should store this name localy, e.g. in its shared preferences. Thus, it would be impossible for a malicious app to derive the folder name without exploiting a major app-specific vulnerability. Consequently, the patch would manage its goal since the attack stems from the fact that the apps' folder names are fixed and consequently "expected", with wrong permissions assigned to them that the apps cannot change. Obviously, to implement the above, the requested changes would only involve changing the connection string to include the prefix in the corresponding files along with adding the necessary queries to generate the initial structure in local databases. Likewise, one could also rename the corresponding files by adding a prefix that would be also stored in shared preferences, given that apps do not have access to the shared preferences of their peers unless they have root privileges.

Acknowledgements The authors would like to thank *ElevenPaths* for their valuable feedback and granting them access to Tacyt.

References

1. Y. Aafer, W. Du, H. Yin, Droidapiminer: mining API-level features for robust malware detection in android, in *International Conference on Security and Privacy in Communication Systems* (Springer, 2013), pp. 86–103
2. E. Alepis, C. Patsakis, Trapped by the UI: the android case, in *International Symposium on Research in Attacks, Intrusions, and Defenses* (Springer, 2017), pp. 334–354
3. E. Alepis, C. Patsakis, *Unravelling security issues of runtime permissions in android* (J. Hardw. Syst, Secur, 2018)
4. Y. Amit, Accessibility clickjacking the next evolution in android malware that impacts more than 500 million devices (2016). https://www.skycure.com/blog/accessibility-clickjacking/
5. Android Developer, Permission changes. https://developer.android.com/about/versions/nougat/android-7.0-changes.html. Accessed 07 Feb 2018
6. Android Developer. Android 7.0 behavior changes (2017). https://developer.android.com/about/versions/nougat/android-7.0-changes.html
7. D. Arp, M. Spreitzenbarth, M. Hubner, H. Gascon, K. Rieck, C.E.R.T. Siemens, Drebin: effective and explainable detection of android malware in your pocket. NDSS **14**, 23–26 (2014)
8. A. Bianchi, J. Corbetta, L. Invernizzi, Y. Fratantonio, C. Kruegel, G. Vigna, What the app is that? deception and countermeasures in the android user interface, in *Proceedings of the 2015 IEEE Symposium on Security and Privacy* (IEEE Computer Society, 2015), pp. 931–948
9. Check Point Mobile Research Team, Android permission security flaw (2017). https://blog.checkpoint.com/2017/05/09/android-permission-security-flaw/. Accessed 09 Sep 2017
10. Q.A. Chen, Z. Qian, Z.M. Mao, *Peeking into your app without actually seeing it: UI state inference and novel android attacks, in 23rd USENIX Security Symposium (USENIX Security 14)* (San Diego, CA, USENIX Association, 2014), pp. 1037–1052
11. ElevenPaths, An innovative tool for the monitoring and analysis of mobile threats. https://www.elevenpaths.com/technology/tacyt/index.html
12. P. Faruki, A. Bharmal, V. Laxmi, V. Ganmoor, M.S. Gaur, M. Conti, M. Rajarajan, Android security: a survey of issues, malware penetration, and defenses. IEEE Commun Surv Tutor **17**(2), 998–1022
13. A.P. Felt, M. Finifter, E. Chin, S. Hanna, D. Wagner, A survey of mobile malware in the wild, in *Proceedings of the 1st ACM workshop on Security and Privacy in Smartphones and Mobile Devices* (ACM, 2011), pp. 3–14
14. Y. Fratantonio, C. Qian, S. Chung, W. Lee, *Cloak and dagger: from two permissions to complete control of the UI feedback loop, in Proceedings of the IEEE Symposium on Security and Privacy (Oakland)* (, San Jose CA, 2017)
15. Google, AOSP source code for filesystem_config. https://android.googlesource.com/platform/system/core/+/master/libcutils/include/private/android_filesystem_config.h
16. M. Grace, Y. Zhou, Q. Zhang, S. Zou, X. Jiang, Riskranker: scalable and accurate zero-day android malware detection, in *Proceedings of the 10th International Conference on Mobile Systems, Applications, and Services* (ACM, 2012), pp. 281–294

17. K. Nohl, J. Lell. *Mind the Gap—Uncovering the Android Patch Gap through Binary-only Patch Analysis* (2018)
18. E.J. Kartaltepe, J.A. Morales, S. Xu, R. Sandhu, Social network-based botnet command-and-control: emerging threats and countermeasures, in *International Conference on Applied Cryptography and Network Security* (Springer, 2010), pp. 511–528
19. Avantika Mathur, Mingming Cao, Suparna Bhattacharya, Andreas Dilger, Alex Tomas, Laurent Vivier, The new ext4 filesystem: current status and future plans. Proc. Linux Symp. **2**, 21–33 (2007)
20. K. Nohl, Mobile self-defense (snoopsnitch), in *Proceedings of Chaos Computer Security Conference* (2014)
21. Ed O'Keefe, https://www.washingtonpost.com/news/post-politics/wp/2013/06/06/transcript-dianne-feinstein-saxby-chambliss-explain-defend-nsa-phone-records-program/?utm_term=.f2e1466faae2 (2013)
22. C. Patsakis, E. Alepis, Knock-knock: the unbearable lightness of android notifications, in *Proceedings of the 4th International Conference on Information Systems Security and Privacy, ICISSP 2018, Funchal, Madeira-Portugal, January 22–24, 2018*, ed. by P. Mori, S. Furnell, O. Camp (SciTePress, 2018), pp. 52–61
23. N. Peiravian, X. Zhu, Machine learning for android malware detection using permission and API calls, in *2013 IEEE 25th International Conference on Tools with Artificial Intelligence (ICTAI)* (IEEE, 2013), pp. 300–305
24. Android Police, https://www.androidpolice.com/2017/11/12/google-will-remove-play-store-apps-use-accessibility-services-anything-except-helping-disabled-users/ (2017)
25. C. Spensky, J. Stewart, A. Yerukhimovich, R. Shay, A. Trachtenberg, R. Housley, R.K. Cunningham, Sok: privacy on mobile devices–it's complicated. Proc. Privacy Enhanc. Technol. **2016**(3), 96–116 (2016)
26. Statista, Global market share held by the leading smartphone operating systems in sales to end users from 1st quarter 2009 to 2nd quarter 2018. https://www.statista.com/statistics/266136/global-market-share-held-by-smartphone-operating-systems/
27. T. Vidas, D. Votipka, N. Christin, All your droid are belong to us: a survey of current android attacks, in *Proceedings of the 5th USENIX Conference on Offensive Technologies* (USENIX Association, 2011), pp. 10–10
28. D.-J. Wu, C.-H. Mao, T.-E. Wei, H.-M. Lee, K.-P. Wu, Droidmat: android malware detection through manifest and API calls tracing, in *2012 Seventh Asia Joint Conference on Information Security (Asia JCIS)* (IEEE, 2012), pp. 62–69
29. L. Ying, Y. Cheng, Y. Lu, Y. Gu, P. Su, D. Feng, Attacks and defence on android free floating windows, in *Proceedings of the 11th ACM on Asia Conference on Computer and Communications Security* (ACM, 2016), pp 759–770
30. Y. Zhou, W. Zhi, W. Zhou, X. Jiang, Hey, you, get off of my market: detecting malicious apps in official and alternative android markets. NDSS **25**, 50–52 (2012)

Part V
Computer Science-Based Technologies in Scheduling and Transportation

Chapter 14
A Fuzzy Task Scheduling Method

Konstantina Chrysafiadi

Abstract Task scheduling is crucial for offering the users of an operating system the impression that the system's response is direct and in real time. However, it is a complicated process that is characterized by vagueness and uncertainty. A solution to this problem is the usage of fuzzy logic. Therefore, in this paper a rule-based fuzzy scheduling method, which considers both the execution time and the waiting time for each task, is presented. Fuzzy sets are used to describe both criteria. The operation of the presented scheduling method is based on a rule-based reasoner that decides dynamically about the priority of the tasks that wait to be executed. This reasoner is triggered each time a change occurs (i.e. the execution of a task ends, a new task arrives etc.). It has been compared with Short Job First (SJF) and First Come First Served (FCFS) scheduling algorithms for a large number of different sets of tasks. The results showed that the presented fuzzy scheduling algorithm in all the cases has the same or almost the same average waiting time with SJF, which is considered as the scheduling algorithm with the best (minimum) average waiting time. However, the presented fuzzy scheduling algorithm ensures that priority is given not only to tasks with short execution time, but also to tasks that remain into the queue waiting to be executed for a long time. In this way, the starvation, which may be caused by SJF algorithm, is eliminated.

Keywords Task scheduling · Fuzzy logic · Rule-based fuzzy reasoner · SJF · FCFS

14.1 Introduction

An operating system is a software, whose goal is to run user programs/applications and manage and utilize effectively the computer's resources. Therefore, an operating system performs multiple tasks that vary from basic tasks like recognizing input from keyboard or displaying output to the computer's screen, to more complex tasks like

K. Chrysafiadi (✉)
Department of Informatics, University of Piraeus, Piraeus, Greece
e-mail: kchrysafiadi@unipi.gr

G. A. Tsihrintzis and M. Virvou (eds.), *Advances in Core Computer Science-Based Technologies*, Learning and Analytics in Intelligent Systems 14,
https://doi.org/10.1007/978-3-030-41196-1_14

file manipulation, device controlling and memory managing. An operating system has to ensure the correct working of the computer system and, simultaneously, it has to offer computer's users the impression that the system's response is direct and in real time. That is succeeded through task scheduling.

Task scheduling is a process, in which the operating systems selects each current time a task, among others that wait in a queue, that has to be executed and allocates the appropriate computer's resources for the execution. Task scheduling is a complicated process and has to take into consideration a variety of criteria, such as the mean waiting time, the execution time, the response time, the throughput, the turnaround time etc. have in order to decide which task has to be scheduled for execution. That is the reason for the existence of a variety of task scheduling algorithms [1].

Furthermore, the complex process of scheduling is characterized by vagueness and uncertainty. A solution to this problem is fuzzy logic, which was introduced by Zadeh [2], as a methodology for computing with words that cannot be done equally well with other methods. Fuzzy logic can be used to handle the uncertainty of scheduling describing the criteria of task scheduling in a more realistic way through fuzzy sets. Also, fuzzy rules over the defined fuzzy sets can be used in order to define the priority of execution for the tasks that wait in a queue.

Considering the above, in this paper, a proposed fuzzy scheduling method is presented. It considers two criteria for each task: (i) the execution time and (ii) the waiting time. Both criteria are described using fuzzy sets. The operation of the presented scheduling method is based on a rule-based reasoner that decides dynamically about the priority of the tasks that wait to be executed. This reasoner is triggered each time a change occurs (i.e. the execution of a task ends, a new task arrives etc.). The presented scheduling algorithm can operate for both pre-emptive and non pre-emptive scheduling. It has been compared with two well-known scheduling algorithms: Short Job First (SJF), which is a scheduling algorithm that selects for execution the task with the smallest execution time, and First Come First Served (FCFS), which selects for execution the queued tasks by the order of their arrival (i.e. the task that arrives first, gets executed first). We applied the presented scheduling algorithm, SJF and FCFS scheduling algorithms to a large number of different sets of tasks. The results showed that the presented fuzzy scheduling algorithm in all the cases has the same or almost the same average waiting time with SJF, which is considered as the scheduling algorithm with the best (minimum) average waiting time. However, the presented fuzzy scheduling algorithm ensures that priority is given not only to tasks with short execution time, but also to tasks that remain into the queue waiting to be executed for a long time. In this way, the starvation, which may be caused by SJF algorithm, is eliminated.

The remainder of this paper is organized as follows. In Sect. 14.2, the related work in task scheduling and fuzzy techniques is presented. In Sect. 14.3, the presented fuzzy scheduling method (criteria and rules) is described. In Sect. 14.4, the evaluation results are presented. In Sect. 14.5, the conclusions drawn from this work and the future work are presented.

14.2 Related Work

Task scheduling is the method by which a task (process, thread) is assigned to resources, such as processor, that execute and complete the task. It is a complicated process and the goal of task scheduling can differ according to each particular case. Also, a scheduling method can have one or more goals. Some of these goals can be: (i) minimizing the waiting time, (ii) minimizing the response time, (iii) maximizing the total amount of work completed per time unit, (iv) minimizing the turnaround time etc. There is a variety of task scheduling algorithms [3, 4]. Two well-known task scheduling algorithms are: Shortest Job First (SJF) scheduling algorithm and First Come First Served (FCFS) scheduling algorithm. A short description of these scheduling algorithms follows:

- Shortest Job First (SJF) scheduling algorithm. It is considered the best approach to minimize the waiting time of tasks. According to this approach, the task with the shortest execution time (duration) is served first. Therefore, the execution time of the tasks has to been known to the processor in advance.
- First Come First Served (FCFS): It is the simplest scheduling algorithm. It queues tasks in the order that they arrive in the ready queue. According to this algorithm, the task that arrives first will be executed first and the next task will be started to be executed only after the previous task gets fully executed.

An error that can be caused during task scheduling is starvation. Starvation is a problem encountered in concurrent computing where a process is perpetually denied necessary resources to process its work [5]. It is a phenomenon associated with priority task scheduling algorithms, in which a task can wait indefinitely for execution due to low priority. For example, in SJF scheduling algorithm, a shorter process, which arrives later in the waiting queue than a task with longer execution time, will be chosen for execution firstly. So, the longer task has to wait for a long time until all the tasks with shorter execution time to be served. Therefore, there is the possibility the task with longer execution time will never be able to get the share of CPU. Researchers have made attempts to minimize starvation [6, 7]. A possible solution to starvation is to use a scheduling algorithm with priority queue that also uses the aging technique. Aging is a technique of gradually increasing the priority of tasks that wait in the system for a long time [6]. For avoiding a task to wait in the system for a long time, we can use fuzzy logic.

In literature review there is a significant amount of works that consider the use of fuzzy logic in task scheduling. Particularly, in [8] the authors propose a new improved scheduling algorithm technique based on Fuzzy Logic. In [9, 10] the authors have incorporated fuzzy techniques for enhancing a priority based scheduling algorithm. Furthermore, Shanmugasundaram and Venkatesh [11] have studied the priority queueing model under fuzzy environment. Also, Behera et al. [12] have proposed an improved fuzzy-based CPU scheduling algorithm for reducing the waiting time and turnaround time in relation to others scheduling algorithms. In addition, an intuitionistic fuzzy rule-based decision-making system for an operating system process scheduler has been proposed [13]. Also, in [14] the idea of achieving an ideal CPU time slice using fuzzy logic was given. However, none of them have combined execution time with the waiting time in a fuzzy scheduling rule-based method in order to ensure that priority is given not only to tasks with short execution time, but also to tasks that remain into the queue waiting to be executed for a long time, minimizing, in this way, the phenomenon of starvation.

14.3 The Proposed Fuzzy Scheduling Algorithm

14.3.1 Criteria of Scheduling and Fuzzy Sets

The presented fuzzy scheduling algorithm is based on the execution time (ET) (or remaining execution time for pre-emptive scheduling) and the waiting time (WT) of the tasks that are waiting in the queue for serving. For both criteria we use the following four fuzzy sets: low (L), medium (M), high (H) and very high (VH). The ET (and WT) of a task is characterized by the quadruplet (μ_L, μ_M, μ_H, μ_{VH}) that defines which fuzzy set (or sets) is (are) active each time for the particular task. The membership functions and their partition (Fig. 14.1) are depicted below.

$$\mu_L(x) = \begin{cases} 1, & x \leq 40 \\ 1 - \frac{x-40}{45-40}, & 40 < x < 45 \\ 0, & x \geq 45 \end{cases}$$

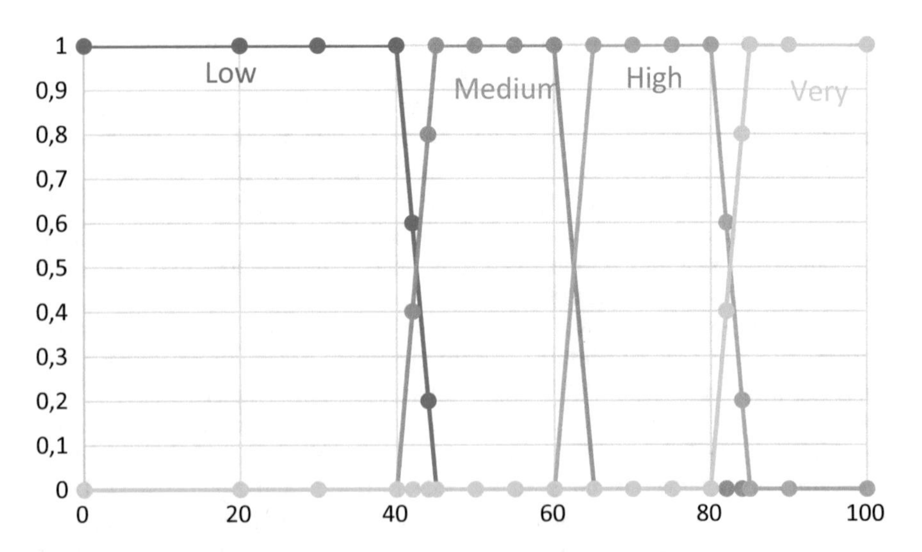

Fig. 14.1 The partition of the fuzzy sets

$$\mu_M(x) = \begin{cases} \frac{x-40}{45-40}, & 40 < x < 45 \\ 1, & 45 \le x \le 60 \\ 1 - \frac{x-60}{65-60}, & 60 < x < 65 \\ 0, & x \le 40 \text{ or } x \ge 65 \end{cases}$$

$$\mu_H(x) = \begin{cases} \frac{x-60}{65-60}, & 60 < x < 65 \\ 1, & 65 \le x \le 80 \\ 1 - \frac{x-80}{85-80}, & 80 < x < 85 \\ 0, & x \le 60 \text{ or } x \ge 85 \end{cases}$$

$$\mu_{VH}(x) = \begin{cases} \frac{X-80}{85-80}, & 80 < x < 85 \\ 1, & 85 \le x \le 100 \\ 0, & x \le 80 \end{cases}$$

X is EX (or WT) calculated as percentage. Particularly, x ε [0, 100]. To calculate x as percentage, we consider that the ET (or WT) of the task with the maximum duration (or waiting time) is 100 and proportionally we calculate the ET (or WT) of the rest tasks that are in the queue waiting to be executed. For example, lets the tasks T_0, T_1, T_2 and T_3 with execution times: 5, 3, 8 and 6 accordingly. Then ET for T_0 is 62.5, ET for T_1 is 37.5, ET for T_2 is 100 and ET for T_3 is 75.

14.3.2 *Fuzzy Rules*

The operation of the presented scheduling method is based on a rule-based reasoner that decides dynamically about the priority of the tasks that wait to be executed. This reasoner is triggered each time a change occurs (i.e. the execution of a task ends, a new task arrives etc.). The reasoner takes as input both the duration (EX) and the waiting time (WT) in the queue for each task and decides which task has to be executed each time, calculating the priority (Pr) of each task. Pr can take one of the following four values: 'Very High' (VH), 'High' (H), 'Medium' (M) and 'Low' (L). The defined fuzzy rules are the following:

1. If EX = 'Low' then Pr = 'Very High'.
2. If EX = 'Medium' and WT = 'Very High' then Pr = 'Very High'.
3. If EX = 'Medium' and WT = 'High' then Pr = 'High'.
4. If EX = 'Medium' and WT = 'Medium' or 'Low' then Pr = 'Medium'.
5. If EX = 'Very High' or 'High' and WT = 'Very High' then Pr = 'High'.
6. If EX = 'Very High' or 'High' and WT = 'High' then Pr = 'Medium'.
7. If EX = 'Very High' or 'High' and WT = 'Medium' or 'Low' then Pr = 'Low'.
8. After the calculation of the tasks' priority by applying the above rules, if we have two tasks T_a and T_b with the same priority, then we calculate $z_a = EX_b + WT_a$ and $z_b = EX_a + WT_b$ and the task with the maximum z is executed first.

14.3.3 Example of Operation

In this section, we present two examples (a pre-emptive and a non pre-emptive scheduling example) of application of the fuzzy scheduling rule-based method to a set of tasks. In Table 14.1, a set of tasks is presented. The table gives us information about the arrival time and the execution time of each task. We apply the presented fuzzy rule-based scheduling algorithm on these tasks for non pre-emptive scheduling.

At time $t = 0$: At the particular time, only T_0 is available for execution. So, T_0 is starting to be executed.

At time $t = 5$: The execution of T_0 finished and T_1, T_2 and T_3 are waiting in the queue. In Table 14.2, we can see the results of the fuzzy calculation and the definition of the tasks' priority according to the rules that are presented above. We notice that T_1 is being scheduled to be executed.

At time $t = 8$: The execution of T_1 finished and T_2 and T_3 are waiting in the queue. In Table 14.3, we can see the results of the fuzzy calculation and the definition of the tasks' priority according to the rules that are presented above. We notice that T_2 and T_3 have the same priority. Therefore, we calculate z for both tasks. For T_2 is $z_2 = EX_3 + WT_2 = 6 + 6 = 12$ and for T_3 is $z_3 = EX_2 + WT_3 = 8 + 5 = 13$. Consequently, according to the rules T_3 is being executed firstly.

In Fig. 14.2 the sequence of task execution according to the presented fuzzy mechanism is presented.

Below, a pre-emptive operation of the presented fuzzy rule-based scheduling method is presented. In Table 14.4, the example's set of tasks is presented. The table gives us information about the arrival time and the execution time of each task.

At time $t = 0$: At the particular time, only T_0 is available for execution. So, T_0 is starting to be executed.

At time $t = 1$: T_1 is arriving and the execution of T_0 is stopping. In Table 14.5, we can see the results of the fuzzy calculation and the definition of the tasks' priority according to the rules that are presented above. We notice that T_1 is being scheduled to be executed, now.

At time $t = 2$: T_2 is arriving and the execution of T_1 is stopping. In Table 14.6, we can see the results of the fuzzy calculation and the definition of the tasks' priority according to the rules that are presented above. We notice that T_1 and T_2 have the same priority. Therefore, we calculate z for both tasks. The value of z for T_1 is 6 and for T_2 is 2. Consequently, according to the rules T_1 is continuing to be executed.

Table 14.1 A set of four tasks

Task	Arrive time	Execution time
T_0	0	5
T_1	1	3
T_2	2	8
T_3	3	6

Table 14.2 The tasks in the waiting queue at t = 5

Task	Arrive time	Exec. time	Waiting time	Exec. time (%)	Waiting time (%)	ET	WT	Rule	Pr
T_0	–	–	–	–	–	–	–	–	–
T_1	1	3	4	37.5	100	L	VH	1	VH
T_2	2	8	3	100	75	VH	H	6	M
T_3	3	6	2	75	50	H	M	7	L

Table 14.3 The tasks in the waiting queue at $t = 8$

Task	Arrive time	Exec. time	Waiting time	Exec. time (%)	Waiting time (%)	ET	WT	Rule	Pr
T_0	–	–	–	–	–	–	–	–	–
T_1	–	–	–	–	–	–	–	–	–
T_2	2	8	6	100	100	VH	VH	5	H
T_3	3	6	5	75	83.33	H	VH	5	H

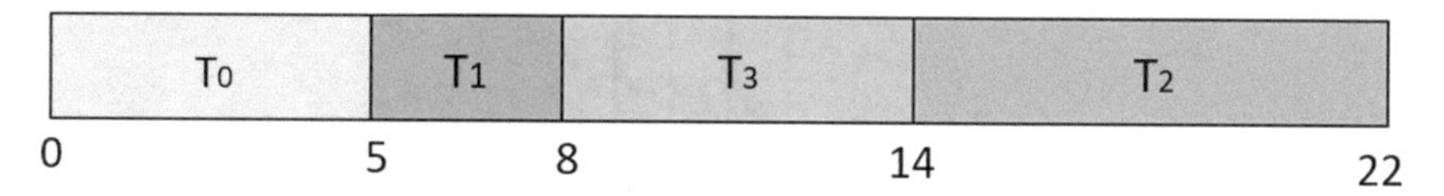

Fig. 14.2 The sequence of task execution

Table 14.4 Another example of a set of tasks

Task	Arrive time	Execution time
T_0	0	21
T_1	1	3
T_2	2	6
T_3	3	2

At time t = 3: T_3 is arriving and the execution of T_1 is stopping. In Table 14.7, we can see the results of the fuzzy calculation and the definition of the tasks' priority according to the rules that are presented above. We notice that T_1, T_2 and T_3 have the same priority. Therefore, we calculate z for the three tasks. For T_1 and T_2 is $z_1 = EX_2 + WT_1 = 6 + 0 = 6$ and $z_2 = EX_1 + WT_2 = 1 + 1 = 2$. So, T_1 has to be executed before T_2. Then, we calculate z for T_1 and T_3. For T_1 $z_1 = EX_3 + WT_1 = 2 + 0 = 2$ and for T_3 is $z_3 = EX_1 + WT_3 = 1 + 0 = 1$. Consequently, T_1 is continuing to be executed.

At time t = 4: The execution of T_1 finished. In Table 14.8, we can see the results of the fuzzy calculation and the definition of the tasks' priority according to the rules that are presented above. We notice that T_2 and T_3 have the same priority. Therefore, we calculate z for both tasks. For T_2 is $z_2 = EX_3 + WT_2 = 2 + 2 = 4$ and for T_3 is $z_3 = EX_2 + WT_3 = 6 + 1 = 7$. Consequently, T_3 is being scheduled to be executed.

At time t = 6: The execution of T_3 finished. In Table 14.9, we can see the results of the fuzzy calculation and the definition of the tasks' priority according to the rules that are presented above. We notice that T_2 is being scheduled to be executed.

In Fig. 14.3 the sequence of task execution according to the presented fuzzy mechanism is depicted.

14.4 Evaluation

For the evaluation of the presented fuzzy rule-based scheduling method, it was compared with First Come First Served (FCFS) scheduling algorithm and with Shortest Job First (SJF) scheduling algorithm (for its non pre-emptive and pre-emptive operation) for a variety of different sets of tasks. After the application of the above referenced scheduling algorithms to the sets of tasks, we calculated the mean waiting time, the mean turnaround time and the mean waiting time for tasks with large duration. The results are presented below.

Table 14.5 The tasks in the waiting queue at t = 1

Task	Arrive time	Exec. time	Waiting time	Exec. time (%)	Waiting time (%)	ET	WT	Rule	Pr
T_0	0	20	0	100	0	VH	L	7	L
T_1	1	3	0	15	0	L	L	1	VH

Table 14.6 The tasks in the waiting queue at t = 2

Task	Arrive time	Exec. time	Waiting time	Exec. time (%)	Waiting time (%)	ET	WT	Rule	Pr
T_0	0	20	1	100	100	VH	VH	5	H
T_1	1	2	0	15	0	L	L	1	VH
T_2	2	6	0	30	0	L	L	1	VH

Table 14.7 The tasks in the waiting queue at t = 3

Task	Arrive time	Exec. time	Waiting time	Exec. time (%)	Waiting time (%)	ET	WT	Rule	Pr
T_0	0	20	2	100	100	VH	VH	5	H
T_1	1	1	0	5	0	L	L	1	VH
T_2	2	6	1	30	50	L	M	1	VH
T_3	2	2	0	10	0	L	L	1	VH

Table 14.8 The tasks in the waiting queue at t = 4

Task	Arrive time	Exec. time	Waiting time	Exec. time (%)	Waiting time (%)	ET	WT	Rule	Pr
T_0	0	20	3	100	100	VH	VH	5	H
T_1	–	–	–	–	–	–	–	–	–
T_2	2	6	2	30	66.67	L	H	1	VH
T_3	2	2	1	10	33.33	L	L	1	VH

Table 14.9 The tasks in the waiting queue at t = 6

Task	Arrive time	Exec. time	Waiting time	Exec. time (%)	Waiting time (%)	ET	WT	Rule	Pr
T_0	0	20	5	100	100	VH	VH	5	H
T_1	–	–	–	–	–	–	–	–	–
T_2	2	6	4	30	80	L	H	1	VH
T_3	–	–	–	–	–	–	–	–	–

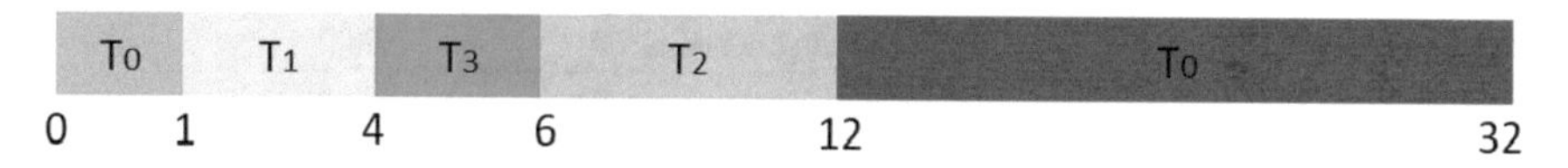

Fig. 14.3 The sequence of task execution

Table 14.10 First set of tasks

Task	Arrive time	Execution time
T_0	0	2
T_1	1	15
T_2	4	4
T_3	4	12
T_4	5	6

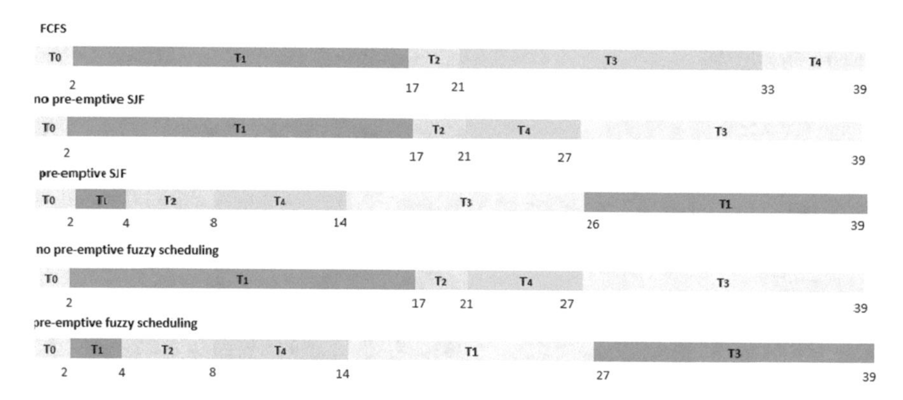

Fig. 14.4 The sequence of tasks execution for the first set of tasks

No 1 set of tasks: In Table 14.10, the first set of tasks is presented. The table gives us information about the arrival time and the duration of each task. In Fig. 14.4 the sequence of task execution according to FCFS, SJF and the presented fuzzy scheduling mechanism are depicted. The comparative results are presented in Table 14.11.

No 2 set of tasks: In Table 14.12, the second set of tasks is presented. The table gives us information about the arrival time and the duration of each task. In Fig. 14.5 the sequence of task execution according to FCFS, SJF and the presented fuzzy scheduling mechanism are depicted. The comparative results are presented in Table 14.13.

No 3 set of tasks: In Table 14.14, the third set of tasks is presented. The table gives us information about the arrival time and the duration of each task. In Fig. 14.6 the sequence of task execution according to FCFS, SJF and the presented fuzzy scheduling mechanism are depicted. The comparative results are presented in Table 14.15.

Table 14.11 The comparative results for the first set of tasks

		Mean waiting time	Mean turnaround time	Waiting time of the task with the maximum duration
FCFS		11.8	19.6	1
SJF	No pre-emptive	10.6	18.4	1
	Pre-emptive	7.2	15	23
Fuzzy scheduling	No pre-emptive	10.6	18.4	1
	Pre-emptive	7.4	15.2	11

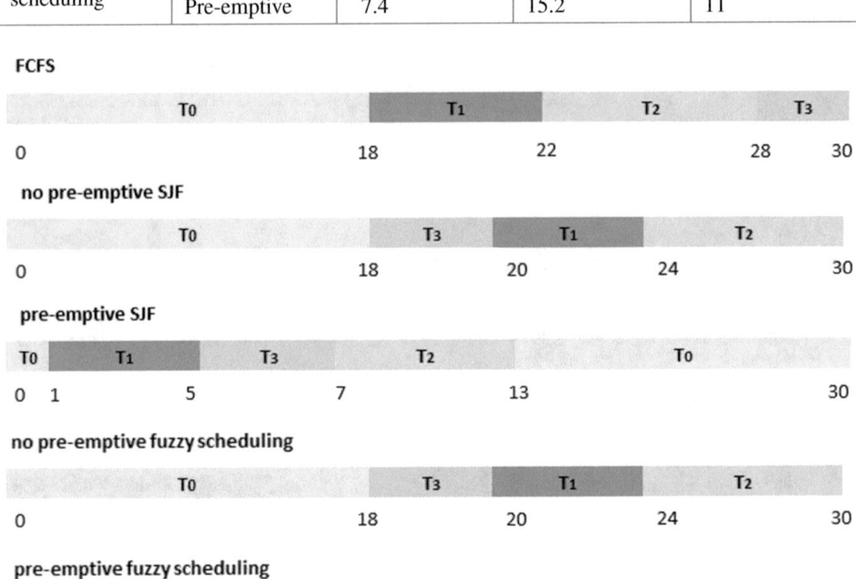

Fig. 14.5 The sequence of tasks execution for the second set of tasks

Table 14.12 Second set of tasks

Task	Arrive time	Execution time
T_0	0	18
T_1	1	4
T_2	1	6
T_3	3	2

Table 14.13 The comparative results for the second set of tasks

		Mean waiting time	Mean turnaround time	Waiting time of the task with the maximum duration
FCFS		15.75	23.25	0
SJF	No pre-emptive	14.25	21.75	0
	Pre-emptive	5	12.5	12
Fuzzy scheduling	No pre-emptive	14.25	21.75	0
	Pre-emptive	5	12.5	12

Table 14.14 Third set of tasks

Task	Arrive time	Execution time
T_0	0	2
T_1	1	15
T_2	3	4
T_3	3	12

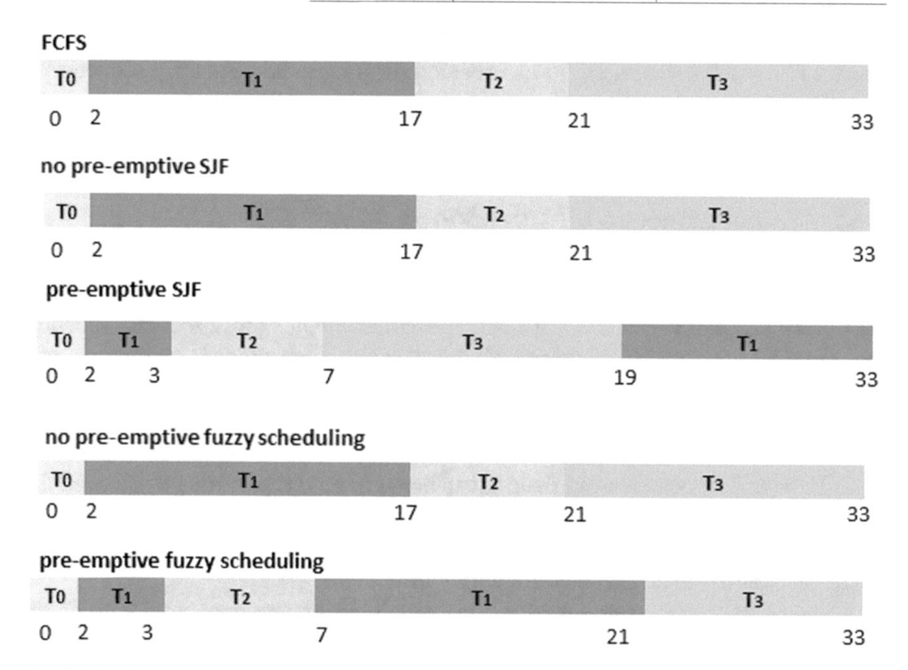

Fig. 14.6 The sequence of tasks execution for the third set of tasks

Table 14.15 The comparative results for the third set of tasks

		Mean waiting time	Mean turnaround time	Waiting time of the task with the maximum duration
FCFS		8.5	16.5	1
SJF	No pre-emptive	8.5	16.5	1
	Pre-emptive	5.25	13.5	17
Fuzzy scheduling	No pre-emptive	8.5	16.5	1
	Pre-emptive	5.75	14	5

Taking into consideration the above, we conclude that the proposed fuzzy rule-based scheduling algorithm has the same or almost the same results with SJF algorithm, concerning the mean waiting time and the mean turnaround time. However, the waiting time of the task with the longer execution time is lower when we apply the presented fuzzy scheduling algorithm rather than when we apply SJF scheduling algorithm. Therefore, the presented fuzzy rule-based scheduling algorithm ensures that priority is given not only to tasks with short execution time, but also to tasks that remain into the queue waiting to be executed for a long time. As a consequence, it can be used to eliminate starvation.

14.5 Conclusions and Future Work

In this paper we presented a fuzzy rule-based scheduling method with aim to handle the uncertainty and vagueness that characterize the complex process of task scheduling. We used fuzzy sets to describe both the execution time and the waiting time of a task, which is waiting to the queue for execution. Also, we used a set of fuzzy rules over the defined fuzzy sets in order to define the priority of execution. The presented fuzzy rule-based scheduling method tries to merge the FCFS and SJF scheduling algorithms and to benefit from their advantages. In this way, it tries to succeed the minimum average waiting time, like SJF algorithm, and at the same time, to give priority not only to tasks with short execution time, but also to tasks that remain into the queue waiting to be executed for a long time.

The presented fuzzy rule-based scheduling algorithm was compared with FCFS and SJF algorithms for a variety of sets of tasks. The evaluation results showed that the presented fuzzy scheduling method has the same or almost the same results, concerning the average waiting time and the average turnaround time, with SJF algorithm, but it succeeds lower waiting time for the task with the longer execution time. Therefore, the presented fuzzy rule-based scheduling algorithm can be used to eliminate starvation.

In future work, a more thorough evaluation is going to be conducted. Also, in our future plans we aim to investigate how artificial techniques other than fuzzy logic can be used in the complex process of task scheduling for improving it.

References

1. J. Wang, Task scheduling, in *Real-Time Embedded Systems* (Wiley Inc., 2017, pp. 53–98)
2. L.A. Zadeh, Fuzzy logic = Computing with words. IEEE Trans. Fuzzy Syst. **4**(2), 103–111 (1996)
3. N. Goel, R.B. Garg, A comparative study of CPU scheduling algorithms. Int. J. Graph. Image Process. **2**(4), 245–251 (2012)
4. P. Singh, V. Singh, A. Pandey, Analysis and comparison of CPU scheduling algorithms. Int. J. Emerg. Technol. Adv. Eng. **4**(1), 91–95 (2014)
5. A. Tanenbaum, *Modern Operating Systems* (Prentice Hall, 2001, pp. 184–185). ISBN 0-13-092641-8.
6. S. Zhang, Z. Qian, H. Wu, S. Lu, Efficient data center flow scheduling without starvation using expansion ratio. IEEE Trans. Parall. Distr. **28**(11), 3157–3170 (2017)
7. S. Tyagi, S. Choudhary, A. Poonia, Enhanced priority scheduling algorithm to minimize process starvation. Int. J. Emerg. Technol. Adv. Eng. **2**(10), 288–294 (2012)
8. P.K. Varshney, N. Akhtar, M.F.H. Siddiqui, Efficient CPU scheduling algorithm using fuzzy logic, in *Proceedings of 2012 International Conference on Computer Technology and Science (ICCTS 2012)* (2012)
9. H. Nirmala, H.A. Grammar, Priority based scheduling algorithm using fuzzy technique, in *International Conference on Computational Systems and Information Systems for Sustainable Solutions, Karnataka, India* (2016)
10. B. Alamo, M.N. Doja, R. Biswas, M. Alam, Fuzzy priority CPU scheduling algorithm. Int. J. Comput. Sci. Issues **8**, 386–390 (2011)
11. S. Shanmugasundaram, B. Venkatesh, Fuzzy retrial queues with priority using DSW algorithm. Int. J. Comput. Eng. Res. (IJCER) **06**, 18–23 (2016)
12. H.S. Behera, R. Pitahaya, P. Mallick, An improved fuzzy-based CPU scheduling (IFCS) algorithm for real time systems. Int. J. Soft Comput. Eng. **2**, 326–331 (2012)
13. M.A. Butt, M. Akram, A new intuitionistic fuzzy rule-based decision-making system for an operating system process scheduler. SpringerPlus. **5**, 1547 (2016). https://doi.org/10.1186/s40064-016-3216-z
14. S. Varshney, N. Akhtar, Efficient CPU scheduling algorithm using fuzzy logic, in *Proceedings of International Conference on Computer Technology Science* (2012)

Chapter 15
Computational Intelligence and Combinatorial Optimization Problems in Transportation Science

Manolis N. Kritikos and Pantelis Z. Lappas

Abstract The purpose of this chapter is to highlight the use of computational intelligence algorithms for solving a special class of combinatorial optimization problems in transportation science called routing problems. Classical routing problems, such as the Traveling Salesman Problem and the Vehicle Routing Problem, as well as highly relevant extensions of classical routing problems like Vehicle Routing Problem with Time Windows and Inventory Routing Problem have received a great deal of attention from academics, consultants and practitioners in the field of Supply Chain Management. The contribution of this study is fourfold: (i) it provides a comprehensive review of various solution algorithms that have been proposed in literature as possible solutions to many of the complex issues surrounding routing problem management, (ii) it presents formulation schemes related to basic routing problems and their extensions, (iii) it promotes the use of selected computational algorithms called meta-heuristics and sim-heuristics by providing various graphical presentation formats so as to simplify complicated issues and convey meaningful insights into the routing problems and (iv) it points out interesting research directions for further development in routing problems.

Keywords Computational intelligence · Combinatorial optimization · Routing problem · Genetic algorithm · Variable neighborhood search algorithm · K-means clustering algorithm · Sim-heuristic algorithm

---

M. N. Kritikos (✉)
Management Science Laboratory, Department of Management Science and Technology, Athens University of Economics and Business, Athens, Greece
e-mail: kmn@aueb.gr

P. Z. Lappas
Stochastic Modeling and Applications Laboratory, Department of Statistics, Athens University of Economics and Business, Athens, Greece
e-mail: pzlappas@aueb.gr

G. A. Tsihrintzis and M. Virvou (eds.), *Advances in Core Computer Science-Based Technologies*, Learning and Analytics in Intelligent Systems 14,
https://doi.org/10.1007/978-3-030-41196-1_15

15.1 Introduction

Supply Chain Optimization (ScO) is one of the most successful applications of operations research modeling in Supply Chain Management (ScM). ScM can be presented as a continuous cycle of activities related to supply, demand, product and information management. Nowadays, ScM aims at integrating and coordinating different supply chain activities within the Supply Chain Network (ScN) so as to maximize the economical benefits of a firm. According to Tan [161] there are three alternative perspectives on ScM: (a) the purchasing and supply perspective of the industrial buyers, (b) the transportation and logistics perspective of the merchants and (c) the unified/integrated ScM strategy that integrates production scheduling, inventory allocation and distribution activities. In particular, ScM can be viewed as a three-level decision-making process including strategic, tactical and operational decisions [148]. The strategic level describes the supply chain network design problem where facility locations, production technologies and plant capacities should be determined. The tactical level reflects the production planning and inventory allocation problems where material flow management policies, assembly policies, inventory policies and lot sizes should be defined. The operational level depicts the distribution planning problem where operations should be scheduled to assure in-time delivery of final products to demand points.

Routing problems are central combinatorial optimization problems in operations research. They have received a great deal of attention from academics, consultants and practitioners in transportation community. Routing problem (RP) or transportation problem is the generic name given to a whole class of problems in which transportation is necessary. The result of an analysis of the scientific and managerial literature led to the identification of four main routing problems in the overall field of ScM where tactical and operational decisions are taken into account on a case-by-case basis: (1) the Traveling Salesman Problem (TSP), (2) the Vehicle Routing Problem (VRP), (3) the Inventory Routing Problem (IRP) and more rarely (4) the Production Routing Problem (PRP). Analytically, TSP and VRP are associated with operational level, while IRP and PRP are related to tactical and operational levels. In practice, routing problems are modeled based on their ScN topology. The number of plants or suppliers and customers may vary, and therefore the structure can be (a) one-to-one, (b) one-to-many, (c) many-to-one or (d) many-to-many.

Assuming a one-to-many ScN topology, the TSP is the most basic routing problem where a vehicle starts from a depot, visits all customers exactly once, and returns to the depot [66]. If the depot consists of m vehicles, then the Multiple TSP (m-TSP) can be assumed as the generalization of the TSP (i.e., TSP is the m-TSP where m is equal to 1) [28]. Usually, customers have a deterministic or stochastic demand, whereas the depot contains a fleet of vehicles with limited capacity. This is the VRP which generalizes the m-TSP [56]. Each of the m vehicles starts from a depot, serves a unique cluster of customers exactly once, and returns to the depot. Each vehicle, based on its capacity, can satisfy the total demand occurs from the customers of the corresponding cluster. The IRP is an extension of the VRP and arises from the application of the Vendor Managed Inventory (VMI) concept, where the supplier (vendor) integrates routing and inventory control decisions [8, 33]. In a typical IRP, a single-product type has to be delivered by a fleet of capacitated vehicles to a number of geographically dispersed customers for a given planning horizon. Therefore, as opposed to VRP, the IRP deals with a longer horizon (e.g., a sequence of days) and the supplier monitors the inventory levels of the customers so as to determine the delivery times, the delivery quantities and the set of routes. Involving the added complexity that every customer should be served within a given time window, the VRP with Time Windows (VRPTW) and the IRP with Time Windows (IRPTW) are generalizations of the VRP and IRP, respectively [112, 154]. The PRP can be assumed to be the generalization of the IRP where production planning, inventory allocation and routing decisions have to be made simultaneously [3].

Due to the NP-hard nature of the routing problems [132] computational intelligence algorithms have been proposed in the literature so as to solve large-scale problems within a reasonable computation time. Small-scale routing problems are usually addressed using exact solutions (exacts) such as branch-and-bound, branch-and-cut, branch-and-cut-and-price algorithms, dynamic programming, column generation and Lagrange relaxation-based methods. In general, computational intelligence algorithms can be categorized into four levels: (a) classic heuristic algorithms (heuristics) (b) meta-heuristic algorithms (meta-heuristics), (c) math-heuristic algorithms (math-heuristics) and (d) sim-heuristic algorithms (sim-heuristics).

This study provides a comprehensive review of various solution algorithms (i.e., exacts, heuristics, meta-heuristics, math-heuristics and sim-heuristics) that have been proposed in literature as possible solutions to many of the complex issues surrounding routing problem management. It presents formulation schemes related to TSP, m-TSP, VRP, VRPTW, IRP and IRPTW. In addition to this, it promotes the use of selected computational algorithms such as the Genetic Algorithm (GA), the K-means clustering Algorithm (K-means) and the Sim-heuristic algorithm (SH). Various graphical presentation formats are provided so as to simplify complicated issues and convey meaningful insights into the routing problems. Finally, it points out interesting research directions for further development in routing problems. The most promising areas are related to (a) the PRP with Time Windows (PRPTW), (b) the Location Production Routing Problem (LPRP) with operational and disruption risks and (c) the Pollution Inventory Routing Problem (PIRP).

The reminder of the chapter is organized as follows. Problem descriptions and mathematical formulations related to TSP, m-TSP, VRP, VRPTW, IRP and IRPTW are presented in Sect. 15.2. Section 15.3 provides a comprehensive summary of the state of the art in research on solution approaches for solving routing problems, whereas selected meta-heuristics and sim-heuristics are promoted for solving TSP and IRPTW. K-means clustering algorithm is described in order to model m-TSP. In addition, an indicative computational study takes place by establishing testing instances. Finally, in Sect. 15.4, conclusions and future directions are given.

15.2 Routing Problem Descriptions and Mathematical Formulations

This section presents modeling frameworks for formulating the following routing problems: TSP, m-TSP, VRP, VRPTW, IRP and IRPTW. The ScN, used for transportation of products, is generally described through a directed or undirected complete graph [13, 74, 164]. If $G = (N, A)$ is a complete directed graph where $N = \{0, n+1\} \cup \{1, \ldots, n\}$ is the set of nodes and $A = \{(i, j) : i, j \in N, i \neq j\}$ is the set of arcs, the corresponding routing problem is called Asymmetric Routing Problem (ARP). Nodes $1, \ldots, n$ correspond to the customers, whereas 0 and $n+1$ represent the single-depot (depot as origin-depot or destination-depot). If $G = (V, E)$ is a complete undirected graph where $V = \{0, \ldots, n\}$ is the set of vertices and $E = \{(i, j) : i, j \in V, j > i\}$ is the set of edges, the corresponding routing problem is called Symmetric Routing Problem (SRP). Vertices $1, \ldots, n$ correspond to the customers, whereas vertex 0 corresponds to the single-depot (supplier). In the following subsections, asymmetric versions of TSP, m-TSP, VRP, VRPTW and IRPTW as well as, the symmetric version of IRP are given.

15.2.1 *Traveling Salesman Problem (TSP)*

The Asymmetric TSP can be defined on a complete directed graph $G = (N, A)$, where $N = \{1, \ldots, n\}$ is the set of n nodes and $A = \{(i, j) : i, j \in N, i \neq j\}$ is the set of arcs. Nodes are associated with points of the plane having the given coordinates $\left(x_i^{coord}, y_i^{coord}\right) \forall i \in N$. Let c_{ij} the travel cost for each arc $(i, j) \in A$. The travel cost is defined as the Euclidean distance between the two nodes $i, j \in N$. Therefore, $c_{ij} = \sqrt{\left(x_i^{coord} - x_j^{coord}\right)^2 + \left(y_i^{coord} - y_j^{coord}\right)^2}$. Generally, the usage of the loop arc, (i, i), is not allowed, and this is imposed by defining $c_{ii} = +\infty$ for all $i \in N$. Moreover, the cost matrix satisfies the triangle inequality: $c_{ir} + c_{rj} \geq c_{ij}$ for all $i, j, r \in N$. The problem is to find a directed tour (i.e., a Hamiltonian cycle) with a minimal distance. Let a binary variable $\psi_{ij} = 1$ if the arc $(i, j) \in A$ is in the optimal TSP tour (i.e., Hamiltonian cycle), and 0 otherwise. The problem is then formulated as follows:

$$\min\left(\sum_{i\in N}\sum_{j\in N} c_{ij}\psi_{ij}\right) \tag{15.1}$$

$$s.t.$$

$$\sum_{i\in N} \psi_{ij} = 1, \forall j \in N \tag{15.2}$$

$$\sum_{j \in N} \psi_{ij} = 1, \forall i \in N \quad (15.3)$$

$$\sum_{i \in S} \sum_{j \in S} \psi_{ij} \leq |S| - 1, |S| = 2, 3, \ldots, n - 1 \quad (15.4)$$

$$\psi_{ij} \in \{0, 1\} \forall i, j \in N \quad (15.5)$$

The objective function (15.1) expresses the total distance cost. In general, constraints (15.1)–(15.3) and (15.5) correspond to the classical assignment problem (Hiller and Lieberman, 2010). Constraints (15.2) ensure that each node $j \in N$ must be entered exactly once. Constraints (15.3) ensure that each node $i \in N$ must be exited exactly once. Constraints (15.5) are the binary constraints. Constrains (15.4) are known as the Dantzig-Fulkerson-Johnson (DFJ) sub-tour elimination constraints [13]. According to sub-tour elimination scheme, the main goal is to prevent from forming sub-tours and allow forming a tour. To achieve this, the number of arcs must be less than the number of nodes for every subset that consists of 2 to $n - 1$ nodes. Mathematically, S represents any non-empty subset of N and $|S|$ symbolizes the size of set S (i.e., the number of nodes in S).

Instead of using (15.4) constraints (i.e., DFJ sub-tour elimination constraints), TSP can be formulated based on Miller-Tucker-Zemlin (MTZ) sub-tour elimination constraints [13]. This sub-tour elimination scheme considers that a tour is just a sequence of all nodes. Let u_j be the sequence number of node $j \in N$ in a tour. The following set of sub-tour elimination constraints can be obtained:

$$u_i - u_j + n\psi_{ij} \leq n - 1 \forall (i, j) \in A, i \neq 1, j \neq 1, j \neq i \quad (15.6)$$

Table 15.1 represents an illustrative example of non-completed directed graph $G = (N, A)$ with 6 nodes so as to demonstrate key modeling features of DFJ and MTZ sub-tour elimination schemes.

15.2.2 Multiple Traveling Salesman Problem (m-TSP)

The Asymmetric m-TSP can be defined on a complete directed graph $G = (N, A)$ where $N = \{1, \ldots, n\}$ is the set of nodes and $A = \{(i, j) : i, j \in N, i \neq j\}$ is the set of arcs. Node 1 represents the depot (supplier), while nodes of set $C = \{2, \ldots, n\}$ represent the customers that must be visited by m vehicles. The set of identical vehicles is denoted by $K = \{1, \ldots, m\}$. c_{ij} is the travel cost (i.e., Euclidean distance) associated with arc $(i, j) \in A$. Let ψ_{ijk} a binary variable equal to 1 if vehicle $k \in K$ visits customer $j \in N$ immediately after customer $i \in N$, and 0 otherwise. The m-TSP consists of finding tours for all m vehicles, which all start and end at the depot, such that each intermediate customer is visited exactly once and the total cost of visiting all customers is minimized. As a consequence, the m-TSP is a generalization

Table 15.1 Illustrative example for DFJ and MTZ sub-tour elimination constraints

G = (N, A)	Non-empty subsets of N		
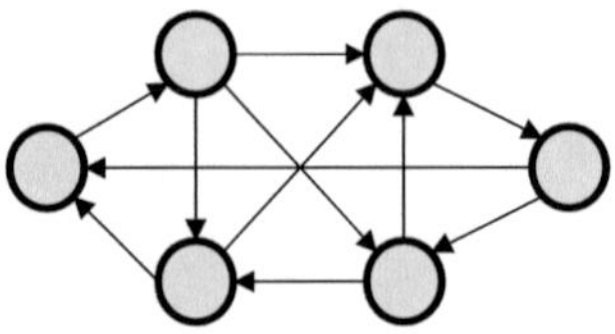	Size of set S	Number of DFJ constraints	Number of MTZ constraints
	$\lvert S\rvert = 2$	0	0
	$\lvert S\rvert = 3$	2	4
	$\lvert S\rvert = 4$	3	4
	$\lvert S\rvert = 5$	2	1

$\lvert S\rvert = 3$	Mathematical constraints
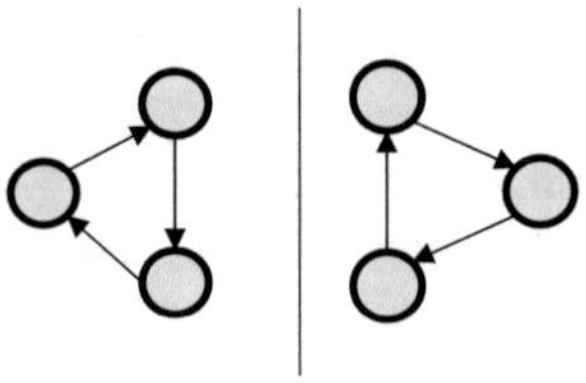	(DFJ) • $\psi_{12} + \psi_{25} + \psi_{51} \leq 2$ • $\psi_{36} + \psi_{64} + \psi_{43} \leq 2$ (MTZ) • $(2, 5) : u_2 - u_5 + 6\psi_{25} \leq 5$ • $(3, 6) : u_3 - u_6 + 6\psi_{36} \leq 5$ • $(6, 4) : u_6 - u_4 + 6\psi_{64} \leq 5$ • $(4, 3) : u_4 - u_3 + 6\psi_{43} \leq 5$
$\lvert S\rvert = 4$	Mathematical constraints
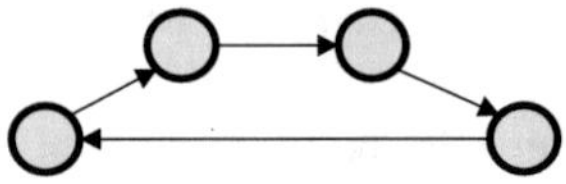	(DFJ) • $\psi_{12} + \psi_{23} + \psi_{36} + \psi_{61} \leq 3$ (MTZ) • $(2, 3) : u_2 - u_3 + 6\psi_{23} \leq 5$
$\lvert S\rvert = 4$	Mathematical constraints
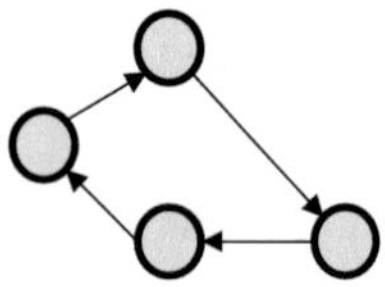	(DFJ) • $\psi_{12} + \psi_{24} + \psi_{45} + \psi_{51} \leq 3$ (MTZ) • $(2, 4) : u_2 - u_4 + 6\psi_{24} \leq 5$ • $(4, 5) : u_4 - u_5 + 6\psi_{45} \leq 5$
$\lvert S\rvert = 4$	Mathematical constraints
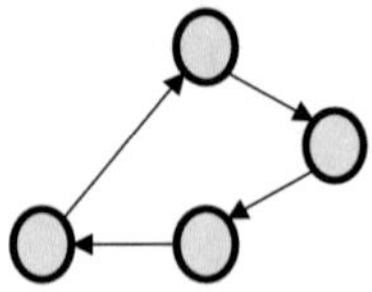	(DFJ) • $\psi_{36} + \psi_{64} + \psi_{45} + \psi_{53} \leq 3$ (MTZ) • $(5, 6) : u_5 - u_6 + 6\psi_{56} \leq 5$
$\lvert S\rvert = 5$	Mathematical constraints
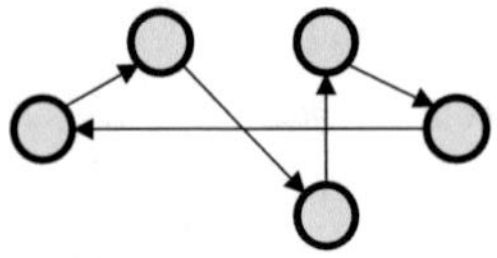	(DFJ) • $\psi_{12} + \psi_{24} + \psi_{43} + \psi_{36} + \psi_{61} \leq 4$
$\lvert S\rvert = 5$	Mathematical constraints
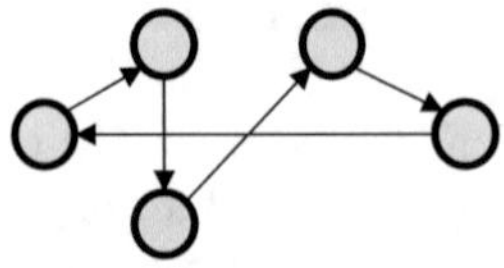	(DFJ) • $\psi_{12} + \psi_{25} + \psi_{53} + \psi_{36} + \psi_{61} \leq 4$ (MTZ) • $(5, 3) : u_5 - u_3 + 6\psi_{53} \leq 5$

of the TSP with a single vehicle ($m = 1$) where all customers should be visited exactly once. The vehicle starts and ends at the depot, such that the total cost of visiting all nodes is minimized (i.e., find the Hamiltonian cycle with a minimal distance).

The Asymmetric m-TSP can be formulated as follows:

$$\min\left(\sum_{i\in N}\sum_{j\in N} c_{ij} \sum_{k\in K} \psi_{ijk}\right) \tag{15.7}$$

$$s.t.$$

$$\sum_{i\in N}\sum_{k\in K} \psi_{ijk} = 1, \forall j \in N \tag{15.8}$$

$$\sum_{i\in N} \psi_{igk} - \sum_{j\in N} \psi_{gjk} = 0, \forall g \in N, \forall k \in K \tag{15.9}$$

$$\sum_{j\in N} \psi_{1jk} = 1, \forall k \in K \tag{15.10}$$

$$u_i - u_j + n \sum_{k\in K} \psi_{ijk} \leq n - 1, \forall i, j \in C, i \neq j \tag{15.11}$$

$$\psi_{ijk} \in \{0, 1\} \forall i, j \in N, \forall k \in K \tag{15.12}$$

The objective function (15.7) minimizes the total transportation costs. Constraints (15.8) state that each customer should be visited exactly once. Constraints (15.9) ensure that once a vehicle visits a customer, then it must also depart from the same customer. Constraints (15.10) ensure that each vehicle is used exactly once. Constraints (15.11) are the extensions of MTZ sub-tour elimination constraints to a three-index decision variable ψ_{ijk} for all $i, j \in N$ and for all $k \in K$. Constraints (15.12) are the binary constraints.

15.2.3 Vehicle Routing Problem (VRP)

The Asymmetric VRP generalizes the Assymetric m-TSP. It can be defined on a complete directed graph $G = (N, A)$ where $N = \{0, n + 1\} \cup \{1, \ldots, n\}$ is the set of nodes and $A = \{(i, j) : i, j \in N, i \neq j\}$ is the set of arcs. Nodes $i = 1, \ldots, n$ correspond to the customers, whereas node 0 corresponds to the depot (supplier). Sometimes in order to simplify modeling the depot is associated with node $n + 1$. Therefore, node 0 and $n + 1$ symbolize the origin-depot and the destination-depot, respectively. The set of arcs represents connections between the depot and the customers and among customers. No arc terminates in node 0, and no arcs originate from $n + 1$. c_{ij} is the transportation cost (i.e., Euclidean distance) associated with arc $(i, j) \in A$. There are m homogenous vehicles, each with capacity Q, which are located at the depot. The fleet of homogenous vehicles is denotes by set $K = \{1, \ldots, m\}$. Each customer $i \in C = N \backslash \{0, n+1\} = \{1, \ldots, n\}$ is associated with a non-negative demand $d_i \forall i \in C$, where $d_0 = 0$. It should be noticed that the total demand of all customers on any single route may not exceed the vehicle's capacity. Binary variables ξ_{ik} are used to assign customers to vehicles, with value 1 indicating that customer $i \in C$ will be visited by vehicle $k \in K$ (0 otherwise). Let ψ_{ijk} a binary variable equal to 1 if vehicle $k \in K$ traverse arc $(i, j) \in A$, and 0 otherwise. The VRP consists of finding a set of K cost-minimal routes, each starting and ending at the depot, such that every customer is visited exactly once and its demand is fully satisfied. Moreover, it is assumed that the supplier has a sufficient supply of items that can cover all customers' demand.

The Asymmetric VRP can be formulated as follows:

$$\min \left(\sum_{i \in N} \sum_{j \in N} c_{ij} \sum_{k \in K} \psi_{ijk} \right) \tag{15.13}$$

$$s.t.$$

$$\sum_{k \in K} \xi_{ik} = 1, \forall i \in C \tag{15.14}$$

$$\sum_{k \in K} \xi_{0k} = m \tag{15.15}$$

$$\sum_{j \in N} \psi_{ijk} = \xi_{ik}, \forall i \in N, \forall k \in K \tag{15.16}$$

$$\sum_{j \in N} \psi_{jik} = \xi_{ik}, \forall i \in N, \forall k \in K \tag{15.17}$$

$$u_{ik} - u_{jk} + Q\psi_{ijk} \leq Q - d_j, \forall i, j \in C, i \neq j, such that d_i + d_j \leq Q, \forall k \in K \tag{15.18}$$

$$d_i \leq u_{ik} \leq Q \forall i \in C, \forall k \in K \tag{15.19}$$

$$\psi_{ijk} \in \{0, 1\} \forall i, j \in N, \forall k \in K \tag{15.20}$$

$$\xi_{ik} \in \{0, 1\} \forall i \in N, \forall k \in K \tag{15.21}$$

The objective function (15.13) expresses the total cost. Constraints (15.14)–(15.17) impose that each customer is visited exactly once, that m vehicles leave the depot, and that the same vehicle enters and leaves a given customer, respectively. Constraints (15.18) and (15.19) are the (MTZ) sub-tour elimination constraints and capacity constraints, respectively. Constraints (15.20) and (15.21) are the binary constraints.

15.2.4 *Vehicle Routing Problem with Time Windows (VRPTW)*

The asymmetric VRPTW can be formulated as an extension of the asymmetric VRP involving the added complexity that every customer should be served within a given time window. Analytically, each customer $i \in C$ is associated with a time interval (i.e., a time window) $[e_i, l_i]$ where $e_i \leq l_i \ \forall i \in C$. The service of each customer must start within a time window, and the vehicle must stop at the demand point (i.e., the customer) for s_i time instants, where $0 \leq s_i \leq l_i - e_i \forall i \in C$. The depot has also time windows $[e_o, l_o]$ and $\left[e_{n+1}, l_{n+1}\right]$ where $e_0 = e_{n+1}$ and $l_0 = l_{n+1}$. Regarding the service time of the depot: $s_0 = s_{n+1} = 0$. It should be noticed that, a vehicle $k \in K$ must arrive at the customer $i \in C$ before l_i. In case of early arrival at the location of customer $i \in C$ (i.e., before e_i), the customer will not be serviced before, while the vehicle generally is allowed to wait until time instant e_i. The time windows associated with the depot represent the earliest possible departure from the depot as well the latest possible return time at the depot, respectively. An extra decision variable α_{ik} is defined for each customer $i \in C$ and each vehicle $k \in K$ for denoting the time instant that a vehicle $k \in K$ starts to service customer $i \in C$. With each arc $(i, j) \in A$, where $i \neq j$, is associated a travel cost c_{ij} and a travel time t_{ij} satisfying the triangle inequality. Therefore, the VRPTW consist of finding a set of K cost-minimal routes, each starting and ending at the depot, such that every customer is visited exactly once, its demand is fully satisfied and the time windows are observed.

The Asymmetric VRPTW can be formulated as follows:

$$\min\left(\sum_{i\in N}\sum_{j\in N} c_{ij}\sum_{k\in K}\psi_{ijk}\right) \tag{15.22}$$

$$s.t.$$

$$\sum_{k\in K}\xi_{ik} = 1, \forall i \in C \tag{15.23}$$

$$\sum_{k\in K}\xi_{0k} = m \tag{15.24}$$

$$\sum_{k\in K}\xi_{n+1,k} = m \tag{15.25}$$

$$\sum_{j\in N}\psi_{jik} = \xi_{ik}, \forall i \in N\backslash\{0\}, \forall k \in K \tag{15.26}$$

$$\sum_{j \in N} \psi_{ijk} = \xi_{ik}, \forall i \in N \backslash \{n+1\}, \forall k \in K \tag{15.27}$$

$$\sum_{i \in N} d_i \xi_{ik} \leq Q, \forall k \in K \tag{15.28}$$

$$\alpha_{ik} + s_i + t_{ij} \leq \alpha_{jk} + M\big(1 - \psi_{ijk}\big), \forall i, j \in N, \forall k \in K \tag{15.29}$$

$$\alpha_{ik} \geq e_i \xi_{ik} \forall i \in \mathrm{N}, \forall k \in K \tag{15.30}$$

$$\alpha_{ik} \leq l_i \xi_{ik} \forall i \in \mathrm{N}, \forall k \in K \tag{15.31}$$

$$\psi_{ijk} \in \{0, 1\} \forall i, j \in N, \forall k \in K \tag{15.32}$$

$$\xi_{ik} \in \{0, 1\} \forall i \in N, \forall k \in K \tag{15.33}$$

The objective function (15.22) expresses the total transportation cost. Constraints (15.23)–(15.28) and (15.32), (15.33) are analogous to the VRP model presented before. However, due to the fact the sub-tour elimination constraints are no longer needed, constraints (15.29) ensure the feasibility in terms of the time necessary when traveling from node i to node j, $\forall i, j \in N$, where M is a large scalar. It is worth mentioning that in case the time constraints should be ignored, the problem becomes a VRP by setting $e_i = 0$ and $l_i = M \ \forall i \in C$. Furthermore, constraints (15.30) and (15.31) impose that service may only starts within a given time interval.

15.2.5 Inventory Routing Problem (IRP)

Whereas the VRP typically deals with a single time instance (e.g., one day), IRP have to deal with a sequence of time instances (e.g., a sequence of days). In the context of VRP, customers place orders, a day before, and the supplier assign the orders (for the next day) to routes based on his fleet of vehicles and vehicles' capacity. IRP arises from the application of the Vendor Managed Inventory (VMI) concept and therefore there are no customer orders. Supplier monitors the inventory levels of the customers and determines the delivery times, the delivery quantities and the suitable set of routes. The typical IRP model deals with the repeated distribution of a single product from a single depot to a set of geographically dispersed customers over a given time horizon. As a consequence, the IRP enlarges the set of decisions to be taken with respect to the VRP adding the time variant to the decision variables.

The Symmetric IRP is defined on a complete undirected graph $G = (V, E)$ where $V = \{0, \ldots, n\}$ is the set of vertices and $E = \{(i, j) : i, j \in V, j > i\}$ is the set of edges. Let $C = V \backslash \{0\} = \{1, \ldots, n\}$ the set of customers, whereas vertex 0 corresponds to the supplier. The set of time horizons is denoted by $T = \{1, \ldots, H\}$, where H is the length of planning horizon. Each customer $i \in C$ faces a different demand d_i^t per time period $t \in T$, maintains his own inventory up to capacity U_i and incurs an inventory holding cost of h_i per period per unit. It is assumed that the supplier has a sufficient supply of product items so as to cover all customers' demands throughout the planning horizon (i.e., $U_0 = +\infty$). A non-negative cost c_{ij} is associated with each edge $(i, j) \in E$ and represents the travel cost between the two vertices $i, j \in V$. The travel cost is defined as the Euclidean distance and satisfies the triangle inequality. An unlimited fleet of identical vehicles denoted by the set $K = \{1, 2, \dots\}$ with capacity Q is available for the distribution of the product. Let ω_{ik}^t the amount of delivery to customer $i \in C$ in period $t \in T$ participating in vehicle route $k \in K$. The three-index decision variable ψ_{ijk} of the VRP model is extended to ψ_{ijk}^t representing the number of times the edge $(i, j) \in E$ is traversed by vehicle $k \in K$ in period $t \in T$. In addition, the two-index decision variable ξ_{ik} is extended to ξ_{ik}^t binary variable and it is used to assign customers to vehicle routes. Value 1 indicates that customer $i \in C$ will be visited by vehicle $k \in K$ in period $t \in T$ (0 otherwise). Finally, let θ_i^t a nonnegative variable indicating the inventory level at customer $i \in C$ at the end of period $t \in T$. At the beginning of the planning horizon, each customer $i \in C$ has an initial inventory level of $\theta_i^0 = 0 \forall i \in C$ of product.

The Symmetric IRP can be formulated as follows:

$$\min\left(\sum_{k\in K}\sum_{t\in T}\sum_{i\in V}\sum_{j\in V, j>i} c_{ij}\psi_{ijk}^t + \sum_{t\in T}\sum_{i\in C} h_i\theta_i^t\right) \tag{15.34}$$

$$s.t.$$

$$\theta_i^t = \theta_i^{t-1} + \sum_{k\in K}\omega_{ik}^t - d_i^t, \forall i \in C, \forall t \in T \tag{15.35}$$

$$\theta_i^t \geq 0, \forall i \in C, \forall t \in T \tag{15.36}$$

$$\sum_{k\in K}\omega_{ik}^t + \theta_i^{t-1} \leq U_i, \forall i \in C, \forall t \in T, \forall k \in K \tag{15.37}$$

$$\sum_{i\in C}\omega_{ik}^t \leq Q\xi_{ik}^t, \forall k \in K, \forall t \in T \tag{15.38}$$

$$\omega_{ik}^t \leq U_i\xi_{ik}^t, \forall i \in C, \forall k \in K, \forall t \in T \tag{15.39}$$

$$\sum_{k\in K}\xi_{ik}^t \leq 1, \forall i \in C, \forall t \in T \tag{15.40}$$

$$\sum_{j\in V, j>i}\psi_{ijk}^t + \sum_{j\in V, j<i}\psi_{jik}^t = 2\xi_{ik}^t, \forall i \in C, \forall k \in K, \forall t \in T \tag{15.41}$$

$$\sum_{i \in S} \sum_{j \in S, j>i} \psi_{ijk}^t \leq \sum_{i \in S} \xi_{ik}^t - \xi_{sk}^t, \forall S \subseteq C, \forall s \in S, \forall k \in K, \forall t \in T \quad (15.42)$$

$$\omega_{ik}^t \geq 0, \forall i \in C, \forall k \in K, \forall t \in T \quad (15.43)$$

$$\psi_{ijk}^t \in \{0, 1\}, \forall i, j \in C, j > i, \forall k \in K, \forall t \in T \quad (15.44)$$

$$\psi_{0jk}^t \in \{0, 1, 2\}, \forall j \in C, \forall k \in K, \forall t \in T \quad (15.45)$$

$$\xi_{ik}^t \in \{0, 1\}, \forall i \in V, \forall k \in K, \forall t \in T \quad (15.46)$$

The objective function (15.34) expresses the transportation and inventory holding costs. Constraints (15.35) are the inventory balance equations for all customers. Constraints (15.36) guarantee that no stock-outs occur at any customer $i \in C$ during the planning horizon, while constraints (15.37) limit the inventory level of the customers to their corresponding maximum inventory levels. Constraints (15.38) ensure that the vehicle capacities are not exceeded in any period $t \in T$ of the planning horizon. Constraints (15.39) ensure that it is impossible to deliver to customer $i \in C$ quantity more than his defined capacity U_i. Constraints (15.40) guarantee that each customer $i \in C$ will be visited exactly once in each period $t \in T$ of the planning horizon. Constraints (15.41) and (15.42) are the routing constraints [164], while constraints (15.43)–(15.46) are the domain constraints.

15.2.6 Inventory Routing Problem with Time Windows (IRPTW)

The IRPTW is a generalization of the standard IRP involving the added complexity that every customer should be served within a given time window.

Using the decision variables of the IRP formulation, presented in Sect. 15.2.5, and the time window domain of the VRPTW, presented in Sect. 15.2.4 (Table 15.2), the Asymmetric IRPTW can be formulated as follows:

$$\min \sum_{i \in N} \sum_{j \in N} c_{ij} \sum_{k \in K} \sum_{t \in T} \psi_{ijk}^t + \sum_{t \in T} \sum_{i \in C} h_i \theta_i^t \quad (15.47)$$

$$s.t.$$

$$\theta_i^0 = U_i, \forall i \in C \quad (15.48)$$

$$\theta_i^{t-1} - \theta_i^t + \sum_{k \in K} \omega_{ik}^t = d_i^t, \forall i \in C, \forall t \in T \quad (15.49)$$

Table 15.2 Decision variables and parameter for formulating IRPTW

Decision variables	Parameters
$\boldsymbol{\omega}^t_{ik}$: the amount of delivery to customer $i \in C$ in period $t \in T$ by vehicle $k \in K$	$\boldsymbol{G} = (\boldsymbol{N}, \boldsymbol{A})$: complete directed graph where $N = \{0, n+1\} \cup \{1, \ldots, n\}$ is the set of nodes and $A = \{(i, j), i, j \in N, i \neq j\}$ is the set of arcs. The set of customers is denoted by $C = \{1, \ldots, n\}$
$\boldsymbol{\psi}^t_{ijk}$: a binary variable that is equal to 1 if vehicle $k \in K$ drives from node i to node $j \forall (i, j) \in A, i \neq j, j \neq n+1, j \neq 0$	$\boldsymbol{T} = \{1, \ldots, \boldsymbol{H}\}$: given planning horizon
$\boldsymbol{\xi}^t_{ik}$: a binary variable that is equal to 1 if customer $i \in C$ is visited by vehicle $k \in K$ in period $t \in T$	$\boldsymbol{d}^t_i$: demand of customer $i \in C$ per time period $t \in T$ $\boldsymbol{c}_{ij}$: travel cost (Euclidean distance) from node i to node $j \forall i, j \in N$ $\boldsymbol{t}_{ij}$: travel time associated with $(i, j) \in A$
$\boldsymbol{\alpha}^t_{ik}$: the time vehicle $k \in K$ starts to service customer $i \in C$ in period $t \in T$	$\boldsymbol{U}_i$: inventory capacity of customer $i \in C$ $\boldsymbol{U}_0 = +\infty$ $\boldsymbol{h}_i$: holding cost $\forall i \in C$
$\boldsymbol{\theta}^t_i$: a nonnegative variable indicating the inventory level at customer $i \in C$ at the end of period $t \in T$.At the beginning of the planning horizon each customer $i \in C$ has an initial inventory level $\theta^0_i = U_i$	$\boldsymbol{K} = \{1, \ldots, \boldsymbol{m}\}$: fleet of m homogenous vehicles with capacity $\boldsymbol{Q}$
$\boldsymbol{\varphi}^t_k$: a binary variable that is equal to 1 if vehicle $k \in K$ is used in period $t \in T$	$[\boldsymbol{e}_i, \boldsymbol{l}_i]$: time window $\forall i \in C$ $[\boldsymbol{e}_0, \boldsymbol{l}_0]$: time window for origin depot $[\boldsymbol{e}_{n+1}, \boldsymbol{l}_{n+1}]$: time window for destination depot $\boldsymbol{s}_i$: service time associated with each customer $i \in C$ $\boldsymbol{s}_0 = \boldsymbol{s}_{n+1} = 0$

$$\theta^t_i \leq U_i, \forall i \in C, \forall t \in T \tag{15.50}$$

$$\sum_{i \in C} \omega^t_{ik} \leq Q\varphi^t_k, \forall k \in K, \forall t \in T \tag{15.51}$$

$$\varphi^t_k \leq \sum_{i \in C} \xi^t_{ik}, \forall k \in K, \forall t \in T \tag{15.52}$$

$$\varphi^t_k n \geq \sum_{i \in C} \xi^t_{ik}, \forall k \in K, \forall t \in T \tag{15.53}$$

$$\sum_{k \in K} \xi^t_{ik} \leq 1, \forall i \in C, \forall t \in T \tag{15.54}$$

$$\sum_{k \in K} \xi_{0k}^t = m, \forall t \in T \tag{15.55}$$

$$\sum_{k \in K} \xi_{n+1,k}^t = m, \forall t \in T \tag{15.56}$$

$$\sum_{j \in N} \psi_{jik}^t = \xi_{ik}^t, \forall i \in N \backslash \{0\}, \forall k \in K, \forall t \in T \tag{15.57}$$

$$\sum_{j \in N} \psi_{ijk}^t = \xi_{ik}^t, \forall i \in N \backslash \{n+1\}, \forall k \in K, \forall t \in T \tag{15.58}$$

$$\alpha_{ik}^t + s_i + t_{ij} \leq \alpha_{jk}^t + M\left(1 - \psi_{ijk}^t\right), \forall i, j \in N, \forall k \in K, \forall t \in T \tag{15.59}$$

$$\alpha_{ik}^t \geq e_i \xi_{ik}^t \forall i \in \mathrm{N}, \forall k \in K, \forall t \in T \tag{15.60}$$

$$\alpha_{ik}^t \leq l_i \xi_{ik}^t \forall i \in \mathrm{N}, \forall k \in K, \forall t \in T \tag{15.61}$$

$$\psi_{ijk}^t \in \{0, 1\}, \forall i, j \in N, j > i, \forall k \in K, \forall t \in T \tag{15.62}$$

$$\theta_i^t \geq 0 \forall i \in C, \forall t \in T \tag{15.63}$$

$$\xi_{ik}^t \in \{0, 1\}, \forall i \in N, \forall k \in K, \forall t \in T \tag{15.64}$$

$$\omega_{ik}^t \geq 0, \forall i \in C, \forall k \in K, \forall t \in T \tag{15.65}$$

$$\alpha_{ik}^t \geq 0, \forall i \in C, \forall k \in K, \forall t \in T \tag{15.66}$$

$$\varphi_k^t \in \{0, 1\}, \forall k \in K, \forall t \in T \tag{15.67}$$

The objective function (15.47) minimizes the total transportation and inventory holding costs. Constraints (15.48) indicate that each customer i $\in$ C has an initial inventory level equal to his inventory capacity. Constraints (15.49) are the inventory balance equations for the customers, while constraints (15.50) guarantee that the available inventory level is equal or less than the maximum inventory level for each customer. Constraints (15.51)–(15.53) ensure that the vehicle capacities are not exceeded during the planning horizon. Constraints (15.54)–(15.61) are the constraints related to the classic VRPTW (see Sect. 15.2.4) including the time variant of the decision variables. Finally, constraints (15.62)–(15.67) are the domain constraints.

15.3 Computational Intelligence Algorithms for Routing Problems

This section aims at presenting a comprehensive review of the RP literature as far as the most important solution approaches are concerned. In addition to this, selected computational intelligence algorithms are promoted for solving several variants of routing problems.

15.3.1 State of the Art

Several solution approaches have been developed in the RP literature. The result of an analysis of the scientific literature led to the identification of two main paths of development in the overall field of the RP: (1) optimization-based or exact algorithms (exacts) and (2) approximation-based or computational intelligence algorithms. Exacts generate optimal (exact) solutions for complex small-scale RPs, while computational intelligence algorithms are applied to obtain near-optimal feasible solutions for large-scale RPs. In recent years, computational intelligence algorithms have received a great deal of attention from researchers. Due to the NP-hard nature of RPs, it is very difficult to develop an exact algorithm that can solve large-scale RPs within a reasonable computation time and as a result many computational intelligence algorithms have been proposed in the literature.

The most popular exacts are related to branch-and-bound, branch-and-cut, branch-and-cut-and-price algorithms, dynamic programming, column generation and Lagrange relaxation-based methods [38, 81]. The following table (Table 15.3) presents the most popular exacts that have been proposed for solving TSP, m-TSP, VRP, VRPTW, IRP and IRPTW.

Computational intelligence algorithms can be divided into six groups of algorithms: (1) construction heuristics, (2) local improvement heuristics, (3) single-point search meta-heuristics, (4) population-based search meta-heuristis, (5) math-heuristics and (6) sim-heuristics [76, 87, 116]. Construction heuristics are used to build initial feasible solutions to RPs, while local improvement heuristics start with an initial solution (obtained from a construction heuristic) and iteratively improve on it by considering neighboring solutions. Tables 15.4 and 15.5 provide the most popular construction and local improvement heuristics, respectively.

Since local improvement heuristics may terminate at "local" optimum solutions, meta-heuristics were proposed so as to escape from these "local" optimums in the solution space. Population-based search meta-heuristics consist of a population of candidate solutions to RPs. At each iteration of the algorithm, the population evolves to a better solution to the RP. However, single-point search meta-heuristics consist of only a single candidate solution at each iteration of the algorithm. In recent years, several meta-heuristic concepts have been proposed regarding TSP, VRP, IRP and their variants. The most popular single-point search meta-heuristics are Tabu Search

Table 15.3 Routing problems: exacts

Branch-and-Bound algorithms	TSP and m-TSP	Little et al. [111], Miliotis [118, 119], Volgenant and Jonker [166], Gavish and Strikanth [69], Pekny and Miller [134]
	VRP	Christofides and Eilon [46], Christofides [44], Christofides et al. [47, 48], Fisher [63], Hadjiconstantinou et al. [79]
Branch-and-Cut algorithms	TSP & m-TSP	Laporte and Nobert [100], Padberg and Rinaldi [131, 130], Clochard and Naddef [50], Naddef [125]
	VRP	Naddef and Rinaldi [126], Letchford et al. [105], Baldacci et al. [24], Lysgaard et al. [115], Hokama et al. [82]
	IRP	Archetti et al. [16], Solyali and Süral [155], Coelho and Laporte [53, 54]
Branch-and-Cut-and-price algorithms	TSP	Applegate et al. [13, 14]
	VRP	Fukasawa et al. [67], Baldacci et al. [23]
Column generation	m-TSP	Desrosiers et al. [58]
	VRP	Agarwal et al. [4], Baldacci and Mingozzi [25]
	VRPTW	Desrosiers et al. [58], Desrochers et al. [57]
Lagrange relaxation-based methods	m-TSP	Gromicho et al. [77]
	VRPTW	Fisher et al. [65], Kohl and Madsen [94], Kallehauge et al. [91]
	IRP	Shen et al. [149], Kamarul et al. [92]
Dynamic programming	VRP	Christofides and Beasley [45]
	VRPTW	Kolen et al. [95]
	IRP	Campbell et al. [36], Zeng and Zhao [170]

Algorithm (TS), Variable Neighborhood Search Algorithm (VNS), Adaptive Large Neighborhood Search Algorithm (ALNS), Greedy Randomized Adaptive Search Procedure (GRASP) and Simulated Annealing Algorithm (SA). Table 15.6 presents selected papers according to RPs.

Table 15.4 Routing problems: construction heuristics

Routing problem	Construction heuristic
TSP and m-TSP	Nearest neighbor [145]
	Sweep [73]
	Convex hull [9]
	Savings heuristic [49]
VRP	Savings heuristic [49]
	Sweep: cluster-first, route-second heuristic [64]
	Sweep: route-first, cluster-second heuristic [27]
VRPTW	Giant-tour heuristic [154]
	Savings heuristic [21, 22, 99, 154]
	Time-oriented nearest neighbor [22, 154]
	Time-oriented sweep heuristic [154]
	Parallel insertion heuristic [140]
	Insertion heuristic [85]
IRP	ETCH: estimated transportation cost heuristic [2]

Table 15.5 Routing problems: local improvement heuristics

Routing problem	Local improvement heuristic
TSP	Two-exchange or 2-opt [66]
	Three-exchange or 3-opt [55]
	Chain-exchange or chain-opt [109]
	Lin-Kernighan heuristic [110]
m-TSP	Effective heuristic for m-tour TSP [146]
	k-opt exchange procedure [138]
VRP	Intra-route improvement heuristics including 2-opt, 3-opt and chain-opt [74]
	Inter-route improvement heuristics including move, swap and cyclic exchange [74]
VRPTW	Intra-route improvement heuristics [139]
	Intra& inter improvement heuristics [35]
IRP	Multi-start two phase heuristic [143]
	CARE: clustering, allocation, routing, extended [127]
	ETCH improvement heuristic [2]

As far as the population-based search meta-heuristics are concerned, the most promising algorithms for RPs are related to evolutionary optimization algorithms such as Genetic Algorithm (GA), Ant Colony Optimization Algorithm (ACO), Particle Swarm Optimization (PSO) and Bee Colony Optimization Algorithm (BCO) (Table 15.7). It should be noted that, evolutionary optimization algorithms usually

Table 15.6 Routing problems: single-point search meta-heuristics

Routing problem	Single-point search meta-heuristic
TSP	TS [62, 97, 147, 169]
	SA [71, 171]
VRP	TS [70, 159, 160, 163]
	VNS [123]
	ALNS [135]
	SA [42]
	GRASP [96]
IRP	TS [15, 106, 142]
	GRASP [78]
	VNS [120–122]
	ALNS [10, 51, 150]

Table 15.7 Routing problems: population-based search meta-heuristics

Routing problem	Population-based search meta-heuristic
TSP	BCO [165]
	HEOA: ACO and SA [114]
	ACO [60]
	GA [137]
	PSO [108]
VRP	GA [20, 98, 141]
	ACO [144]
	PSO [7, 39, 117]
IRP	ACO [84, 162]
	HEOA – PSO & LNS [113]
	PSO [168]
	GA [1, 19, 43, 124, 133, 151, 152]
	HEOA—GA and SA [101]

are combined with single-point search meta-heuristics constructing new approaches called Hybrid Evolutionary Optimization Algorithms (HEOAs).

Finally, math-heuristics combine mathematical programming techniques (i.e., exacts) with meta-heuristics, whereas sim-heuristics combine Monte-Carlo simulation techniques with meta-heuristic approaches (Table 15.8). It should be mentioned that sim-heuristics usually are applied to address RPs with stochastic demands.

Table 15.8 Routing problems: matheuristics and sim-heuristics

Matheuristics	
VRP	Leggieri and Haouari [104], Keskin and Çatay [93]
IRP	Coelho et al. [52], Bertazzi et al. [30], Bertazzi et al. [31], Hemmati et al. [80], Archetti et al. [17], Bertazzi et al. [32]
Sim-heuristics	
VRP	Juan et al. [89], Juan et al. [88], Grasas et al. [75]
IRP	Juan et al. [90]

Table 15.9 Routing problems: real-life applications

Field	Research papers
School bus routing	m-TSP [12]
Maritime transportation	IRP [5, 6, 156]
Industrial gas distribution	VRP [37, 41]
	IRP [11, 72, 153]
Distribution of perishable products	IRP [59, 61, 157, 158]
Fuel delivery	VRP [18]
	IRP [136]
Medical waste collection and medical drug distribution	IRP [128, 129]
Distribution of agriculture products and groceries	IRP [68, 107]

Several real-life applications of the RPs have been found using TSP, VRP or IRP concepts. Compare to the TSP, VRP or IRP are more adequate to model real-life applications, since they are capable of handling more than one parameter (e.g., demand, inventory capacity, holding costs, etc.). Table 15.8 lists some real-life applications in the following fields: (1) school bus routing, (2) maritime transportation, (3) industrial gas distribution, (4) distribution of perishable products, (5) fuel delivery, (6) medical waste collection and medical drug distribution and (7) distribution of agriculture products and groceries (Table 15.9).

15.3.2 Selected Computational Intelligence Algorithms

The purpose of this section is to promote selected computational intelligence algorithms for solving RPs through application examples. In addition, it provides various graphical presentation formats so as to simplify complicated issues and convey meaningful insights into the RPs. To begin with, a population-based search metaheuristic, that is GA, is used to solve TSP, while K-means Clustering algorithm is proposed to decompose m-TSP into m TSPs. Moreover, a sim-heuristic approach is discussed for solving the single-period IRPTW with stochastic demands. It should be mentioned that the goal of this section is to present some computational intelligence approaches for solving large-scale RPs through illustrative examples and not to compare the approaches from a computational perspective according to the best available computational results in the literature. The algorithms that will be presented were developed in the MATLAB programming language (MATLAB R2016a) and executed on a DELL personal computer with an Intel ® CoreTM i3-2120, clocked at 3.30 GHz, a microprocessor with 4 GB of RAM memory under the operating system M/S Windows 7 Professional.

15.3.2.1 Genetic Algorithm for Solving the TSP

A population-based search meta-heuristic such as the GA is presented for solving TSP. GA is a biologically inspired optimization approach that was proposed by Holland [83] based on some of the mechanisms of evolution in nature. An initial population of individuals (i.e., a population of TSP solutions) is randomly generated. Individual's representation (i.e., TSP representation) corresponds to the chromosome or genotype. Path representation is used for representing a TSP tour. A chromosome consists of genes that represent the nodes of the graph. One iteration of creating a new population through an evolutionary process (optimization algorithm) is called generation. The fitness of each individual is related to the objective function (i.e., minimum total routing cost) and used for evaluation purposes. A selection process, allows parent solutions with high fitness to be selected from the current population. The goal is to select every time two parents to produce two offspring so as to keep the population size constant across generations. From generation to generation the creation of new population is achieved by applying crossover and mutation genetic operators. Crossover operators allow two parent-chromosomes (current solutions) to exchange genes (nodes) to each other so as to produce two new offspring-chromosomes (new solutions). In addition, mutation operator represents a slight change to a single individual. Algorithm 1 provides a pseudo-code listing of the GA for minimizing the total routing cost.

```
Algorithm1. GA for TSP
Inputs: p_m, pop_size, term_criterion, TSP_instance

gen ← 0
P(gen) ← NearestNeighborInitialization(pop_size, TSP_instance)
F(gen) ← PopulationEvaluation(P(gen))
Tour_opt ← FindBestTour(P(gen), F(gen))
while not term_criterion do
        gen ← gen + 1
        O_length ← 0
        while O_length ≤ pop_size do
                [P_1, P_2] ← RouletteWheelSelection(P(gen − 1))
                P_selected ← [P_1, P_2]
                [O_1, O_2] ← Crossover(P_selected)
                for 1:2 do
                        rand_No ← U(0,1)
                        if rand_No < p_m then
                                O_i,updated ← Mutation(O_i)
                                O(gen) ← AddToNewPopulation(O_i,updated)
                        else
                                O(gen) ← AddToNewPopulation(O_i)
                        end-if
                end-for
                O_length ← length(O_gen)
        end-while
        Merge(gen) ← O(gen) ∪ P(gen − 1)
        F(gen) ← PopulationEvaluation(Merge(gen))
        P(gen) ← SelectBestIndividuals(Merge(gen), F(gen))
        if FindBestTour(P(gen), F(gen)) is better than Tour_opt then
                Tour_opt ← FindBestSolution(P(gen), F(gen))
        end-if
end-while

Outputs: Tour_opt
```

In order to initialize a candidate solution, a constructive heuristic such as Nearest Neighbor [145] is applied based on the problem instance (i.e.,$TSP_{instance}$). By selecting randomly the starting node, Nearest Neighbor can be performed more than once so as to obtain different candidate solutions (initial population, $P(gen)$). For each individual (i.e., TSP tour) its fitness is evaluated so as to find the initial best solution (i.e.,$Tour_{opt}$). Since TSP is a minimization problem the fitness of each chromosome (i.e., $F(gen)$) is defined as follows: $fitness = \frac{1}{\sum_{i \in N} \sum_{j \in N} c_{ij} \psi_{ij}}$. Therefore, each individual has a probability of being selected that is proportional to its fitness. The higher the individual's fitness is, the more likely it is to be selected. This process depicts the Roulette-Wheel selection process. While a well-defined termination criterion (i.e., $term_{criterion}$) is not satisfied an evolution process starts where two parents are selected (i.e., P_1, P_2) to produce two offspring (i.e., O_1, O_2). Since the population size (i.e., pop_{size}) should be constant from generation to generation (i.e., gen), crossover and mutation genetic operators are applied in order to produce a new population (i.e., $O(gen)$) of the same size (i.e.,O_{length}). The two populations are merged (i.e., $Merge(gen)$) so as to select pop_{size} best solutions and find a better optimal tour per generation. Regarding crossover, three approaches are applied randomly: (a) single-point crossover, (b) order crossover and (c) cycle crossover. As far as the mutation is concerned, 2-opt and chain-opt local improvement heuristics are used.

The algorithm is applied to the following instances of the TSP, taken from TSPLIB[1]: (1) Ulysses16, (2) Ulysses22, (3) Berlin52 and (4) Gr96. For each of the above instances, 10 independent runs were performed for 1000–2000 generations. Table 15.10 shows the best and worst results obtained by the GA implementation for TSP, the average solutions, the best known solutions reported and the relative errors from the best known solutions. Furthermore, for each TSP instance the corresponding Hamiltonian cycle is provided.

To visually verify the above calculations, the following figures illustrate the solutions of the reported TSP instances as well as some typical graphs of the minimum TSP routing cost (distance), in the population, as a function of generation number (Fig. 15.1). The entirety direction of evolution during convergence indicates improvement with respect to the minimization of TSP routing costs.

It can be observed that when the number on nodes related to TSP instance (large-scale instances) is increased, more independent runs are needed so as to obtain better solutions (Fig. 15.2).

A recent application of GA for solving the Inventory Routing Problem has been proposed by Lappas et al. [101]. An emphasis is given to how the GA can be used

Table 15.10 TSP instances: computational results

TSP instance	Number of nodes	Worst solution	Best solution	Average solution	Best known solution	Relative error (%)
Ulysses16	16	6870	6859	6863.4	6859	**0.00**
Ulysses22	22	7163	7013	7046	7013	**0.00**
Berlin52	52	7848	7542	7740.2	7542	**0.00**
Gr96	96	62,018	57,307	58,996.2	55,209	3.80
Lin105	105	16,545	15,378	15,960.8	14,379	6.94
Best TSP tours found						
Ulysses16	**1** 14 13 12 7 6 15 5 11 9 10 16 3 2 4 8 **1**					
Ulysses22	**1** 14 13 12 7 6 15 5 11 9 10 19 20 21 16 3 2 17 22 4 18 8 **1**					
Berlin52	**1** 22 31 18 3 17 21 42 7 2 30 23 20 50 29 16 46 44 34 35 36 39 40 37 38 48 24 5 15 6 4 25 12 28 27 26 47 13 14 52 11 51 33 43 10 9 8 41 19 45 32 49 **1**					
Gr96	**1** 30 31 32 36 37 35 34 33 44 45 11 46 47 48 25 24 23 63 62 61 60 58 54 53 55 51 52 50 49 43 42 41 40 39 38 79 80 83 81 82 70 69 57 56 59 71 72 73 76 75 74 84 85 86 87 91 90 89 88 78 77 92 93 95 94 96 65 66 68 67 64 27 28 26 22 21 19 18 20 17 16 15 14 13 12 10 9 8 7 6 5 4 3 2 29 **1**					
Lin105	**1** 2 6 7 10 11 15 103 21 22 29 30 31 32 104 40 49 45 48 50 55 56 59 67 68 71 78 82 83 84 85 91 92 96 97 101 102 93 98 99 100 95 94 87 88 90 89 81 75 74 70 69 73 76 80 86 79 77 72 64 105 57 54 51 47 44 41 42 43 46 52 53 58 62 63 66 65 61 60 39 38 35 34 14 13 4 5 9 8 17 16 18 25 26 37 36 33 28 23 20 27 24 19 12 3 **1**					

[1] https://wwwproxy.iwr.uni-heidelberg.de/groups/comopt/software/TSPLIB95/index.html.

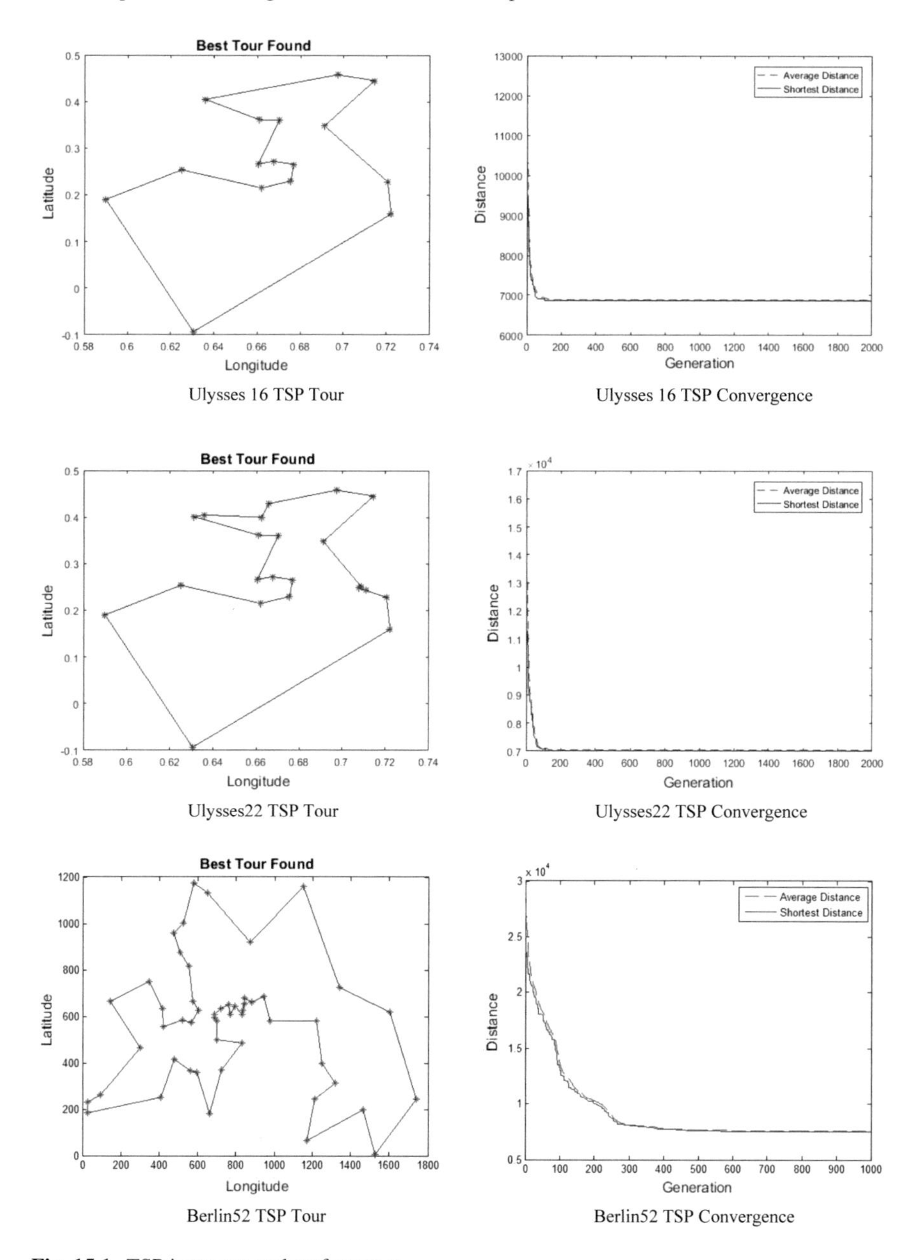

Fig. 15.1 TSP instances and performance

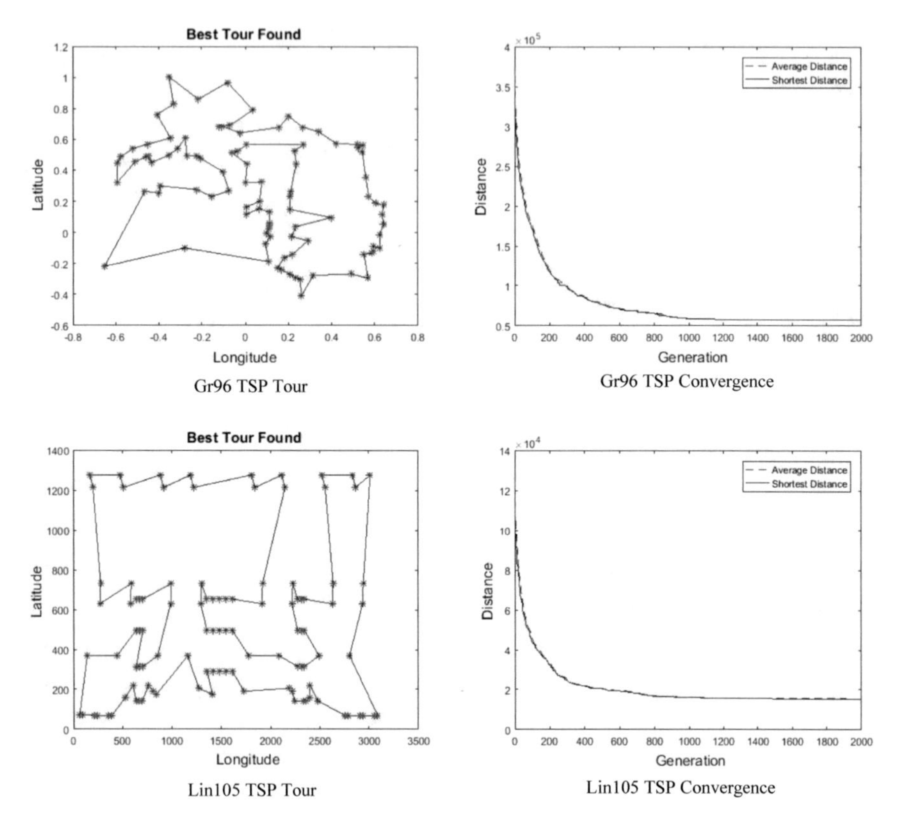

Gr96 TSP Tour

Gr96 TSP Convergence

Lin105 TSP Tour

Lin105 TSP Convergence

Fig. 15.1 (continued)

in hybrid synthesis with SA. In particular, the GA is related to the planning phase, while the SA is associated with the routing phase of the IRP. A repetitive procedure, combining characteristics from both meta-heuristics, is applied so as to obtain a near-optimal feasible solution. The proposed algorithm has been tested on a newly introduced set of 18 IRP benchmark instances.[2]

15.3.2.2 K-means Clustering Algorithm for Modeling the M-TSP

K-means Clustering (KmC) is an algorithm of unsupervised machine learning domain, which is used to divide a set of objects into homogenous groups or clusters [167]. Based on the application domain and the selected evaluation criteria, two arbitrary objects belonging to the same cluster are more similar to each other than

[2]https://www.msl.aueb.gr/files/GaSaIRP.zip.

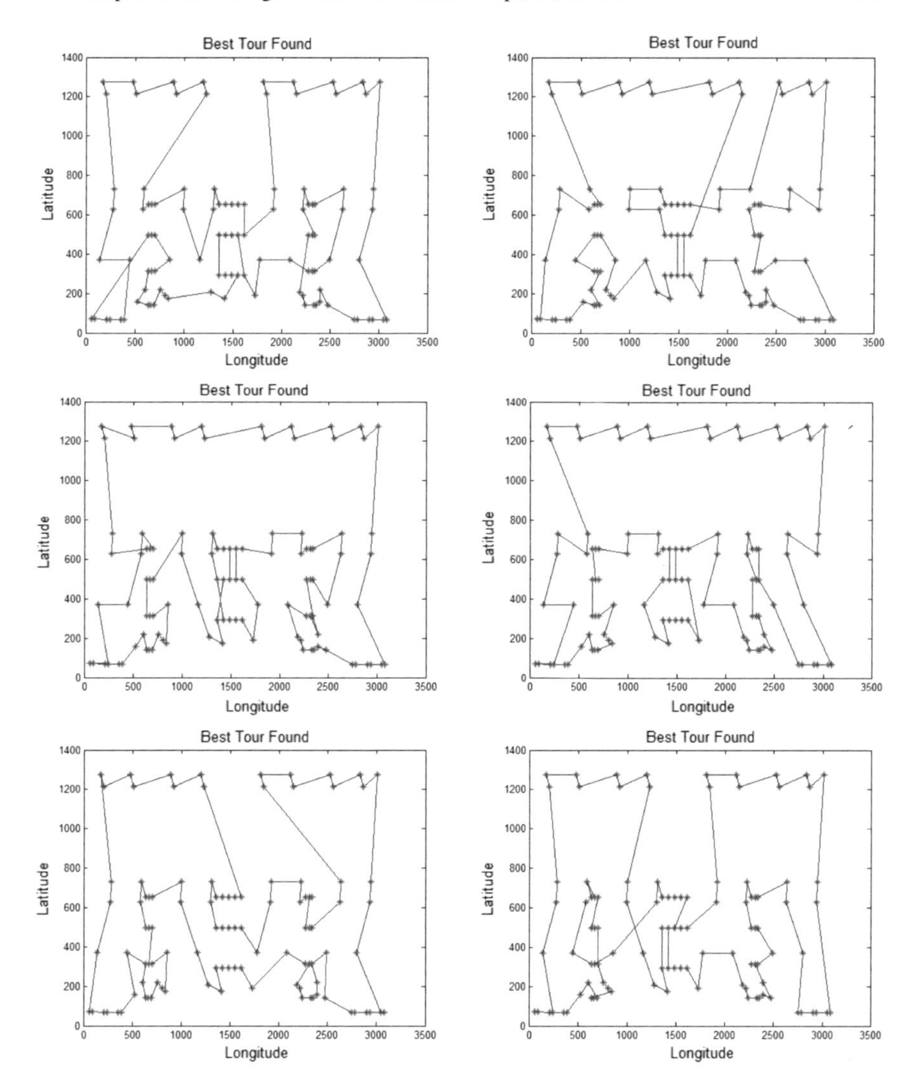

Fig. 15.2 Multiple runs for Lin105 TSP instance

two arbitrary objects belonging to different clusters. KmC could be applied for modeling m-TSP in the application domain of RPs taking into account, as an evaluation criterion, the Euclidean distance between the nodes of a graph (Algorithm 2).

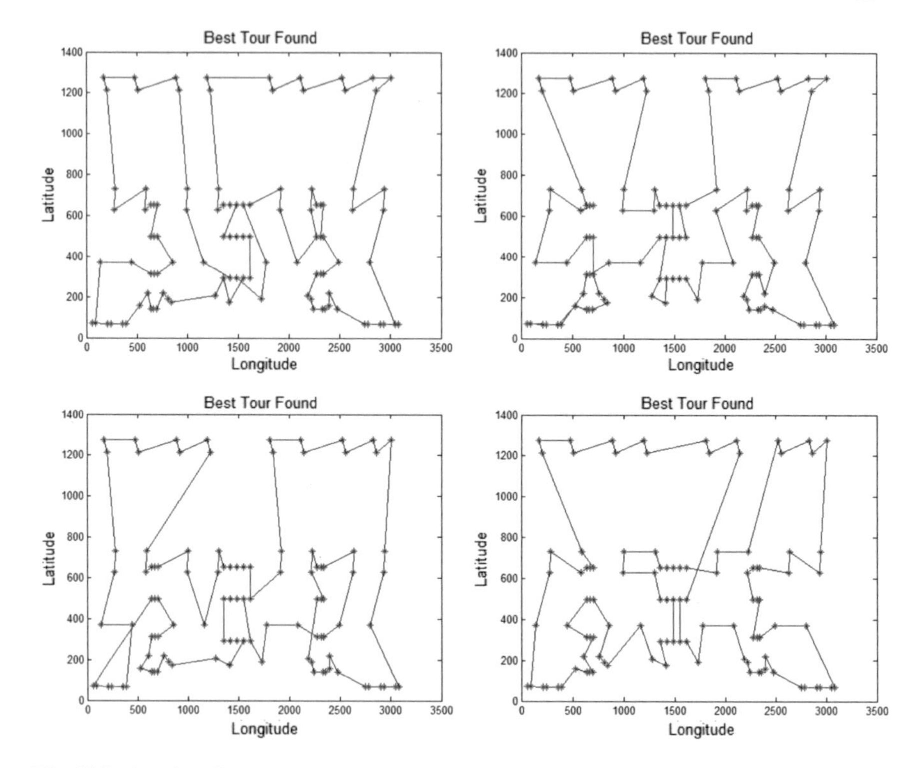

Fig. 15.2 (continued)

The KmC algorithm inputs are the number of clusters K and the dataset. K corresponds to the m vehicles (i.e., salesmen), while dataset corresponds to the TSP problem instance which is modeled on a complete directed graph $G = (N, A)$ where $N = \{0\} \cup \{1, \ldots, n\}$ is the set of nodes and $A = \{(i, j) : i, j \in N, i \neq j\}$ is the set of arcs. Therefore, objects are related to the nodes of the graph and the data are associated with points of the plane having the given coordinates $(x_i, y_i) \forall i \in N$.

Algorithm 2. KmC for modeling m-TSP

Inputs: $K, TSP_{instance}, term_{criterion}$

```
N ← size(TSP_instance)
[c_1, c_2, ..., c_K] ← RandomSelectionOf.K.Centroids(TSP_instance)
while not term_criterion do
        for i = 1:N do
                for g = 1:K do
                        dist(i, c_g) ← EuclideanDistance(i, c_g)
                end-for
        end-for
        for i = 1:N do
                val ← min(dist(i,:))
                ClusterAllocation(i) ← RelatedCentroid(val, dist(i,:))
        end-for
        g ← 1
        cluster_index ← g
        while g ≤ K do
                for i = 1:N do
                        if ClusterAllocation(i) = cluster_index then
                                c_g ← AddNode(i)
                        end-if
                end-for
                g ← g + 1
                cluster_index ← g
        end-while
end-while
```

Outputs: $c_1, c_2, \ldots, c_k$

KmC starts with initial estimates for the K centroids, which can be randomly selected from the dataset. Then an iterative procedure starts. Each centroid defines one of the K clusters. Each data point is assigned to its nearest centroid, based on the Euclidean distance. In the next step, centroids are recomputed by taking the mean of all data points assigned to that centroid's cluster. The above procedure is repeated until a stopping criterion is satisfied (data points do not change clusters or the sum of distances is minimized or some maximum number of iterations is reached, etc.).

An illustrative example is used to demonstrate key modeling features of the KmC. A small sample problem of 8 nodes can be considered, while K is defined equal to 3. Dataset is based on the first 8 nodes of the Ulysses16 TSP instance. Suppose that nodes 1,4 and 7 are selected randomly as the initial cluster centers (i.e., centroids: $c_1 = (x_1, y_1), c_2 = (x_4, y_4), c_3 = (x_7, y_7)$). The following tables show the first two iterations of the algorithm. Each cluster could be assumed as a TSP problem. Summing the routing cost obtained from each TSP, an m-TSP problem can be modeled and solved so as to optimize the total routing costs.

Table 15.11 shows the clustering schema obtained from the first iteration of the algorithm. After computing the Euclidean distances between each node and the defined centroids, the cluster in which the selected node belongs corresponds to the column with the shortest distance. Before the second iteration of the algorithm, centroids should be recomputed based on the nodes of each cluster. Analytically, the new centroid is a point with x-coordinate equal to the average of the nodes' x-coordinates and y-coordinate is equal to the average of the nodes' y-coordinates. Therefore, for the first iteration:

Table 15.11 First iteration of KmC

Node i	Point (x_i, y_i)	(38.24, 20.42)	(36.26, 23.12)	(38.42, 13.11)	Selected cluster
		$dist(i, c_1)$	$dist(i, c_2)$	$dist(i, c_3)$	
1	(38.24, 20.42)	**0**	3.3482	7.3122	1
2	(39.57, 26.15)	5.8823	**4.4874**	13.0906	2
3	(40.56, 25.32)	5.4215	**4.8301**	12.3961	2
4	(36.26, 23.12)	3.3482	**0**	10.2404	2
5	(33.48, 10.54)	10.9669	12.8835	**5.5685**	3
6	(37.56, 12.19)	8.2580	11.0070	**1.2594**	3
7	(38.42, 13.11)	7.3122	10.2404	**0**	3
8	(37.52, 20.44)	**0.7203**	2.9614	7.3850	1
Clustering schema					
$Cluster1 = \{1, 8\}$ $Cluster1 = \{2, 3, 4\}$ $Cluster1 = \{5, 6, 7\}$			$c_1 = (37.8800, 20.4300)$ $c_2 = (38.7967, 24.8633)$ $c_3 = (36.4867, 11.9467)$		

$$c_1 = \left(\frac{x_1 + x_8}{2}, \frac{y_1 + y_8}{2}\right)$$
$$= \left(\frac{38.24 + 37.52}{2}, \frac{20.42 + 20.44}{2}\right) = (37.8800, 20.4300)$$

$$c_2 = \left(\frac{x_2 + x_3 + x_4}{3}, \frac{y_2 + y_3 + y_4}{3}\right)$$
$$= \left(\frac{39.57 + 40.56 + 36.26}{3}, \frac{26.15 + 25.32 + 23.12}{3}\right) = (38.7967, 24.8633)$$

$$c_3 = \left(\frac{x_5 + x_6 + x_7}{3}, \frac{y_5 + y_6 + y_7}{3}\right)$$
$$= \left(\frac{37.56 + 38.42 + 37.52}{3}, \frac{12.19 + 13.11 + 20.44}{3}\right) = (36.4867, 11.9467)$$

Actually, clustering is stopped when cluster allocation from one iteration to the next does not change. Table 15.12 shows the second iteration of the algorithm. As it can be observed the clustering schema is the same with the clustering schema of the first iteration. As a result, there is no need for third iteration. It should be mentioned that the number of iterations depends on the starting centroid-points of the algorithm. In our example, due to the fact that nodes 1, 4 and 7 were selected as the initial cluster centers, only two iterations are needed so at to allocate the nodes to three clusters.

15.3.2.3 Sim-Heuristic Algorithm for the Single-Period IRPTW with Stochastic Demands

Juan et al. [90] presented a sim-heuristic algorithm for the single-period stochastic IRP. They combined Monte-Carlo simulation with a multi-start randomized heuristic based on Savings heuristic so as to solve the IRP with stochastic demands. A myopic solution approach that combines simple simulation and VNS approach has been presented by Lappas et al. [102] in order to solve a deterministic version of the multi-period IRPTW. Inspired by the paper of Juan et al. [90], this section promotes a sim-heuristic approach for the single-period Inventory Routing Problem with Time Windows and Stochastic Demands (IRPTWSD).

The proposed sim-heuristic focuses on a scenario where a single-product type has to be delivered by a fleet of capacitated homogenous vehicles from a single depot to a number of geographically dispersed customers. The demand is stochastic and as a result it is not fully available to the decision maker at the beginning of the planning horizon. The length of the planning horizon is equal to a single period in which vendor has to make inventory allocation and routing decisions, simultaneously. Each customer is associated with a time window and a service time, while he maintains his own inventory up to maximum inventory capacity. Therefore, the service of each customer must start within the associated time window and inventory holding costs

Table 15.12 Second iteration of KmC

Node i	Point (x_i, y_i)	(37.8800, 20.4300)	(38.7967, 24.8633)	(36.4867, 11.9467)	Selected cluster
		$dist(i, c_1)$	$dist(i, c_2)$	$dist(i, c_3)$	
1	(38.24, 20.42)	**0.3601**	4.4780	8.6528	1
2	(39.57, 26.15)	5.9644	**1.5012**	14.5341	2
3	(40.56, 25.32)	5.5762	**1.8215**	13.9799	2
4	(36.26, 23.12)	3.1401	**3.0780**	11.1756	2
5	(33.48, 10.54)	10.8246	15.2782	**3.3195**	3
6	(37.56, 12.19)	8.2462	12.7335	**1.1005**	3
7	(38.42, 13.11)	7.3399	11.7593	**2.2563**	3
8	(37.52, 20.44)	**0.3601**	4.6039	8.5559	1

Clustering schema

$Cluster1 = \{1, 8\}$	$c_1 = (37.8800, 20.4300)$
$Cluster1 = \{2, 3, 4\}$	$c_2 = (38.7967, 24.8633)$
$Cluster1 = \{5, 6, 7\}$	$c_3 = (36.4867, 11.9467)$

are taken into account. In addition to this, for each customer vendor is allowed to apply different inventory replenishment policies such as (a) full refill policy, (b) ¾ refill policy, (c) ½ refill policy, (d) ¼ refill policy and (e) no refill policy [90]. As in the paper of Juan et al. [90], the main purpose here is to find the best personalized inventory replenishment policies and optimal routes that minimize the total costs related to inventory allocation and routing decisions during a planning horizon of a single period. The sim-heuristic algorithm combines Monte-Carlo simulation with insertion heuristics and single-point search meta-heuristic algorithm such as the VNS [103]. The algorithm can be described by the following two phases. Phase 1 aims at finding an initial feasible solution in three steps:

1. Monte-Carlo simulation is used so as to estimate the expected inventory holding costs associated with each customer-inventory policy combination.
2. Push Forward Insertion Heuristic (PFIH) of Solomon [154] is applied in order to obtain a good solution for the associated VRPTW related to each customer-inventory policy combination.
3. Total inventory holding and routing costs are estimated related to each of the inventory replenishment policies. The solution with the lowest total cost is set as the best solution found so far.

Phase 2 depicts a repetitive procedure with the following steps:

1. A solution is found by applying a full refill policy to all customers.
2. For each customer the reduction in total cost with respect to each non full-refill policy is estimated. PFIH is used to solve one VRPTW for each customer and for each non full-refill policy combination.
3. The best inventory replenishment policy for each customer that corresponds to the one with the lowest total cost is selected, whereas a single-point search meta-heuristic, that is VNS, is used to solve the corresponding VRPTW. The VNS approach is based on the application of PFIH and the systematically usage of 2-interchange and cross-exchange mechanisms [102].

Six classes of IRPTWSD dataset have been developed by generalizing the well-known dataset C1, C2, R1, R2, RC1 and RC2 of Solomon [154] with 25, 50 and 100 customers.[3] The dataset are named in the form of "SIRPTW-N-X" strings, where "N" stands for the number of customers and "X" stands for the class related to a specific dataset of Solomon [154]. Nodes coordinates were modified in such a way that the depot is located at the origin (0, 0). As far as the probabilistic demands and their variance scenarios, the structure of the inventory cost function, the initial inventory levels and the maximum inventory capacities are concerned, the same assumptions are followed as in the paper of Juan et al. [90]. Regarding the inventory holding cost, for each customer the holding cost is randomly generated and could be belong to a low holding cost interval, [0.01, 0.05], or a high holding cost interval, [0.1, 0.5].

A sample problem of a distribution system that consists of a single supplier and 25 customers can be considered to illustrate the behavior of the proposed sim-heuristic

[3]https://www.msl.aueb.gr/files/SIRPTW-DataSet.zip.

algorithm for the IRPTW with stochastic demands under an intermediate, a high and a low variance scenario (Table 15.13).

As it can be observed, under a different variance scenario the inventory cost is influenced. Due to the demand variations, the vendor should apply different inventory replenishment policies (e.g., see customers 1, 23 and 24). However, the routing cost is maintained. The minimum total cost is obtained under low variance scenario. The proposed sim-heuristic algorithm has been tested on 18 benchmark instances under intermediate variance scenario with 500 observations with respect to the stochastic demand. The following table (Table 15.14) shows some computational results related to the benchmark instances. Computation times in bold show the maximum and the minimum computation times observed for same size of problem instances (i.e., 25 or 50 or 100 customers).

15.4 Conclusion and Future Work

An initial objective of this chapter was to present several formulation schemes related to asymmetric and symmetric versions of RPs. A second objective was to provide a comprehensive review of exacts, heuristics, meta-heuristics, math-heuristics and sim-heuristics that have been proposed in the literature for solving RPs, as well as, real-life applications. Selected computational intelligence algorithms such as the GA, the KmC and the SH were promoted. Therefore, a third objective of the chapter was to provide numerical examples and various graphical presentation formats so as to simplify complicated issues and convey meaningful insights into the routing problems. Analytically, the GA was discussed in the context of TSP where several TSP instances including 16–105 nodes were solved. Kmc was explained in order to model the m-TSP. The main idea was to decompose m-TSP into m TSPs so as to allocate nodes in m clusters. The SH approach was developed for solving the single-period IRPTW with stochastic demands. New benchmark instances including 25–100 nodes were provided so as investigate algorithm's performance.

In terms of future work, three future research directions can be discussed in the context of RPs. To begin with, PRP, which has not been excessively researched in the literature, is a generalization of the IRP including decisions in tactical and operational levels [3]. Promising computational intelligence algorithms, such as a hybrid GA approach and a Tabu Search approach, have been developed by Boudia and Prins [34] and Bard and Nananukul [26], respectively. These algorithms could be extended by involving the added complexity that every customer should be served within a given time window.

As mentioned in Sect. 15.1, most of the routing problems deal with decisions in operational (TSP, VRP, IRP, PRP) and tactical levels (IRP, PRP). As a result, it would be very challenging the extension of RPs so as to include decisions in strategic level. This can be achieved by taking into account supply chain network design parameters (strategic decisions). In the direction of supply chain integration and coordination, an interesting approach is to combine PRP with Facility Location Problem (FLP). This

Table 15.13 IRPTWSD illustrative example

*: full-refill policy ◇: ¾ refill policy ▷: ½ refill policy	Routing problem details
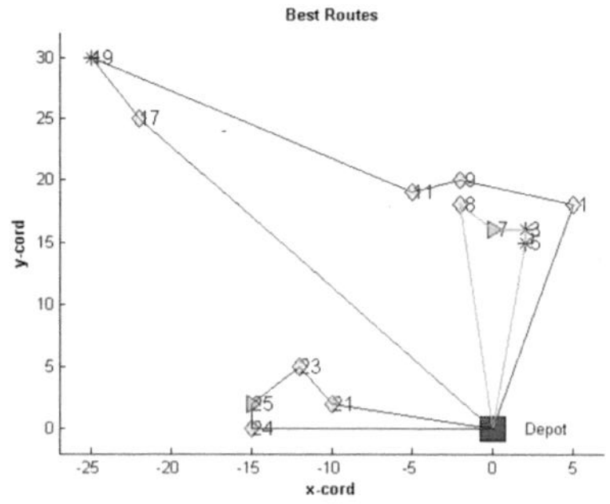	Intermediate variance scenario • Total cost = 466.8731 • Inventory cost = 301.6730 • Routing cost = 165.2001 Routes • Route 1 = {0-17-19-11-9-1-0} • Route 2 = {0-5-3-7-8-0} • Route 3 = {0-24-25-23-21-0}

(continued)

Table 15.13 (continued)

*: full-refill policy ◇: ¾ refill policy ▷: ½ refill policy	Routing problem details
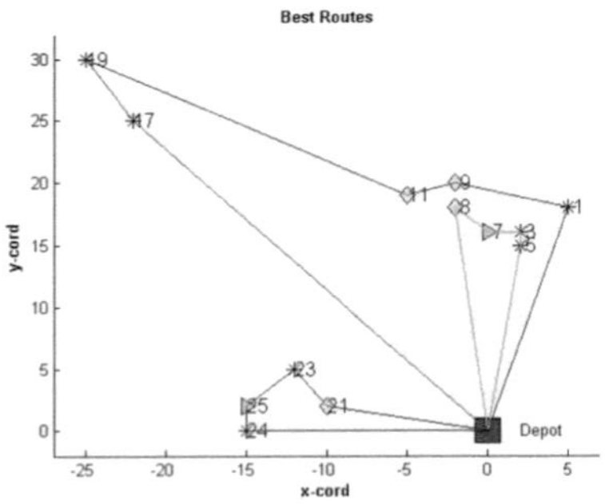	High variance scenario • Total cost = 472.3044 • Inventory cost = 307.1043 • Routing cost = 165.2001 Routes • Route 1 = {0-17-19-11-9-1-0} • Route 2 = {0-5-3-7-8-0} • Route 3 = {0-24-25-23-21-0}
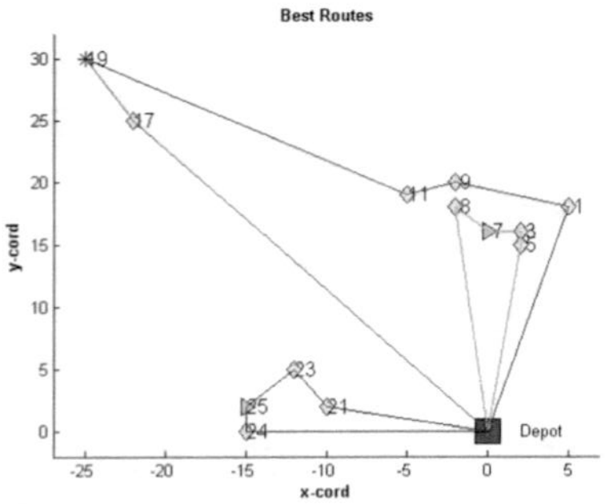	Low variance scenario • Total cost = 458.4902 • Inventory cost = 293.2901 • Routing cost = 165.2001 Routes • Route 1 = {0-17-19-11-9-1-0} • Route 2 = {0-5-3-7-8-0} • Route 3 = {0-24-25-23-21-0}

Table 15.14 Computational results

Problem instance	Routing cost	Inventory cost	Total cost	Computation time (in seconds)
SIRPTW-25-C101	165.2001	301.6730	466.8731	24.5442
SIRPTW-25-C201	196.9326	72.4444	269.3770	19.5303
SIRPTW-25-R101	427.1743	167.4126	594.5869	**17.2404**
SIRPTW-25-R201	360.6851	168.9611	529.6462	30.8168
SIRPTW-25-RC101	353.1109	153.7742	506.8850	**33.0239**
SIRPTW-25-RC201	323.8963	121.6999	445.5962	29.9826
SIRPTW-50-C101	298.6178	636.2788	934.8966	**84.4473**
SIRPTW-50-C201	327.4662	268.9006	596.3677	229.5602
SIRPTW-50-R101	845.8014	303.3265	1.1491e + 003	128.2360
SIRPTW-50-R201	665.5252	181.6128	847.1380	**294.8992**
SIRPTW-50-RC101	815.7262	693.8653	1.5096e + 003	139.4052
SIRPTW-50-RC201	668.0280	174.9085	842.9365	129.4450
SIRPTW-100-C101	772.3109	952.3588	1.7247e + 003	**372.4813**
SIRPTW-100-C201	752.5103	602.3987	1.3549e + 003	1.5470e + 003
SIRPTW-100-R101	1.1908e + 003	506.4760	1.6973e + 003	764.6141
SIRPTW-100-R201	906.9003	553.6929	1.4606e + 003	**2.6422e + 003**
SIRPTW-100-RC101	1.2877e + 003	1.2581e + 003	2.5458e + 003	886.0844
SIRPTW-100-RC201	1.2854e + 003	878.6101	2.1640e + 003	1.4165e + 003

can be called Location Production Routing Problem (LPRP) which can be addressed in the field of Supply Chain Risk Management (ScRM). Therefore, a second future research direction could be to develop computational intelligence algorithms for solving the LPRP with supply chain risks. It should be mentioned that supply chain risks may include operational risks that are associated with the inherent supply chain uncertainties, like uncertain supply or demand, etc. [153], as well as, disruption risks with reference to the major disruptions caused by natural and man-made disasters, such as earthquakes, floods, terrorist attacks, etc. [86].

Finally, a third future research direction is related to the pollution (green) variant of RPs. Bektaş and Laporte [29] presented the Pollution-Routing Problem (P-RP) as an extension of the classical VRP. The proposed objective function consists of transportation costs, as well as, costs related to green-house emissions, fuel and travel times. The literature offers very little in terms of modeling Pollution Inventory Routing Problem (PIRP) [40]. As a consequence, PIRP can be seen as a promising research area.

References

1. T. Abdelmaguid, M. Dessouky, A genetic algorithm approach to the integrated inventory-distribution problem. Int. J. Prod. Res. **44**(21), 4445–4464 (2006)
2. T. Abdelmaguid, M. Dessouky, F. Ordóñez, Heuristic approaches for the inventory-routing problem with backlogging. Comput. Ind. Eng. **56**(4), 1519–1534 (2009)
3. Y. Adulyasak, J.F. Cordeau, R. Jans, The production routing problem: a review of formulations and solution algorithms. Comput. Oper. Res. **55**, 141–152 (2015)
4. Y. Agarwal, K. Mathur, M. Salkin, A set-partitioning-based exact algorithm for the vehicle routing problem. Networks **19**(7), 731–749 (1989)
5. A. Agra, M. Christiansen, A. Delgado, L.M. Hvattum, A maritime inventory routing problem with stochastic sailing and port times. Comput. Oper. Res. **61**, 18–30 (2015)
6. A. Agra, M. Christiansen, A. Delgado, L. Simonetti, Hybrid heuristics for a short sea inventory routing problem. Eur. J. Oper. Res. **236**(3), 924–935 (2014)
7. T.J. Ai, V. Kachitvichyanukul, A particle swarm optimization for the vehicle routing problem with simultaneous pickup and delivery. Comput. Oper. Res. **36**(5), 1693–1702 (2009)
8. M. Akhbari, Y. Mehrjerdi, H. Zare, A. Makui, VMI-type supply chains: a brief review. J. Optim. Ind. Eng. **7**(14), 75–87 (2014)
9. S. Akl, G. Toussaint, A fast convex hull algorithm. Inf. Process. Lett. **7**(5), 219–222 (1978)
10. D. Aksen, O. Kaya, F.S. Salman, Ö. Tüncelb, An adaptive large neighborhood search algorithm for a selective and periodic inventory routing problem. Eur. J. Oper. Res. **239**(2), 413–426 (2014)
11. H. Andersson, M. Christiansen, G. Desaulniers, A new decomposition algorithm for a liquefied natural gas inventory routing problem. Int. J. Prod. Res. **54**(2), 564–578 (2016)
12. R. Angel, W. Caudle, R. Noonan, A. Whinston, Computer assisted school bus scheduling. Manag. Sci. **18**(6), 279–288 (1972)
13. D. Applegate, R. Bixby, V. Chvátal, W. Cook, *The Traveling Salesman Problem: A Computational Study* (Princeton University Press, New Jersey, 2006)
14. D. Applegate, R. Bixby, V. Chvátal, W. Cook, Implementing the Dantzig-Fulkerson-Johnson algorithm for large scale traveling salesman problems. Math. Program. **97**(1–2), 91–153 (2003)
15. C. Archetti, L. Bertazzi, A. Hertz, M.G. Speranza, A hybrid heuristic for an inventory routing problem. INFORMS J. Comput. **24**(1), 101–116 (2012)
16. C. Archetti, L. Bertazzi, G. Laporte, M.G. Speranza, A branch-and-cut algorithm for a vendor-managed inventory-routing problem. Transp. Sci. **41**(3), 382–391 (2007)
17. C. Archetti, N. Boland, M.G. Speranza, A matheuristic for the multivehicle routing problem. INFORMS J. Comput. **29**(3), 377–387 (2017)
18. P. Avella, M. Boccia, A. Sforza, Solving a fuel delivery problem by heuristic and exact approaches. Eur. J. Oper. Res. **152**(1), 170–179 (2004)
19. N.A.B. Aziz, N.H. Moin, Genetic algorithm based approach for the multi product multi period inventory routing problem. Paper presented at the proceedings of 2007 IEEE international conference on industrial engineering and engineering management (Asia, Singapore, 2007), pp. 1619–1623
20. B. Baker, M. Ayechew, A genetic algorithm for the vehicle routing problem. Comput. Oper. **30**(5), 787–800 (2003)
21. E. Baker, J. Schaffer, Solution improvement heuristics for the vehicle routing and scheduling problem with time windows constraints. Am. J. Math. Manag. Sci. **6**(3–4), 261–300 (1989)
22. N. Balakrishnan, Simple heuristics for the vehicle routeing problem with soft time windows. J. Oper. Res. Soc. **44**(3), 279–287 (1993)
23. R. Baldacci, N. Christofides, A. Mingozzi, An exact algorithm for the vehicle routing problem based on the set partitioning formulation with additional cuts. Math. Program. **115**(2), 351–385 (2008)

24. R. Baldacci, E. Hadjiconstantinou, A. Mingozzi, An exact algorithm for the capacitated vehicle routing problem based on a two-commodity network flow formulation. Oper. Res. **52**(5), 723–738 (2004)
25. R. Baldacci, A. Mingozzi, A unified exact method for solving different classes of vehicle routing problems. Math. Program. 120–347 (2009)
26. J. Bard, N. Nananukul, The integrated production-inventory-distribution-routing problem. J. Sched. **12**, 257–280 (2009)
27. J. Beasley, Route first-cluster secosnd methods for vehicle routing. Omega **11**(4), 403–408 (1983)
28. T. Bektaş, The multiple traveling salesman problem: an overview of formulations and solution procedures. Omega **34**(3), 209–219 (2006)
29. T. Bektaş, G. Laporte, The pollution-routing problem. Transp. Res. Part B: Methodol. **45**(8), 1232–1250 (2011)
30. L. Bertazzi, A. Bosco, F. Guerriero, D. Laganà, A stochastic inventory routing problem with stock-out. Transp. Res. Part C: Emerg. Technol. **27**, 89–107 (2013)
31. L. Bertazzi, A. Bosco, D. Laganà, Managing stochastic demand in an inventory routing problem with transportation procurement. Omega **56**, 112–121 (2015)
32. L. Bertazzi, L. Coelho, A. Maio, D. Laganà, A matheuristic algorithm for the multi-depot inventory routing problem. Transp. Res. Part E: Logist. Transp. Rev. **122**, 524–544 (2019)
33. L. Bertazzi, M. Savelsbergh, G. Speranza, Inventory routing, in *The Vehicle Routing Problem: Latest Advances and New Challenges*, ed. by B. Golden, S. Raghavan, E. Wasil (Springer, New York, 2008), pp. 49–72
34. M. Boudia, C. Prins, A memetic algorithm with dynamic population management for an integrated production-distribution problem. Eur. J. Oper. Res. **195**(3), 703–715 (2009)
35. O. Bräysy, Fast local searches for the vehicle routing problem with time windows. INFOR: Inf. Syst. Oper. Res. **40**(4), 319–330 (2003)
36. A. Campbell, L. Clarke, A. Kleywegt, M. Savelsbergh, The inventory routing problem, in *Fleet Management and Logisticsi*, ed. by T. Crainic, G. Laporte (Springer, Berlin, 1998), pp. 95–113
37. A. Campbell, L. Clarke, M. Savelsebergh, in *The Vehicle Routing Problem*, ed. by P. Toth, D. Vigo (SIAM, Philadelphia, 2002), pp. 53–81
38. D.S. Chen, R. Batson, Y. Dang, *Applied Integer Programming: Modeling and Solution* (Willey, New Jersey, 2010)
39. A.L. Chen, G. Yang, Z. Wu, Hybrid discrete particle swarm optimization algorithm for capacitated vehicle routing problem. J. Zhejiang Univ.—Sci. A **7**(4), 607–614 (2006)
40. C. Cheng, P. Yang, M. Qi, L.M. Rousseau, Modeling a green inventory routing problem with a heterogeneous fleet. Transp. Res. Part E: Logist. Transp. Rev. **97**, 97–112 (2017)
41. W.C. Chiang, R. Russell, Integrating purchasing and routing in a propane gas supply chain. Eur. J. Oper. Res. **154**(3), 710–729 (2004)
42. W.C. Chiang, R. Russell, Simulated annealing metaheuristics for the vehicle routing problem with time windows. Ann. Oper. Res. **63**(1), 3–27 (1996)
43. W.D. Cho, Y.H. Lee, Y.T. Lee, M. Gen, An adaptive genetic algorithm for the time dependent inventory routing problem. J. Intell. Manuf. **25**(5), 1025–1042 (2013)
44. N. Christofides, The vehicle routing problem. RAIRO **10**(1), 55–70 (1976)
45. N. Christofides, N. Beasley, The period routing problem. Networks **14**(2), 237–241 (1984)
46. N. Christofides, S. Eilon, An algorithm for the vehicle dispatching problem. J. Oper. Res. Soc. **20**(3), 309–318 (1969)
47. N. Christofides, A. Mingozzi, P. Toth, Exact algorithms for the vehicle routing problem, based on spanning tree shortest path relaxations. Math. Program. **20**(1), 255–282 (1981)
48. N. Christofides, A. Mingozzi, P. Toth, Space state relaxation procedures for the computation of bounds to routing problems. Networks **11**(2), 145–164 (1981)
49. G. Clarke, J. Wright, Scheduling of vehicles from a central depot to a number of delivery points. Oper. Res. **12**(4), 568–581 (1964)

50. J.M. Clochard, D. Naddef, Using path inequalities in a branch and cut code for the symmetric traveling salesman problem. Paper presented at the proceedings of the 3rd IPCO conference (Italy, Erice, 29 April—1 May 1993), pp. 291–311
51. L. Coelho, J.F. Cordeau, G. Laporte, The inventory-routing problem with transhipment. Comput. Oper. Res. **39**(11), 2537–2548 (2012)
52. L. Coelho, J.F. Cordeau, G. Laporte, Consistency in multi-vehicle inventory-routing. Transp. Res. Part C: Emerg. Technol. **24**, 270–287 (2012)
53. L. Coelho, G. Laporte, A branch-and-cut algorithm for the multi-product multi-vehicle inventory-routing problem. Int. J. Prod. Res. **51**(23–24), 7156–7169 (2013)
54. L. Coelho, G. Laporte, The exact solution of several classes of inventory-routing problems. Comput. Oper. Res. **40**(2), 558–565 (2013)
55. G. Croes, A method for solving large scale symmetric traveling salesman problems. Oper. Res. **6**(6), 791–812 (1958)
56. G. Dantzig, J. Ramser, The truck dispatching problem. Manag. Sci. **6**(1), 80–91 (1959)
57. M. Desrochers, J. Desrosiers, M. Solomon, A new optimization algorithm for the vehicle routing problem with time windows. Oper. Res. **40**(2), 342–354 (1992)
58. J. Desrosiers, F. Soumis, M. Desrochers, Routing with time windows by column generation. Networks **14**(4), 545–565 (1984)
59. A. Diabat, T. Abdallah, T. Le, A hybrid tabu search based heuristic for the periodic distribution inventory problem with perishable goods. Ann. Oper. Res. **242**(2), 373–398 (2016)
60. B. Escario, J. Jimenez, J. Giron-Sierra, Ant colony extended: Experiments on traveling salesman problem. Expert. Syst. Appl. **42**(1), 390–410 (2015)
61. A. Federgruen, G. Prastacos, P.H. Zipkin, An allocation and distribution model for perishable products. Oper. Res. **34**(1), 75–82 (1986)
62. C.N. Fiechter, A parallel tabu search algorithm for large traveling salesman problems. Discret. Appl. Math. **3**(6), 243–267 (1994)
63. M. Fisher, Optimal solution of vehicle routing problems using minimum k-trees. Oper. Res. **42**(4), 626–542 (1994)
64. M. Fisher, R. Jaikumar, A generalized assignment heuristic for vehicle routing. Networks **11**(2), 109–124 (1981)
65. M. Fisher, K. Jornsteen, O. Madsen, Vehicle routing with time windows: two optimization algorithms. Oper. Res. **45**(3), 488–492 (1997)
66. M. Flood, The travelling-salesman problem. Oper. Res. **4**(1), 61–75 (1956)
67. R. Fukasawa, H. Longo, J. Lysgaard, M. Aragão, M. Reis, E. Uchoa, R. Werneck, Robust branch-and-cut-and-price for the capacitated vehicle routing problem. Math. Program. **106**(3), 491–511 (2006)
68. V. Gaur, M.L. Fisher, A periodic inventory routing problem at a supermarket chain. Oper. Res. **52**(6), 813–822 (2004)
69. B. Gavish, K. Strikanth, An optimal solution method for large-scale multiple traveling salesman problems. Oper. Res. **34**(5), 698–717 (1986)
70. M. Gendreau, A. Hertz, G. Laporte, A tabu search heuristic for the vehicle routing problem. Manag. Sci. **40**(10), 1276–1290 (1994)
71. X. Geng, Z. Chen, W. Yang, D. Shi, K. Zhao, Solving the traveling salesman problem based on an adaptive simulated annealing algorithm with greedy search. Appl. Soft Comput. **11**(4), 3680–3689 (2011)
72. Y. Ghiami, T. Van Woensel, M. Christiansen, G. Laporte, A combined liquefied natural gas routing and deteriorating inventory management problem, in *Computational Logistics*, ed. by F. Corman, S. Voβ, R. Negenborn (Springer, Switzerland, 2015), pp. 91–104
73. B. Gillett, L. Miller, A heuristic algorithm for the vehicle dispatch problem. Oper. Res. **22**(2), 340–349 (1974)
74. M. Goetschalckx, *Supply Chain Engineering* (Springer, New York, 2011)
75. A. Grasas, A. Juan, H. Lourenço, SimILS: a simulation-based extension of the iterated local search metaheuristic for stochastic combinatorial optimization. J. Simul. **10**(1), 69–77 (2014)

76. S.E. Griffis, J.E. Bell, D.J. Closs, Metaheuristics in logistics and supply chain management. J. Bus. Logist. **33**(2), 90–106 (2012)
77. J. Gromicho, J. Paixão, I. Branco, Exact solution of multiple traveling salesman problems, in *Combinatorial Optimization*, ed. by M. Akgül, H. Hamacher, S. Tüfekçi (Springer, Berlin, 1992), pp. 291–292
78. O. Guemri, A. Bektar, B. Beldjilali, D. Trentesaux, GRASP-based heuristic algorithm for the multi-vehicle inventory routing problem. 4OR **14**(4), 377–404 (2016)
79. E. Hadjiconstantinou, N. Christofides, A. Mingozzi, A new exact algorithm for the vehicle routing problem based on q-paths and k-shortest paths relaxations. Ann. Oper. Res. **61**(1), 21–43 (1995)
80. A. Hemmati, L.M. Hvattum, M. Christiansen, G. Laporte, An iterative two-phase hybrid matheuristic for a multi-product short sea inventory-routing problem. Eur. J. Oper. Res. **252**(3), 775–788 (2016)
81. F. Hillier, G. Lieberman, *Introduction to Operations Research* (McGraw-Hill Education, Singapore, Asia, 2010)
82. P. Hokama, F.K. Miyazawa, E. Xavier, A branch-and-cut approach for the vehicle routing problem with loading constraints. Expert. Syst. Appl. **47**(3), 1–13 (2016)
83. J. Holland, *Adaptation in Natural and Artificial Systems* (University of Michigan Press, Ann Arbor, 1975)
84. S.H. Huang, P.C. Lin, A modified ant colony optimization algorithm for multi-item inventory routing problems with demand uncertainty. Transp. Res. Part E: Logist. Transp. Rev. **46**(5), 598–611 (2010)
85. G. Ioannou, M. Kritikos, G. Prastacos, A greedy look-ahead heuristic for the vehicle routing problem with time windows. J. Oper. Res. Soc. **52**(5), 523–537 (2001)
86. A. Jabbarzadeh, S.G.J. Naini, H. Davoudpour, N. Azad, Designing a supply chain network under the risk of disruptions. Math. Probl. Eng. **2012**, 1–23 (2012)
87. A. Juan, J. Faulin, S. Grasman, M. Rabe, G. Figueira, A review of simheuristics: extending metaheuristics to deal with stochastic combinatorial optimization problems. Oper. Res. Perspect. **2**, 62–72 (2015)
88. A. Juan, J. Faulin, J. Jorba, J. Caceres, J.M. Marques, Using parallel & distributed computing for real-time solving of vehicle routing problems with stochastic demands. Ann. Oper. Res. **207**(1), 43–65 (2013)
89. A. Juan, J. Faulin, S. Grasman, D. Riera, J. Marull, C. Mendez, Using safety stocks and simulation to solve the vehicle routing problem with stochastic demands. Transp. Res. Part C: Emerg. Technol. **19**(5), 751–765 (2011)
90. A. Juan, S. Grasman, J. Caceres Cruz, T. Bektaş, A simheuristic algorithm for the single-period stochastic inventory-routing problem with stock outs. Simul. Model. Pract. Theory **46**, 40–52 (2014)
91. B. Kallehauge, J. Larsen, O. Madsen, Lagrangian duality applied to the vehicle routing problem with time windows. Comput. Oper. Res. **33**(5), 1464–1487 (2006)
92. M. Kamarul, I.A. Rahim, Y. Zhong, E. Aghezzaf, T. Aouam, Modelling and solving the multiperiod inventory-routing problem with stochastic stationary demand rates. Int. J. Prod. Res. **52**(14), 4351–4363 (2014)
93. M. Keskin, B. Çatay, A matheuristic method for the electric vehicle routing problem with time windows and fast chargers. Comput. Oper. Res. **100**, 172–188 (2018)
94. N. Kohl, O. Madsen, An optimization algorithm for the vehicle routing problem with time windows based on Lagrangian relaxation. Oper. Res. **45**(3), 395–406 (1997)
95. N. Kolen, A. Rinnooy Kan, H. Trienkens, Vehicle routing with time windows. Oper. Res. **35**(2), 266–273 (1987)
96. G. Kontoravdis, J. Bard, A GRASP for the vehicle routing problem with time windows. ORSA J. Comput. **7**(1), 10–23 (1995)
97. J. Knox, Tabu search performance on the symmetric traveling salesman problem. Comput. Oper. Res. **21**(8), 867–876 (1994)

98. N. Labadi, C. Prins, M. Reghioui, A memetic algorithm for the vehicle routing problem with time windows. RAIRO **42**(3), 415–431 (2008)
99. A. Landeghem, A bi-criteria heuristic for the vehicle routing problem with time windows. Eur. J. Oper. Res. **36**(2), 217–226 (1988)
100. G. Laporte, Y. Nobert, A cutting planes algorithm for the m-salesmen problem. J. Oper. Res. Soc. **31**(11), 1017–1023 (1980)
101. P. Lappas, M. Kritikos, G. Ioannou, A hybrid evolutionary optimization algorithm for the inventory routing problem. Int. J. Oper. Quant. Manag. **24**(2), 75–115 (2018)
102. P. Lappas, M. Kritikos, G. Ioannou, A two-phase solution algorithm for the inventory routing problem with time windows. J. Math. Syst. Sci. **7**(9), 237–247 (2017)
103. P. Lappas, M. Kritikos, G. Ioannou, A. Burnetas, A combination of Monte-Carlo simulation and a VNS meta-heuristic algorithm for solving the stochastic inventory routing problem with time windows. Paper presented at the proceedings of the international congress on optimization and decision science, Italy, Sorrento, 4–7 September 2017b
104. V. Leggieri, M. Haouari, A matheuristic for the asymmetric capacitated vehicle routing problem. Discret. Appl. Math. **234**, 139–150 (2016)
105. A. Letchford, R. Eglese, J. Lysgaard, Multistars, partial multistars and the capacitated vehicle routing problem. Math. Program. **94**(1), 21–40 (2002)
106. K. Li, B. Chen, A. Lyer Sivakumar, Y. Wu, An inventory-routing problem with the objective of travel time minimization. Eur. J. Oper. Res. **236**(3), 936–945 (2014)
107. L. Liao, J. Li, Y. Wu, Modeling and optimization of inventory-distribution routing problem for agriculture products supply chain. Discret. Dyn. Nat. Soc. (2013). https://doi.org/10.1155/2013/409869
108. Y.F. Liao, D.H. Yau, C.L. Chen, Evolutionary algorithm to traveling salesman problems. Comput. Math. Appl. **64**(5), 788–797 (2012)
109. S. Lin, Computer solutions of the traveling salesman problem. Bell. Syst. Tech. J. **44**(10), 2245–2269 (1965)
110. S. Lin, B. Kernigham, An effective heuristic algorithm for the traveling salesman problem. Oper. Res. **21**(2), 498–516 (1973)
111. J. Little, K. Murty, D. Sweeney, C. Karel, An algorithm for the traveling salesman problem. Oper. Res. **11**(6), 972–989 (1963)
112. S.C. Liu, W.T. Lee, A Heuristic method for the inventory routing problem with time windows. Expert Syst. Appl. **38**(10), 13223–13231 (2011)
113. S.C. Liu, M.C. Lu, C.H. Chung, A hybrid heuristic method for the periodic inventory routing problem. Int. J. Adv. Manuf. Technol. **85**(9–12), 2345–2352 (2016)
114. M. López-Ibáñez, C. Blum, J. Ohlmann, B. Thomas, The traveling salesman problem with time windows: Adapting algorithms from travel-time to makespan optimization. Appl. Soft Comput. **13**, 3806–3815 (2013)
115. J. Lysgaard, A. Letchford, R. Eglese, A new branch-and-cut algorithm for the capacitated vehicle routing problem. Math. Program. **100**(2), 423–445 (2004)
116. V. Maniezzo, T. Stützle, S. Voβ, *Matheuristics: Hybridizing Metaheuristics and Mathematical Programming* (Springer, New York, 2009)
117. Y. Marinakis, G.R. Iordanidou, M. Marinaki, Particle swarm optimization for the vehicle routing with stochastic demands. Appl. Soft Comput. **13**(4), 1693–1704 (2013)
118. P. Miliotis, Using cutting planes to solve the symmetric travelling salesman problem. Math. Program. **15**(1), 177–188 (1978)
119. P. Miliotis, Integer programming approaches for the travelling salesman problem. Math. Program. **10**(1), 367–378 (1976)
120. A. Mjirda, B. Jarboui, R. Macedo, S. Hanafi, A variable neighborhood search for the multi-product inventory routing problem. Electron. Notes Discret. Math. **39**, 91–98 (2012)
121. A. Mjirda, B. Jarboui, R. Macedo, S. Hanafi, N. Mladenović, A two phase variable neighborhood search for the multi-product inventory routing problem. Comput. Oper. Res. **52**, 291–299 (2014)

122. A. Mjirda, B. Jarboui, J. Mladenović, C. Wilbaut, S. Hanafi, A general variable neighborhood search for the multi-product inventory routing problem. IMA J. Manag. Math. **27**, 39–54 (2016)
123. N. Mladenović, P. Hansen, Variable neighborhood search. Comput. Oper. Res. **24**(11), 1097–1100 (1997)
124. N.H. Moin, S. Salhi, N.A.B. Aziz, An efficient hybrid genetic algorithm for the multi-product multi-period inventory routing problem. Int. J. Prod. Econ. **133**(1), 334–343 (2011)
125. D. Naddef, Polyhedral theory and branch-and-cut algorithms for the symmetric TSP, in *The Traveling Salesman Problem and Its Variations*, ed. by G. Gutin, A. Punnen (Springer, New York, 2002), pp. 29–116
126. D. Naddef, G. Rinaldi, Branch-and-cut algorithms for the capacitated VRP, in *The Vehicle Routing Problem*, ed. by P. Toth, D. Vigo (SIAM, Philadelphia, 2002), pp. 53–81
127. R. Nambirajan, A. Mendoza, S. Pazhani, T.T. Narendran, K. Ganesh, CARE: heuristics for two-stage multi-product inventory routing problems with replenishments. Comput. Ind. Eng. **97**, 41–57 (2016)
128. P. Nolz, N. Absi, D. Feillet, A stochastic inventory routing problem for infectious medical waste collection. Networks **63**(1), 82–95 (2014)
129. P. Nolz, N. Absi, D. Feillet, A bi-objective inventory routing problem for sustainable waste management under uncertainty. J. Multi-Criteria Decis. Anal. **21**(5–6), 299–314 (2014)
130. M. Padberg, G. Rinaldi, A branch-and-cut algorithm for the resolution of large-scale symmetric travelling salesman problems. SIAM Rev. **33**(1), 60–100 (1991)
131. M. Padberg, G. Rinaldi, Optimization of a 532-city symmetric traveling salesman problem by branch and cut. Oper. Res. Lett. **6**(1), 1–7 (1987)
132. C. Papadimitriou, K. Steiglitz, *Combinatorial Optimization: Algorithms and Complexity* (Dover Publications, New York, 1998)
133. Y.B. Park, J.S. Yoo, H.S. Park, A genetic algorithm for the vendor-managed inventory routing problem with lost sales. Expert Syst. Appl. **53**, 149–159 (2016)
134. J. Pekny, D. Miller, A parallel branch and bound algorithm for solving large asymmetric traveling salesman problems. Math. Program. **55**(1–3), 17–33 (1992)
135. D. Pisinger, S. Ropke, A general heuristic for vehicle routing problems. Comput. Oper. Res. **34**(8), 2403–2435 (2007)
136. D. Popović, M. Vidović, G. Radivojević, Variable neighborhood search heuristic for the inventory routing problem in fuel delivery. Expert Syst. Appl. **39**(18), 13390–13398 (2012)
137. J.Y. Potvin, Genetic algorithms for the traveling salesman problem. Ann. Oper. Res. **63**(3), 337–370 (1996)
138. J.Y. Potvin, G. Lapalme, J. Rousseau, A generalized k-opt exchange procedure for the MTSP. INFOR: Inf. Syst. Oper. Res. **27**(4), 474–481 (1989)
139. J.Y. Potvin, J. Rousseau, An exchange heuristic for routeing problems with time windows. J. Oper. Res. Soc. **46**(12), 1433–1446 (1995)
140. J.Y. Potvin, J. Rousseau, A parallel route building algorithm for the vehicle routing and scheduling problem with time windows. Eur. J. Oper. Res. **66**(3), 331–340 (1993)
141. C. Prins, A simple and effective evolutionary algorithm for the vehicle routing problem. Comput. Oper. Res. **31**(12), 1985–2002 (2004)
142. L. Qin, L. Miao, Q. Ruan, Y. Zhang, A local search method for periodic inventory routing problem. Expert Syst. Appl. **41**(2), 765–778 (2014)
143. B. Raa, Fleet optimization for cyclic inventory routing problems. Int. J. Prod. Econ. **160**, 172–181 (2015)
144. M. Reimann, K. Doerner, F. Hartl, D-ants: Savings based ants divide and conquer the vehicle routing problem. Comput. Oper. Res. **31**(4), 563–591 (2004)
145. D. Rosenkrantz, R. Stearns, P. Lewis, An analysis of several heuristics for the traveling salesman problem. SIAM J. Comput. **6**(3), 563–581 (1977)
146. R. Russel, An effective heuristic for the m-tour traveling salesman problem with some side conditions. Oper. Res. **25**(3), 517–524 (1977)

147. J. Ryan, T. Bailey, J. Moore, W. Carlton, Reactive tabu search in unmanned aerial reconnaissance simulations. Paper presented at the proceedings of the 1998 winter simulation conference (United States of America, Washington, 13–16 December, 1998), pp. 873–880
148. G. Schmidt, W. Wilhelm, Strategic, tactical and operational decisions in multi-national logistics networks: A review and discussion of modeling issues. Int. J. Prod. Res. **38**(7), 1501–1523 (2000)
149. Q. Shen, F. Chu, H. Chen, A Lagrangian relaxation approach for a multi-mode inventory routing problem with transshipment in crude oil transportation. Comput. Chem. Eng. **35**(10), 2113–2123 (2011)
150. V. Shirokikh, V. Zakharov, Dynamic adaptive large neighborhood search for inventory routing problem, in *Modeling, Computation and Optimization in Information Systems and Management Sciences*, ed. by H.A.L. Thi, T.P. Dinh, N.T. Nguyen (Springer, New York, 2015), pp. 231–241
151. N. Shukla, M.K. Tiwari, D. Ceglarek, Genetic-algorithms-based algorithm portfolio for inventory routing problem with stochastic demand. Int. J. Prod. Res. **51**(1), 118–137 (2013)
152. D. Simić, S. Simić, Evolutionary approach in inventory routing problem, in *Advances in Computational Intelligence*, ed. by I. Rojas, G. Joya, J. Cabestany (Springer, New York, 2013), pp. 395–403
153. T. Singh, J. Arbogast, N. Neagu, An incremental approach using local-search heuristic for inventory routing problem in industrial gases. Comput. Chem. Eng. **80**, 199–210 (2015)
154. M. Solomon, Algorithms for the vehicle routing and scheduling problem with time window constraints. Oper. Res. **35**(2), 254–265 (1987)
155. O. Solyali, H. Süral, A branch-and-cut algorithm using a strong formulation and an a priori tour based heuristic for an inventory-routing problem. Transp. Sci. **45**(3), 335–345 (2011)
156. J.H. Song, K. Furman, A maritime inventory routing problem: practical approach. Comput. Oper. Res. **40**(3), 657–665 (2013)
157. M. Soysal, J. Bloemhof-Ruwaard, R. Haijema, J. Van der Vorst, Modeling a green inventory routing problem for perishable products with horizontal collaboration. Comput. Oper. Res. **89**, 168–182 (2016)
158. M. Soysal, J. Bloemhof-Ruwaard, R. Haijema, J. Van der Vorst, Modeling an inventory routing problem for perishable products with environmental considerations and demand uncertainty. Int. J. Prod. Econ. **164**, 118–133 (2015)
159. É. Taillard, Parallel iterative search methods for vehicle routing problems. Networks **23**(8), 661–673 (1993)
160. É. Taillard, P. Badeau, M. Gendreau, F. Guertin, J.Y. Potvin, A tabu search heuristic for the vehicle routing problem with soft time windows. Transp. Sci. **31**(2), 170–186 (1997)
161. K.C. Tan, A framework of supply chain management literature. Eur. J. Purch. Supply Manag. **7**(1), 39–48 (2001)
162. V. Tatsis, K. Parsopoulos, K. Skouri, I. Konstantaras, An ant-based optimization approach for inventory routing, in *EVOLVE—A Bridge Between Probability, Set Oriented Numerics, and Evolutionary Computation IV*, ed. by M. Emmerich, A. Deutz, O. Schütze, T. Bäck, E. Tantar, A. Tantar, P. Moral, P. Legrand, P. Bouvry, C. Coello (Springer, New York, 2013), pp. 107–121
163. P. Toth, D. Vigo, The granular tabu search and its application to the vehicle routing problem. INFORMS J. Comput. **15**(4), 333–346 (2003)
164. P. Toth, D. Vigo, *The Vehicle Routing Problem* (SIAM, Philadephia, 2002)
165. P. Venkatesh, A. Singh, Two metaheuristic approaches for the multiple traveling salesperson problem. Appl. Soft Comput. **26**, 74–89 (2015)
166. T. Volgenant, R. Jonker, A branch and bound algorithm for the symmetric traveling salesman problem based on the 1-tree relaxation. Eur. J. Oper. Res. **9**(1), 83–89 (1982)
167. S. Wierzchoń, M. Kłopotek, *Modern Algorithms of Cluster Analysis* (Springer, Cham, 2018)
168. Z. Yang, M. Emmerich, T. Bäck, J. Kok, Multicriteria inventory routing by cooperative swarms and evolutionary algorithms, in *Bioinspired computation in artificial systems*, ed. by J.M. Ferrández Vicente, J.R. Álvarez- Sánchez, F. de la Paz López, J. Toledo-Moreo, H. Adeli (Springer, New York, 2015), pp. 127–137

169. M. Zachariasen, M. Dam, Tabu search on the geometric traveling salesman problem, in *Meta-Heuristics*, ed. by I. Osman, J. Kelly (Springer, Boston, 1996), pp. 571–587
170. W. Zeng, Q. Zhao, Study of stochastic demand inventory routing problem with soft time windows based on MDP, in *Advances in Neural Network Research and Applications*, ed. by Z. Zeng, J. Wang (Springer, Shanghai, 2010), pp. 193–200
171. S. Zhan, J. Lin, Z. Zhang, Y. Zhong, List-based simulated annealing algorithm for traveling salesman problem. Comput. Intell. Neurosci. **2016**, 1–12 (2016)

Part VI
Computer Science-Based Technologies in Medicine and Biology

Chapter 16
Homeodynamic Modelling of Complex Abnormal Biological Processes

Athanasios Sofronis and Panagiotis Vlamos

Abstract Biological systems are defined by their complexity and nonlinearity and thus provide fertile ground for the development of nonlinear deterministic models for predicting aspects of their behavior. This approach motivated the introduction of the concept named Homeodynamics which we will outline and then proceed to present a case study on the application of bifurcation theory and stability analysis, both topological approaches of dynamical systems, on three biological mechanisms of great importance in the context of Homeodynamics. These will be protein folding, protein dynamics and epigenetics.

16.1 Biological Processes

Proteins are highly dynamic polymers which carry out most of the functions inside and outside a cell. They are made out of 20 possible amino-acids strung together by peptide bonds and fold in 3D shapes tailored for carrying their specific function or functions. In certain diseases the proteins most implicated in their pathogenesis have folded in incorrect shapes resulting in either loss of function or worse in the gain of toxic function when the misfolded proteins form oligomers or aggregates disrupting the normal functioning of the cell.

It was assumed that there exists one unique functional conformation which is termed the native state and trying to predict it given the linear sequence of the particular amino-acids which make up the protein has been a longstanding problem called "the protein folding problem". There has been some progress lately with the use of Machine Learning and other computational approaches but generally remains far from solved. But relatively recent discoveries are showing that the original formulation of the problem was built on a generally incorrect premise. Because it seems that in reality there is not even a unique, "native fold" to speak of in the first

A. Sofronis · P. Vlamos (✉)
Bioinformatics and Human Electrophysiology Lab, Department of Informatics, Ionian University, Corfu, Greece
e-mail: vlamos@ionio.gr

G. A. Tsihrintzis and M. Virvou (eds.), *Advances in Core Computer Science-Based Technologies*, Learning and Analytics in Intelligent Systems 14,
https://doi.org/10.1007/978-3-030-41196-1_16

place since most proteins contain large regions (>30 amino-acids) which fold into an ensemble of different conformations with correspondingly different functions, called Intrinsically Disordered Proteins (IDPs). There is a broad spectrum in the amount of possible disorder, from fully unstructured proteins to partially structured ones. And to make matters even more complicated in some cases it has been found even the amount of disorder of a particular protein is not always fixed. Some contain a small intrinsically disordered region called molecular recognition feature which undergoes disorder-to-order upon binding to their molecular partners while others retain their disorder during interaction forming what are called fuzzy complexes.

Intrinsically disordered proteins with their flexibility play a very important role in protein dynamics, a term denoting the highly dynamic structure-function landscape of protein-protein interactions and enzyme catalysis. One such highly important example of protein dynamics in which IDPs are involved is allosteric regulation which is the process in which the binding of a molecule to a site different from the active site of an enzyme where catalysis occurs, causes conformational changes that affect the active cite and hence the catalysis of a reaction. It comes as no surprise that the majority of proteins implicated in diseases are IDPs.

One last biological process we will discuss is epigenetics. This refers to highly complex networks which exert control and regulation of gene expression and activity without directly changing the DNA sequence and which give the cell its ability to respond to the vast number of different scenarios it is confronted with despite the relatively small number of genes in the DNA.

This for the most part is accomplished by what are called transcription factors, which can bind to specific non-coding regions of DNA and either cause either an increase or a decrease in the expression of a particular gene.

Structures of an IDP (Thylakoid soluble phosphoprotein Protein Data Base id: 2FFT) and a well-folded (ubiquitin PDP id: 1D3Z).

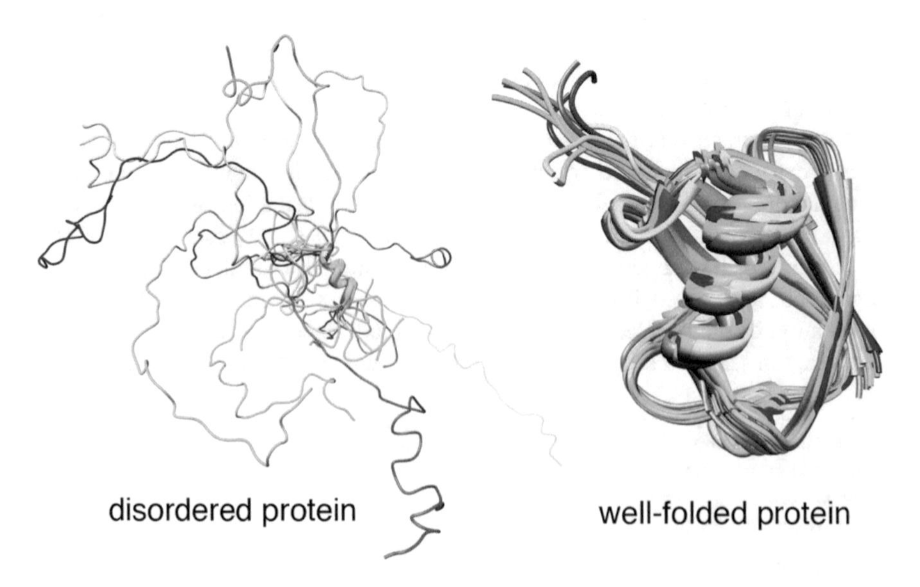

16.2 Nonlinear Dynamics in Mathematics and Physics

It is significant to note that the great power in terms of predictive ability of the physical laws which Physics enjoys stems largely from the linearity of the mathematical formulations in which they are cast. Most, if not all, of physical processes are modeled as differential equations and even the notoriously counter-intuitive quantum mechanics as well as Maxwell's laws of electromagnetism are in fact linear partial differential equations. Linear deterministic dynamical systems, defined as those systems whose future states are fully determined by their initial state plus an evolution law or differential equation, obey the superposition principle which states that if some independent solutions have been found, their linear combination is also a solution of the equation. It allows one therefore, to completely determine a linear system if in possession of a very small subset of possible solutions as long as they are (linearly) independent. However, most real-life physical processes are not linear and linear approximations to them can in some cases fall seriously short of modeling them within acceptable ranges of accuracy. Analytic solutions of their differential equations, that is solutions easily obtained in linear systems which are expressed as arithmetic operations and typical functions are in most cases impossible to obtain but Poincare, in the 19th century, while grappling with the three-body problem came up with a qualitative and topological approach which shifted the focus away from trying to derive precise predictions of future states of a system and instead inquired about the various possible types of stability and long term behavior that the trajectories of the system can display as time goes from zero to infinity. To that end what becomes the new objects of study are the phase spaces and the phase portraits associated with a dynamical system or differential equation. The phase space represents the space whose number of dimensions stand for the degrees of freedom of the system (for example 3 for position in 3D space) and where all possible trajectories of it are located. The phase portrait is used to gauge the more prominent features of a dynamical system's trajectories such as the stability of its various equilibrium (or fixed) points.

Those represent states which if the system ever reaches, will remain at as time approaches infinity. A more formal approach to dynamical systems is to view them as maps or transformations and so fixed points can be represented as those points x^* such that $f(x^*) = x^*$ where f represents the map. One can then proceed to classify the different stabilities associated with each fixed point according to whether it repels or attracts nearby trajectories or, viewed from a different angle, according to whether small perturbations away from them will return back to the fixed point or whether they will diverge away as times runs forward. There is a further classification on different types of stability of equilibrium points and the most commonly encountered are Lyapunov stable points where solutions starting near the fixed points will forever stay near them and asymptotically stable which is a stronger condition where fixed points are Lyapunov stable and in addition nearby trajectories converge to them as time goes to infinity. In linear autonomous systems this stability analysis is carried out by inspection of the eigenvalues associated with the coefficient matrix of the system.

This stability analysis can be carried out on nonlinear systems as well, by linearizing near the equilibrium points which involves constructing the so-called Jacobian matrix whose entries are the first order Taylor polynomials of the vector function and serves as the linearized version of the previously nonlinear system.

One suggestive example showcasing the richness of behavior that nonlinear systems can exhibit is a bifurcation. By a bifurcation we mean that a very abrupt change in the topological features of an equilibrium point via a small change in the value of a parameter, the bifurcation parameter that occurs at a bifurcation point. Just as with stabilities of equilibrium points, bifurcations are classified as well. A saddle-note or fold bifurcation denotes the mechanism by which equilibrium points are created and destroyed.

In a transcritical bifurcation the fixed point changes its stability at the bifurcation point. The following diagram depicts one such case with its associated equation. The bifurcation parameter is r and the two fixed points switch their stability at the bifurcation point.

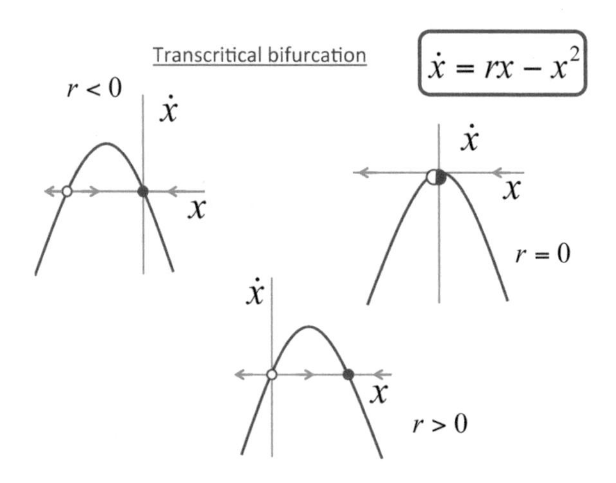

Other than fixed points the next most encountered type of behavior that dynamical systems exhibit with unique qualitative features and whose stability we can analyze is the limit cycle. This is a closed curve to which either neighboring trajectories spiral in where it is called a stable limit cycle or away in the case of the unstable limit cycle. Many oscillatory phenomena in nature can be modeled by limit cycles.

Another type of bifurcation which involves both fixed points and limit cycles is the Hopf bifurcation. One case is when stable fixed point transforms into a stable limit cycle surrounding an unstable fixed point at the bifurcation point, which is called supercritical Hopf bifurcation and in the other case an unstable fixed point turns into an unstable limit cycle surrounding a stable fixed point. Below is depicted a supercritical Hopf bifurcation with parameter β:

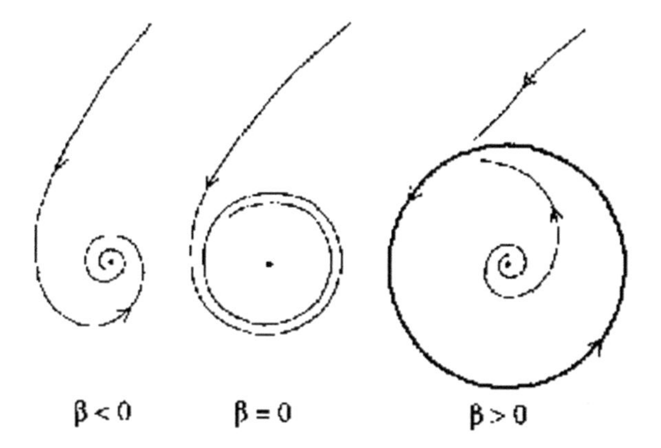

Lastly one commonly encountered bifurcation shown below is a period-doubling bifurcation where at each bifurcation parameters the period of the system doubles encountered in the logistic map which is a discrete dynamical system as opposed to differential equations which are continuous.

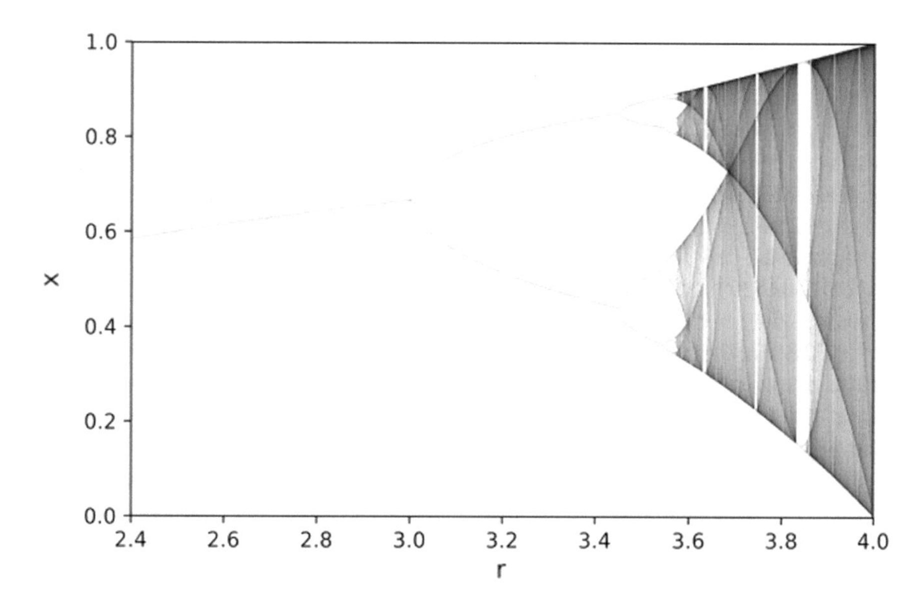

A more general concept under which the concept of a fixed point is subsumed, is that of an attractor which is defined as any subset of the phase space where trajectories sufficiently close to the basin of attraction, the basin being a neighborhood of the attractor, get attracted inside the subspace that is the attractor and stay there forward in time. Attractors initially were simple points or lines but now attractors with more exotic geometry have been included such as manifolds. Perhaps the most prominent and exotic example of exotic geometry is that of fractal set which are called strange attractors. Strange attractors are the attractors of a chaotic system, a system that is non periodic and distinguished by its sensitive dependence on initial conditions.

The study of nonlinear systems has evidently given rise to a broad field with wide modelling ranges by expanding on its topological methods for stability analysis coupled with tried and tested linearization approximations whenever those are admissible.

And no better source of exciting and challenging new nonlinear systems for this field to bring its powerful methods to bear on, than what biological systems offer in plenitude. We will pursue this line of thought in more detail.

16.3 Homeodynamics

Physiology, the field that studies the normal functioning of an organism, has traditionally been operating under one unifying principle, that of homeostasis introduced in 1929 by Walter Cannon. The idea is that all organisms strive to keep their internal environment regulated within a functional range of certain important parameters such as temperature or electrolyte concentrations and maintain a steady state of internal conditions. But the highly dynamic and non-linear nature of biological systems and processes that has been emerging from their study called for a new concept to reflect this shift in our understanding. That concept was called homeodynamics and was introduced by F. Eugene Yates as early as 1994 and it comprises a far richer and suitable framework upon which we can develop more realistic and thus useful biomedical models with the hope that a more powerful new Medicine can be deployed as a result.

Most of the new ideas and tools which homeodynamics used to build on top of homeostasis before it, were an amalgamation of concepts from nonlinear dynamical systems and non-equilibrium thermodynamics, a field we will discuss shortly, and applied them in the context of biological systems. It no longer saw one stable steady state resistant to perturbations like one complicated thermostat. In its place it introduced the concept of bifurcations covered previously. An organism can indeed display drastic qualitative transitions in both the stability status of its fixed points and limit points as well as potentially becoming chaotic for certain bifurcation parameters.

The other field mentioned above from which homeodynamics borrowed ideas is so-called non-equilibrium thermodynamics. Much in the same way that the study of nonlinear systems became a field of its own long after the study of linear ones despite nonlinear systems being the norm in Nature, non-equilibrium thermodynamics arose as an object of study long after traditional equilibrium-oriented thermodynamics became a field, and despite ourselves being living creatures operating away from equilibrium states in a state of order. Unlike nonlinear dynamics though there is no unifying framework as of yet, under which all such systems can be studied and there are various different approaches to it. As it pertains to biological systems though the dissipative systems idea introduced by Prigogine as a prototypical open system, which all biological organisms are, served as a scaffolding upon which the notion of self-organization was developed. Self-organization described all phenomena where order arises spontaneously out of disorder in certain contexts and is thought to be a type of phase transition, like the boiling of water, which can be modeled with the use of certain nonlinear dynamical models. In the image below is shown the Belouzov–Zhabotinsky reaction which serves as both an example of non-equilibrium thermodynamics and also of a chaotic system.

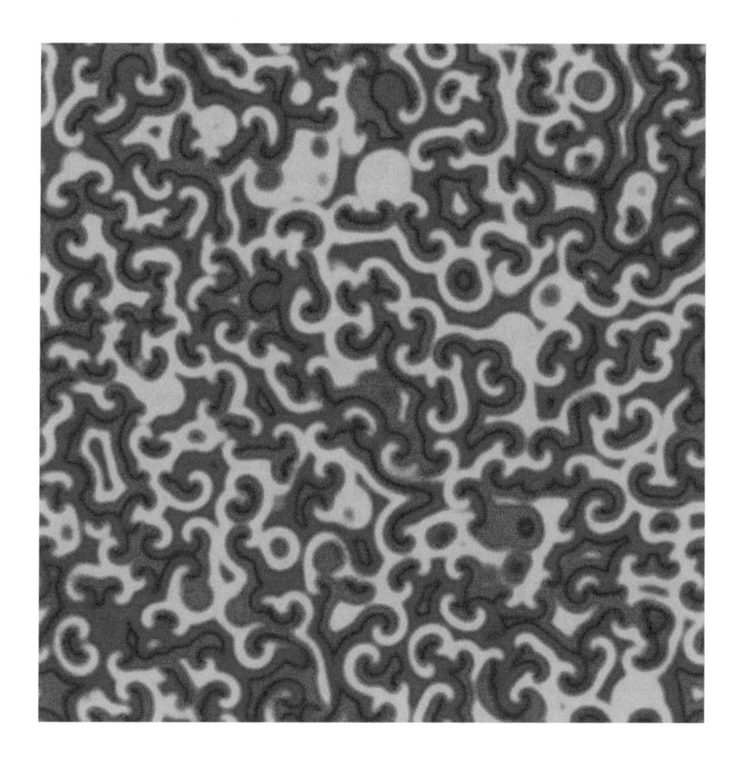

16.4 Homeodynamic Modeling of Protein Folding/Binding

Two such possible and particularly attractive processes where homeodynamic approaches can be applied is protein folding and protein binding interactions. Firstly folding, as was touched upon briefly earlier in this paper, constitutes a process that still evades our understanding and the recent discoveries on IDPs only serve to further obfuscate our understanding and hinder our attempts at modeling it. Secondly, binding interactions is so widespread a process and integral to every single molecular event taking place in the trillions of cells we have and in all of life, that it is hardly considered a separate biological process. Also called molecular binding, it comprises of all possible binding forces and interactions known, with covalent being the strongest, hydrophobic forces and hydrogen bonds being the most widespread in use for molecular recognition and thus all vital cellular systems such as communication (between cells for example) and London forces, salt bridges and disulfide bridges being the rest of them with important functions as well.

The two key features which those two processes possess and make them highly suitable for a homeodynamic approach are bistability and cooperativity. Bistability refers to a dynamical system possessing two stable equilibrium points with one unstable in between them and corresponds to a pitch-fork bifurcation.

Bistability is a widespread phenomenon among biological processes but in this particular case we make use of the fact that amyloidogenic proteins have been found to be able to switch spontaneously between their amyloid and non-amyloid conformation and thus it is not unreasonable to hypothesize the existence of a bistable topology whose bifurcation parameter if found could shed light on the misfolding process in general.

Another pertinent line of research discovered bistability that seems to occur when a particular chaperone—a protein tasked with dealing with misfolded proteins by either refolding them or tagging them for degradation if the previous attempts prove unsuccessful—vary in its concentrations, due to naturally occurring fluctuations. A threshold transition arises then between a state of aggregation of the misfolded proteins and a state with no aggregation present, induced by the chaperone whose concentration plays the role of the bifurcation parameter.

This type of bistability could help explain certain nonlinear kinetic processes associated with aggregations of misfolded proteins as that found in the development of prion disease whose precise mechanics still elude us.

This aforementioned threshold effect is being increasingly found in various cellular processes and in particular those involved in transcriptional control and metabolic regulation. For example there is a threshold effect associated with the number of mitochondrial DNAs that are defective in a particular tissue or organ which determines whether or not they are defective. It is this defectiveness that is a causative factor in aging and a multitude of degenerative diseases and thus poses another attractive candidate for the nonlinear approach. A last but equally important process exhibiting bistability with important implications for disease and aging is Nrf-1, a nuclear transcriptional factor whose function is to upregulate genes involved in mitochondrial biogenesis, an all around health inducing action, as well as other genes which can help the cell in oxidative conditions. The bistable nature of the process becomes evident when one considers that the function of Nrf-1 is dependent upon its redox state. When oxidized, in other words when a single electron gets pulled away from its outer shell by a free radical, it is in its active form and when it gets reduced by dedicated enzymes it loses its activity. There are more transcription factors involved in the regulation of the redox state of the cell which in return when outside of certain parameters is perhaps the most significant causative factor in all degenerative diseases and aging. Modeling all these threshold effects as bifurcation systems and deriving the relevant parameters therefore evidently holds great promise as a potential therapeutic strategy against the most debilitating diseases.

The second key feature mentioned above is cooperativity. This refers to various phenomena all related to binding interactions among molecules where the affinity of multiple (nearly) identical ligands for a particular molecule is not independent as would be expected in that if more binding sites are occupied by ligands the affinity of the remaining sites is greater than it would be if the other sites were unoccupied where we have positive cooperativity and a correspondingly less than expected affinity in the case of negative cooperativity. Both behaviors are displayed by haemoglobin positive cooperativity when Oxygen binds to it and negative cooperativity is induced by binding of CO_2. This whole system (haemoglobin, oxygen, CO_2) could be modeled as a limit cycle. Cooperativity as is clear by now exemplifies a nonlinear system and its modelling could shed light on the highly intricate interconversions between states that underpin both the functions of enzymes and receptors, molecules whose roles in both health and disease are of primal importance.

16.5 Conclusion

In this paper we sought to delineate the developments associated with the highly promising field of nonlinear dynamical systems as well as an almost parallel development of ideas inside the fields of Biology and Physiology taking the shape of homeodynamics. Additionally, we provided a few noteworthy examples of possible biomedical applications of these ideas, both past works as well as some new ideas for potential future research in the field. Although far from a comprehensive coverage of the various topics, we hope it will suffice to give a birds-eye view of its contours.

References

1. R. Rosen, *Dynamical Systems Theory in Biology* (Wiley Interscience, New York, 1970)
2. A.K. Dunker, J.D. Lawson, C.J. Brown, R.M. Williams, P. Romero, J.S. Oh, C.J. Oldfield, A.M. Campen, C.M. Ratliff, K.W. Hipps, J. Ausio, M.S. Nissen, R. Reeves, C. Kang, C.R. Kissinger, R.W. Bailey, M.D. Griswold, W. Chiu, E.C. Garner, Z. Obradovic, Intrinsically disordered protein. J. Mol. Graph. Model. (2001)
3. G. Boeing, Visual analysis of nonlinear dynamical systems: chaos, fractals, self-similarity and the limits of prediction (2016)
4. M.J. Korenberg, I.W. Hunter, The identification of nonlinear biological systems: Volterra kernel approaches. Ann. Biomed. Eng. (1996)
5. S. Lichter, T.L. Friesz, *Networks and Dynamics: The Structure of the World We Live In. Network Science, Nonlinear Science and Infrastructure Systems* (2007)
6. A. Marathe, R. Govindarajan, Nonlinear dynamical systems, their stability, and chaos. Appl. Mech. Rev. (2014)
7. S. Strogatz, *Non-linear Dynamics and Chaos: With applications to Physics, Biology, Chemistry and Engineering* (CRC PRESS, 2000)
8. I. Prigogine, *Introduction to Thermodynamics of Irreversible Processes*, 3rd edn. (Wiley Interscience, New York, 1955/1961/1967)

9. F.E. Yates, Order and complexity in dynamical systems: homeodynamics as a generalized mechanics for biology. Math. Comput. Model. (1994)
10. T.R. Rieger, R.I. Morimoto, V. Hatzimanikatis, Bistability explains threshold phenomena in protein aggregation both in vitro and in vivo. Biophys. J. (2006)
11. N.D. Lee Know, *Mitochondria and the Future of Medicine* (Chelsea Green Publishing, 2014)

Chapter 17
Metadata Web Searching EEG Signal

Marios Poulos and Sozon Papavlasopoulos

Abstract In this paper, the problem of developing appropriate information search and retrieve mechanisms and tools in the web environment, is investigated. This problem is of great interest to those in information technology, since a vast amount of heterogeneous data are available, end so, are not interoperable on the Web to researchers or other interest groups. The problem is addressed here using, as, effective encoding for locating and sharing a very specific class of data, that of uniform diagnostic EEG features. In this study is proposed a suitable metadata schema, based on knowledge of medical EEG signal processing. The defined schema tries to initiate a dialog for further development of metadata specific formats of EEG recordings. The final aim of this study is to offer a web searching tool for data recorded and stored in a different operational structure or using several software and hardware systems, in a uniform EEG data collection for research and research purposes.

Keywords XML schema · EEG · Biomedical research · Spectral features · Diagnostic factors · Metadata · Information interoperability

17.1 Introduction

The EEG has a long history in clinical evaluations of cerebro-vascular diseases. EEG brain mapping is a term commonly used for several quantitative (computerized) EEG techniques [1] and can help highlight or identify regional features of the EEG. The classification of specific EEG features can also help in determining whether some features are present to an abnormal degree. EEG data has been traditionally recorded using flat file formats, such as the MIT-BIH file library [2, 3].

M. Poulos (✉) · S. Papavlasopoulos
Laboratory of Information Technology-Faculty of Information Sciences and Informatics, Ionian University, Corfu, Greece
e-mail: mpoulos@ionio.gr

S. Papavlasopoulos
e-mail: sozon@ionio.gr

G. A. Tsihrintzis and M. Virvou (eds.), *Advances in Core Computer Science-Based Technologies*, Learning and Analytics in Intelligent Systems 14,
https://doi.org/10.1007/978-3-030-41196-1_17

Interoperability problems between heterogeneity information systems for the exchange of complex medical data have grown significantly in recent years. For this reason, the HL7 committee has been actively cooperating with the World Wide Web Consortium (W3C) to define XML guidelines to represent medical information (XML) [4]. Other XML-based initiatives for the representation and distribution of biomedical information have been developed such us the ASTM E31.25 subcommittee [5], the CEN/TC251 Task Force on XML Applications in Healthcare [6] and the Clinical Data Interchange Standards Consortium (CDISC) [7]. However, those efforts have not focused on EEG data on patient records as well as the administrative and financial transactions associated with a clinical workplace environment. Taking this into account all the above mentioned we can summarize that these are addressed in the following three factors:
First: The clinical application of EEG brain mapping is still very limited. Most scientific reports on these techniques concerned research applications rather than clinical ones. Techniques, used in EEG brain mapping, vary significantly among different philosophy laboratories regarding the software strategy of management thus, a significant number of technical and clinical problems arise with many medical applications.
Second: There is a problem with interoperability among several EEG databases because they are constructed using different philosophies. Thus, the collection of uniform EEG data concerning a specific disease case and using web technology is impossible.

Current research [2] has developed an initial set of information for the storage and representation of the EEG, encoded into an XML-vocabulary. Although, this model is promising for the flexible exchange of data and the analysis of EEG information on the Web platform, itis based on recorded EEG data independent of the number of channels or type of experiments involved. On the one hand, this approach does provide a solution, in theory, to the problems of the interoperability of EEG data.
Third: The application of the XML provides a significant solution to the problem of information interoperability. However, there is universal disagreement concerning metadata models, especially in the medical area and this puts considerable constraint on the design and development of new search systems, capable of profiting benefiting from the new information attached to digital resources, initiatives, like the Open Archives Initiative [8]. Data models have proved themselves able to offer effective search and retrieval attributes, which involve a vast set of resources, by efficiently employing new technologies. Thus, taking into account that heterogeneity is a predetermined problem and not yet, efficiently solved the development of the proposed XML schema was undertake in this studying.
Our Study
The present work is a continuation of our previous work on the same subject entitled 'An XML schema for the sharing and communication of heterogeneous EEG data for diagnostic and research purposes' [9] and the study [10] which introduces the new metadata language entitled "EEGML", which is based on in an XML format.

This paper presents an XML strategy which targets by which to be exploited on medical and collaboration mechanisms and that would allow web information resources, to be searched and specific EEG data retrievable in the same way as data from digital libraries or other areas with well-defined search and retrieval standards. The achievement of this goal means facing the challenge of medical information interoperability, which covers numerous areas, including metadata formats, document model extraction and the use of statistical models for EEG analysis. This also means taking into consideration information extensibility or community/industry standards. The abstract structure of the paper is as follows: The paper begins with an introduction to the XML Schema and its classification structure though does not pretend to be a complete state-of-the-art review of Web information search and retrieval technologies. After the initial analysis, we begin the construction of the schema by classifying EEG factors for each EEG processing stage.

Our aim being to promote, the application of standards for message exchange and EEG data integration. In this way, the most significant clinical diagnostic EEG features will be transformed into a common metadata XML schema in order to answer, at any given time, queries and inquiries derived from a set of scenarios.

Generally, these definitions of the above scenarios are implemented in four main branches. In the first branch (EXPERIMENTAL_ VARIABLES)", is the discovery of the EEG's original data according to those significant factors which influence its background activity 0 such as, demographic and medication variables and symptoms. In the second branch, (EEG_RECORDING_PARAMETERS), we determine describing the recording characteristic of each EEG data and this is divided into seven (7) main elements "ARTIFACT_REJECTION, CHANELL_TYPE, SAMPLE_SIZE_SELECTION, SAMPLING_RECORDING, SLEEPING_CONDITIONS, ELECTRODE_MONTAGE and MATRIX_DATA_COLLECTION. In the third branch can be found the variable elements of EEG_ANALYSIS and those are: Primary Analysis and Feature extraction. Finally, in the fourth branch of the proposed XML schema are the elements of CLINICAL_ DIAGNOSIS.

All these stages are also agreed with the conclusions of who ese stages. The general descriptive schedule of the proposed schema is presented in Fig. 17.1.

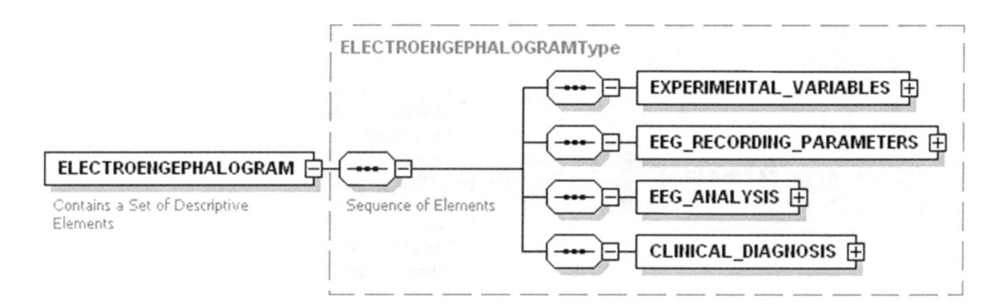

Fig. 17.1 The general descriptive schedule EEG schema

In general, the workflow developed focuses, in EEG analysis, on using, mainly, the global element of matrix data collection, as a compatibility tool for the various EEG data and the others of CLINICAL_ DIAGNOSIS and EXPERIMENTAL_ VARIABLES optionally.

17.2 The Structure of the Proposed Schema

In all cases regarding schema design and value representation (see Fig. 17.1) we consider a series of attributes attached to the end elements of the schema. These attributes are attached to the element in order to express decision that also characterized by a "use" predicate which denotes the usage on the representation of data.

17.2.1 The Experimental Variables

The experimental variables of an EEG considered as a significant factor of the influence of the EEG analysis [10, 11]. In this way, we included three child elements that are depicted in Fig. 17.2.

Also, in Table 17.1 the attributes of the elements attached to the EXPERIMENTAL_VARIABLES complex type are summarized.

In more details the analysis of the DEMOGRAPHICS element [12] is presented in Fig. 17.3.

Furthermore, PHARMACOLOGICAL_VARIABLES issue is influenced by EEG activity and thus, two major categories are adopted: NON_MEDICATION and MEDICATION. More specifically, the medication variables are divided into PSYCHOTROPIC and ANTIEPILEPTIC categories where the Psychotropic category is then divided into PSYCHO_STIMULANT, ANTIPSYCHOTIC, ANTIDEPRESSANT and TRANQUILIZERS categories of medication (see Fig. 17.4).

Finally, the SYPTOMS issue is divided in nine (9) cases as shown in Fig. 17.5.

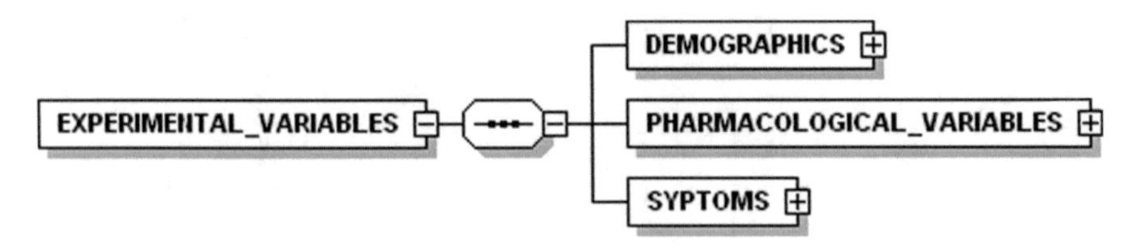

Fig. 17.2 The schema of the experimental variables of EEG

Table 17.1 Attributes attached to the elements of the Experimental_Variables complex type

Attribute name	Type	Elements
YEARS	xs:integer	AGE
PRESENT	xs:boolean	GENTER_MALE
PRESENT	xs:boolean	GENTER_FEMALE
PRESENT	xs:boolean	HANDEDNESS_LEFT
PRESENT	xs:boolean	HANDEDNESS_RIGHT
DAYS	xs:integer	DAYS_IN_HOSPITAL
PRESENT	xs:boolean	ANTIDEPRESSANTS
PRESENT	xs:boolean	TRANQUILIZERS
PRESENT	xs:boolean	ANTIPSYCHOTIC
PRESENT	xs:boolean	PSYCHO_STIMULANTS
PRESENT	xs:boolean	ANTIEPILEPTIC
PRESENT	xs:boolean	NON_MEDICATION
PRESENT	xs:boolean	SYNCOPE
PRESENT	xs:boolean	NARCOLEPSY
PRESENT	xs:boolean	CATAPLEXY
PRESENT	xs:boolean	SLEEP_DISSORDERS
PRESENT	xs:boolean	LOSS_CONSCIOUSNESS
PRESENT	xs:boolean	ABNORMAL
PRESENT	xs:boolean	TRAUMATIC_INJURY
PRESENT	xs:boolean	DYSLEXIC_SYMPTOMS

17.2.2 The EEG Recording Parameters

The target of the main strategy of the proposed schema is to create a flexible global element, which can connect the specific and multi-complexity factors of an EEG recording with the dominant branch of "EEG_ANALYSIS". Thus, in the second branch of the schema "EEG_RECORDING_ PARAMETERS", a global element, MATRIX_DATA_COLLECTION, is created in which a string type attribute is included (see Fig. 17.4). In this way, this type of global element describes all the recording characteristics "ELECTRODE_MONTAGE, SLEEPING_CONDITIONS, SAMPLING_RECORDING, SAMPLE_SIZE_SELECTION, CHANELL_TYPE, ARTIFACT_REJECTION. Lab SOFTWARE" of each EEG. The usability of this element is in evidence from the fact that it is used **as an input element** in all sub-case children elements of the main EEG_ANALYSIS branch (see Fig. 17.6).

17.2.2.1 Artifact Rejection

According to reference [13] the ARTIFACT_REJECTION is analyzed in Fig. 17.7.

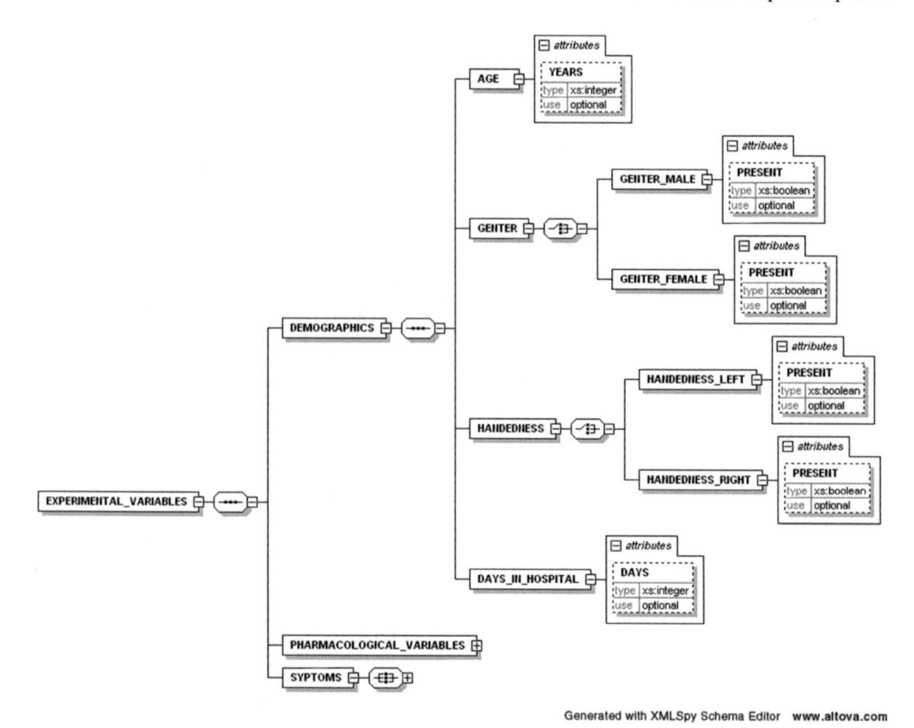

Fig. 17.3 The schema of the demographics variables of EEG

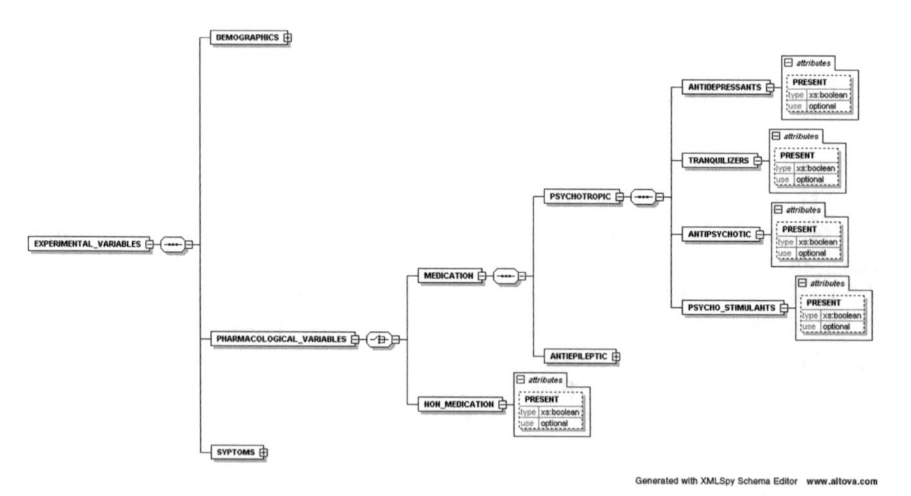

Fig. 17.4 The schema of the pharmacological variables of EEG

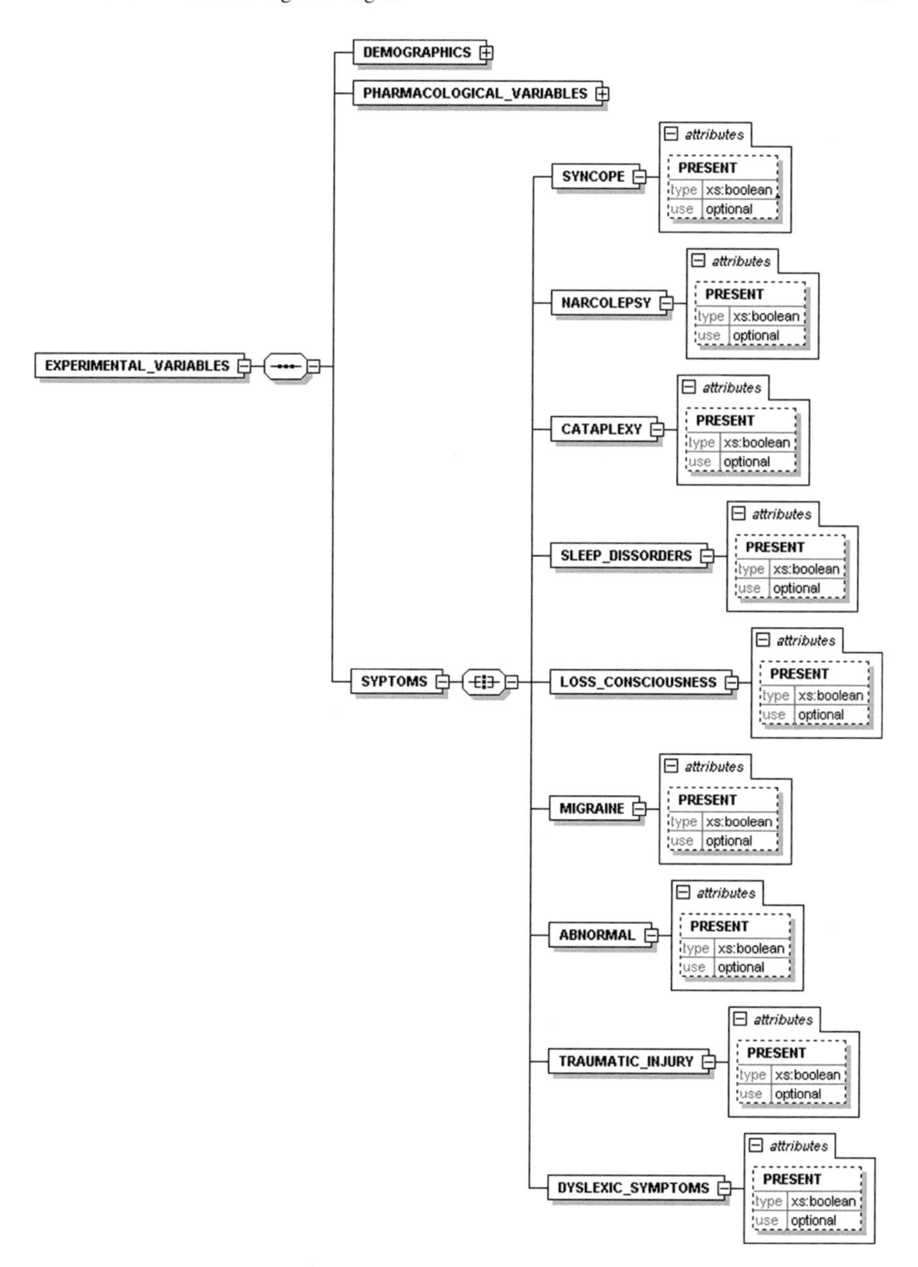

Fig. 17.5 The schema of the symptom's variables of EEG

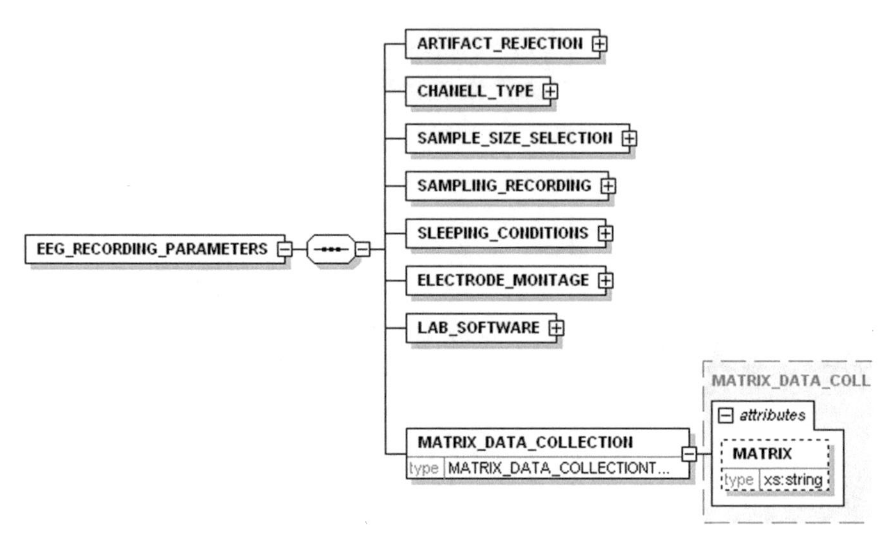

Fig. 17.6 The schema of the EEG recording parameters

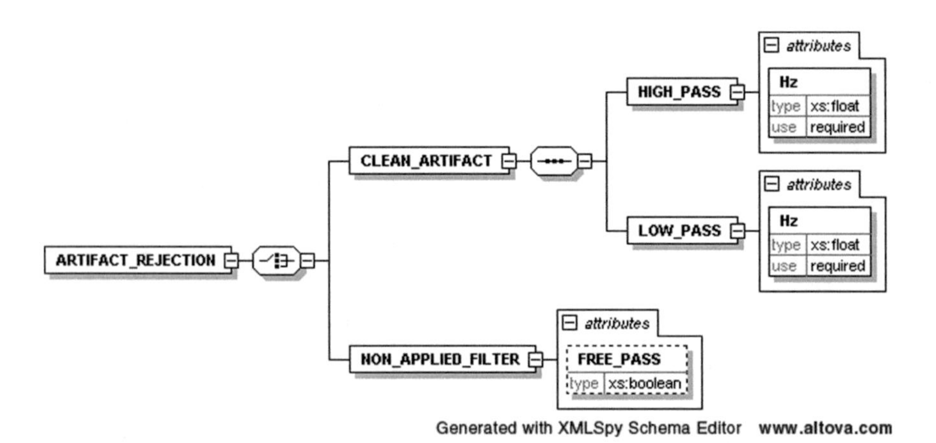

Fig. 17.7 The schema of the EEG artifact rejection

In more details, this schema is developed in a hierarchical way which consisted, by numerical values of filters (low and high) and these are described by an attached float type attribute. Finally, the selection of the case of filter is archived by an attached Boolean type attribute (see Fig. 17.7).

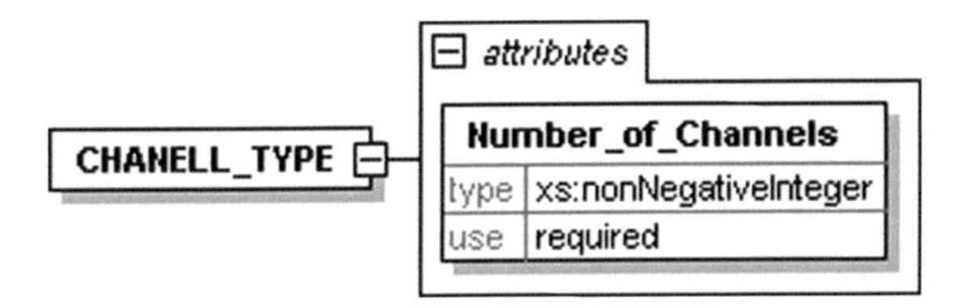

Fig. 17.8 The schema of the EEG channel type

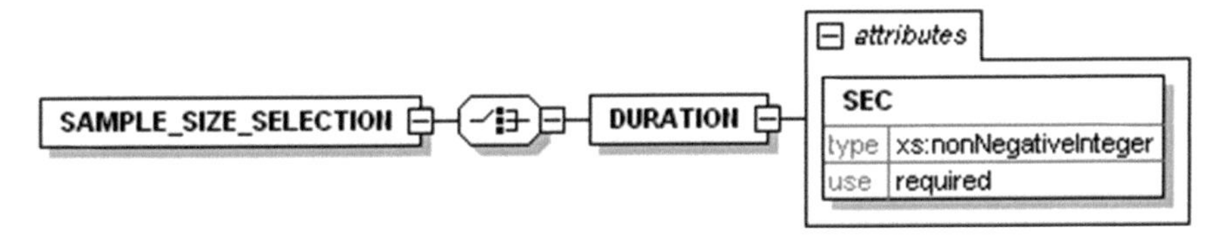

Fig. 17.9 Sample sizes attached to a sequence on the duration element of the schema

17.2.2.2 Channel Type

We adopt the attribute (CHANNEL_TYPE) with a non-negative integer value in order to Number_of_Channels to be described arithmetically (see Fig. 17.8).

17.2.2.3 Sample Size Selection

With regard to the problem of correct sample size selection (SAMPLE_SIZE_SELECTION) studies by a number of investigators have found that the duration of each case is variable [9]. Thus, in our case regarding the schema design and value representation, we consider a series of attributes attached to the end elements of the schema. These attributes are also characterized by a "use" predicate which denotes usage regarding the representation of data (Fig. 17.9).

Where there is a numeric attribute (size) describing the duration (integer form) having also the "required" predicate which yields that a validated XML data file has to include this attribute.

17.2.2.4 Sampling Recording

With regards to the sampling rate frequency, the most commercial software supports multi main sampling frequencies 128 Hz (Hz_128), 256 Hz (Hz_256) and 512 Hz (Hz_512) or higher. Thus, we adopt an element (SAMPLING_RECORDING) with an attached "non negative integer" type attribute, which denotes usage regarding the representations of data. (see Fig. 17.10).

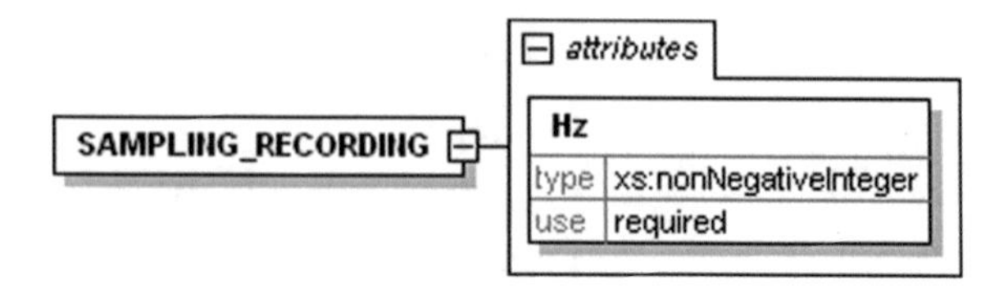

Fig. 17.10 The sampling frequency element followed by a group of types

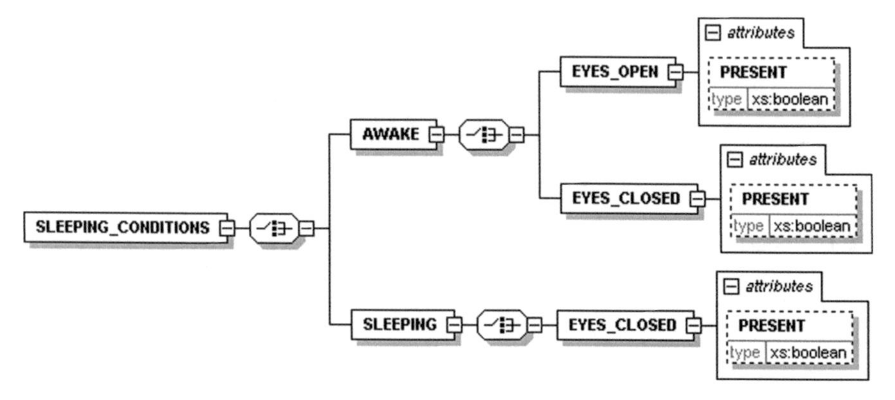

Fig. 17.11 The Sleeping_Conditions complex type expanded

17.2.2.5 Sleeping Conditions

There are two major types of experiment, which differ in their approach regarding control of patients' state of consciousness. The simplest type is where no special effort is made to control a patient's state of consciousness. The traditional conditions of such a routine clinical EEG examination are representatively awake (AWAKE). which includes "eye-open" (EYES_OPEN) and "eye-closed" (EYES_CLOSED) and sleeping (SLEEPING),which includes "eye-closed" (EYES_CLOSED) (see Fig. 17.11). After an experiment design has run, the data must be checked for relevant differences between clinical categories or experimental conditions, and data contributing to these differences should be eliminated.

17.2.2.6 Electrode Montage

There has been an increasing preference in recent years for data collection using online digitalization. In this case, the continuous voltage EEG waveform is converted into a set of binary values. In our case, we chose to make the schema 10–20 specific due to the large amount of already available EEG recordings recorded with the 10–20 system. The 10–20 system is based on the relationship between the location of an electrode and the underlying area of cerebral cortex. Each point on this Fig. 17.12 to the left indicates a possible electrode position. Each site has a letter (to identify the

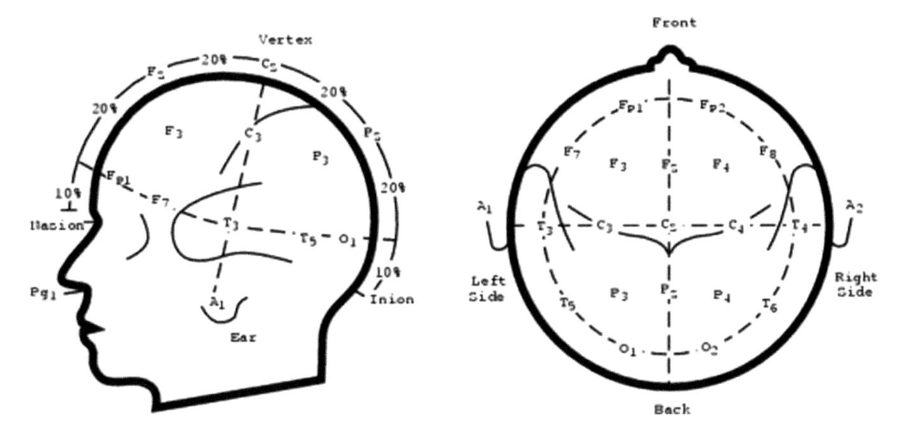

Fig. 17.12 The nomenclature of the electrode location according to international 10–20 system

lobe) and a number or another letter to identify the hemisphere location. The letters F, T, C, P, and O stand for Frontal, Temporal, Central, Parietal and Occipital. (Note that there is no "central lobe", but this is just used for identification purposes.) Even numbers (2, 4, 6, 8) refer to the right hemisphere and odd numbers (1, 3, 5, 7) refer to the left hemisphere. The z refers to an electrode placed on the midline. Also, note that the smaller the number, the closer the position is to the midline [14].

However, in the current state we define an attribute (series_number) of an ELECTROD_PAIR element where the selection of the montage of the electrodes is described in Fig. 17.13.

Thereinafter, the montage of each electrode pair is determined by the combination of two elements {(ELECTRODE_TYPE_1 and ELECTRODE_TYPE_2)} with string types attributes. Each of these elements is analyzed by the following string attributes of elements, which are analytically presented in Fig. 17.14 and describe the montage locations of sequence of electrodes.

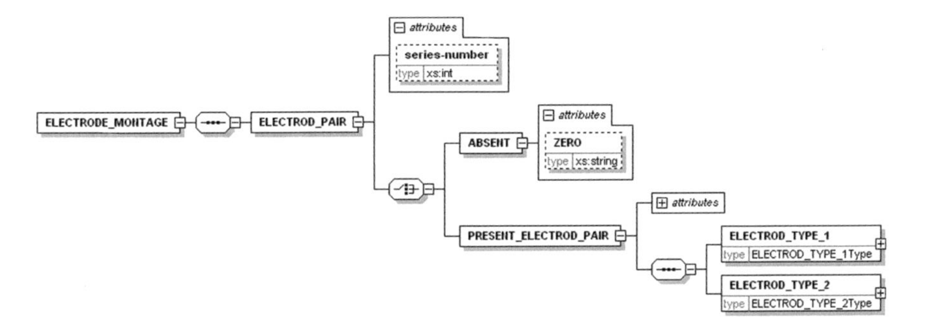

Fig. 17.13 The electrode montage elements

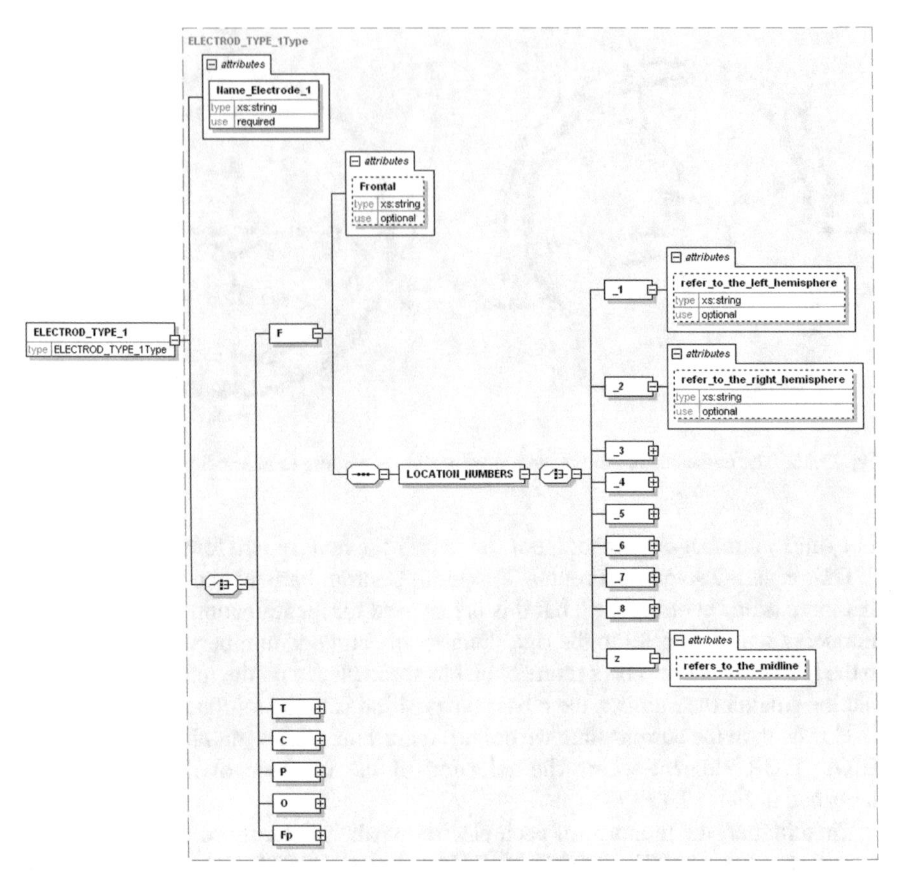

Fig. 17.14 The location montage elements

17.2.2.7 Lab Software

However, the EEG signal considered that presents is one of the major problems in the incompatibility [15] of biological signals. It almost limits the exchange of data and processing algorithms to clinics and laboratories using equipment and software from the same manufacturer, which is one of the causes of the lack of coherent progress in clinical encephalography in recent decades [16]. For this reason, an appropriate flexible schema is added to the basic schema. Since the proposed schema aims to be interoperable the inclusion of software specific characteristics would be against the interoperability target. Therefore, we only include an optional element denoting the name of the software which can help categorize EEG recordings and we attach a string value in order to describe the name of this software, see Fig. 17.15.

Fig. 17.15 The interpretation of the lab-software on the schema

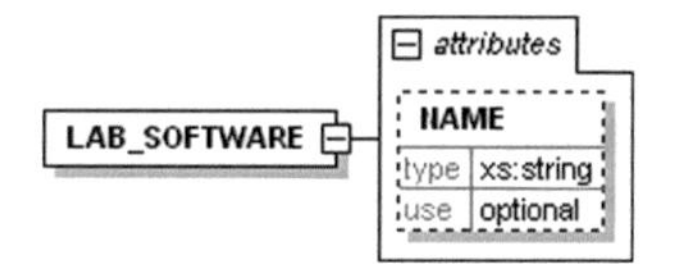

17.2.2.8 Matrix Data Collection

Finally, all these settings are described by a string attribute (MATRIX) of a global element (MATRIX_DATA_COLLECTION) (see Fig. 17.6). Furthermore, in Table 17.2 the attributes of the elements attached to the EEG_RECORDING_PARAMETERS complex type are summarized.

17.2.3 EEG Analysis

EEG analysis comprises the most important part of the proposed schema. In this part all the most significant statistical variables and features are presented in a flexible and interoperable way. To be more specific, this branch is divided into two significant sub-branches which are described by the two main elements of FEATURE_EXTRACTION and PRIMARY_ANALYSIS as presented in

Table 17.2 Attributes attached to the elements of the EEG_RECORDING_PARAMETERS complex type

Attribute name	Type	Elements
Hz	xs:float	LOW_PASS
Hz	xs:float	HIGH_PASS
FREE_PASS	xs:boolean	NON_APPLIED_FILTER
MONTAGE	xs:nonNegativeInteger	CHANELL_TYPE
SEC	xs:nonNegativeInteger	DURATION
Hz	xs:nonNegativeInteger	SAMPLING_RECORDING
PRESENT	xs:boolean	EYES_OPEN
PRESENT	xs:boolean	EYES_CLOSED
series-number	xs:int	ELECTROD_PAIR
ZERO	xs:string	ABSENT
series-number	xs:int	PRESENT_ELECTROD_PAIR
Name_ Electrod_Type_1	xs:string	ELECTROD_TYPE_1
Frontal	xs:string	F
refer_to_the_left_hemisphere	xs:string	_1
MATRIX	xs:string	MATRIX_DATA_COLLECTION

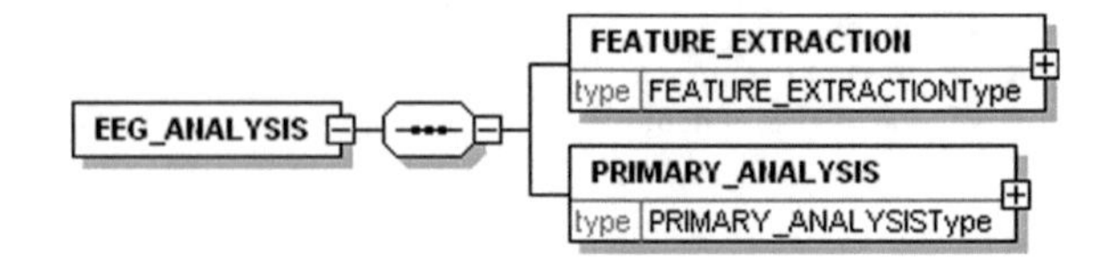

Fig. 17.16 The hierarchical interpretation of EEG analysis

Fig. 17.16. Furthermore, in Table 17.3 the attributes of the elements attached to the EEG_RECORDING_PARAMETERS complex type are summarized.

Table 17.3 Attributes of the elements attached to the EEG_ANALYSIS

Attribute name	Type	Elements
MATRIX	xs:string	INPUT_DATA_F_A
Hz	xs:float	DOMINANT_ALPHA
NOT_DIFFER_SIGNIFICANTLY	xs:boolean	ANOVA
Hz	xs:float	DOMINANT_BETA
MATRIX	xs:string	INPUT_DATA_WAVELET_ANALYSIS
MATRIX_1X9	xs:string	WAVELET_COEFFICIENTS
MATRIX_2XN	xs:string	MORLET_WAVELETS__1_2
STRENGTH_VARIATION	xs:float	MATRIX_2XN
MATRIX	xs:string	MOCK_AVERAGE
MATRIX	xs:string	MEDIAN_AVERAGES
MATRIX	xs:string	SELECTIVELY_FORMING_AVERAGES_ WITHOUT_OUTLIERS
MATRIX	xs:string	WITH_OUTLIERS_DOWN_ WEIGHTED_AVERAGES
MATRIX	xs:string	ALTERNATE
MATRIX	xs:string	MMSE_FILTERING
Cross_Coherence_Magnitude	xs:float	ICA_COMPONENTS_PAIRS
MATRIX	xs:string	MATRIX_DATA_COLLECTION_BILINEAR
MARTIX_1X8	xs:string	BILINEAR_COEFFICIENT
MATRIX	xs:string	MATRIX_DATA_COLLECTION_WAVELET
MARTIX_1X8	xs:string	WAVELET_TRANSFORMATION
MATRIX	xs:string	MATRIX_DATA_COLLECTION_M_V
MARTIX_1X8	xs:string	Maximal_Value
DEGREE_OF_VALIDATION	xs:float	LEAVE_OUT_JACKNIFE
MATRIX	xs:string	MATRIX_DATA_COLLECTION_P_C
MARTIX_1X8	xs:string	MATRIX

17.2.3.1 Feature Extraction

Feature extraction is a crucial but difficult step in a complete analysis. Its purpose is to concentrate the data generated by the primary analysis by forming summary indices, called features, which characterize the signal properties most relevant to the hypothesis under consideration. Given the limited number of patients and observations in an experiment, this number of variables is far too large, and must be substantially reduced to a smaller number of essential variables prior to statistical hypothesis testing. There are ad hoc (heuristic) (HEURISTIC), and classifier-directed methods (CLASSIFIER_DIRECTED) of feature extraction (see Fig. 17.17).

Heuristic Methods (HEURISTIC)

Heuristic methods of time shift measurement are often used in the analysis of EEG signals and especially in the analysis of epileptic discharges and seizure spread. Numerous works have been published using the following methods:
I. Linear cross-correlation
The linear cross-correlation [17–19] method, in which the position of maximal value (Maximal_Value) of the cross-correlation function is determined by the time delay between signals from a particular electrodes pair source. These are described by the global element (MATRIX_DATA_COLLECTION) and represented by (MATRIX_DATA_ COLLECTION_M_V) the element with the attached string type attribute (MATRIX_1X8). This factor is selected for medical diagnostic purposes (such as epilepsy) see Fig. 17.18.
II. Phase/coherence

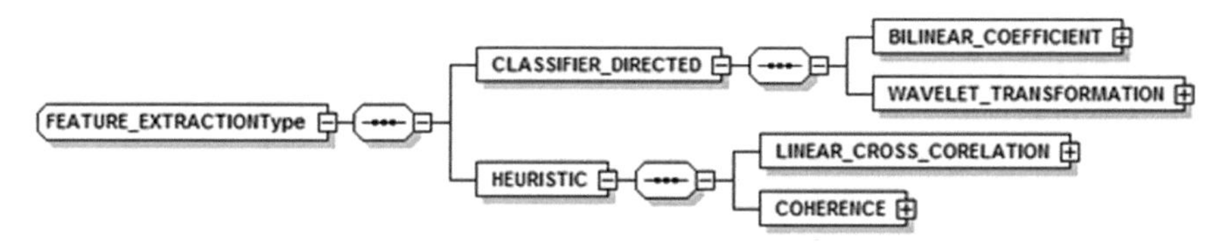

Fig. 17.17 The representation of feature extraction in the schema

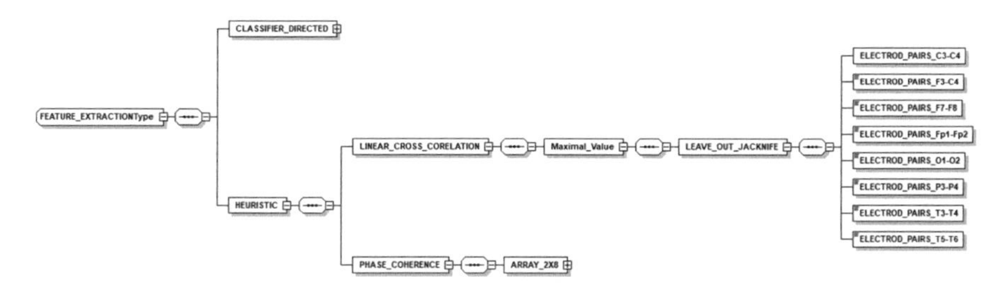

Fig. 17.18 The representation of the linear cross correlation method in the schema

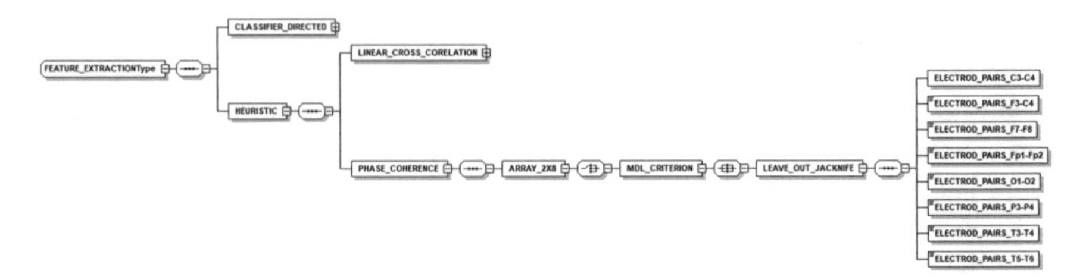

Fig. 17.19 The representation of the phase-coherence method in the schema

The phase/coherence method [20–27] is, in principle, similar to the previous method, but in this case, the cross-spectral phase is estimated by autoregressive modelling. Using auto- and cross- correlation coefficients, the Yule Walker equations are compiled and then the cross-phase spectrum is derived. The time shift is also computed from the slope of the regression line and fitted to the selected part of the phase. An important feature in this approach is the selection of an adequate model order. A method with an underestimated model order does not describe all the signal details and an overestimated order causes false peaks in the spectrum with no relation to a signal. The model order is estimated using Minimal Description Length (MDL) criterion [28] (MDL_CRITERION). Usually, this yields an 8-order model or 8 coefficients (ARRAY_1X8) [29]. This is described by the global element (MATRIX_DATA_COLLECTION) and is represented by the (MATRIX_DATA_COLLECTION_P_C) element with the attached string type attribute (MATRIX_1X8) see Fig. 17.19.

Validation

The next step in a complete analysis is validation, which is of paramount importance whenever extensive analysis has been performed on data. During validation (see Fig. 17.25), it is determined whether the results are significant, reliable on replication and physiology or medically meaningful. The most convincing validation is obtained with an entirely independent data sample taken from a new set of patients. However, since this is not always feasible, especially during the early exploratory phase of research, one of several types of statistical validation scheme is most often used. In these procedures, a portion of the data (one-third, one-half, etc.) is reserved for testing the results obtained against the rest of the data: the average significance of "independent" validations can then be determined. The significance determined in this way will probably not be the same as the significance obtained on an independent set of patients, but it can provide a reasonable estimate of disease instances.

The proposed method is called "leave-out- jackknife validation" [30] and is used to validate cross-statistical results of heuristic methods. This method assumes that each EEG recording is segmented into at least two (2) equal segments, in order to compare the validation of the proposed heuristic methods.

The extracted data of the above methods is collected in a two-dimensional array format described by a string type attribute (MATRIX_1X8) and attached by the (Maximal_Value) element in the one case and by a string type attribute (MATRIX_1X8)

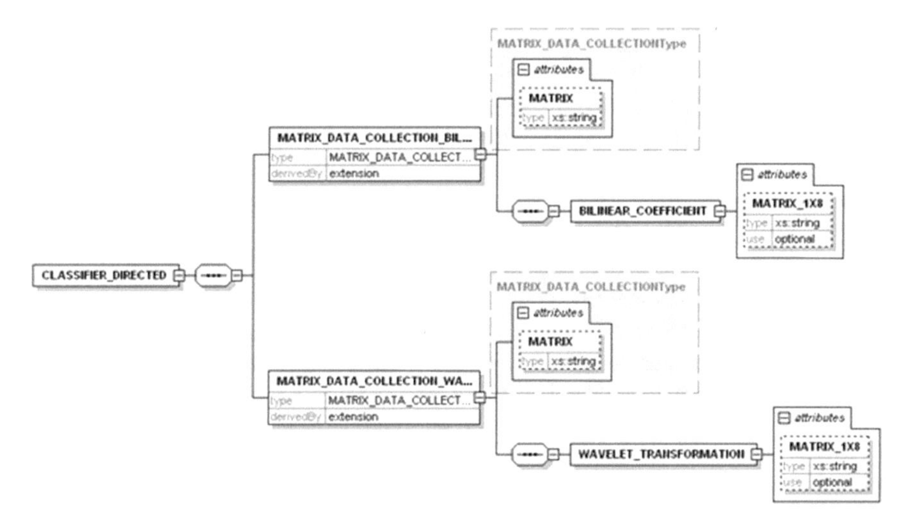

Fig. 17.20 The representation of the Classifier-directed methods (CLASSIFIER_DIRECTED) in the schema

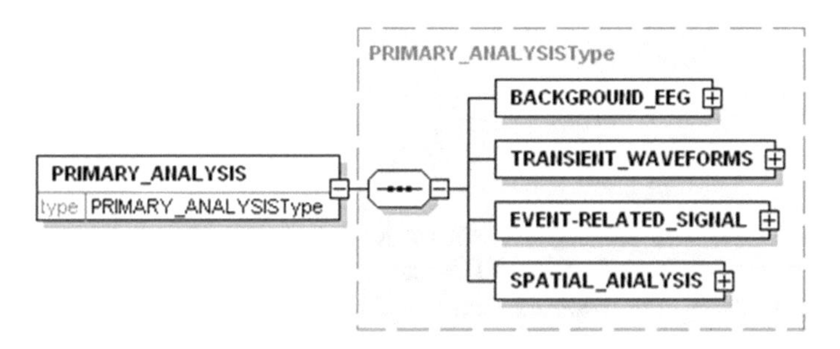

Fig. 17.21 The hierarchical interpretation of the primary analysis

and attached by the (MDL_CRITERION) element in the other case. Then, the Jack-nife method extracts the degree of validation (LEAVE_OUT_JACKNIFE) which is attached by a (DEGREE_OF_VALIDATION), float type attribute in a cross-correlation procedure by calculating the array values for each of the above two cases (see Figs. 17.20, 17.21).

Classifier-Directed Methods (CLASSIFIER_DIRECTED)

Statistical methods of feature extraction suffer from a fundamental limitation. Feature extraction by the maximization of some criteria, such as the amount of variance accounted for by the new features, frequently does not derive the measures which best distinguish the clinical categories or conditions of the experiment. As an alternative, statistical pattern recognition algorithms have successfully

been used to choose features and feature combinations, which are relatively good, or even optimal, for hypothesis testing. Many techniques offer specific classifier features for the neural network training procedure. These are: the Wavelet transformation (WAVELET_TRANSFORMATION) and Bilinear Coefficients (BILINEAR_COEFFICIENT) see Fig. 17.20.
I. Wavelet transformation (WAVELET_TRANSFORMATION)
Wavelet transformation [31] happens in the case where the resultant of a five (5) component feature vector for each channel is extracted, or more simply the definition of the most significant electrode pairs which are determined by the global element (MATRIX_DATA_COLLECTION Type). These vectors may be characterized as significant vector classifiers for diagnostic classification purposes (see Fig. 17.20). The final components are represented via the element (BILINEAR_COEFFICIENT) and are described by the string type's attached attribute (MATRIX_1X8), see Fig. 17.20.
II. Bilinear Coefficients (BILINEAR_COEFFICIENT)
The Coefficients of a 12th order bilinear model, based on a non-linear EEG analysis, is a promising method [32] for feature extraction for diagnostic classification purposes using an independent neural network classifier. The described data is determined by a recording of k sec duration for each channel. All these settings (montage time) are determined by the global element (MATRIX_DATA_COLLECTIONType) (see Fig. 17.20).

17.2.3.2 Primary Analysis

Signal conditions, digitalization, artifact rejection, formation of data sets and the examination of distribution, using different types of primary analysis are applied as a first step in extracting meaning from the data. The four (4) major objectives of primary analysis are:

(1) Analysis of the background EEG (BACKGROUND_EEG),
(2) Transient waveforms in the EEG (TRANSIENT_WAVEFORMS),
(3) Event-related signals (EVENT-RELATED_SIGNAL) and
(4) Spatial processes (SPATIAL_ANALYSIS).

It must be noted that in this section all the factors of significant statistical values are taken from the total number of EEG channels and for this reason we have avoided specifically referring to each separately (see Fig. 17.21).

Analysis of the Background EEG (BACKGROUND_EEG)

The background analysis (BACKGROUND_EEG) may be divided into two main types of analysis: frequency analysis (FREQUENCY_ANALYSIS) and non-stationary analysis (NON_STATIONARY_ANALYSIS) (see Fig. 17.22).
I. Frequency analysis (FREQUENCY_ANALYSIS)

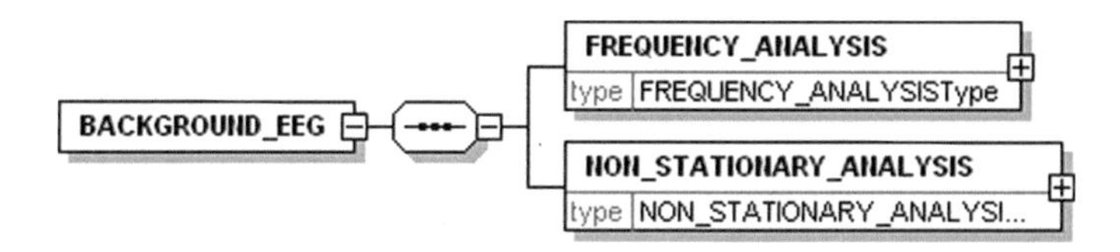

Fig. 17.22 The hierarchical interpretation of the background EEG analysis

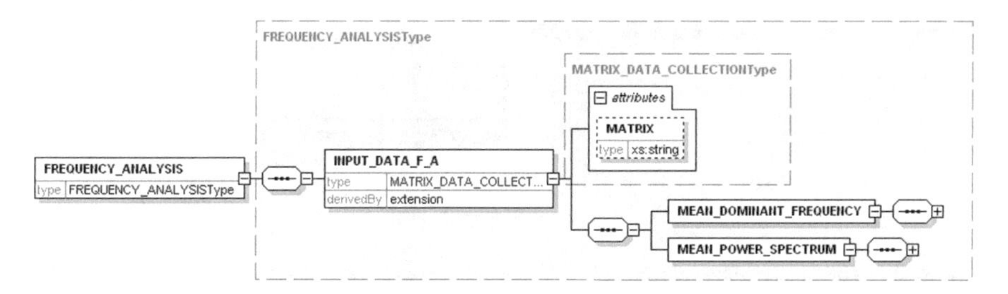

Fig. 17.23 The hierarchical interpretation of the frequency analysis

In Frequency Analysis (element: Frequency_Analysis) (see Fig. 17.23) the automated EEG analysis is based on the parameters' mean dominant frequency (MEAN_DOMINANT_FREQUENCY) and mean power spectrum (MEAN_POWER_SPECTRUM) parameters. In this case, a global element with an attached string type attribute is introduced as input data (INPUT_DATA_F_A). These types of global element and the attached string type attribute (MATRIX) describe all the features of the EEG recording setting.

In the mean dominant frequency (MEAN_DOMINANT_FREQUENCY) the relative powers of the delta and theta bands, which are suitable for the objective classification of hepatic encephalopathy in individual patients, is another significant diagnostic factor [14]. The most significant dominant frequency is the highest of alpha activity, in other words, dominant alpha activity (DOMINANT_ALPHA) and dominant beta activity (DOMINANT_BETA). These data are described by the attached attributes (Hz) of float type elements DOMINANT_ALPHA and DOMINANT_BETA correspondingly (see Fig. 17.24).

The mean power spectrum is divided into 6 bands: delta (2–4 Hz) (MEAN_POWER_SPECTRUM), theta (4–8 Hz) (THETA), alpha (8-13 Hz) (ALPHA), beta 1 (13–20 Hz) (BETA1), beta 2 (20–30 Hz) (BETA2) and gamma (30–40 Hz) (GAMMA). The power spectrum of any given band is averaged across all the electrodes, which are described by the global element MATRIX_DATA_COLLECTIONType. For the five data sets (see Fig. 17.27), the following descriptors [15] are used and may be considered to be significant diagnostic factors especially in Traumatic brain injury [16]. These data are described by the attached attributes (Hz) of float type elements THETA, ALPHA, BETA1, BETA2, and GAMMA correspondingly (see Fig. 17.25).
Decision (hypothesis testing)

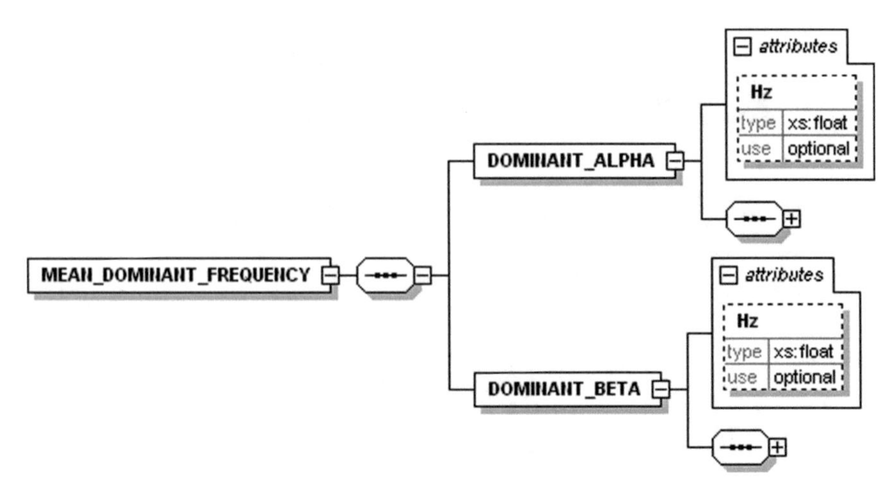

Fig. 17.24 The hierarchical interpretation of the mean dominant frequency

Furthermore, a significant factor of an EEG analysis specifically in the case of the repeated measures EEGs is statistical analysis. Thus, we adopt the ANOVA method as optimum for this purpose. In the XML schema these are set at the lowest level, "leaf nodes and specifically as a simple numerical element.

The most important step in any analysis is deciding whether the data obtained is sufficient to support the hypothesis or hypotheses under investigation. This may be accomplished with a myriad of different statistical methods: parametric or non-parametric, unvariate or multivariate. Sometimes the results of feature extraction are so clear that statistical tests are applied to prove the obvious conclusion. For example, if a major change in the number and duration of 3 s EEG spike and wave discharges is observed because of an alteration of an anticonvulsant drug regime, the obvious conclusion would be drawn. Of course, standards for such changes must previously have been compiled from a large group of patients in order to determine whether the observed change was significant. However, in many instances, the results of primary analysis and feature extraction are not so obviously related to the experimental or clinical condition under investigation. For example, in attempting to predict the onset of a grand mal seizure 600 s or more prior to its occurrence, no obvious relation between EEG spectral features and the subsequent seizure onset may be apparent. In these instances, it is necessary to employ statistical analyses of primary analysis (PRIMARY_ANALYSIS Type) and data of frequency analysis (FREQUENCY_ANALYSIS) such as analysis of variance ANOVA (ANOVA) (see Fig. 17.26). More specifically, the Anova analysis is performed in order to investigate the differences between the group means of several EEG mean value indexes, extracted from each channel recorded, such as MEAN_DOMINANT_FREQUENCY and MEAN_POWER_SPECTRUM testing if the means of the groups formed by the values of the independent variable (or combinations of values for multiple independent variables) are different enough not to have occurred by chance. If the group

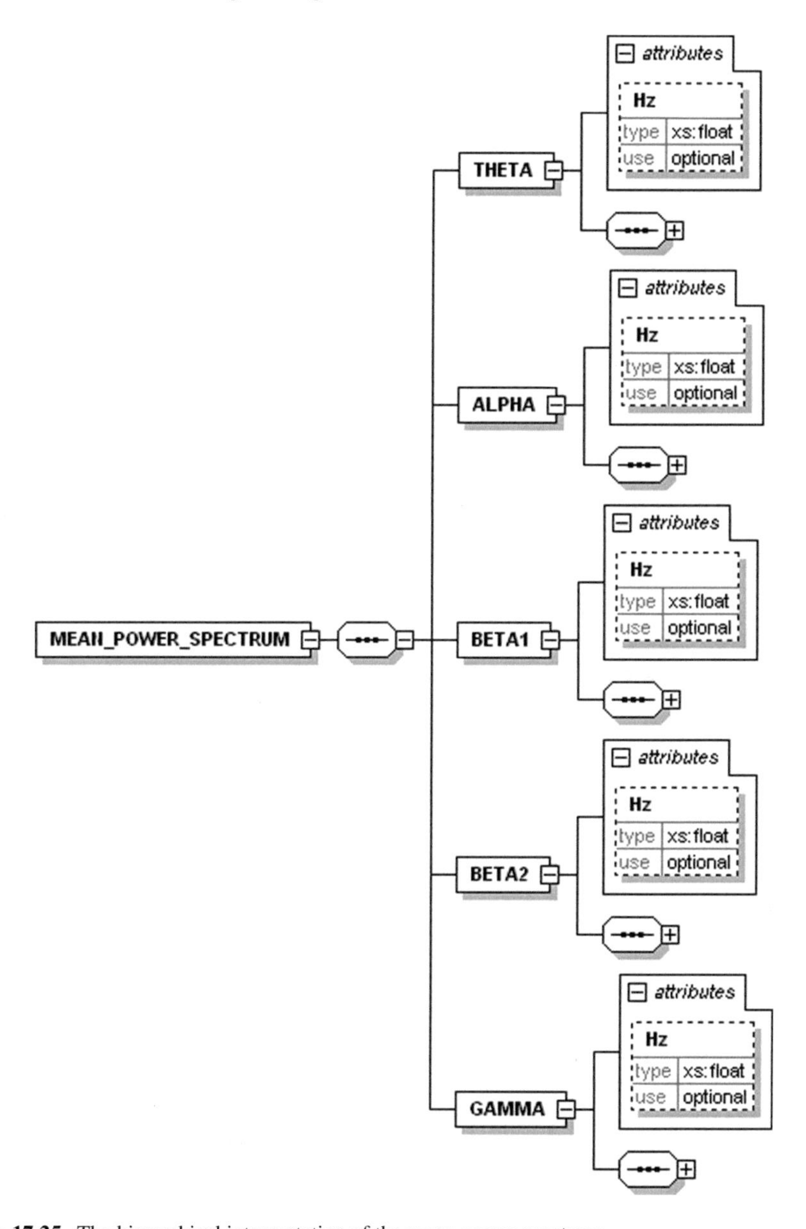

Fig. 17.25 The hierarchical interpretation of the mean power spectrum

means do NOT_DIFFER_ SIGNIFICANTLY then it is inferred that the independent variable(s) did not have an effect on the dependent variable. This is implemented using as the query the statistical criterion (null hypothesis) where the difference of the above mean population is zero $H = 0$ with $p <= 0.01$.
II. Non-Stationary Analysis (NON_STATIONARY_ANALYSIS)

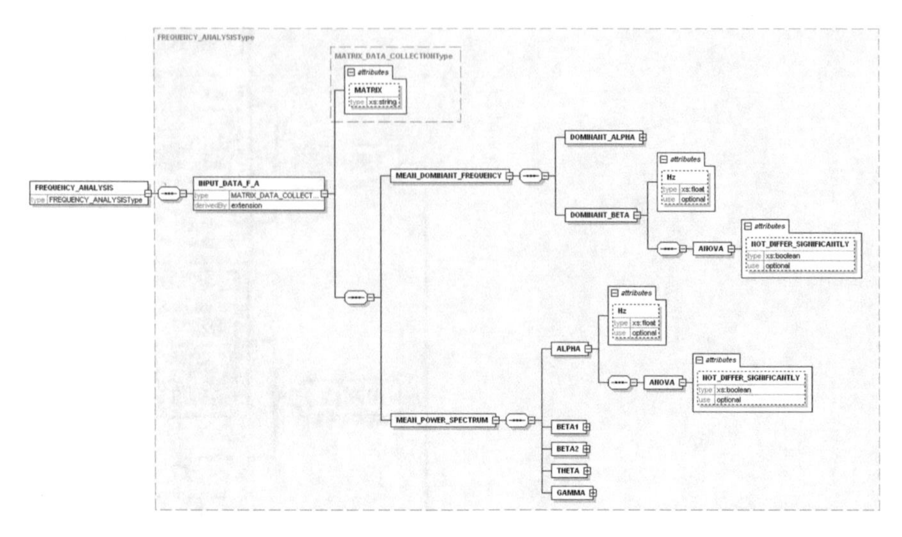

Fig. 17.26 The decision variables involved in the schema

One important trend in research is concerned with characterizing non-stationary or time-varying processes. The term "non- stationary" means that characteristics of a time series, such as its mean and spectral characteristics, are changing within data segments chosen for analysis. Recent statistical tests of stationary characteristics showed that the electrical signal from the brain, recorded on the scalp (above the occipital cortex), in response to a visual stimulus, was a significant factor in clinical disease diagnosis. More specifically, the motion-onset Visual Evoked Potential (VEP) elicited by stationary gratings served as a control condition [33].

Wavelet representation is a recent development in the analysis of non-stationary EEG signals [34]. Wavelet analyses of EEGs obtained from a population of patients can potentially suggest the physiological processes that the brain is undergoing with the onset of epilepsy [35]. More specifically, nine (9) wavelet coefficients (WAVELET_COEFFICIENTS) are extracted from each channel and this is the decomposition in wavelet sub bands and reconstruction from nine (9) wavelet coefficients, statistically differentiating Evoked Potentials (EP) from an on-going EEG [36]. More simply, we test the recorded data of the most significant electrode pairs, which are described in the MATRIX_DATA_COLLECTION string type global element (see Fig. 17.27).

Transient Waveforms in the EEG (TRANCIENT_WAVEFORMS)

Accordingly, a great deal of energy has been expended over the years in efforts to search long recordings for epileptic phenomena, but with mixed results [37].

More recently, an approach has been proposed in which the temporal sharpness is measured in different "spans of observation", involving different amounts

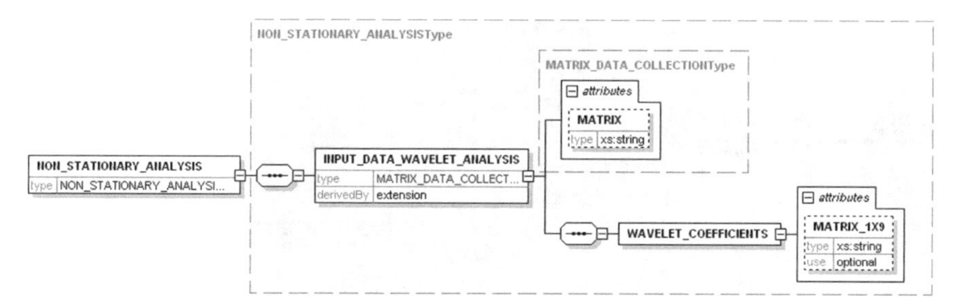

Fig. 17.27 The non-stationary Analysis is composed of a sequence of elements

of temporal context. Alternatively, this can be seen as a type of template matching of important signal characteristics at different scales (dilations) while maintaining the fundamental morphology of the wavelet. There are many suitable wavelets which can be used such as those developed by Mallet, Daubechies, and Morlet [38, 39] (see Fig. 17.30), though the wavelet function is, in fact, an offspring of "Morlet's wavelet" (MORTLETS_WAVELET) and presents the most significant factors in the detection of interictal spikes. The characterization of infrequent, morphologically variable transient events, especially those associated with the epilepsies ('spikes', 'spikes and waves' etc.) is an essential component of a traditional clinical EEG. Pathological synchronization is a main mechanism responsible for an epileptic seizure [40]. Since many features of EEG signals cannot be generated by linear models, it is generally argued that non-linear measures are likely to give more information, such as MORLET_WAVELETS_TRANSFORMATION [26, 27] than that obtained with conventional linear approaches [41]. Thus, according to recent studies [41] the PHASE_ SYNCHRONIZATION measure is a significant factor for comparing two series (2 channels) of EEGs. In particular, for two given time series (channel recording data) the Morlet wavelet coefficients are extracted separately for each time series, as it is known that they include the phase information of the signal [41]. Finally, the measure of synchronization strength (STRENGTH_VARIATION) varies from 0 to 1and is implemented by the coherence procedure COHERENCE_PHASE_SYNCHRONIZATION between wavelets coefficients (see Fig. 17.28). All the above elements depend on the introduced global element and all are attached to a float type element (STRENGTH_VARIATION).

Event-Related Signals (EVENT-RELATED_SIGNAL)

The aim of analyzing Event-Related Signals (ERSs) in neural neuro-physiological studies is to explore information processing in the human brain. In most cases stimulus synchronous averaging takes place in order to improve the Signal-to-Noise Ratio (SNR) though this does not allow for the variability between individual (ERSs). For that reason each single ERS must be analyzed for loss of information and present day research into signal processing procedures includes the estimation of the single ERS

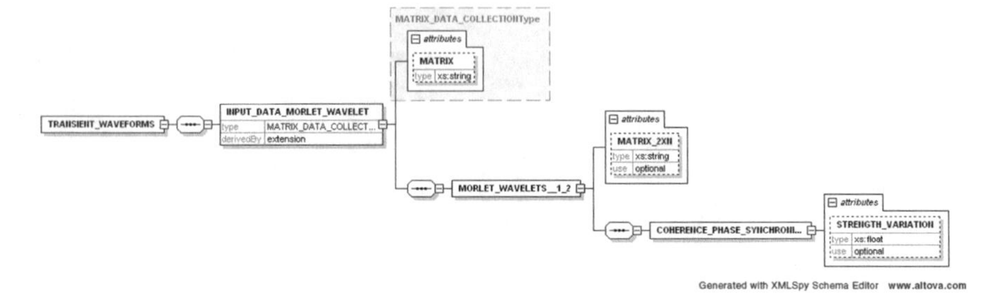

Fig. 17.28 The representation of the Mortlet's wavelet in the schema

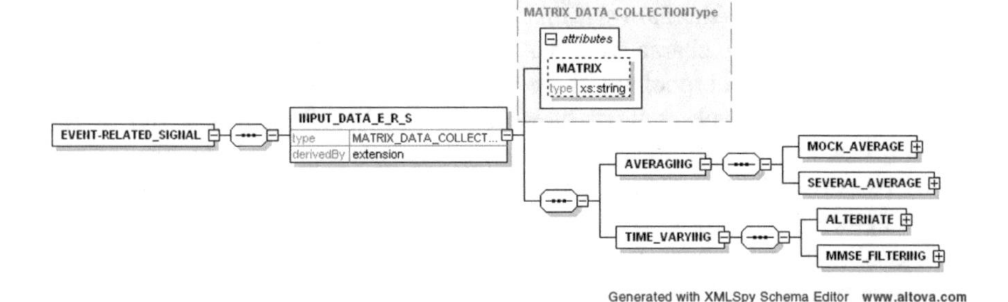

Fig. 17.29 The representation of the event-related signals in the schema

parameters of amplitude and latency. This branch is fed by the INPUT_DATA_E_R_S element where it is of the global element MATRIX_DATA_COLLECTIONType (see Fig. 17.29).

I. Averaging (AVERAGING)

A very simple, but surprisingly effective method of primary analysis consists of averaging a number of brief EEG time series, which are time-registered to an expected or actual stimulus or response. Because the event-related brain signal tends to occur at about the same time after the stimulus on each trial, while background EEG waves have a random relation to stimulus, the event related signals are enhanced when many trials are averaged together, and the unrelated background EEG waves are suppressed. The signal –to-noise ratio is thus improved in proportion to the square root of the number of trials averaged. Statistical tests have been devised to determine the amount of variability of event-related processes comprising the Mock average (MOCK_AVERAGE) [42] and several average methods (SEVERAL_ AVERAGE) have been directed at reducing the effect of trials such as computing the median averages (MEDIAN_AVERAGES) or selectively forming averages without outliers (SELECTIVELY_FORMING_ AVERAGES_WITHOUT_OUTLIERS) or outliers with down weighted averages (WITH_OUTLIERS_DOWN_ WEIGHTED_AVERAGES) [43] (see Fig. 17.30).

All these elements are described by an attached string type element baptized as "MATRIX".

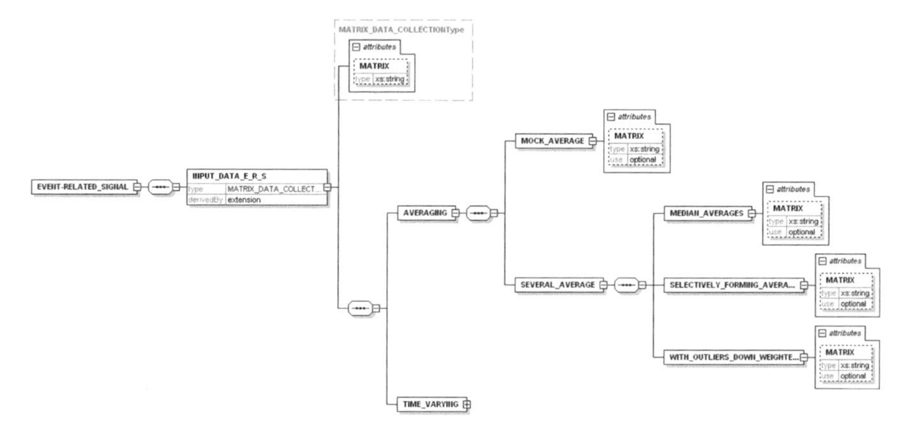

Fig. 17.30 The representation of the averaging method in the schema

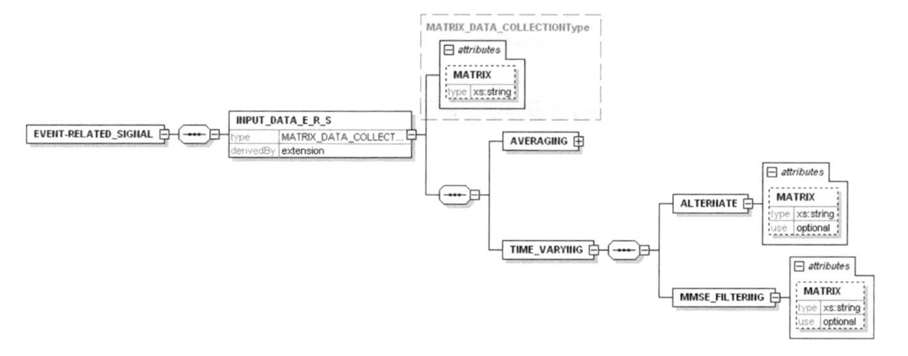

Fig. 17.31 The representation of the time varying method in the schema

II. Time–varying Filtering (TIME_VARYING)

Brain electrical activity resulting from sensory stimulation is referred to as an evoked potential (EP) or an event-related potential (ERP) in the case of an electrical signal, and as an evoked field (EF) or an event-related field (ERF) in the case of magnetic signals. Since the timing and shape of the individual peaks of the ERP can shift independently, it is obvious that the ERP is a time –varying process. For example, de Weerd's [44] contribution is an economical procedure that generalizes the method of MMSE filtering (MMSE_FILTERING) to the time-varying case. In recent research this filtering is implemented via the "Alternate Sub ensemble Averaging" filter (ASA) (ALTERNATE) [45], which is based on the original signal and noise scalogram estimates (see Fig. 17.31).

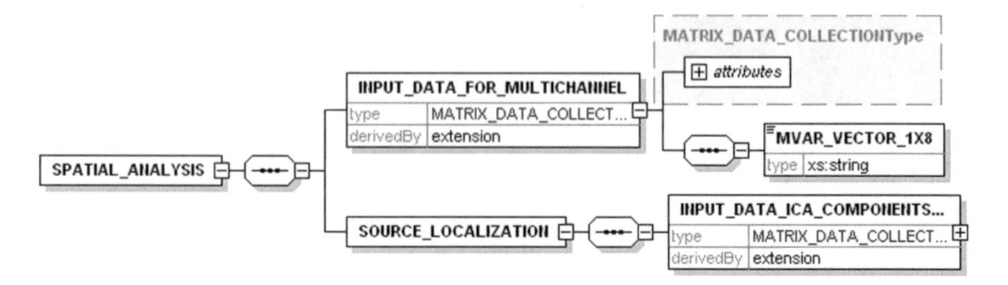

Fig. 17.32 The representation of spatial analysis in the schema

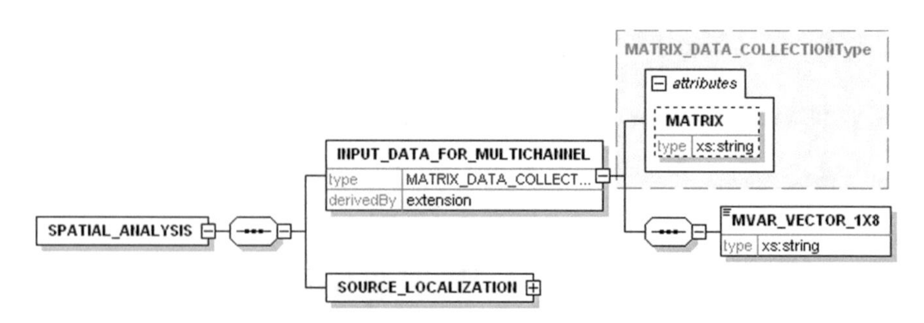

Fig. 17.33 The representation of multi-channel analysis in the schema

Spatial Analysis (SPATIAL_ANALYSIS)

In the case of spatial analysis of brain activity, the electrical field is also analyzed and resembles a field map of distributed electrical potential on the scalp. The two (2) special type analyse (see Fig. 17.32) involved use the global element (MATRIX_DATA_COLLECTION) as described input data and are:

(a) multi-channel processing (MULTICHANNEL) and
(b) source location analysis (SOURCE _ LOCALIZATION)

I. Multi-channel and inter-impendency analysis (MULTICHANNEL)
In most applications, EEG or ERP time series are first analyzed one channel at a time, and then comparisons made among them. In one of the first multi-dimensional time series analyses [46], multiple-coherence measures are applied in an attempt to measure the linear predictability of EEGs. However, this category of method becomes problematic when applied to very short durations in which there are too few points available to estimate the measures. In recent studies, this approach has been improved by calculating the Mean of a Multi-Variate Autoregressive model (MVAR) [47], which is applied to k MVAR_COEFFICIENTS data where k is usual 8. (see Fig. 17.33). Each coefficient represents a Multi-Variate of data, which are sourced by particular electrode pairs, and these are described in the global element (MATRIX_DATA_COLLECTION). This factor characterizes the sensitivity [48] of

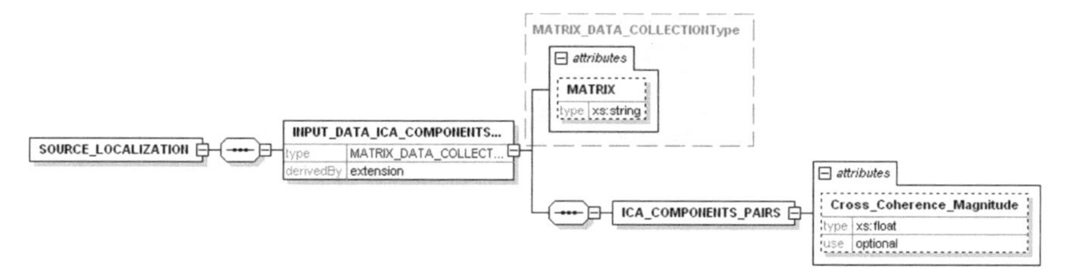

Fig. 17.34 The structure of source localization element in the schema

the multiple instantaneous EEG and this may be valuable for the clinical diagnosis of many diseases such as epilepsy, encephalopathy etc.
II. Source Localization (SOURCE_LOCALIZATION)
While observation and analysis of spatial patterns is worthwhile, the basic but more difficult problem is to determine the source in the brain of the patterns observed at the scalp. In recent studies [49, 50] the sensitivity of EEG electrodes to the location and direction of sources in the brain was determined, and electrodes spaced 5 cm apart were found to be about ten (10) times more sensitive to nearby cortical sources deep within the brain; the more widely spaced the electrodes, the greater the relative sensitivity to deeper sources. Computing a source from a scalp distribution requires the imposition of somewhat arbitrary assumptions in the form of a priori model of the character and number of sources. The most common model is that scalp field distribution arises from an equivalent current dipole source. Each source was located via the standardized "Talairach" maps. More specifically, the general EEG inverse problem can be stated as follows: Given a time dependent set of electric potentials on the surface of the head and the associated positions of those measurements, as well as the geometry and conductivity of the different regions within the head, calculate the locations and magnitudes of the electric current sources within the brain. The solution to this inverse problem can be formulated by finding a least square fit of a set of current dipoles to the observed data for a single time step, or minimization with respect to the model parameters. The ICA uses an "infomax" algorithm that attempts to minimize the common information among the temporal projections derived from single component weights and their accompanying activations, and maximize the information in each component. [51]. Thus, the evaluation of the calculation of ICA components from the electrode channels, which are determined (required) in the EEG_RECORDING_PARAMETERS element (montage attribute) is achieved by indicating the amount of synchronization between two components (a significant factor for the diagnosis of diseases such as Alzheimer case). This control is implemented using the Cross_Coherence_Magnitude (from 0 to 1) procedure with significant statistical confidence $p < 0.01$ which satisfied the hull hypothesis that the investigated pairs of components are synchronized. The above stages are implemented in the following part of the original schema (see Fig. 17.34).

Table 17.4 Attributes of the elements attached to the CLINICAL_DIAGNOSIS complex type

Attribute name	Type	Elements
PRESENT	xs:boolean	MULTIPLE_SCLEROSIS
PRESENT	xs:boolean	MULTIPLE_INJURY
PRESENT	xs:boolean	DYSLEXIA
PRESENT	xs:boolean	ATTENTION_DEFICIT
PRESENT	xs:boolean	SCHIZOPHRENIA
PRESENT	xs:boolean	DEPRESSION
PRESENT	xs:boolean	DRUG_ABUSE
PRESENT	xs:boolean	EPILEPSY
PRESENT	xs:boolean	TUMORS

17.2.4 Clinical Diagnosis

lIn this stage, we adopt clinical diagnosis in order to compare the signal analysis diagnostic decision with the experimental clinician neurologist decision. In this way, the above distribution takes places in order to clearly determine the discrimination between the experimental part, the significant features of EEG analysis and the clinical diagnosis factors. Furthermore, in Table 17.4 the attributes of the elements attached to the EEG_RECORDING_PARAMETERS complex type is summarized.

The clinical diagnostic evaluation of individual patients for possible tumors (TUMORS), multiple sclerosis (MULTIPLE_SCLEROSIS), minor head injury (MINOR_HEAD_INJURY), dyslexia (DYSLEXIA), attention deficit disorder (ATTENTION_DEFICIT), schizophrenia (SCHIZOPHRENIA), depression and alcoholism (DEPRESSION) drug abuse (DRUG_ABUSE) and epilepsy (EPILEPSY), may continue to be considered an important factor. In the proposed schema these elements are described by a Boolean type attribute (PRESENT/NO) (see Fig. 17.35).

17.3 Conclusions

Our study reflects an attempt to contribute further to the standardization of medical signal encoding, storage and retrieval research. Furthermore, our proposed schema of metadata was done having in mind the EEG factors for enrichment its reusability in a clinical context. Similar studies such as [16, 52] have orientated on signal related features. Furthermore, the construction of this research from the schema point of view is based on in the international 10–20 recording system. Additionally, the adaption of this schema in other systems such as 3D or magnetic context could be achieved by a revised version.

The implementation of this schema in a real clinical context yields a useful meta-analysis tool valuable for the clinical decision. In our study we implement a metadata

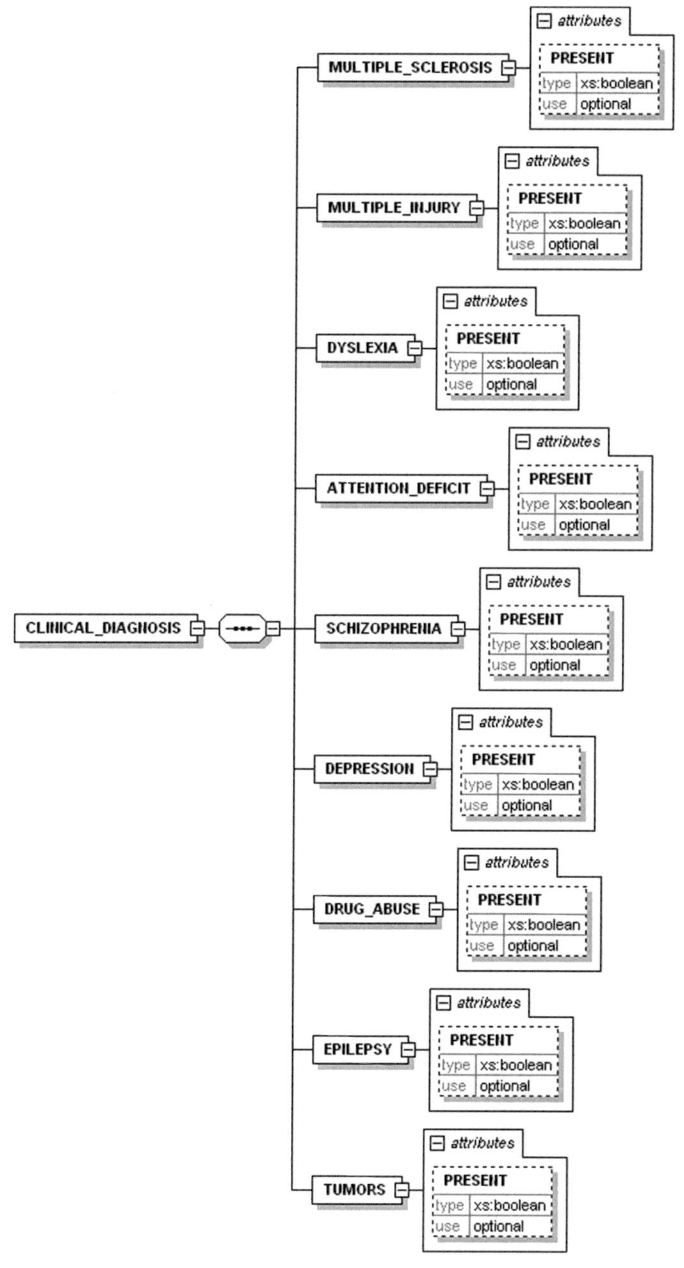

Fig. 17.35 The clinical diagnosis element

web searching schema making interoperable EEG formats. Undoubtedly, the schema may avoid elements that describe different methods to investigating and understanding the EEG. For that reason, we present a web searching schema which aims to be extensible through the metadata standards and medical procedures. The construction of this schema rests mainly on the feasibility of enabling the signal production software to produce an interoperable XML file. Objectively, problems such as programming issues and industry standardization emerge and are topics to nonstop dialogue with the industry investors. It is known, that this manner is a high priority axis from the industry investors because this could be solve problems of communication of medical standards.

In future work, this study can be focused to other web searching tools such as the deep learning machine [53] on an XML basis [54, 55]. Additionally, the creation of such an associated bioinformatics framework could assist the parting of both operational and research databases and maximize the utilization and effective of current, scattered bio-medical databases.

References

1. J.L. Willems, P. Arnaud, J.H. van Bemmel, R. Degani, P.W. Macfarlane, C. Zywietz, Common standards for quantitative electrocardiography: goals and main results. Methods Inf. Med. **29**, 263–271 (1990)
2. H. Wang, F. Azuaje, B. Jung, N. Black, A markup language for electrocardiogram data acquisition and analysis (ecgML). BMC Med. Inform. Decis. Mak. **3**, 4 (2003). https://doi.org/10.1186/1472-6947-3-4
3. MIT-BIH Arrhythmia Database. http://www.physionet.org/physiobank/database/mitdb/
4. Health Level Seven XML Patient Record Architecture. http://xml.coverpages.org/hl7PRA.html
5. ASTM, subcommittee E31.25. http://www.astm.org/COMMIT/COMMITTEE/E31.htm
6. J. Dudeck, TC 251 task force on XML application in healthcare. CEN/TC251 Task Force XML-Final Report (1999). http://www.centc251.org/TCMeet/Doclist/TCdoc99/N99-067.doc
7. Clinical Data Interchange Standards Consortium. http://www.cdisc.org/
8. Open Archives Initiative. http://www.openarchives.org/
9. M. Poulos, S. Papavlasopoulos, G. Bokos, A. Evangelou, An XML schema for the sharing and communication of heterogeneous EEG data for diagnostic and research purposes. J. Inf. Technol. Healthc. **4**, 253–273 (2006)
10. X. Zhang, B. Hu, L. Zhou, J. Chen, P. Moore, An XML format for electroencephalogram data presentation (EEGML), in *2013 IEEE International Conference on Bioinformatics and Biomedicine (BIBM)* (IEEE, 2013), pp. 584–588
11. N.K. Kasabov, Brain disease diagnosis and prognosis based on EEG data, in *Time-Space, Spiking Neural Networks and Brain-Inspired Artificial Intelligence* (Springer, Berlin, Heidelberg, 2019), pp. 339–359
12. S. Mahato, S. Paul, Electroencephalogram (EEG) signal analysis for diagnosis of major depressive disorder (MDD): a review, in *Nanoelectronics, Circuits and Communication Systems* (Springer, Singapore, 2019), pp. 323–335
13. C.J. James, O.J. Gibson, Temporally constrained ICA: an application to artifact rejection in electromagnetic brain signal analysis. IEEE Trans. Biomed. Eng. **50**(9), 1108–1116 (2003)
14. The "10–20 System" of Electrode Placement. http://faculty.washington.edu/chudler/1020.html
15. J. Olivan, Formats in clinical neurophysiology: the point of view of a user (2003). http://neurotraces.com/views/formats.html

16. J. Durka, D. Ircha, SignalML: metaformat for description of biomedical time series. http://eeg.pl/SignalML/SignalML/SignalML.html
17. R. Cohn, H.S. Leader, Synchronization characteristics of paroxysmal EEG activity. Electroencephalogr. Clin. Neurophysiol. **22**, 421–428 (1967)
18. B.R. Tharp, The penicillin focus: a study of field characteristics using cross-correlation analysis. Electroencephalogr. Clin. Neurophysiol. **31**, 45–55 (1971)
19. T. Matsuzaka, K. Ono, H. Baba, M. Matsuo, S. Tanaka, Y. Tsuji, S. Sugai, Interhemispheric correlation analysis of EEGs before and after corpus callosotomy. Jpn. J. Psychiatry Neurol. **47**, 329–330 (1993)
20. A. Medvedev, L. Mackenzie, J.J. Hiscock, J.O. Willoughby, Frontal cortex leads other brain structures in generalised spike-and-wave spindles and seizure spikes induced by picrotoxin. Electroencephalogr. Clin. Neurophysiol. **98**, 157–166 (1996)
21. S.H. Papavlasopoulos, M.S. Poulos, G.D. Bokos, A. Evangelou, Classification control for discrimination between interictal epileptic and non-epileptic pathological EEG events. Int. J. Biomed. Sci. **1**(1), 34–41 (2007)
22. B. Zhang, T. Lei, H. Liu, H. Cai, EEG-based automatic sleep staging using ontology and weighting feature analysis. Comput. Math. Methods Med. (2018)
23. F.H. Lopes da Silva, J.P. Pijn, P. Boeijinga, Interdependence of EEG signals: linear versus nonlinear associations and the significance of time delays and phase shifts. Brain Topogr. **2**, 9–18 (1989)
24. N.J. Mars, F.H. Lopes da Silva, Propagation of seizure activity in kindled dogs. Electroencephalogr. Clin. Neurophysiol. **56**, 194–209 (1983)
25. M.A. Brazier, Spread of seizure discharges in epilepsy: anatomical and electrophysiological considerations. Exp. Neurol. **36**, 263–272 (1972)
26. J. Gotman, Interhemispheric relations during bilateral spike-and-wave activity. Epilepsia **22**(4), 453–466 (1981)
27. K. Kobayashi, Y. Ohtsuka, E. Oka, S. Ohtahara, Primary and secondary bilateral synchrony in epilepsy: differentiation by estimation of interhemispheric small time differences during short spike-wave activity. Electroencephalogr. Clin. Neurophysiol. **83**(2), 93–103 (1992)
28. R. Cmejla, Criteria for autoregressive model order estimation in analysis of speech signals. (In Czech) Acoust. Lett. **22**, 4–7 (2000)
29. M. Poulos, M. Rangousi, N. Alexandris, A. Evangelou, Person identification from the EEG using nonlinear signal classification. Methods Inf. Med. **41**, 64–75 (2002)
30. B. Efron, *The Jackknife, the Bootstrap, and Other Resampling Plans* (SIAM, Philadelphia, 1982)
31. F. Karameh, M.A. Dahleh, Automated classification of EEG signals in brain tumor diagnostics, in *June 2000 Proceedings of the American Control Conference ACC2000* (Chicago IL, 2000), pp. 4169–4173
32. N. Hazarika, A.C. Tsoi, A.A. Sergejew, Nonlinear considerations in EEG signal classification A.A. IEEE Trans. Signal Process. **45**(4), 829–836 (1997)
33. M. Poulos, F. Geogiacodis, V. Chrissicopoulos, A. Evangelou, Diagnostic test for the discrimination between interictal epileptic and non-epileptic pathological EEG events using auto-cross-correlation methods. Am. J. Electroneurodiagnostic Technol. **43**, 228–264 (2003)
34. I. Clark, R. Biscay, M. Echeverria, T. Virues, Multiresolution decomposition of non-stationary EEG signals: a preliminary study. Comput. Biol. Med. **25**(4), 373–382 (1995)
35. S. Papavlasopoulos, M. Poulos, A. Evangelou, Feature extraction from interictal epileptic and non-epileptic pathological EEG events for diagnostic purposes using LVQ1 neural network, in *Mathematical Methods in Scattering Theory and Biomedical Engineering* (2006), pp. 390–398
36. P. Durka, From wavelets to adaptive approximations: time-frequency parameterization of EEG. BioMedical Eng. Online **2**, 1 (2003)
37. F. Jose Maria et al., What does an epileptiform spike look like in MEG? comparison between coincident EEG and MEG spikes. J. Clin. Neurophysiol. **22**(1), 68–73 (2005)
38. I. Daubechies, *Ten Lectures on Wavelets* (SIAM, Philadelphia, 1992)

39. S. Mallat, A Theory for multiresolution signal decomposition: the wavelet representation. IEEE Trans. Pattern Anal. Mach. Intell. **14**, 710–732 (1992)
40. E. Niedermeyer, Epileptic seizure disorders, in *Electroencephalography: Basic Principles, Clinical Applications and Related Fields*, ed. by E. Niedermeyer (Williams and Wilkins, LdSFBM, 1999), pp. 476–585
41. R. Quian Quiroga, A. Kraskov, T. Kreuz, P. Grassberger, Performance of different synchronization measures in real data: a case study on electroencephalographic signals. Phys. Rev. E **65**, 041903 (2002)
42. J. Mocks, T. Gasser, How to select epochs of the EEG at open eyes for quantitative analysis. Electroencephalogr. Clin. Neurophysiol. **58**, 89–92 (1984)
43. A.S. Gevins, B.C. Cutillo, Signals of cognition, in *Clinical Applications of Computer Analysis of EEG and other Neurophysiological Signals. Handbook of Electroencephalography and Clinical Neurophysiology,* vol. 2, ed. by F. Lopes da Silva, W. Storm van Leeuwen, A. Remond (Elsevier, Amsterdam, 1986), pp. 335–381
44. J.P. De Weerd, J.I. Kap, A posteriori time-varying filtering of averaged evoked potentials II. Mathematical and computational aspects. Biol. Cybern. **41**, 223–234 (1981)
45. X.H. Yu, Z.Y. He, Y.S. Zhang, Time-varying adaptive filters for evoked potential estimation. IEEE Trans. Biomed. Eng. **41**(11), 1062–1071 (1994)
46. D.O. Walter, W. Adey: Is the brain linear?, in *Technical and biological Problems of Control-a Cybernic Viwe,* vol. 41, ed. by A.S. Iberall, J.B. Reswick (Instrument Society of America, Pittsburgh, P.A, 1968), pp. 11–22
47. M. Poulos, S. Papavlasopoulos, N. Alexandris, E. Vlachos, Comparison between auto-cross-correlation coefficients and coherence methods applied to the EEG for diagnostic purposes. Med. Sci. Monit. **10**(10), MT99-MT108 (2004)
48. D.L. Gilbert et al., Meta-analysis of EEG test performance shows wide variation among studies. Neurology **60**, 564–570 (2003)
49. S. Rush, D.A. Driscoll, EEG electrode sensitivity—an application of reciprocity. IEEE Trans, biomed. Eng. (BME-16), 15–22 (1969)
50. L. Zhukov, D. Weinstein, C. Johnson, Statistical analysis for FEM EEG source localization in realistic head models. Technical report–techreports-2000. http://www.cs.utah.edu/techreports/2000/pdf/UUCS-00-003.pdf
51. J.E. Richards, Recovering dipole sources from scalp-recorded event-related-potentials using component analysis: principal component analysis and independent component analysis. Int. J. Psychophysiol. **54**, 201–220 (2004)
52. D. Gardner et al., Common data model for neuroscience data and data model exchange. J. Am. Med. Inform. Assoc **8**(1), 17–33 (2001)
53. M.P.G. Bokos, N.K.S. Papavlasopoulos, M. Avlonitis, Specific selection of FFT amplitudes from audio sports and news broadcasting for classification purposes. J. Graph Algorithms Appl. **11**(1), 277–307 (2007). http://jgaa.info/vol
54. P. Ježek, R. Moucek, EEG/ERP portal–semantic web extension: generating ontology from object oriented model, in *2010 Second WRI Global Congress on Intelligent Systems* (IEEE, 2010), pp. 392–395
55. N. Mukherjee, S. Neogy, S. Chattopadhyay, *Big Data in ehealthcare: Challenges and Perspectives* (CRC Press, 2019)

Part VII
Theoretical Advances in Computer Science with Significant Potential Applications in Technology

Chapter 18
Algorithmic Methods for Computing Bounds for Polynomial Roots

Doru Ştefănescu

Abstract We present some basic results concerning the evaluation of the absolute values of roots of univariate polynomials with complex or real coefficients. There are discussed classical and modern algorithmic methods and their computational efficiency.

18.1 Introduction

One of the main problems in the study of *univariate polynomials* (called also *polynomials in one indeterminate*) is to find their roots. Because the exact computation of the roots with real or complex coefficients is possible only in some particular cases (namely for degrees smaller than 4 or for polynomials having some special expressions), it is necessary to have efficient methods for the approximation of these roots. This is the so called *numerical computation of roots* and the general methods for the numerical computation of zeros of nonlinear functions can be applied. However, before using such methods as the bisection, the secant or the tangent, it is useful to find real intervals, respectively complex domains, that contain all the roots.

A first step is the computation of upper and lower bounds for the absolute values of the roots. We review some algorithms for the computation of such upper bounds and we compare their efficiency. Some new bounds will be also described. By considering the reciprocal polynomial one can easily derive lower bounds.

To Professor Nikolaus Alexandris.

D. Ştefănescu (✉)
University of Bucharest, Bucharest, Romania
e-mail: stef@rms.unibuc.ro

G. A. Tsihrintzis and M. Virvou (eds.), *Advances in Core Computer Science-Based Technologies*, Learning and Analytics in Intelligent Systems 14,
https://doi.org/10.1007/978-3-030-41196-1_18

18.2 Complex and Real Roots of Polynomials

If we consider a nonconstant univariate polynomial

$$P(X) = a_0 X^d + \cdots + a_{d-1} X + a_d$$

with real or complex coefficients, the fundamental theorem of algebra certifies that there exist unique complex numbers $z_1, \ldots, z_s$ and unique natural exponents $m_1, \ldots, m_s \geq 1$ such that

$$P(X) \; = a_o (X - z_1)^{m_1} \cdots (Z - z_s)^{m_s}, \quad \text{with} \quad m_1 + \cdots + m_s = d.$$

If the polynomial has real coefficients it is not sure that the roots are real. In this case it is possible that all the roots are complex or, in many cases, that some roots are complex and the others are real.

The ideal solution for a polynomial equation $P(x) = 0$ is to find exactly all its roots. This can be done only for polynomials of small degrees. For degrees $d \leq 4$ there exist explicit formulas for the roots using radicals. Évariste Galois has proved that this is not possible for all polynomials of degree $d \geq 5$. However, for quintic and sixtic polynomials there exist expressions for the roots using special power series. In practice, even for cubic or quartic polynomials, it is not practical to use exact solutions.

But the applications of mathematics needs concrete solutions. The approach is to compute approximations of solutions, the so called numerical solution of algebraic equations. In many practical instances sharp approximations are sufficient for solving specific problems that make use of polynomials.

18.3 Rule of Signs of Descartes

In the case of univariate polynomials with real coefficients one of the first problem is to know how many positive and how many negative solutions exist (if the polynomial has also real solutions). The answer to this question was given by Descartes in his *Geometry*. He gave a rule that allows to estimate the number of positive roots, and also that of negative roots. We will see that this result is also useful for computing bounds for the absolute values of the polynomial roots. We give here a short proof of the Rule of signs of Descartes.

18.3.1 A Historical Note on Descartes' Rule of Signs

René Descartes stated in his *Geometry* (1637, p. 373), a rule for counting the number of positive and negative roots for univariate polynomials with real coefficients, by counting the sign variations in the sequence of the coefficients. In particular he stated that the number of positive roots is equal to that of sign variations, which is true only for some particular classes of polynomials, for example for those having only real roots.

The statement of Descartes was criticized by some of his contemporaries but it was restated clearly and proved by several authors since the publication of his *Geometry*. We give a short proof of the complete version of the rule of signs of Descartes.

18.3.2 Sign Variations of the Coefficients

We consider a nonconstant polynomial F with real coefficients

$$F(X) = a_n X^n + a_{n-1} X^{n-1} + \cdots + a_1 X + a_0.$$

We associate to the polynomial F the sequence of its coefficients

$$a_0, a_1, \ldots, a_{n-1}, a_n. \tag{18.1}$$

We say that there is a sign variation between two members a_i and a_j, where $i < j$, if they are nonzero and have opposite signs, provided the terms $a_{i+1}, \ldots, a_{j-1}$ are equal to 0.

Example 18.1 In the sequence

$$1, -1, 2, 3, -5, -2, 4, 6, -6, -2$$

there are five sign variations.

We will denote by $var(a_0, a_1, \ldots, a_n)$ the number of sign variations of the sequence (18.1).

Lemma 18.1 *We have*

1. *The number $var(a_0, a_1, \ldots, a_n)$ is even if and only if $a_0 a_n > 0$.*
2. *The number $var(a_0, a_1, \ldots, a_n)$ is odd if and only if $a_0 a_n < 0$.*

Proof Without loss of generality we may suppose that $a_n > 0$. The sequence of signs is

$$+ - + - \cdots + - \pm$$

Before the last sign we have an even number of sign changes. As the last one is the sign of a_0, this proves the lemma. □

18.3.3 Sign Variations Versus Number of Positive Roots

We consider first two examples.

Example 18.2 Let

$$P(x) = x^7 + 2x^6 - 28x^5 - 70x^4 + 119x^3 + 308x^2 - 92x - 240 \in \mathbb{R}[x]$$

and

$$\begin{aligned} Q(x) &= (x^2 + x + 1)P(x) \\ &= x^9 + 3x^8 - 25x^7 - 96x^6 + 21x^5 + 357x^4 + 335x^3 - 24x^2 - 332x - 240. \end{aligned}$$

We note that the polynomials P and Q have 3 sign variations and 3 positive roots.

Example 18.3 Let

$$S(x) = x^3 - x^2 + 2x - 12.$$

The polynomial S has 3 sign variations and one positive root.

Lemma 18.2 *The number of sign variations of the sequence of the coefficients and the number of positive roots have the same parity.*

Proof We suppose that $z_1, z_2, \ldots, z_n$ are the (complex) roots of the polynomial F. We have the following relations

$$F(x) \ = \ a_n(x - z_1)\cdots(x - z_n)$$

and

$$a_0 = F(0) = (-1)^n a_n z_1 z_2 \cdots z_n. \tag{18.2}$$

Let $p = p(F)$ be the number of positive roots, $q = q(F)$ be the number of negative roots and $v = v(F)$ be the number of sign variations in the sequence (18.1).

By Lemma 18.1 we have

$$\mathrm{sign}(a_0 a_n) = (-1)^v.$$

We note that if there exist pure complex roots, their product is a positive number. In fact, the complex roots occur in conjugate pairs. We denote by $2d$ the number of complex roots and by $z_1, \ldots, z_m$ the real roots. From (18.2) we have

$$\begin{aligned}
\operatorname{sign}(a_0 a_n) = \operatorname{sign}(\tfrac{a_n}{a_0}) &= (-1)^n \cdot \operatorname{sign}(z_1 \cdots z_m) \\
&= (-1)^n (-1)^q \\
&= (-1)^{2d+p+q} (-1)^q \\
&= (-1)^{2(d+q)} \cdot (-1)^p \\
&= (-1)^p .
\end{aligned}$$

We finally obtain $(-1)^p = (-1)^v$.
This proves that the difference between v and p is an even number. □

18.3.4 The Inequality $v(F) \geq p(F)$

The next step for obtaining the rule of signs is to prove that $v(F) \geq p(F)$. We give a proof that avoids the use of Segner's Lemma (1728).

Proposition 18.3 *We have $v(F) \geq p(F)$.*

Proof We prove the statement by induction on the degree of the polynomial F.
If $\deg(F) = 1$ we have $v(F) = p(F)$. More precisely $v(F) = p(F) = 1$ if $a_0 a_1 < 0$, respectively $v(F) = p(F) = 0$ if a_0 and a_1 have the same sign.
Suppose now the result proved for degrees $\leq n-1$ and that $n = \deg(P) \geq 2$. We consider the derivative of F

$$F'(X) = n a_n X^{n-1} + (n-1) a_{n-1} X^{n-2} + \cdots + 2 a_2 X + a_1$$

and we observe that the signs of the coefficients do not change but the sign of a_n is missing. Therefore

$$v(F') \leq v(F). \tag{18.3}$$

On the other hand, by Fermat's theorem, between two real roots of F there exists one real root of F'. Let $z_1, \ldots, z_k$ be the positive roots of F, having multiplicities $m_1, \ldots, m_k$, with

$$m_1 + \cdots + m_k = p(F).$$

By Fermat, in each interval (z_j, z_{j+1}) there exists a root of F', so we have

$$\begin{aligned}
p(F') &\geq (m_1 - 1) + \cdots + (m_k - 1) + (k - 1) \\
&= m_1 + \cdots + m_k - 1 \\
&= p(F) - 1,
\end{aligned}$$

so

$$p(F) \le p(F') + 1. \tag{18.4}$$

By the induction hypothesis we have $p(F') \le v(F')$. Using also (18.3) and (18.4) we have

$$p(F) \le p(F') + 1 \le v(F') + 1 \le v(F) + 1.$$

If $p(F) = v(F) + 1$ it would follow that $v(F) - p(F)$ should be odd, a contradiction with Lemma 18.1. Therefore

$$p(F) \le v(F).$$

□

18.3.5 The Rule of Signs

We can now state the main result.

Theorem 18.4 (Descartes) *The number of positive roots of a univariate polynomial with real coefficients $F(X)$ is at most equal to the number of sign variations of the coefficients and can differ from it with an even number.*

Proof By Proposition 18.3 we have $v(F) \ge p(F)$ while by Lemma 18.2 $v(F)$ and $p(F)$ have the same parity. Hence the conclusion. □

18.3.6 The Number of Negative Roots

From Theorem 18.4 it follows that the following rule concerning the negative roots holds.

Theorem 18.5 *The number of negative roots of a univariate polynomial with real coefficients $F(X)$ is at most equal to the number of sign variations of the coefficients of the polynomial $F(-X)$ and can differ from it with an even number.*

18.3.7 Descartes' Rule for Hyperbolic Polynomials

Finally we prove the original statement of Descartes, i.e. the rule of signs for polynomials having only real roots. Such polynomials are called *hyperbolic*.

We first need a result concerning the number of signs variations.

Lemma 18.6 *Let $var(a_0, a_1, \ldots, a_n)$ be the number of signs variations in the sequence of real numbers $(a_0, a_1, \ldots, a_i, \ldots, a_n)$. We have*

$$var(a_0, a_1, \ldots, a_i, \ldots, a_n) + var(a_0, -a_1, \ldots, (-1)^i a_i, \ldots, (-1)^n a_n) \leq n.$$

Proof We put

$$s_n := var(a_0, a_1, \ldots, a_i, \ldots, a_n) + var(a_0, -a_1, \ldots, (-1)^i a_i, \ldots, (-1)^n a_n)$$

and we verify the inequality $s_n \leq n$ by induction on n. The case $n = 0$ is verified. We suppose the lemma is true for n and we verify it for $n + 1$. We observe that possible new sign variations could occur only in $var(a_n, a_{n+1})$ and in $var((-1)^n a_n, (-1)^{n+1} a_{n+1})$, and this adds at most one sign variation. So

$$s_{n+1} \leq s_n + 1 \leq n + 1.$$

□

We can state now the original result of Descartes.

Theorem 18.7 (Descartes) *If the polynomial F is hyperbolic, the number of its positive roots is equal to the number of signs variations of $F(X)$, while the number of its negative roots is equal to the number of signs variations of $F(-X)$.*

Proof We put

$$v_1 = var(a_0, a_1, \ldots, a_i, \ldots, a_n),$$

$$v_2 = var(a_0, -a_1, \ldots, (-1)^i a_i, \ldots, (-1)^n a_n).$$

By Theorems 18.3 and 18.5 we have

$$p \leq v_1,$$
$$q \leq v_2.$$

and $v_1 + v_2 \leq n$.

It follows that

$$n = p + q \leq v_1 + v_2 \leq n.$$

Therefore $p = v_1$ and $q = v_2$. □

18.4 Bounds for Complex Roots

It is known, by the fundamental theorem of algebra, that any univariate polynomial with complex coefficients has all the roots in the complex field $\mathbb{C}$.

18.4.1 The Bound of Cauchy

The French mathematician Augustin-Louis Cauchy published in 1822 a smart criterion that allows to estimate the bounds for the moduli of roots of univariate polynomials with complex coefficients.

Theorem 18.8 *Let ρ be the unique positive root of the polynomial* $F(X) = X^n - |a_1|X^{n-1} - \cdots - |a_n|$, *where* $a_1, \ldots, a_n \in \mathbb{C}$.

Then all the roots of the polynomial $P(X) = X^n + a_1X^{n-1} + \cdots + a_{n-1}X + a_n$ *lie in the disk* $\{|z| \leq \rho\}$.

Proof We first note that the polynomial

$$F(X) = X^n - |a_1|X^{n-1} - \cdots - |a_{n-1}|X - |a_n|$$

has only one positive root because all the coefficients are real and in the sequence of the coefficients there is only one sign variation (Descartes' rule of signs).

Let $z \in \mathbb{C}$, $|z| > \rho$. We have

$$|P(z)| \geq |z|^n - \left(|a_1| \cdot |z|^{n-1} + \cdots + |a_{n-1}| \cdot |z| + |a_n|\right) = F(|z|) > 0,$$

therefore $P(z) \neq 0$. □

18.4.2 Applications of the Method of Cauchy

The method described by Theorem 18.8 allows to obtain elementary expressions for the upper bounds of the absolute values of the roots in function of the size of the coefficients and of the degree of the polynomial P. We present some applications of this result for obtaining such bounds in the case of polynomials with complex coefficients. A first result was obtained by Cauchy himself.

Corollary 18.9 (Cauchy, 1822) *The number* $1 + M$, *where* $M = \max_{i=1}^{n} |a_i|$, *is an upper bound for the absolute values of the roots of the polynomial* $P(x) = x^n + a_1x^{n-1} + \cdots + a_n \in \mathbb{C}[x]$.

Proof Using the notation in Theorem 18.8 we have

$$\begin{aligned} F(1+M) &= (1+M)^n - \left(|a_1|(1+M)^{n-1} + \cdots + |a_n|\right) \\ &\geq (1+M)^n - M\left((1+M)^{n-1} + \cdots + 1\right) \\ &= (1+M)^n - M \cdot \frac{(1+M)^n - 1}{1+M-1} = 1 \\ &> 0. \end{aligned}$$

Therefore $1 + M > \rho$, so $1 + M$ is an upper bound for the absolute values of the roots of the polynomial P. □

Example 18.4 We consider the polynomial $P(X) = X^5 - 2X^4 - 2X^3 - X^2 + X - 2$. By the previous result we obtain the upper bound $1 + \max\{2, 1\} = 3$.

Proposition 18.10 *The number $M = 2\max_{s=1}^{n} |a_s|^{1/s}$ is an upper bound for the absolute values of the roots of the polynomial P.*

Proof Because $|a_s| \le (M/2)^s$ we have

$$|a_i| \cdot M^{n-i} \le \frac{M^n}{2^i}$$

so

$$\sum_{i=1}^{n} |a_i| \cdot M^{n-i} \le M^n \left(\frac{1}{2} + \frac{1}{2^2} + \cdots + \frac{1}{2^n}\right) = M^n \left(1 - \frac{1}{2^n}\right).$$

Using the notation from Theorem 18.8 we obtain

$$F(M) = M^n - \sum_{i=1}^{n} |a_i| \cdot M^{n-i} \ge M^n - M^n \left(1 - \frac{1}{2^n}\right) = \left(\frac{M}{2}\right)^n > 0.$$

therefore M is an upper bound for the moduli of the roots.

Example 18.5 We consider again the polynomial

$$P(X) = X^5 - 2X^4 - 2X^3 - X^2 + X - 2$$

and we obtain the upper bound $2 \cdot \max\{2, 1\} = 4$.

The following result allows to obtain many families of upper bounds.

Theorem 18.11 (Fujiwara, [5]) *Let $P(X) = X^n + a_1X^{n-1} + \cdots + a_{n-1}X + a_n$ be a non-constant polynomial with complex coefficients and let $\lambda_1, \ldots, \lambda_n > 0$ be such that*

$$\frac{1}{\lambda_1} + \frac{1}{\lambda_2} + \cdots + \frac{1}{\lambda_n} \le 1.$$

The number

$$\max_{i=1}^{n} (\lambda_i |a_i|)^{1/i}$$

is an upper bound for the absolute values of the roots of the polynomial P.

Proof We put $M = \max_{i=1}^{n} (\lambda_i |a_i|)^{1/i}$.

Therefore $|a_i| \le \dfrac{M^i}{\lambda_i}$ for all i, hence

$$\sum_{i=1}^{n} |a_i|\, M^{n-i} \le \sum_{i=1}^{n} \frac{M^i}{\lambda_i} \cdot M^{n-i} \le M^n \sum_{i=1}^{n} \frac{1}{\lambda_i}.$$

Therefore

$$F(M) \ge M^n - M^n \sum_{i=1}^{n} \frac{1}{\lambda_i} = M^n \cdot \left(1 - \sum_{i=1}^{n} \frac{1}{\lambda_i}\right) > 0,$$

which proves the theorem.

Another application of Theorem 18.8 is

Proposition 18.12 *Let* $\alpha > 0$, $|a_1| \le \alpha$. *Then*

$$\alpha + \max_{i=2}^{n} \left|\frac{a_i}{\alpha}\right|^{\frac{1}{i-1}}$$

is an upper bound for the absolute values of the roots of the polynomial P.

Proof Considering $M = \alpha + \max_{i=1}^{n} \left|\frac{a_i}{\alpha}\right|^{1/i}$, we have $|a_i| \le \alpha(M-\alpha)^{i-1}$. Hence

$$\begin{aligned}
\sum_{i=1}^{n} |a_i|\, M^{n-i} &\le \alpha \sum_{i=1}^{n} (M-\alpha)^{i-1}\, M^{n-i} \\
&= \frac{\alpha\, M^n}{M-\alpha} \sum_{i=1}^{n} \left(\frac{M-\alpha}{M}\right)^i \\
&< \frac{\alpha\, M^n}{M-\alpha} \cdot \frac{M-\alpha}{M} \cdot \frac{1}{1 - \dfrac{M-\alpha}{M}} \\
&= \alpha\, M^{n-1} \cdot \frac{M}{\alpha} \\
&= M^n.
\end{aligned}$$

Therefore

$$F(M) \ge M^n - M^n = 0.$$

Remark If we consider $a_1 = 0$ the number α in Proposition 18.12 can be any positive number.

Example 18.6 We consider

$$P(X) = X^9 - 2X^8 + X^6 - 4X^4 + X - 1.$$

We have $a_1 = 2$, so we can choose any number $a \geq 2$. We obtain

$$\max_{2 \leq i \leq 9} \left| \frac{a_i}{\alpha} \right|^{1/(i-1)} = \max \left\{ \frac{1}{a}, \left(\frac{1}{a} \right)^{1/2}, \left(\frac{4}{a} \right)^{1/4}, \left(\frac{1}{a} \right)^{1/7}, \left(\frac{1}{a} \right)^{1/8} \right\} = \left(\frac{4}{a} \right)^{1/4}.$$

Therefore other upper bounds are given by

$$M(a) = a + \left(\frac{4}{a} \right)^{1/4} \quad \text{for all} \quad a \geq 2.$$

Choosing $a = 2$ we get the upper bound

$$M(2) = 3.414.$$

In fact, the true upper bound of the moduli of the roots is 1.997.

Remark Other bounds for the roots can be obtained using linear recurrent sequences, s. Mignotte and Ştefănescu [12]. For a review of bounds for complex roots, see Hong [7] and Mignotte and Stefanescu [10].

18.5 The Bound $R + \rho$ of Lagrange

J.-L. Lagrange wrote a key memory, followed by a more detailed volume, on the numerical solution of algebraic equations. In this seminal work he stated a result on the upper bounds for positive roots of univariate polynomials with real coefficients (s. [9]. Lagrange has not given a proof of his statement. We restate his theorem in the general case of univariate polynomials with complex coefficients.

We give a proof based on the Theorem 18.11 of Fujiwara, as in [11].

Theorem 18.13 (The bound $R + \rho$) *Let $a_1, a_2, \ldots, a_n$ be complex numbers, $a_n \neq 0$. If $R = |a_j|^{1/j} \geq |a_i|^{1/i} = \rho \geq |a_k|^{1/k}$ for all $k \neq i, \; j$, the number $R + \rho$ is an upper bound vor the absolute values of the roots of the polynomial*

$$F(X) = X^n + a_1 X^{n-1} + \cdots + a_{n-1} X + a_n.$$

Proof By Theorem 18.9 of Cauchy it is sufficient to prove that the unique real root of the polynomial

$$G(X) = X^n - |a_1| X^{n-1} - \cdots - |a_{n-1}| X - |a_n|$$

is smaller than $R + \rho$.

We note that $R = |a_j|^{1/j} \geq |a_i|^{1/i} = \rho \geq |a_k|^{1/k}$ for all $k \neq i, \, j$. We are searching for suitable values in the Theorem 18.11 of Fujiwara.

We would like to have $\lambda_1, \ldots, \lambda_n > 0$ such that

$$\lambda_k |a_k| \le (R+\rho)^k \quad \text{for all} \quad k = 1, 2, \ldots, n.$$

It would be sufficient that the following inequalities be fulfilled

$$\lambda_k \, \rho^k \le (R+\rho)^k \quad \text{for all} \quad k \neq j$$

and

$$\lambda_j \, R^j \le (R+\rho)^j.$$

We choose

$$\lambda_j = \left(\frac{R+\rho}{R}\right)^j, \quad \lambda_k = \left(\frac{R+\rho}{\rho}\right)^k \quad \text{for} \quad k \neq j \quad (1 \le k \le n).$$

Hence

$$\begin{aligned}
\sum_{k=1}^{n} \frac{1}{\lambda_k} &= \sum_{k \neq j} \frac{1}{\lambda_k} + \frac{1}{\lambda_j} \\
&= \sum_{\substack{k=1 \\ k \neq j}}^{n} \left(\frac{\rho}{R+\rho}\right)^k + \left(\frac{R}{R+\rho}\right)^j \\
&= \frac{\rho}{R}\left(1 - \left(\frac{\rho}{R+\rho}\right)^n\right) + \frac{R^j - \rho^j}{(R+\rho)^j} = \frac{\rho}{R} + \frac{R^j - \rho^j}{(R+\rho)^j} - \frac{\rho^{n+1}}{R(R+\rho)^n}.
\end{aligned}$$

We consider now $y = R/\rho$ and we observe that $y \ge 1$ and

$$\begin{aligned}
\frac{\rho}{R} + \frac{R^j - \rho^j}{(R+\rho)^j} - \frac{\rho^{n+1}}{R(R+\rho)^n} &= \frac{1}{y} + \frac{y^j - 1}{(y+1)^j} - \frac{1}{y(y+1)^n} \\
&= \frac{(y+1)^n + y(y^j-1)(y+1)^{n-j} - 1}{y(y+1)^n}.
\end{aligned}$$

Note that the right hand side member of the inequality is subunitary if and only if

$$g(y) = y(y+1)^n - (y+1)^n - y(y^j - 1)(y+1)^{n-j} + 1 \ge 0.$$

We have

$$g(y) = (y+1)^{n-j} \cdot h(y) + 1,$$

where $h(y) = (y-1)(y+1)^j - y(y^j - 1)$. For checking this it is sufficient to have $h(y) \geq 0$ for all $y \geq 1$.

In fact,

$$\frac{h(y)}{y-1} > y^j + \left(\binom{j}{1} - 1\right) y^{j-1} + \left(\binom{j}{2} - 1\right) y^{j-2} + \cdots + \left(\binom{j}{j-1} - 1\right) y + 1$$
$$\geq 0$$

because all parentheses in the last row are positive. Therefore we have $h(y) \geq 0$ for all $y \geq 1$.

It follows that $\sum_{k=1}^{n} \frac{1}{\lambda_k} \leq 1$.

By Theorem 18.11 the number $R + \rho$ is an upper bound for the absolute values of the roots of F.

18.6 Lower Bounds

The information on the upper bounds for the absolute values of the roots of univariate polynomials with complex or real coefficients allows also the computation of lower bounds. This can be known using the reciprocal polynomials, as stated in the following result.

Proposition 18.14 *Let* $P(X) = a_0 X^n + a_1 X^{n-1} + \cdots + a_n \in \mathbb{C}[X]$ *and let* $P^*(X) = a_n X^n + a_{n-1} X^{n-1} + \cdots + a_0$ *be its reciprocal polynomial.*

If K *is an upper bound for the absolute values of the roots of the reciprocal polynomial* P^*, *the number* $1/K$ *is a lower bound for the absolute values of the polynomial* P.

Proof It is sufficient to observe that we have the relation

$$P^*(X) = X^n \cdot P(\frac{1}{X}),$$

which proves that the roots of the reciprocal polynomial are the inverses of the roots of the initial polynomial.

Example 18.7 Let $P(X) = X^7 - X^6 + 2X^4 - 2X^3 + X + 1$.

The reciprocal polynomial is $P^*(X) = X^7 + X^6 - 2X^4 + 2X^3 - X + 1$. By Theorem 18.13 an upper bound for the absolute values of its roots is

$$K = R + \rho = \sqrt[3]{2} + \sqrt[4]{2} \approx 2.449$$

Therefore, a lower bound for the roots of the polynomial P is $m = 0.408$.

18.7 Applications to Complex Roots

The computation of bounds for the absolute values of univariate polynomials with real or complex coefficients does not solve immediately the question of the localization of the roots of such polynomials. In applications we are interested to find bounds that are close to the true ones.

On the other hand it is preferred to choose algorithmic methods that can be performed in real time—if possible using only paper and pencil. As computers today can perform in few minutes—sometimes in few seconds—a huge number of operations, we consider to be efficient those methods that can be implemented on computers.

Example 18.8 Let $P(X) = X^5 - 2X^4 + 2X^2 + X - 2 \in \mathbb{C}[X]$. Then $M = 2$ and by Theorem 18.9 we obtain the upper bound $1 + M = 3$.

On the other hand, Proposition 18.10 and Theorem 18.13 give the bound 4.

Example 18.9 We consider the polynomial

$$P(X) \;=\; X^5 + 4X^3 + 100X + 99.$$

Theorem 18.9 of Cauchy gives the upper bound

$$M_1 = 1 + \max\{4, 100, 99\} = 1 + 100 = 101.$$

On the other hand Proposition 18.10 gives

$$M_2 \;=\; 2 \cdot \max\{4^{1/2}, 100^{1/4}, 99^{1/5}\} \;=\; 2 \cdot 3.163 \;=\; 6.326.$$

Example 18.10 Let

$$P(X) \;=\; X^{11} - 2X^{10} + X^9 - 2X^8 - 8X^4 + X - 1.$$

Using the notation in Theorem 18.13 we have

$$R = 2 \quad \text{and} \quad \rho \approx 1.346.$$

So the bound of Lagrange is $R + \rho = 3.346$.

On the other hand

$$1 + \max\{|a_i|\,;\, 1 \le i \le 11\} \;=\; 1 + 8 = 9$$

We obtain the following upper bounds.

Other upper bounds for the absolute values of the roots can be obtained using Proposition 18.12.

In the case of our polynomial we have $a_1 = 2$, so we can choose any real number $a \ge 2$. This gives

9	Theorem 18.9
4	Proposition 18.10
3.346	Theorem 18.13

$$\max_{2\leq i\leq 11}\left|\frac{a_i}{\alpha}\right|^{1/(i-1)} = \max\left\{\frac{1}{a}, \left(\frac{2}{a}\right)^{1/2}, \left(\frac{8}{a}\right)^{1/6}, \left(\frac{1}{a}\right)^{1/9}, \left(\frac{1}{a}\right)^{1/10}\right\} = \left(\frac{8}{a}\right)^{1/6}.$$

Therefore other upper bounds are given by

$$M(a) = a + \left(\frac{8}{a}\right)^{1/6} \quad \text{for all} \quad a \geq 2.$$

But the function $g : [0, \infty) \longrightarrow \mathbb{R}$, $g(x) = x + (8/x)^{1/6}$ is increasing, so the best upper bound given by Proposition 18.12 is

$$M(2) = 3.259.$$

We conclude that the best upper bounds for the absolute values of the roots of the polynomial P are given by Theorem 18.13 and Proposition 18.12.

Using an efficient package, for example `gp-pari`, we observe that the maximal absolute value of a root is 2.079.

The bounds of the absolute values of univariate polynomials with complex coefficients can be computed in function of the degree and of the size of the coefficients using almost elementary devices. Among the most efficient are the bound $R + \rho$ of Lagrange and that of Fujiwara. The estimation of such bounds is a key step for the numerical solution of algebraic equations.

18.8 Bounds for Real Roots of Polynomials with Real Coefficients

We give new bounds for the real roots of univariate polynomials with real coefficients. Our results refine the bounds of Kioustelidis and those of Lagrange. We consider polynomials which have at least one negative and one positive coefficient. The upper bounds of the real roots are expressed as functions of the first positive coefficients and of the two largest absolute values of the negative ones.

We obtain new upper bounds for the real roots of univariate polynomials with real coefficients. Some other bounds were obtained in [14]. Such bounds are useful for the location of real roots and polynomial real root isolation (cf. Emiris and Tsigaridas [4]). The effective computation of positive roots of univariate polynomials with real coefficients is relevant to iterative numerical processes (Herzberger [6]).

Our results give tighter bounds for the positive roots than those of Lagrange [9].

18.9 Upper Bounds for Positive Roots

Until very recently the most used result concerning the bounding of positive roots was the following theorem obtained by Lagrange in 1767.

Theorem 18.15 (J.-L. Lagrange) *Let* $P(X) = a_0X^d + \cdots + a_mX^{d-m} - a_{m+1}X^{d-m-1} \pm \cdots \pm a_d \in \mathbb{R}[X]$, *with all* $a_i \geq 0$, $a_0, a_{m+1} > 0$. *Denote*

$$A = \max\{a_i\,;\, i = m+1, \ldots, d\}.$$

The number

$$1 + \left(\frac{A}{a_0}\right)^{\frac{1}{m+1}}$$

is an upper bound for the positive roots of P.

Proof We consider $x \in \mathbb{R}$, $x > 1$. We have

$$\begin{aligned} |P(x)| &\geq |a_0x^d + \cdots + a_mx^{d-m}| - |a_{m+1}x^{d-m-1} \mp + \cdots \mp a_d| \\ &\geq a_0x^d - A(x^{d-m-1} + \cdots + 1) \\ &= a_0\, x^d - A\,\frac{x^{d-m} - 1}{x - 1} \\ &= \frac{a_0\, x^d(x-1) - Ax^{d-m} + A}{x - 1}. \end{aligned} \tag{18.5}$$

We would like to have

$$a_0\, x^d(x-1) > Ax^{d-m} - A,$$

which is equivalent to

$$a_0(x-1)x^m > A - Ax^{-d+m}.$$

It is sufficient to have

$$a_0(x-1)x^m > A.$$

Because $x^m > (x-1)^m$ it is sufficient to have

$$a_0(x-1)^{m+1} > A. \tag{18.6}$$

But (18.6) is satisfied for

$$x > 1 + \left(\frac{A}{a_0}\right)^{\frac{1}{m+1}},$$

so the number

$$1 + \left(\frac{A}{a_0}\right)^{\frac{1}{m+1}}$$

is an upper bound for the positive roots. □

Remark We observe that in Theorem 18.15 of Lagrange the bound is always larger than 1. This is not always convenient in practice. For example there exist classes of polynomials (including Legendre and Chebyshev orthogonal polynomials) that have all the roots subunitary.

18.9.1 Positive Roots that Can Be Subunitary

In the last decades there were discovered bounds that can return also subunitary values by Kiostelidis, H. Hong and D. Ştefănescu. At the same time, a forgotten bound of Lagrange was revisited by Mignotte-Ştefănescu, G. E. Collins and Batra-Mignotte-Ştefănescu.

For polynomials with an even number of sign variations, another bound is given by the following theorem of Ştefănescu [13].

Theorem 18.16 (Ştefănescu, 2005) *Let $P(X) \in \mathbb{R}[X]$ be such that the number of variations of signs of its coefficients is even. If*

$$P(X) = c_1X^{d_1} - b_1X^{m_1} + c_2X^{d_2} - b_2X^{m_2} + \cdots + c_kX^{d_k} - b_kX^{m_k} + g(X),$$

with $g(X) \in \mathbb{R}_+[X]$, $c_i > 0$, $b_i > 0$, $d_i > m_i > d_{i+1}$ for all i, the number

$$B_3(P) = \max\left\{\left(\frac{b_1}{c_1}\right)^{1/(d_1-m_1)}, \ldots, \left(\frac{b_k}{c_k}\right)^{1/(d_k-m_k)}\right\}$$

is an upper bound for the positive roots of the polynomial P.

Proof Suppose $x > 0$. We have

$$|P(x)| \geq c_1x^{d_1} - b_1x^{m_1} + \cdots + c_kx^{d_k} - b_kx^{m_k}$$

$$= x^{m_1}(c_1x^{d_1-m_1} - b_1) + \cdots + x^{m_k}(c_1x^{d_k-m_k} - b_k),$$

which is strictly positive for

$$x > \max\left\{\left(\frac{b_1}{c_1}\right)^{1/(d_1-m_1)}, \ldots, \left(\frac{b_k}{c_k}\right)^{1/(d_k-m_k)}\right\}.$$

In fact, for such an x we have

$$c_j x^{d_j} - b_j > 0 \quad \text{for} \quad j = 1, \ldots, k.$$

□

The proof of Theorem 18.16 can be extended also to the general case of polynomials with real coefficients having at least one positive root (that is at least one sign variation in the sequence of the coefficients). Such an extension is the following.

Theorem 18.17 *Let*

$$P(X) = a_1X^{d_1} + a_2X^{d_2} + \cdots + a_sX^{d_s} - b_1X^{e_1} - b_2X^{e_2} - \cdots - b_tX^{e_t} \in \mathbb{R}[X],$$

where $a_i > 0$, $b_j > 0$, $d_1 = \deg(P)$ and $d_1 > d_2 > \cdots > d_s$. An upper bound for the positive roots of P is given by

$$B_6(P) = \max_{\substack{1\le i\le s, 1\le j\le t\\ d_i\ge e_j}} \left(\frac{b_j}{\beta_j a_i}\right)^{\frac{1}{d_i-e_j}}$$

for any $\beta_j > 0$ such that

$$\beta_1 + \cdots + \beta_t \le 1.$$

Proof We suppose $x \in \mathbb{R}$, $x > 0$. We have

$$\begin{aligned}
P(x) &= \sum_{i=1}^{s} a_i x^{d_i} - \sum_{j=1}^{t} b_j x^{e_j} \\
&\ge \sum_{i=1}^{s} (\beta_1 + \cdots + \beta_t) a_i x^{d_i} - \sum_{j=1}^{t} b_j x^{e_j} \\
&\sum_{i=1}^{s} \left(\sum_{j=1}^{t} \beta_j a_i\, x^{d_i}\right) - \sum_{j=1}^{t} b_j x^{e_j} \\
&= \sum_{j=1}^{t} \left(\left(\sum_{i=1}^{s} \beta_j\, a_i x^{d_i}\right) - b_j x^{e_j}\right) \\
&\ge \sum_{j=1}^{t} \left(\left(\sum_{\substack{i=1\\ d_i\ge e_j}}^{s} \beta_j\, a_i x^{d_i}\right) - b_j x^{e_j}\right) \\
&= \sum_{j=1}^{t} \left(\left(\sum_{\substack{i=1\\ d_i\ge e_j}}^{s} \beta_j\, a_i x^{d_i-e_j}\right) - b_j\right) x^{e_j}.
\end{aligned}$$

The last sum is positive if

$$\beta_j\, a_i x^{d_i - e_j} - b_j \; > 0 \quad \text{for all} \quad i,\, j \quad \text{such that} \quad d_i \geq e_j$$

i.e. if

$$x > \left(\frac{b_j}{\beta_j a_i}\right)^{\frac{1}{d_i - e_j}},$$

which proves the result. □

18.9.2 Applications of Theorem 18.17

Corollary 18.18 (D. Ştefănescu [13]) *Let*

$$P(X) = X^d - b_1 X^{d-m_1} - \cdots - b_t X^{d-m_t} + g(X),$$

with $b_1, \ldots, b_k > 0$ *and* $g(X) \in \mathbb{R}_+[X]$.

The number

$$B_1(P) = \max\{(tb_1)^{1/m_1}, \ldots, (tb_t)^{1/m_t}\}$$

is an upper bound for the positive roots of P.

Proof Let $Q(X) = X^d - b_1 X^{d-m_1} - \cdots - b_t X^{d-m_t}$. We observe that a bound for the positive roots of g is also a bound for the positive roots of P.

With the notation from Theorem 18.17 we have $s = 1$ and $a_1 = 1$. We consider

$$\beta_1 = \ldots \beta_t = \frac{1}{t}$$

and it follows that the number

$$B_1(P) = \max\{(tb_1)^{1/m_1}, \ldots, (tb_t)^{1/m_t}\}$$

is an upper bound for the positive roots of Q.

But any upper bound for the positive roots of the polynomial Q is also an upper bound for the positive roots of P. □

Corollary 18.19 (J. B. Kioustelidis [8]) *Let*

$$P(X) = X^d - b_1 X^{d-m_1} - \cdots - b_t X^{d-m_t} + g(X),$$

where $b_1, \ldots, b_k > 0$ *and* $g \in \mathbb{R}_+[X]$.

The number

$$K(P) = 2 \cdot \max\{b_1^{1/m_1}, \ldots, b_k^{1/m_k}\}$$

is an upper bound for the positive roots of P.

Proof As in the proof of the previous result, it is sufficient to check that K is an upper bound for the positive roots of the polynomial $Q(X) = X^d - b_1 X^{d-m_1} - \cdots - b_k X^{d-m_k}$.

We consider

$$\beta_i = \left(\frac{1}{2}\right)^{d-m_i} \quad \text{for all} \quad i.$$

Without loss of generality we may suppose that

$$m_1 < \cdots < m_t$$

and we have

$$\begin{aligned}
\beta_1 + \cdots + \beta_t &= \left(\frac{1}{2}\right)^{d-m_1} + \cdots + \left(\frac{1}{2}\right)^{d-e_t} \\
&= \frac{1}{2^{m_1}} \sum_{m=1}^{t} \left(1 + \frac{1}{2^{m_2-m_1}} + \cdots + \frac{1}{2^{m_t-m1}}\right) \\
&\le \frac{1}{2^{m_1}} \sum_{m=1}^{t} \left(1 + \frac{1}{2} + \cdots + \frac{1}{2^{t-1}}\right) \\
&\le \frac{1}{2^{m_1}} \cdot \frac{1 - \frac{1}{2^t}}{1 - \frac{1}{2}} \\
&< \frac{1}{2^{m_1-1}} \\
&\le 1.
\end{aligned}$$

It follows that the conditions from Theorem 18.17 are fulfilled, so $K(P)$ is an upper bound for the positive roots of Q, then also for those of Q. □

Remark The result of Kioustelidis returns also subunitary bounds, if they exist. The bound B_2 is obtained through the estimation of the unique positive root of the associated polynomial

$$P_{\text{ass}}(X) = X^d - \sum_{j=1}^{k} b_j\, X^{d-m_j}.$$

Note that the estimation of the unique positive root of polynomials with negative coefficients excepting the dominant one have important applications in finance mathematics [6].

18.9.3 Theorem $R + \rho$ of Lagrange, Real Coefficients

We finally give a new (short) proof of the Theorem 18.13 of Lagrange (see [1–3]) for the case of real coefficients.

Theorem 18.20 *Let $F(X) = X^d + a_1X^{d-1} + \cdots + a_{d-1}X + a_d$ be a polynomial with real coefficients, with $d > 0$ and $a_d \neq 0$ and let $\{a_{i_t}\ 1 \leq t \leq s\}$ be the sequence of its negative coefficients. We suppose*

$$|a_{i_1}|^{1/i_1} \geq |a_{i_2}|^{1/i_2} \geq \cdots \geq |a_{i_s}|^{1/i_s}.$$

We put $R = |a_{i_1}|^{1/i_1}$, $\rho = |a_{i_2}|^{1/i_2}$. The number $R + \rho$ is an upper bound for the positive roots of the polynomial F.

Proof By Theorem 18.8 of Cauchy it is sufficient to check that $f(R + \rho) > 0$, where

$$f(X) = X^d - \sum_{t=1}^{s} |a_t|X^{d-t}$$

For simplifying the presentation we put $i_1 = j$. Without loss of generality we may suppose $t = d$.

We observe that $|a_i| \leq \rho^i$ for all $i = 1, 2, \ldots, d, i \neq j$. Then, for $x > \rho$, we have

$$f(x) \geq x^d - (R^j - \rho^j)x^{d-j} - \sum_{k=1}^{d} \rho^k x^{d-k}.$$

We put

$$g(X) = X^d - (R^j - \rho^j)X^{d-j} - \sum_{k=1}^{d} \rho^k X^{d-k}, \quad h(X) = (X - \rho)g(X)$$

and we observe that

$$h(X) = (X - \rho)X^{d+1} - (R^j - \rho^j)(X - \rho)X^{d-j} - \sum_{s=0}^{d-1} \rho^{s+1} X^{d-s} + \sum_{s=1}^{d} \rho^{s+1} X^{d-s}$$

$$= (X - 2\rho)X^d - (X - \rho)(R^j - \rho^j)X^{d-j} + \rho^{d+1}.$$

For $x > 0$ we have $h(x) > (x-2\rho)x^d - (R^j - \rho^j)(x-\rho)x^{d-j} = \big((x-2\rho)x^j - (x-\rho)(R^j-\rho^j)\big)\,x^{d-j}$.

We put $u(X) = (X-2\rho)X^j - (X-\rho)(R^j-\rho^j)$ and we notice that

$$\begin{aligned}
u(R+\rho) &= (R-\rho)(R+\rho)^j - R(R^j-\rho^j)\\
&= (R-\rho)\left(\sum_{k=0}^{j}\binom{j}{k}R^{k-j}\rho^k - (R^j + R^{j-1}\rho + \cdots + R\rho^{j-1})\right)\\
&= (R-\rho)\Bigg((jR^{j-1}\rho + \frac{j(j-1)}{2}R^{j-2}\rho^2 + \cdots + jR\rho^{j-1})\\
&\qquad -(R^{j-1}\rho + R^{j-2}\rho^2 + \cdots + R^2\rho^{j-2} + R\rho^{j-1})\Bigg)\\
&> 0.
\end{aligned}$$

Therefore $f(R+\rho) > 0$, which proves the result. □

18.9.4 Comparisons of Results on Real Roots

We first compare our bound $B_1(P)$ with the bound $B_2(P) =: K(P)$ of J. K. Kioustelidis.

For $k = 1$ we have $B_1 = b_1^{1/m_1} < 2b_1^{1/m_1} = B_2$.

For $k \geq 2$ and $k < 2^{m_j}$ $(1 \leq j \leq k)$ we always have $B_1(P) < B_2(P)$.

If we consider

$$P(X) \;=\; 4\,X^7 - X^6 + 0.0004\,X^5 - X^4 + 0.00004\,X^3 + 0.0000004\,X - 1$$

we obtain

$$B_1(P) \;=\; 0.96, \qquad B_2(P) \;=\; 1.64.$$

Note that the true upper bound for the positive roots of the polynomial P is 0.928.

For polynomials with an even number of signs we also consider the bound $B_3(P)$ given in Theorem 18.16.

Let

$$\begin{aligned}
Q_1(X) &= 3X^4 - X^3 + 7X^2 - 3X + 0.001,\\
Q_2(X) &= X^5 - 1.01\,X^4 + X^3 - 1.1\,X + 0.1,\\
Q_3(X) &= 3X^7 - X^6 + 7X^5 - 3X^2 + 0.001,\\
Q_4(X) &= 10X^9 - 17\,X^5 + 10X^4 - 13\,X + 1.
\end{aligned}$$

We have

The bound $B_3(P)$ gives in many cases better results. For particular polynomials the method used in Theorem 18.16 can be used for deriving better limits for the roots.

	B_1	B_2	B_3	Largest positive root
Q_1	1.256	2	0.428	0.421
Q_2	2.02	2.048	1.024	1.003
Q_3	1.148	2	0.753	0.725
Q_4	1.357	2.283	1.141	1.121

For a given polynomial with real coefficients having at least one negative coefficient there are, generally, several possibilities to choose the positive coefficients $c_1, \ldots, c_k$. If $b_j/c_j > 1$ the optimal choice is for $m_j - d_j$ maximal, while for $b_j/c_j < 1$ the optimal choice is for $m_j - d_j$ minimal.

References

1. P. Batra, M. Mignotte, D. Ştefănescu, Improvements of Lagrange's bound for polynomial roots. J. Symb. Comp. **82**, 19–25 (2017)
2. A.-L. Cauchy, *Exercices de Mathématiques*, t. 4, Paris (1829)
3. G.E. Collins, Krandick's proof of Lagrange's real root bound claim. J. Symb. Comp. **70**, 106–111 (2015)
4. I.Z. Emiris, E.P. Tsigaridas, Univariate polynomial real root isolation: continued fractions revisited, in *Algorithms ESA 2006*, ed. by Y. Azar, T. Erlebach. LNCS, vol. 4168 (Springer, Berlin, Heidelberg, 2006)
5. M. Fujiwara, Über die obere Schranke des absoluten Betrages der Wurzeln einer algebraischen Gleichung. Tôhoku Math. J. **10**, 167–171 (1916)
6. J. Herzberger, Construction of bounds for the positive root of a general class of polynomials with applications, in *Inclusion Methods for Nonlinear Problems with Applications in Engineering, Economics and Physics* (Munich, 2000), Comput. Suppl. **16**, Springer, Vienna (2003)
7. H. Hong, Bounds for absolute positiveness of multivariate polynomials. J. Symb. Comp. **25**, 571–585 (1998)
8. J.B. Kioustelidis, Bounds for positive roots of polynomials. J. Comput. Appl. Math. **16**, 241–244 (1986)
9. J.-L. Lagrange, *Traité de la résolution des équations numériques* (Paris, 1798). (Reprinted in *Œuvres*, t. VIII, Gauthier–Villars, Paris, 1879)
10. M. Mignotte, D. Ştefănescu, *Polynomials: An Algorithmic Approach* (Springer, 1999)
11. M. Mignotte, D. Ştefănescu, On an estimation of polynomial roots by Lagrange, Prépubl. IRMA, 25/2002 (Strasbourg, 2002), 17 pp. HAL Id: hal-00129675, https://hal.archives-ouvertes.fr/hal-00129675
12. M. Mignotte, D. Ştefănescu, Linear recurrent sequences and polynomial roots. J. Symb. Comput. **35**, 637–649 (2003)
13. D. Ştefănescu, New bounds for the positive roots of polynomials. J. Univ. Comput. Sci. **11**, 2125–2131 (2005)
14. D. Ştefănescu, *Inequalities on Upper Bounds for Real Polynomials Roots*. LNCS, vol. 4194 (Springer, 2006), pp. 284–294

Printed in the United States
by Baker & Taylor Publisher Services